VOLUME SEVENTY THREE

ADVANCES IN QUANTUM CHEMISTRY

Electron Correlation in Molecules – ab initio Beyond Gaussian Quantum Chemistry

VOLUME SEVENTY THREE

ADVANCES IN QUANTUM CHEMISTRY

Electron Correlation in Molecules – ab initio Beyond Gaussian Quantum Chemistry

Edited by

PHILIP E. HOGGAN

Institut Pacal, UMR CNRS 6602,
Campus Universitaire des Cézeaux,
Aubiere Cedex, France

TELHAT OZDOGAN

Department of Physics, Faculty of Arts and Sciences,
Amasya University, Amasya, Turkey

ELSEVIER

AMSTERDAM • BOSTON • HEIDELBERG • LONDON
NEW YORK • OXFORD • PARIS • SAN DIEGO
SAN FRANCISCO • SINGAPORE • SYDNEY • TOKYO
Academic Press is an imprint of Elsevier

Academic Press is an imprint of Elsevier
50 Hampshire Street, 5th Floor, Cambridge, MA 02139, USA
525 B Street, Suite 1800, San Diego, CA 92101-4495, USA
The Boulevard, Langford Lane, Kidlington, Oxford OX5 1GB, UK
125 London Wall, London, EC2Y 5AS, UK

First edition 2016

ISBN: 978-0-12-803060-8
ISSN: 0065-3276

For information on all Academic Press publications
visit our website at http://store.elsevier.com/

CONTENTS

PREFACE

This thematic volume of *Advances in Quantum Chemistry* collects expert topical contributions on electronic structure, regarding the correlation methods and exponential type orbitals. In this respect, it covers "Electron Correlation in Molecules: *Ab Initio* Beyond Gaussian Quantum Chemistry."

The first section is on exponential type orbitals (ETO). It therefore covers the eigenfunctions of a one-electron atom (with molecular applications). They show exponential decrease at long range, hence the name. They are a challenge to handle for molecules especially, maintaining this fundamental work at the cutting edge.

Often, an approximate expansion in Gaussian functions is used which facilitates application to quantum chemistry at the expense of the physical representation of atomic orbitals. ETO work is a traditional strong point of the Molecular Electronic Structure (MES) workshop series. One indication of this strong *ab initio* quantum methodology is that 10 contributions constitute this section, beginning with reviews by Ancarani and Gasaneo, and Nakatsuji and Harris and continuing with research work by Pachucki, Ruiz, Perez, Erturk, Yasui, Ozdogan, and Yükçü. Contributions of this section were presented in Amasya University (Turkey) in September 2014 and the next workshop is scheduled for Buenos Aires (Argentina) in September 2016. Another indication of this tradition is the Poster award at Amasya to Magdalena Zienkiewicz (Warsaw) for work entitled: Precise Born–Oppenheimer potential for excited states of H_2 molecule.

Electronic correlation is a major frontier in quantum chemistry nowadays. The methods of *ab initio* quantum chemistry will only give accurate properties for molecules and solids when electron correlation is taken into account. Its treatment ranges from rapid density functional theory (DFT) approaches to almost exact quantum Monte Carlo (QMC) approaches. These methods are covered in the second section.

These theoretical techniques include electron correlation and are introduced with the help of general contributions by Pastorczak, Weatherford, and Musial as far as DFT is concerned and by Toulouse and Umrigar as well as Pederiva regarding fundamental quantum Monte Carlo work and both DFT and QMC have been applied to quantum chemical treatment of reactivity for difficult cases in this section, where correlation needs to be obtained accurately to cater for variation from reactants to products: this is illustrated by Grabowski *et al.* in recent work on the noble gas dimers, evaluating correlation effects in density functional theory along the dissociation path. Early 2014 also saw the first validated *ab initio* benchmark for the H_2 dissociation energy barrier on Cu(111), given using QMC by Hoggan: 14.79 kcal/mol, well within standard error (0.5) of the experimental value 14.48 kcal/mol. (see ArXiv).

A couple of novel research-reviews close this volume, where theory is used as a very direct complementary approach to experiment, one on X-Ray Constrained Wave Functions: Fundamentals and Effects of the Molecular Orbitals Localization by Genoni and the other regarding Electron Impact Ionization by Saha *et al.*

Hopedfully readers enjoy this volume as much as we have enjoyed editing it!

Philip E. Hoggan and Telhat Ozdogan
Editors

CONTRIBUTORS

Selda Akdemir
Department of Science Education, Faculty of Education, Sinop University, Sinop, Turkey

Lorenzo Ugo Ancarani
Equipe TMS, UMR CNRS 7565, ICPM, Université de Lorraine, 57078 Metz, France

Roland Assaraf
Sorbonne Universités, UPMC Univ Paris 06, and CNRS, UMR 7616, Laboratoire de Chimie Théorique, Paris, France

Arun K. Basak
Department of Physics, University of Rajshahi, Rajshahi, Bangladesh

Ahmed Bouferguene
Faculté Saint-Jean, University of Alberta, Alberta, Canada

Adam Buksztel
Institute of Physics, Faculty of Physics, Astronomy and Informatics, Nicolaus Copernicus University, Torun, Poland

Melek Eraslan
Department of Physics, Faculty of Arts and Sciences, Amasya University, Amasya, Turkey

Murat Erturk
Department of Physics, Faculty of arts and sciences, Onsekiz Mart University, Çanakkale, Turkey, and Institute of Physics, Nicholas Copernicus University, Torun, Poland

Alexei M. Frolov
Department of Applied Mathematics, University of Western Ontario, London, Ontario, Canada

Gustavo Gasaneo
Departamento de Física, Universidad Nacional del Sur, 8000 Bahía Blanca, Buenos Aires, and Consejo Nacional de Investigaciones Científicas y Técnicas CONICET, Argentina

Daniel H. Gebremedhin
Department of Physics, Florida A&M University, Florida, USA

Alessandro Genoni
CNRS, and Université de Lorraine, Laboratoire SRSMC, UMR 7565, Vandoeuvre-lès-Nancy, France

Nikitas I. Gidopoulos
Department of Physics, Durham University, Durham, United Kingdom

Ireneusz Grabowski
Institute of Physics, Faculty of Physics, Astronomy and Informatics, Nicolaus Copernicus University, Torun, Poland

Carlos Mario Granados-Castro
Equipe TMS, UMR CNRS 7565, ICPM, Université de Lorraine, 57078 Metz, France, and Departamento de Física, Universidad Nacional del Sur, 8000 Bahía Blanca, Buenos Aires, Argentina

A.K. Fazlul Haque
Department of Physics, University of Rajshahi, Rajshahi, Bangladesh

Frank E. Harris
Department of Physics, University of Utah, Salt Lake City, and Quantum Theory Project, University of Florida, Gainesville, Florida, USA

Philip E. Hoggan
Institut Pacal, UMR CNRS 6602, Campus Universitaire des Cézeaux, Aubiere Cedex, France

Yusaku I. Kurokawa
Quantum Chemistry Research Institute, Kyoto, Japan

Federico Latorre
Department of Theoretical Chemistry, Friedrich-Alexander University Erlangen–Nürnberg, Erlangen, Germany, and Institute of Theoretical Chemistry, University of Vienna, Vienna, Austria

Leszek Meissner
Institute of Physics, Nicholas Copernicus University, Torun, Poland

Benjamin Meyer
CNRS, and Université de Lorraine, Laboratoire SRSMC, UMR 7565, Vandoeuvre-lès-Nancy, France

Dario M. Mitnik
Consejo Nacional de Investigaciones Científicas y Técnicas CONICET, and Instituto de Astronomía y Física del Espacio (IAFE) and Departamento de Física, Universidad de Buenos Aires, C1428EGA Buenos Aires, Argentina

Paolo Mori
Dipartimento di Fisica, Università di Trento, Trento, Italy

Anna Motyl
Institute of Chemistry, University of Silesia, Katowice, Poland

Abishek Mukherjee
ECT*, Strada delle Tabarelle 286, Trento, Italy, and ClusterVision B.V., Nieuw-Zeelandweg 15B, Amsterdam, Netherlands

Monika Musiał
Institute of Chemistry, University of Silesia, Katowice, Poland

Hiroyuki Nakashima
Quantum Chemistry Research Institute, Kyoto, Japan

Hiroshi Nakatsuji
Quantum Chemistry Research Institute, Kyoto, Japan

Telhat Ozdogan
Department of Physics, Faculty of Arts and Sciences, Amasya University, Amasya, Turkey

Emin Öztekin
Department of Physics, Faculty of Science and Arts, Ondokuz Mayıs University, Samsun, Turkey

Krzysztof Pachucki
Faculty of Physics, University of Warsaw, Warsaw, Poland

Ewa Pastorczak
Institute of Physics, Lodz University of Technology, Lodz, Poland

M. Atiqur R. Patoary
Department of Physics, University of Rajshahi, Rajshahi, Bangladesh

Francesco Pederiva
Dipartimento di Fisica, Università di Trento, and INFN-TIFPA, Trento Institute for Fundamental Physics and Applications, Trento, Italy

Jorge E. Pérez
Departamento de Física, Facultad de Ciencias Exactas Fco-Qcas y Naturales, Universidad Nacional de Río Cuarto, Río Cuarto, Argentina

Katarzyna Pernal
Institute of Physics, Lodz University of Technology, Lodz, Poland

Alessandro Roggero
Institute for Nuclear Theory, University of Washington, Seattle, Washington, USA

María Belén Ruiz
Department of Theoretical Chemistry, Friedrich-Alexander University Erlangen–Nürnberg, Erlangen, Germany

Bidhan C. Saha
Department of Physics, Florida A & M University, Tallahassee, Florida, USA

Patrycja Skupin
Institute of Chemistry, University of Silesia, Katowice, Poland

Szymon Śmiga
Institute of Physics, Faculty of Physics, Astronomy and Informatics, Nicolaus Copernicus University, Torun, Poland

Julien Toulouse
Sorbonne Universités, UPMC Univ Paris 06, and CNRS, UMR 7616, Laboratoire de Chimie Théorique, Paris, France

M. Alfaz Uddin
Department of Physics, University of Rajshahi, Rajshahi, Bangladesh

Cyrus J. Umrigar
Laboratory of Atomic and Solid State Physics, Cornell University, Ithaca, New York, USA

Charles A. Weatherford
Department of Physics, Florida A&M University, Florida, USA

Jun Yasui
RIKEN Innovation Center, Hirosawa, Wako, Saitama, and Frontier Research Center, Canon Inc., Ohta-ku, Tokyo, Japan

Niyazi Yükçü
Department of Energy Systems Engineering, Faculty of Technology, Adıyaman University, Adıyaman, Turkey

Magdalena Zientkiewicz
Faculty of Physics, University of Warsaw, Warsaw, Poland

SECTION I

Exponential Type Basis Quantum Chemistry

CHAPTER ONE

A Sturmian Approach to Photoionization of Molecules

Carlos Mario Granados-Castro*,†,1, Lorenzo Ugo Ancarani*,1, Gustavo Gasaneo†,‡, Dario M. Mitnik‡,§

*Equipe TMS, UMR CNRS 7565, ICPM, Université de Lorraine, 57078 Metz, France
†Departamento de Física, Universidad Nacional del Sur, 8000 Bahía Blanca, Buenos Aires, Argentina
‡Consejo Nacional de Investigaciones Científicas y Técnicas CONICET, Argentina
§Instituto de Astronomía y Física del Espacio (IAFE) and Departamento de Física, Universidad de Buenos Aires, C1428EGA Buenos Aires, Argentina
[1]Corresponding authors: e-mail address: carlos.mario-granados.castro@univ-lorraine.fr; ugo.ancarani@univ-lorraine.fr

Contents

Advances in Quantum Chemistry, Volume 73
ISSN 0065-3276
http://dx.doi.org/10.1016/bs.aiq.2015.06.002

Abstract

An accurate theoretical description of photoionization processes is necessary in order to understand a wide variety of physical and chemical phenomena and allows one to test correlation effects of the target. Compared to the case of many-electron atoms several extra challenges occur for molecules. The scattering problem is generally multicenter and highly noncentral. The molecular orientation with respect to the polarization of the radiation field must also be taken into account. These features make the computational task much more cumbersome and expensive than for atomic targets. In order to calculate cross sections, one needs to describe the ejected electron with a continuum wavefunction with appropriate Coulomb asymptotic conditions. Making a number of initial approximations, many different theoretical/numerical methods have been proposed over the years. However, depending on the complexity of the molecule, agreement among them is not uniform and many features of the experimental data are not so well reproduced. This is illustrated through a number of examples. In order to have a global theoretical overview, we present a survey of most of the methods available in the literature, indicating their application to different molecules. Within a Born–Oppenheimer, one-center expansion and single active electron approximation, we then introduce a Sturmian approach to describe photoionization of molecular targets. The method is based on the use of generalized Sturmian functions for which correct boundary conditions can be chosen. This property makes the method computationally efficient, as illustrated with results for H_2O, NH_3, and CH_4.

1. INTRODUCTION

The quantum description of both bound and unbound orbitals is necessary to study collisions with atoms and molecules. The study of single photoionization (PI) provides an indirect tool to test our capacity to describe the target before and after the interaction correctly and thus correlation and many-body effects. PI plays an important role beyond atomic and molecular physics, since it has a wide variety of applications, such as astrophysics,[1–3] planetary,[4–6] atmospheric,[7–8] plasma,[9–11] or medical physics.[12,13] Also PI helps to understand different processes in surfaces, such as structural changes upon surface adsorption, quantifying the relationship between shape resonances and the bond lengths[14–17]; or to characterize the relation between gas, chemisorbed and solid-state phases in surface reactions.[18–21]

In the last few years, a Sturmian approach[22–23] has been introduced to study single and double ionization of atoms induced by electron[24] or photon[25] impact. It is the purpose of this contribution to extend, implement and apply such an approach to the single PI of molecular targets. The Hamiltonian for molecules being generally multicenter and highly noncentral makes the problem more difficult than for atomic targets. Indeed, the absence of any spherical symmetry couples different angular momenta from different atomic orbitals (AOs) that form the molecular orbitals (MOs) and thus convergence of "traditional" methods is considerably more difficult to achieve. Additionally, there are various many-body effects that can be important in ionization processes, such as the relaxation of all MOs, due to the creation of a hole (ionized electron), or the change of the remaining pair correlation energies because of such relaxation. An issue which does not arise in PI of atoms is the orientation of the molecular target. In most experiments, the molecule is randomly oriented and this must be taken into account within the theoretical calculations.

When leaving an atomic or molecular target, an ionized electron needs to be described accurately by a continuum wavefunction, which has well defined boundary conditions. Over the years, quite a few methods have been proposed and applied successfully to atoms. The extension of these methods and their computational codes to molecular targets is not straightforward, as several complications arise beyond the many-body nature of the problem and not all of them can provide the correct asymptotic form. Different approaches have been applied to a large variety of molecules ranging from the smallest one, H_2, up to, e.g., DNA bases. The success of each method depends on the studied molecule and photon energy range, the validity of some approximations and possibly on convergence issues or limitations related to the numerical implementation.

Except for small molecules, experimental data are not so abundant and do not always span the whole photoelectron energy range; they therefore do not permit a full assess to the quality of different theoretical descriptions. As several theories often disagree with each other and with experimental data, especially close to threshold, we made a survey of most existing methods applied to PI in molecules. For each, we briefly indicate the main ingredients, the advantages and possible limitations. We also found useful to draw a list (rather complete to the best of our knowledge) of all molecules for which PI has been investigated theoretically.

In order to calculate the transition amplitudes for single PI in atomic or molecular systems, many considerations must be taken into account. Usually the starting point is the treatment of the ionized electron as a one-electron function, the one-center expansion (OCE). In many cases, the vibrational structure of the molecule can be ignored, especially in high-energy collisions, justifying the use of the Born–Oppenheimer (BO) approximation. Also, in order to simplify the calculations, the frozen core (FC) approximation and the static exchange approximation (SEA) are considered. It is within this frame, together with a model molecular potential, that we implement the generalized Sturmian approach. In the literature, several Sturmian function implementations exist, as reviewed, e.g., in the introductions of Refs. 22 and 23. Similarly to previous publications on scattering studies (see the recent review 22 and references therein), in this contribution we shall name *Generalized Sturmian Functions* (GSF) those defined in Section 5.1; note that other authors use the same terminology to define a different class of Sturmian functions. One of the advantages of such a method is that it ensures that the continuum wavefunction has the correct asymptotic behavior.[22] To assess the validity of our approach, we will compare the calculated PI cross sections for a number of small molecules with theoretical and experimental data found in the literature.

The rest of this paper is organized as follows. We start with some generalities on PI in Section 2; we continue in Section 3 with a brief panorama of what sort of agreement one observes in the literature between theoretical and experimental cross sections. In Section 4, we present a survey of existing theoretical methods used to investigate molecular PI. In Section 5, we introduce the Sturmian approach, and compare our results for PI of H_2O, NH_3, and CH_4 to several theoretical and experimental data.

Atomic units ($\hbar = e = m_e = 1$) are assumed throughout, unless stated otherwise.

2. GENERALITIES

In the study of the interaction of a radiation field (a photon) with a molecular target several processes may occur. Consider a photon of energy $E_\gamma = \hbar\omega$, such that $E_\gamma > I_0$, where I_0 is the ionization potential of the molecule. Once it strikes a polyatomic molecule RA in an initial vibrational state ν_0 (R is the polyatomic radical and A is an individual atom), the different outcomes may be

$$\hbar\omega + \mathrm{RA}(\nu_0) \rightarrow \begin{cases} \mathrm{RA}^+ + \mathrm{e}^-(\ell) & \text{Photoionization,} \\ \mathrm{R}^* + \mathrm{A}^* & \text{Photodissociation,} \\ \mathrm{R}^* + \mathrm{A}^{+*} + \mathrm{e}^-(\ell) & \text{Dissociative photoionization.} \end{cases} \tag{1}$$

If we have a dissociation process, the final products can be in an excited state. If we have an ejected electron, called photoelectron, it has a defined angular momentum ℓ. In this contribution, we will concentrate only on single PI which can be considered as a "half-scattering" processes. It involves a bound-free transition for which one needs to know only the initial state Ψ_0 (energy E_0) of the molecule, usually its ground state, and the final state of the ionized electron. The transition operator, that connects both initial and final states, is described semi-classically via the dipolar approximation; the dipolar operator in both length (L) and velocity (V) gauges reads

$$\widehat{D}^{(L)} = -\hat{\varepsilon}\cdot\mathbf{r}, \tag{2a}$$

$$\widehat{D}^{(V)} = -\hat{\varepsilon}\cdot\mathbf{p}, \tag{2b}$$

where $\hat{\varepsilon}$ gives the polarization of the field. In this work, we consider linear polarization along the z direction.

The major task is to calculate accurately the wavefunction Ψ of the photoelectron, that is an electron in a continuum state of the ionized molecular target, with an energy $E = k^2/2$ defined by the energy of the incident photon $E = E_\gamma - I_0$. Such continuum wavefunctions are more difficult to calculate than the low-lying bound-states as they oscillate up to infinity. They are solutions of the time-dependent Schrödinger equation (TDSE) or the time-independent Schrödinger equation (TISE), with well defined properties. They must be regular at the origin of the coordinate system, and the asymptotic boundary conditions are given by the superposition of an incoming-wave Coulomb function plus an incoming spherical wave, generated by the non-Coulomb part of the molecular potential[26]

$$\lim_{r\to\infty}\Psi^{(-)} \propto \mathrm{e}^{-i\left(kz+\frac{Z}{k}\ln k(r-z)\right)} + f\left(\hat{k},\hat{r}\right)\frac{1}{r}\mathrm{e}^{-i\left(kr-\frac{Z}{k}\ln(2kr)\right)}, \tag{3}$$

where $f\left(\hat{k},\hat{r}\right)$ is the transition amplitude and $Z = -1$ for an initial neutral target.

One quantity that is measurable experimentally is the PI cross section, defined theoretically as

$$\frac{\mathrm{d}\sigma}{\mathrm{d}E} = \frac{\pi e^2}{m^2\hbar^2 c}\omega^{(g)}\left|\left\langle\Psi_0\left|\widehat{D}^{(g)}\right|\Psi\right\rangle\right|^2, \tag{4}$$

where $\omega^{(L)} = E - E_0$ or $\omega^{(V)} = (E - E_0)^{-1}$ and c is the speed of light.

In most experiments, it is difficult to determine the spatial orientation of the molecule in a given laboratory frame. Only a few advanced experimental techniques can perform a full angle-resolved spectroscopy, such as the one based on ultrashort pump-probe laser pulses[27,28] and the full kinematic experiments as COLTRIMS (cold target recoil ion momentum spectroscopy).[29] In most cases, therefore, one must consider a random orientation of the molecule when it interacts with the radiation field. To do that, two different coordinates systems, whose origin coincide with the center of mass of the target, are considered[30]: the laboratory frame, $\mathbf{r}'$, defined by the polarization axis of the electric field, and a molecular-fixed frame, $\mathbf{r}$, defined by the axis of highest symmetry. Let β and α be the polar angles of this molecular axis with respect to the laboratory frame, and let the set of Euler angles $\hat{\mathfrak{R}} = (\alpha, \beta, \gamma)$ denote hereafter the molecular orientation. A rotation $\hat{\mathfrak{R}}$ will bring the molecular fixed frame into coincidence with the laboratory frame. The dipolar operator in length gauge (2a), for a linearly polarized field (axis z), in the laboratory frame is then

$$z' = \left(\frac{4\pi}{3}\right)^{1/2} r \sum_{\mu} Y_1^{\mu}(\hat{r}) \mathcal{D}_{0\mu}^{1}(\hat{\mathfrak{R}}), \tag{5}$$

where $\mathcal{D}_{0\mu}^{1}(\hat{\mathfrak{R}})$ is the rotation matrix[31] that rotates the dipolar operator to the molecular frame. The rotated dipolar operator in velocity gauge follows a similar expression. In order to calculate a cross section for a randomly oriented molecule, we must calculate first the orientation-dependent transition amplitudes in Eq. (4) (see also Section 5.2) and then perform an angular average over $\hat{\mathfrak{R}}$, defined as

$$\int \mathrm{d}\hat{\mathfrak{R}} \equiv \frac{1}{8\pi^2} \int_0^{2\pi} \mathrm{d}\alpha \int_0^{\pi} \sin\beta \, \mathrm{d}\beta \int_0^{2\pi} \mathrm{d}\gamma\,, \tag{6}$$

of the square modulus of such transition amplitudes.

3. EXAMPLES TAKEN FROM THE LITERATURE

As mentioned in Section 1, many different theoretical methods and computational codes have been developed over the years to study PI in multielectron atoms. For molecules, many complications arise. The problem is highly noncentral and generally multicenter so that continuum wavefunctions are quite difficult to calculate. Additionally the vibrational structure can have an important influence on the electronic structure and therefore on the PI itself.

To overcome all these complexities, additional to the "traditional" (FC) or the SEA, one starts to separate the electronic motion from that of the nuclei and this is done using the BO approximation. One may also implement the fixed nuclei (FN) approximation, and it is possible to go further and use the OCE, where all electrons are referred to a common center, usually the center of mass of the molecule. Such variety of approximations (which are needed to deal with molecular systems), together with the choice of basis functions or adopted numerical approach, translates into a considerable non uniformity in the quality of the end product. Except for H_2, for most molecules the PI cross sections obtained using different theoretical or numerical methods show no overall satisfactory agreement, on the one hand between them and, on the other hand, with experimental data. This is illustrated below with four molecules: H_2, N_2, CO_2, and C_6H_6. We emphasize that almost all experimental data presented here have no explicit error bars, either because they are not indicated in the given references or because they are too small, typically smaller than 3%.

3.1 H_2

We start with H_2, the simplest many-electron molecule. Figure 1 shows the PI cross sections obtained using different methods: self-consistent field (SCF,

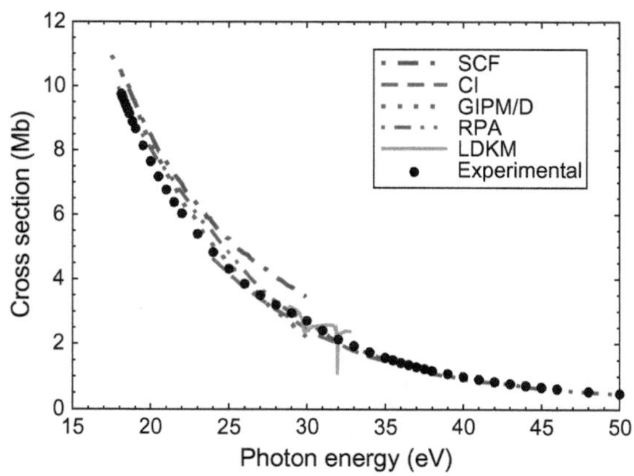

Figure 1 PI cross section in Mb versus photon energy in eV for the ground state of H_2 molecule. We compare the results obtained using SCF[33] (purple (dark gray in the print version), dash-dot); CI[34] (red (gray in the print version), dash); GIPM/D[35] (brown (gray in the print version), dots); RPA[36] (blue (gray in the print version), dash-dot-dot), and LDKM[37] (orange (light gray in the print version), solid) with experimental data[32] (black dots).

see Section 4.2.1), configuration interaction (CI, see Section 4.1), ground state inversion method (GIPM/D, see Section 4.7.2), random-phase approximation (RPA, see Section 4.9) and logarithmic derivative Kohn method (LDKM, see Section 4.11.1). They are further compared with the experimental data of Chung et al.[32] For this example, SCF and CI calculations used OCE, and LDKM the FC approximation. Except for the SCF results, we see an excellent agreement between all theories with experimental data. Indeed, the molecule H_2 is sufficiently simple to allow for a PI study taking into account all interactions. One aspect, though, that remains challenging is to calculate precisely the positions and widths of the doubly excited states that depend on the nuclear motion.

3.2 N_2

We show in Fig. 2 the PI cross sections for the outer valence orbital $3\sigma_g$ of N_2. For such MO, we show calculations performed with CI (Section 4.1), time-dependent density functional theory (TD-DFT, see Section 4.3.2), multiple-scattering Xα (MS Xα, see Section 4.6), Stieltjes–Tchebycheff technique (STT, see Section 4.10) and iterative-Schwinger method (ISM, see Section 4.12.1). Note that the results for CI and TD-DFT were obtained using OCE, and for ISM using the FC approximation. The theoretical cross sections are compared with the experimental results of Plummer et al.[38]

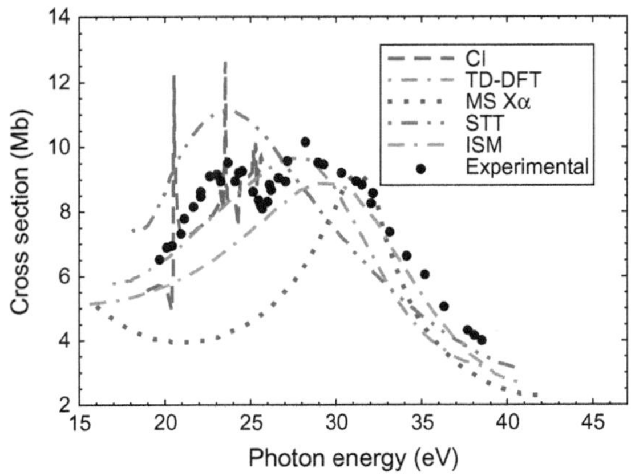

Figure 2 Partial PI cross section in Mb versus photon energy in eV from the MO $3\sigma_g$ of N_2. Results for CI[39] (red (dark gray in the print version), dash); TD-DFT[40] (green (gray in the print version), dash-dot); MS Xα[41] (blue (dark gray in the print version), dots); STT[42] (gray, dash-dot-dot), and ISM[43] (orange (light gray in the print version), dash-dash-dot) are compared with experimental data[38] (black dots).

The situation changes drastically when moving from H_2 to a more complex molecule such as N_2. The figures show that the agreement between different theories and experimental data is basically lost, especially for energies close to the threshold. Moreover, only a partial agreement for higher energies is observed. Except for the CI results, none of the other calculations reproduces the different series of resonances located between 20 and 25 eV.

3.3 CO_2

The PI cross sections for CO_2 are shown in Fig. 3 for the MO $1\pi_g$. We compare the results obtained with GIPM/D (Section 4.7.2), STT (Section 4.10), ISM (Section 4.12.1) and *R*-matrix method (RMM, see Section 4.8). The experimental data are taken from Brion and Tan.[44]

Here, results for ISM and RMM used both the FC and the FN approximations. Depending on the energy range, the different theoretical calculations present again only a partial agreement, and even if they cannot reproduce completely the experimental data, they perform rather well beyond 25 eV. Although the center of mass of CO_2 is close to the C atom because of its linear geometry, this molecule is particularly difficult to describe: the density of charge is completely delocalized around the molecule and only the use of multicenter wavefunctions yields acceptable PI results, as in the GIPM/D case.

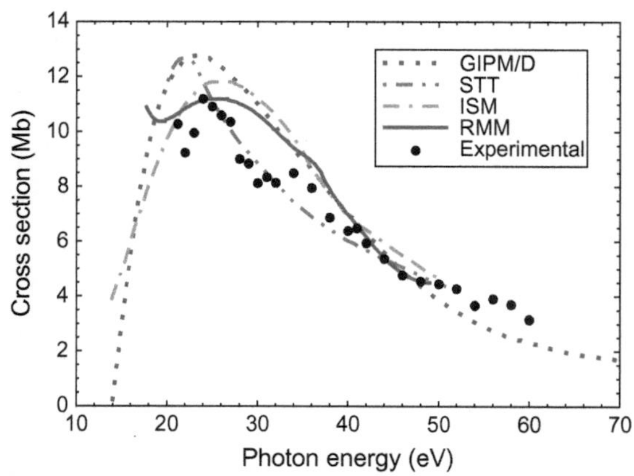

Figure 3 Partial PI cross section in Mb versus photon energy in eV from the MO $1\pi_g$ of CO_2. Results for GIPM/D[45] (brown (dark gray in the print version), dots); STT[42] (gray, dash-dot-dot); ISM[46] (orange (light gray in the print version), dash-dash-dot), and RMM[47] (blue (dark gray in the print version), solid) are compared with experimental data[44] (black dots).

3.4 C_6H_6

Finally, for benzene (C_6H_6), PI cross sections for the outer valence orbital $1e_{1g}$ are shown in Fig. 4. The theoretical results obtained using DFT (Section 4.3.1), TD-DFT (Section 4.3.2), GIPM/D (Section 4.7.2) and LDKM (Section 4.11.1) are compared with the experimental data of Carlson et al.[48]

This is a rather complex molecule to describe theoretically and the difficulties show up in the PI spectra. None of the calculations accurately reproduce the resonances (neither their energy position nor their intensity), let alone the overall cross section magnitude except at rather high photoelectron energies.

As evidenced from Figs. 1 to 4, except for H_2, for all other molecules we can draw similar conclusions: (1) major disagreement between methods are clearly seen when comparing PI cross sections; (2) experimental data (in particular near threshold) are generally not well reproduced (other features of the continuum spectra are also difficult to reproduce). This is also true for H_2O, NH_3 or CH_4 molecules; the cross sections will be presented in Section 5.3, where we shall compare different theoretical calculations including ours obtained with the Sturmian approach. To have an overview of most methods that have been proposed to describe molecular PI, we present in the next section a survey and indicate to which molecules they have been applied (a rather complete list is presented in Appendix).

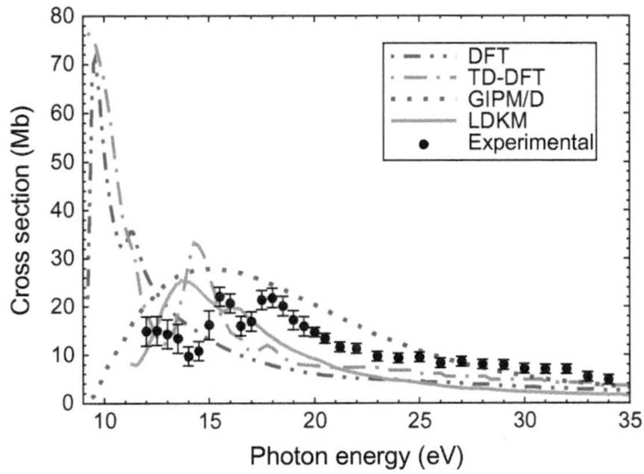

Figure 4 Partial PI cross section in Mb versus photon energy in eV from the MO $1e_{1g}$ of C_6H_6 (benzene). Results for DFT[49] (blue (dark gray in the print version), dash-dot-dot); TD-DFT[50] (green (gray in the print version), dash-dot); GIPM/D[51] (brown (dark gray in the print version), dots), and LDKM[52] (orange (gray in the print version), solid) are compared with experimental data[48] (black dots).

4. SURVEY OF THEORETICAL METHODS

4.1 CI

One of the "classical" methods used to study electronic structure in atoms and molecules is configuration interaction (CI); a description of its use for the study of PI of molecules can be found in Ref. 53.

Some results obtained using the CI method are the studies by Daasch et al.[54] for CO_2, van Dishoeck et al.[55] for HCl and Decleva et al.[56] for O_3. Using B-splines[57] as a basis set, Apalategui and Saenz[58] studied multiphoton ionization of H_2; Vanne and Saenz[59] studied HeH^+; Fojón et al.[60] also studied H_2; Sanz–Vicario et al.[34] studied PI of H_2 by ultrashort laser pulses and Sansone et al.[28] for H_2 and D_2; Dowek et al.[61] studied circular dichroism in H_2. Using the so-called time-dependent CI,[62] we find the work of Klinkusch et al.[63] for LiCN, and of Sonk and Schlegel[64] for C_4H_6 (butadiene). Finally, using the multichannel CI complete-active-space,[65] we can find the work of Stratmann et al.[39] for N_2, and of Stratmann and Lucchese[66] for O_2.

4.2 Hartree–Fock Methods

4.2.1 Self-Consistent Field

Among the studies that have used the Hartree–Fock (HF) method and the self-consistent field (SCF) to study PI of molecules, we find the work of Dalgarno[67] for CH_4; Kelly[33] studied H_2; Schirmer et al.,[68] together with the Green's function formalism, studied the inner-valence PI of N_2 and CO; Padial et al.,[1] using Gaussian-type orbitals (GTOs), studied C_2. For calculations performed with the relaxed-core HF approximation, we have the results of Larkins and Richards[69] for Li_2; the studies of Saito et al. on the *K*- shell photoelectron angular distribution from CO_2[70] and from NO_2[71]; Semenov et al.[72] studied the PI from the *K*-shell of CO. We should also mention the review of different applications of the SCF by Ågren et al.[73]

4.2.2 Multiconfiguration Time-Dependent Hartree–Fock

In general, it is difficult to describe with high precision highly excited states and nonadiabatic dynamics in molecules, especially if one is interested in studying ionization by high-intensity radiation fields. The multiconfiguration time-dependent Hartree–Fock (MCTDHF) approach is a method that uses a linear combination of determinants of time-dependent orbitals, and is flexible enough to describe the response of a molecule to short and intense laser pulses. The formalism can be found in Refs. 74–76.

The MCTDHF has been used by Kato and Kono[76] and by Haxton et al.[77] to study PI of H_2 by intense laser fields, and also by Haxton et al.[78] for HF.

4.3 Density Functional Theory

Density functional theory (DFT) is widely used in quantum chemistry. It allows for easy determination of the electronic structure for given systems (an atom, a molecule, a crystal, etc.), regardless of its extension or the number of particles that constitute it. While "standard" quantum mechanics works directly with the many-body wavefunctions of the different particles in a given system, DFT uses the one-electron electronic density $n(\mathbf{r})$ and is based on two theorems, called the Hohenberg–Kohn theorems.[79]. In different implementations of the DFT to study PI of molecules, $n(\mathbf{r})$ is calculated using a conventional linear combination of AOs (LCAO).[49]

4.3.1 Kohn–Sham DFT

In the Kohn–Sham DFT (KS DFT),[80] the Hamiltonian of the molecular system is determined by the density of the occupied orbitals in the ground state and in terms of the Hartree potential, the electron–nuclei interaction, and the so-called exchange-correlation potential which contains all the "unknowns" of the system. Different potentials are available in the literature for different atomic and molecular systems (see, for instance Refs. 81 and 82), based, for example, on the local density approximation or on the generalized gradient approximation.

The KS DFT has been used by Venuti et al.[49] to study PI in C_6H_6; by Stener and Decleva, using the OCE approximation, to study HF, HCl, H_2O, H_2S, NH_3, and PH_3 (Ref. 83), and CH_4, SiH_4, BH_3, and AlH_3 (Ref. 84). Toffoli et al.,[85] using the multicenter expansion, calculated cross sections for Cl_2, $(CO)_2$, and $Cr(CO)_6$. Woon and Park[86] also studied C_6H_6 (benzene), $C_{10}H_8$ (naphthalene), $C_{14}H_{10}$ (anthracene) and $C_{16}H_{10}$ (pyrene). Stranges et al.[87] studied the dynamics in circular dichroism of the C_3H_6O (methyl-oxirane). Toffoli et al.[88] studied the PI dynamics in $C_4H_4N_2O_2$ (uracil).

4.3.2 Time-Dependent DFT

Time-dependent DFT (TD-DFT)[89] constitutes another line of development of the DFT methods. In the first order time-dependent perturbative scheme, where the zeroth order is equivalent to the KS DFT,[90] the linear response of the electronic density $n(\mathbf{r})$ to an external weak time-dependent electromagnetic field can be described by a SCF potential, given by Zangwill and Soven.[91]

The TD-DFT has been used by Levine and Soven[40] to calculate photoemission cross sections and asymmetry parameters of N_2 and C_2H_2. Stener, Decleva and coworkers, using B-splines[57] and the OCE, studied PI for different molecules: Stener and Decleva[90] calculated the cross sections for N_2 and PH_3; Stener et al.[92] for CH_4, NH_3, H_2O, and HF; Stener et al.[93] for CO and also from the *K*-shell[94]; Fronzoni et al.[95] for C_2H_2; Stener et al.[50] for CS_2 and C_6H_6; Toffoli et al.[96] and Patanen et al.[97] for CF_4, and Holland et al.[98] for pyrimidine and pyrazine. We also find the work of Russakoff et al.[99] for C_2H_2 and C_2H_4, and by Madjet et al.[100] for C_{60}. Different results for molecular PI have been reviewed by Stener et al.[101]

For the sake of completeness, we also mention some studies of molecular PI that use a slightly different approach, the static-exchange DFT: Plésiat et al.[102] investigated PI of N_2 and CO, and Kukk et al.[103] from the inner-shells of CO.

4.4 Complex Methods

4.4.1 Complex Scaling

The complex scaling (CS) method[104–105] has been used extensively to study ionization and, mainly, resonance phenomena in atoms and molecules. The idea behind this method is to scale the coordinates of all particles in the Hamiltonian by a complex-valued scale factor: $r \to re^{i\theta}$. One variant of the CS is the so-called exterior complex scaling (ECS),[106–108] whereby the coordinates scale only outside a fixed radius R_0

$$r \to R(r) = \begin{cases} r & \text{for } r \leq R_0, \\ R_0 + (r - R_0)e^{i\theta} & \text{for } r > R_0. \end{cases} \quad (7)$$

The ECS method has been applied to study general scattering problems using L^2 basis set representations. It is especially well suited to study ionization processes in molecules, since the definition of the exterior scaling (7) avoids complicated scaling expressions in the nuclear attraction terms of the Hamiltonian[107] when R_0 is large enough to enclose all the molecular nuclei.

The ECS has been used mainly by McCurdy, Rescigno, Martín and coworkers to study different ionization processes in atoms and molecules: McCurdy and Rescigno[109–110] used Cartesian Gaussian-type orbitals (CGTOs) to calculate PI cross sections of H_2^+; Vanroose et al.,[111–112] using

B-splines,[57] studied double PI (DPI) of H_2; Rescigno et al.[113] performed *ab initio* DPI calculations of H_2; Tao et al., using discrete variable representation (DVR),[114] calculated PI cross sections for H_2^+ [115–116] and angular distribution for DPI of H_2.[117]

4.4.2 Complex Basis Functions

In the complex basis function (CBF) technique[109,110,118] (the CS method can be considered a particular case of the CBF where the basis functions are defined in terms of the physics of the problem), the continuum scattering information is extracted from a finite-matrix representation of the electronic Hamiltonian in a set of complex square-integrable basis functions. The resulting matrix elements necessary to obtain the cross section, can be calculated efficiently using a discrete basis set approximation to the spectrum of the Hamiltonian.[109]

The CBF technique, together with complex GTO, has been used by McCurdy and Rescigno[109] to calculate PI cross section of H_2^+; by Yu et al.[118] for valence- and *K*-shell ionization of N_2, and by Morita and Yabushita[119] for H_2^+ and H_2.

4.5 Linear Algebraic Method

The linear algebraic method (LAM), developed by Collins and Schneider,[120–121] has been applied successfully to study molecular excitation and ionization by electron collisions. The adaptation of the method to study PI in molecules is given in Ref. 122. The LAM presents the advantage of including explicitly an effective optical potential in order to introduce correlation effects into the scattering solution.[122]

While the initial state is treated separately, usually in terms of GTO or CGTO,[120] the method is used to calculate directly the ejected electron unbound wavefunction that satisfies the TISE. To do so, the configuration space is divided into two regions, with the boundary at $r=a$: (1) for $r \geq a$, where nonlocal effects are negligible, the wavefunction can be calculated by standard propagation procedures; (2) for $r < a$, where exchange and correlation effects are important, the wavefunction is expanded in two terms: one as a linear combination of the wavefunctions of the molecular-ion target and the scattering wavefunction, and the other in a set of "correlation" functions that are added for completeness.[122]

In the LAM, one obtains a set of differential equations in the SEA, that can be converted into a set of radial integro-differential equations using an expansion in partial waves of the electronic wavefunctions. Then, this set of scattering equations is further transformed into a set of coupled integral equations using Coulomb Green's functions.[120] Finally, by introducing a discrete quadrature to evaluate the integrals, one obtains a set of linear-algebraic equations that can be solved with standard linear systems routines. This solution must be matched at $r=a$ with the result of the propagation scheme to the asymptotic region. More details on the effective optical potential are given in the Refs. 120, 122 and 123.

To our knowledge, this method has been used only by Collins and Schneider[122] to calculate cross sections for H_2, N_2, NO, and CO_2.

4.6 Multi-Scattering

The multiple-scattering method (MSM) has been developed in different physics fields, as in nuclear physics,[124] solid state physics,[125] and also in atomic and molecular physics (see, for example, Refs. 126 and 127, and references therein). The idea behind the MSM is to represent the molecular field, that in general is highly noncentral in the molecular core region, by a set of three potentials V_{I}, V_{II}, and V_{III}, defined in different nonoverlapping spheres (called muffin-tin partitioning): (I) defined by the $\{\mathrm{I}_i\}$ spherical regions containing the different atomic nuclei at their center $r_i=0$, and with radii $\{\rho_i\}$; (II) defined by $r_i>\rho_i$ and $r_0<\rho_{\mathrm{III}}$, where r_0 is the radial coordinate from the center of the molecule and ρ_{III} is the outer sphere radius, measured from the molecular center. In general, the potential V_{II} is considered constant; (III) defined by $r_0\geq\rho_{\mathrm{III}}$. The potential V_{III} has a spherical symmetry. One can construct the photoelectron continuum wavefunction taking into account the continuity conditions between all three regions, and imposing the incoming boundary conditions (3) in the external region. The total wavefunction is written as $\Psi=\sum_i\Psi_i+\Psi_{\mathrm{II}}+\Psi_{\mathrm{III}}$, where each term is a solution to the potential of the corresponding region of the molecular field, and obeys the adequate asymptotic boundary conditions.

The MSM or, equivalently, multiple-scattering with an undetermined factor α (MS Xα),[128] have been widely used to study ionization of molecules by photon and electron impact. For example, Davenport calculated cross sections for N_2 and CO,[41,129] and for H_2[129]; Dehmer and Dill calculated the K-shell PI of N_2[130]; Grimm[131] calculated the cross section for C_2H_4 and Grimm et al.[132] for N_2, CO, CO_2, COS, and CS_2; Rosi et al.[133] studied PI in CH_4 and CF_4; Tse et al.[134] investigated the photoabsorption spectra in $SiCl_4$; Ishikawa et al.[135] studied, implementing a DVR[114] method, SiH_4,

SiF_4, and $SiCl_4$; Powis studied PI in PF_3[136], CH_3I[137], and CF_3Cl.[138] Finally, Jürgensen and Cavell[139] compared directly experimental results with the MS Xα for NF_3 and PF_3.

4.7 Plane-Wave-Based Methods

4.7.1 Plane-Wave and Orthogonalized Plane-Wave Approximations

The simplest description of an ionized electron is the plane-wave approximation (PWA), but it is not expected to give accurate results near threshold.[140] To our knowledge, the first implementations of the PWA are due to Kaplan and Markin,[141–142] Lohr and Robin,[143] and to Thiel and Schweig.[144–145]

The final state of the molecule describes one electron that has been excited from a given initial MO to a continuum normalized plane-wave orbital.[140] This plane-wave is not necessarily orthogonal to any of the occupied MOs; if orthonormality is imposed, we have the orthogonalized PWA.

The PWA and the orthogonalized PWA, together with Slater-type orbitals (STOs) to describe AO, have been used by Rabalais et al.[146] and by Dewar et al.[147] to calculate PI cross sections for H_2, CH_4, N_2, CO, H_2O, H_2S, and H_2CCH_2. Huang et al.[148] used the orthogonalized PWA to calculate angular asymmetry parameters for H_2, N_2, and CH_4. Beerlage and Feil[149] calculated cross sections for HF, $(CN)_2$, CaHCN, $C_2(CN)_2$, N_2, CO, H_2O, furan, pyrole and tetrafluoro-pyrimidine. Schweig and Thiel[150] calculated the relative band intensity of N_2, CO, H_2O, H_2S, NH_3, PH_3, CH_4, $(CH_3)_2S$, C_6F_6, among others. Hilton et al.[151] have used the so-called effective PWA to calculate cross sections for H_2, CO, H_2O, and C_2H_4. Finally, Deleuze et al.[152] used the orthogonalized PWA, together with a many-body Green's function framework, to calculate PI cross sections for CH_4, H_2O, C_2H_2, N_2, and CO.

4.7.2 Ground Inversion Potential Method

The so-called ground state inversion potential method (GIPM) has been developed by Hilton, Hush, Nordholm and coworkers[151,153] with the aim of obtaining a chemical theory of PI intensities.[154] This method uses the standard one-electron PWA, the orthogonalized PWA or the energy shifted PWA[151] in order to calculate the electronic continuum final wavefunction. The cross section is obtained from an atomic summation theory together with a plane wave analysis of diffraction effects from photoelectron amplitudes from different atoms that interfere with each other.[35,154] The main difference of GIPM with a standard PWA is that the potential felt by an electron when leaving an atomic center in a molecule is calculated directly by inversion of the ground state HF orbital.[153–154] The GIPM theory can include three important effects: change in the atomic orbitals nature upon formation of the molecule, diffraction effects[154] and exchange in an exact sense.

The GIPM has been used by Hilton et al. to calculate PI cross sections for H_2O[155] and for H_2, N_2, and CO.[35] Also Kilcoyne et al. calculated cross sections for H_2, HF, and N_2[154]; H2O, NH_3, and CH_4[156]; CO, CO_2, and N_2O,[45] and for C_2H_4 and C_6H_6.[51]

4.8 *R*-Matrix Method

Originally introduced in nuclear physics, the *R*-matrix method (RMM) has been adapted to atomic and molecular physics by Burke and coworkers (see Ref. 157 and references therein). Applications of this method, in particular for electron collisions, have been reviewed elsewhere.[158–160] The idea behind the RMM is to enclose the scattering particles and the target within a sphere of radius *a*, so that it should be possible to characterize the system using the eigenenergies and the eigenstates computed within the sphere. Then by matching them to the known asymptotic solutions, one can extract all the scattering parameters. The *R*-matrix is defined as the matrix that connects the two regions in which the space is divided into. They are: (1) an internal region, where all the particles are close to one another, so that the short-range interactions and exchange are important; (2) an external region, where all particles are still interacting, but the forces are direct and could have a multipolar character. In the most conventional use of the RMM, the Hamiltonian of the internal region is diagonalized in order to obtain the *R*-matrix eigenenergies and eigenfunctions, generally using the nonadiabatic formalism.[161] The initial and final states are expanded in terms of these eigenstates. The corresponding coefficients for the initial state are usually obtained by performing an all-channels-closed scattering calculation, and in this case the problem is reduced to find the zeros of a determinant.[162–163] To obtain the coefficients for the final state, calculations of electron scattering by the corresponding molecule can be made and the resulting *R* matrices represent the result of a full nonadiabatic treatment of the internal region of the scattering problem,[160] and provides the solution in the external region.[164] Finally, with both sets of coefficients, it is possible to calculate the required transition dipole moments, and thus the PI cross section (4).

Since the corresponding formalism is relatively new, the RMM has not been used for molecules as much as for atoms. However, we have the works by Tennyson et al.[165] for H_2, and by Tennyson[166] for H_2 and D_2. The so-called R-matrix Floquet theory[167–168] has been used by Burke et al.[168] and by Colgan et al.[169] to study multiphoton processes in H_2. Saenz,[170] using STOs, studied PI in HeH^+. Tashiro[171] calculated cross sections for N_2 and NO. Finally, Harvey et al.[47] recently studied CO_2, using GTOs combined with Coulomb and Bessel functions.

4.9 Random Phase Approximation

The random phase approximation (RPA) is a method that has been applied with success to study PI in atoms and molecules.[172–173] One advantage is that PI cross sections calculated in length or velocity gauges coincide. Additionally, the computational effort required in the RPA implementation is comparable to calculations in the single active electron (SAE) approximation, since the RPA uses only two-electron integrals involving two occupied and two unoccupied orbitals.[174–175]

In the standard procedure of the RPA, the ground state and the one-electron wavefunctions for the excited and continuum states of the molecule are calculated at HF level. With these, all required matrix elements and in particular the Coulomb and dipole matrix elements, can be calculated directly. Next, the RPA dipole matrix elements are calculated solving the corresponding equation, and the results are used to obtain directly PI cross sections or the required observables.[174–177]

The RPA has been used to study PI of H_2 by Martin et al.,[36] by Schirmer and Mertins[178] and by Semenov and Cherepkov.[177,179] For N_2, we can find calculations performed by Lucchese and Zurales[180]; by Semenov and Cherepkov[176,181]; by Yabushita et al.,[175] using complex functions; and by Montuoro and Moccia,[182] using mixed L^2 basis sets (STOs and B-splines[57]). For H_2S, we have the results of Cacelli et al.[183] For LiH, calculations were performed by Carmona-Novillo et al.[184] For C_2H_2, we have the results of Yasuike and Yabushita,[185] who used complex basis functions (see Section 4.4.2) and by Montuoro and Moccia, using the mixed L^2 basis sets. We can find also calculations for the K shell of N_2 by Cherepkov et al.[186]; for the ion C_{60}^+ by Polozkov et al.,[187] or for the fullerenes C_{20} and C_{60} by Ivanov et al.[188] Extensive calculations have been performed by Cacelli et al.,[189] to study PI in CH_4, NH_3, H_2O and HF, and by Amusia et al.,[172] who used the RPA with exchange to calculate PI cross sections of CH_4, C_2H_6, C_3H_8, C_2H_4, C_2H_2, NH_3, H_2O, CN^-, N_2, CO, CO_2, N_2O, and NO_2^-.

4.10 Stieltjes–Tchebycheff Technique

The Stieltjes–Tchebycheff technique (STT), developed by Langhoff and coworkers (see, for example, Ref. 190 and 191 and references therein), is based on theorems from the theory of moments[192]; its flexibility allows the use of different type of basis sets.[191,193,194] The technique has been widely and successfully used to study ionization processes in different atomic and molecular systems.

The strength of the interaction of unpolarized radiation with a target gas can be described by Kramers–Heisenberg expression of the polarizability (frequency dependent) of the constituent molecules. This strength can be written as a Stieltjes integral over the appropriate oscillator strength distribution[190,194] or, alternatively, by the use of the cumulative oscillator-strength distribution which can be approximated by an histogram (Stieltjes procedure). Even if such an histogram cannot represent adequately the continuum of the molecule, it can give good approximations to the related power moments, and it rigorously bounds the correct distribution through Tchebycheff inequalities.[192] Technical details about the direct computational implementation of the STT are provided in Ref. 191.

The STT has been used to study PI in CH, using STOs, by Barsuhn and Nesbet[195]; in H_2, using GTOs in a CI method (see Section 4.1), by ONeil and Reinhardt[196]; in N_2, together with GTOs, by Rescigno et al.[194] and using LCAO with optimized STOs by Stener et al.[42] In H_2O by Williams et al.[197] and by Delaney et al.[198] in the SEA, both using GTOs; by Diercksen et al.,[199] using Cartesian Gaussian basis sets and by Cacelli et al.[200] using STOs in the independent-channel approximation. By Cacelli et al., we also find calculations for NH_3,[200] HF,[201] HCl,[202] H_2S,[203] and CH_4.[201] For CO, we mention the work by Görling and Rösch,[204] who used GTOs. For F_2, Orel et al.[205] used contracted Gaussians. For C_6H_6, Gokhberg et al.[206] used the STT together with the Lanczos algorithm. Finally, Stener et al.[42] have performed calculations using LCAO with optimized STOs for CO_2, N_2O, SF_6, C_2N_2, $TiCl_4$, and $Cr(CO)_6$.

4.11 The Kohn Variational Method

Among different approximate methods used to determine the energy spectra and the corresponding wavefunctions, we have the perturbation theory and the standard Ritz variational method,[26] where approximate solutions of the TDSE or the TISE for a given problem are found in a subspace of the real problem. Besides the standard Ritz method, there is also, e.g., the Kohn variational method (KVM).[207] The idea behind the latter is to find a variational expression that allows one to calculate the wavefunction with a correct asymptotic behavior. This is dictated by two arbitrary $f_\ell(r)$ and $g_\ell(r)$ functions, that behave asymptotically as the regular $F_\ell(kr)$ and, respectively, irregular $G_\ell(kr)$ Coulomb functions. The trial wavefunction can be written as

$$\psi_\ell^{\mathrm{t}}(r) = f_\ell(r) + \lambda^{\mathrm{t}} g_\ell(r) + \sum_i c_i \varphi_i(r), \tag{8}$$

where $\{\varphi_i\}$ is a set of L^2 functions and λ^{t} is a trial parameter. The Kato identity[208] is used to find a stationary λ^{s} value. We can distinguish two options for the trial wavefunction (8): (1) if f_ℓ and g_ℓ are the regular and irregular Coulomb functions, then we have $\lambda = \tan\delta_\ell$, where δ_ℓ is the phase shift related to a short range potential; (2) if g_ℓ is an outgoing function $h_\ell^{(-)}$, called "regularized" irregular Coulomb function (defined as $h_\ell^{(-)}(r) = ik^{-1/2}[F_\ell(kr) - ic(r)G_\ell(kr)]$, where $c(r)$ is a cutoff function) then $\lambda = e^{i\delta_\ell}\sin\delta_\ell$, i.e., the T-matrix (transition matrix). In this case, we have the complex Kohn method.[208–209]

Two different implementations of the KVM in the study of PI of molecules are separately hereafter described.

4.11.1 Logarithmic Derivative Kohn Method

The logarithmic derivative Kohn method (LDKM),[210–211] and its variant, the finite-volume variational method,[212] were originally proposed to generate a translational basis for reactive scattering, using Lobatto shape functions.[211,213,214] In this method, all the required radial integrals are performed explicitly over a finite volume V, usually a sphere. The main difference between the LDKM and the standard KVM is that different coefficients are added to the functions f_ℓ and g_ℓ in (8); these coefficients can be determined by matching the wavefunction and their derivatives with the exact Coulomb functions across the surface that delimits the integration volume V.[52,214] In many of the implementations of the LDKM, Lobatto shape functions, referred usually as free-type functions, are used as the basis set $\{\varphi_i\}$ in (8).

The LDKM has been used to calculate PI cross sections for H_2^+ by Le Rouzo and Raşeev[212] and by Rösch and Wilhelmy[214]; Raşeev[37] studied autoionization in H_2; and Wilhelmy et al.[52,215] calculated cross sections and asymmetry parameters for N_2, CO, and C_6H_6.

4.11.2 Complex Kohn Method

The complex Kohn method (CKM), developed by McCurdy, Rescigno and coworkers to study excitation and ionization of molecules by electron collisions,[209,216,217] have proved to be very effective, in particular in the first-order calculation of dipolar transition moments.[216] The adaptation of the method to study PI in molecules has been described by Lynch and Schneider.[218] Different elections of the arbitrary cutoff function $c(r)$ or the irregular function $g_\ell(r)$ have been tested.[218–219]

The CKM has been used Lynch and Schneider[218] to study PI of H_2 and N_2; by Rescigno et al.[220] to study CO, examining the effects of the inter-channel coupling; Orel and Rescigno[221] to study NH_3 and, more recently, Jose et al.[222] to study PI of SF_6 also adding inter-channel coupling effects.

4.12 The Schwinger Variational Method

While many variational methods are based on the TISE (a differential equation), several others, as the Schwinger variational method (SVM)[223] are based on the equivalent integral equation, the Lippmann–Schwinger equation (LSE).[224–225]

The advantage of the LSE over the TISE to study collisions processes is that the correct boundary conditions of the problem are automatically incorporated through the use of the corresponding Green function G^C. The SVM is a powerful formulation of the scattering problem that can provide highly accurate solutions without requiring expansions in very large basis sets.[226–227] The idea behind this method is to obtain a stationary variational condition over the T-matrix. In general, one can obtain better converged results using the SVM compared to the KVM results.

The implementation of the SVM has been developed along two methods, named the Schwinger multichannel method[228] and the iterative-Schwinger method (ISM). The latter, and a variant using continued fractions, are now briefly described.

4.12.1 Iterative Schwinger

The ISM is an iterative approach to the solution of collisions problems using the SVM[226] to solve the LSE. The first implementation of ISM[227,229,230] was the study of scattering of low-energy electrons by atoms and molecules. In the case of molecules, the fixed-nuclei approximation was used together with the assumption that the interaction between the ionized electron and the molecular ion is described by the static-exchange potential.[230–231] The description of the ISM implementation to study PI is given with details in Refs. 43 and 46.

In the ISM instead of solving the associated LSE for each partial-wave of the scattering function, one solves equivalently a LSE for the T matrix[226,231]: $T = U + UG^C T$. The iterative method begins by approximating the short-range potential U by a separable potential $\tilde{U}$, using an initial set of expansion functions R; then the scattering solutions for the approximate potential $\tilde{U}$ are obtained from the corresponding LSE. The iterative procedure is continued by augmenting the original set of functions with those obtained with the approximated potential. Using this augmented set of functions, the first iteration is completed by calculating a new T matrix. A second iteration is begun by constructing a new set of solutions and combining them with the initial trial functions set; this will yield a new T matrix. The iterative procedure is continued until the wavefunctions converge, yielding the LSE solutions for the exact potential U.[230–231]

This method has been widely used to study PI of molecular systems. Using CGTO as the initial set of functions R, we find calculations by Lucchese et al.[43] for N_2; using spherical GTOs, Lucchese et al.[46] calculated PI cross section for CO_2 and Lynch et al.[232] for C_2H_2. Natalense et al. presented results for SF_6[233] and for CH_4, CF_4, and CCl_4[234]; Machado et al. for H_2O[235] and SiH_4[236]; Machado and Masili[237] studied H_2; Stephens and McKoy[238] for OH; Braunstein et al.[239] for CH_4; Wells and Lucchese[240] for C_2H_2; for C_{60} we find the results by Gianturco and Lucchese[241]; and Wiedmann et al.[242] calculated the rotationally resolved PI cross section for CH_3, H_2O, H_2S, and H_2CO.

4.12.2 Continued Fractions

The method of continued fractions (MCF) was originally proposed by Horáček et Sasakawa[243–244] for the study of elastic scattering of fast electrons by atoms; subsequently, Lee et al. adapted it to study scattering of slow electrons by atoms[245] and by linear molecules,[246] and extended it to study ionization by electron collisions in polyatomic molecules.[247–248] The extension of the MCF to the PI study of molecules is explained with details in Ref. 249. The idea is to represent the scattering matrix as a continued fraction. The continuum wavefunction is obtained from the solution of the LSE using the static-exchange potential, with the long-range Coulomb potential of the ionic core removed. The MCF does not require basis functions and it is characterized by rapid convergence.

The application of the MCF starts with the definition of a nth-order weakened potential operator $U^{(n)}$, from which the reactance matrix $\mathbf{K}$ is expressed in the form of a continued fraction. The nth-order correction to $\mathbf{K}$, as well as to the wavefunction, can be approximated successively. The operator $U^{(n)}$ becomes weaker and weaker as n increases, and the procedure can be stopped after a few steps. The converged $\mathbf{K}$ matrix corresponds to the exact solution for a given potential U in LSE.[249]

To our knowledge, the MCF has been used only to study PI of NH_3.[249]

4.13 Crank–Nicolson

The Crank–Nicolson (CN) method[250] was originally developed to solve differential equations of heat-conduction type numerically using a combination of backward/forward finite difference of all variables involved. It is correct up to the second order in $\hat{H}\Delta t$, and is numerically stable. The CN scheme can be used to propagate an initial wavefunction with an imaginary time evolution operator, in which, by the Wick rotation, the time t is replaced by $-i\tau$. In such a way, any initial arbitrary state can converge directly to a particular desired state (bound or continuum), just by adjusting the time step of the propagator.

The CN scheme has been used to study general PI features by Goldberg and Shore[251]; we can find also different studies in PI of H_2^+ by Picón et al.,[252] Yuan et al.,[253] Silva et al.,[254] and Bian[255]. The ion HeH^{2+} has been studied by Bian[255] and the angular distributions for H_2 by Yuan et al.[253]

5. STURMIAN APPROACH

5.1 Generalized Sturmian Functions

In the literature, we can find different approaches to Sturmian functions, depending on the type of problem to be solved. There are essentially two lines, one associated to bound states and another to scattering problems. The first line initiated by Shull and Löwdin,[256] has been formalized by Goscinski[257] and impulsed later on by Aquilanti and coworkers.[258–259] It is within this line that the generalized Sturmian functions were introduced by Avery and coworkers[260–261] to deal with many electron atoms and chemical systems. On the scattering line, the work was initiated by Rawitscher[262–263] and continued by Macek, Ovchinnikov and coworkers.[264–265] Some of us extended the scattering functions proposed by Rawitscher and started to use them in scattering studies with the name Generalized Sturmian Functions

(GSFs) to indicate that the basis functions are solving general atomic potentials.

Details on the GSF used here are given in Refs. 22 and 23 (and references therein) and only the essentials are recalled below. GSF are solutions of a Sturm–Liouville problem, from which Rotenberg[266–267] took the name. Noted $\mathcal{S}_n^{(\ell,\,E)}(r)$, they are regular at the origin and satisfy the two-body non-homogeneous Schrödinger equation

$$\left[-\frac{1}{2}\frac{\mathrm{d}^2}{\mathrm{d}r^2}+\frac{\ell(\ell+1)}{2r^2}+\mathcal{U}(r)-E\right]\mathcal{S}_n^{(\ell,\,E)}(r)=-\beta_n^{(\ell,\,E)}\mathcal{V}(r)\mathcal{S}_n^{(\ell,\,E)}(r), \quad (9)$$

where E is an externally fixed parameter and $\beta_n^{(\ell,\,E)}$ are the eigenvalues for a given angular momentum ℓ. In general, the generating potential $\mathcal{V}(r)$, a short-range potential, dictates the size of the inner region in which most of the dynamics is supposed to occur, while the auxiliary potential $\mathcal{U}(r)$ determines the common asymptotic behavior of all GSFs. This property is illustrated in Fig. 5, for functions with a fixed energy $E=0.5$ a.u., an auxiliary Coulomb potential with charge $Z=-1$ and a Yukawa generating potential. The outgoing asymptotic behavior of the GSFs with an auxiliary Coulomb potential is given by (see second term of (3))

$$\lim_{r\to\infty}\mathcal{S}_n^{(\ell,E)}(r)\propto \mathrm{e}^{-i\left(kr-\frac{Z}{k}\ln(2kr)-\ell\frac{\pi}{2}+\delta_\ell\right)}, \quad (10)$$

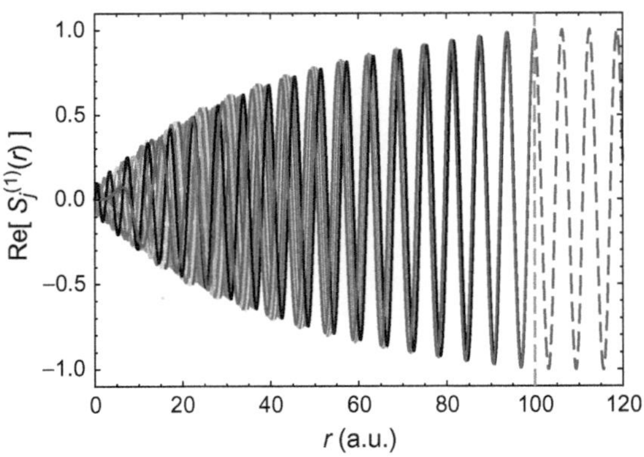

Figure 5 Real part of 10 generalized Sturmian functions, with a fixed energy $E=0.50$ a.u. and $\ell=1$, obtained solving Eq. (9), for $r\in[0,100]$, together with a Coulomb auxiliary potential with charge $Z=-1$, and a Yukawa potential as a generating potential, with parameter $\alpha_{\mathrm{short}}=0.0219$. The exact Coulomb (analytic) regular function (blue (gray in the print version), dash) is also shown.

where $\delta_\ell = \arg[\Gamma(l + 1 + iZ/k)]$ is the Coulomb phase shift. Additionally, all the solutions form a complete basis set, with the potential-weighted orthogonality relation

$$\int_0^\infty dr\, \mathcal{S}_{n'}^{(\ell,\,E)}(r)\mathcal{V}(r)\mathcal{S}_{n}^{(\ell,\,E)}(r) = \delta_{n'n}. \tag{11}$$

Note that the integral is defined without taking the complex conjugate of the function $\mathcal{S}_{n'}^{(\ell,\,E)}(r)$.

5.2 Sturmian Approach to Photoionization Process

We shall present in this section the theoretical formalism developed within a Sturmian approach for molecules. We start with a brief description of the molecular model potentials used, then we derive the driven equation of the TISE and provide the necessary formula to calculate the PI cross section. As a simple illustration of the numerical implementation in the atomic case, we show results for the hydrogen atom.

5.2.1 *Molecular Model Potentials*

To study PI of molecules, we shall use the SAE approximation[268] for the initial state wavefunction. We then need a molecular model potential that plays the role of a scattering potential. Consider the active electron placed in the MO i of the ground state and denote $\phi_i(\mathbf{r})$ the corresponding wavefunction. The molecular model potential we shall use is the following[269]

$$V_{i\,\mathrm{mol}}(\mathbf{r},\mathbf{R}) = -\sum_{n=1}^{M} \frac{Z_n}{|\mathbf{r}-\mathbf{R}_n|} + \sum_{j=1}^{N_{\mathrm{MO}}} N_{ij} \int d\mathbf{r}' \frac{\left|\phi_j(\mathbf{r}')\right|^2}{|\mathbf{r}-\mathbf{r}'|}, \tag{12}$$

where M is the number of nuclei in the molecule, Z_n is the charge of each nucleus, $\mathbf{R}_n$ is the position of each nucleus with respect to the molecule center of mass, N_{MO} is the number of MOs and $N_{ij} = 2 - \delta_{ij}$. This potential is the direct term within the SEA. For the sake of simplicity, the nuclei $\mathbf{R}_n$ dependence, collectively represented as $\mathbf{R}$, is omitted hereafter.

We shall take the MO i given by Moccia; they are expressed as

$$\phi_i(\mathbf{r}) = \sum_{j=1}^{N} A_{ij}\mathcal{R}_j(r) S_{\ell_j}^{m_j}(\hat{r}), \tag{13}$$

where $S_{\ell_j}^{m_j}(\hat{r})$ are the real spherical harmonics,[31] and the N radial wavefunctions are given as Slater type-orbitals (STOs)

$$\mathcal{R}_j(r) = \left[\frac{(2\zeta_j)^{2n_j+1}}{(2n_j)!}\right]^{1/2} r^{n_j-1} e^{-\zeta_j r}, \tag{14}$$

with tabulated integers n_j and exponents ζ_j. These MOs allow one to calculate analytically, in a partial-wave expansion, the molecular model potential.

As mentioned before, in a typical experiment the molecules are randomly oriented. Although this is not the proper way to proceed, we may consider as starting point an angular average of the model potential (12), i.e., a central potential

$$U_{i\,\text{mol}}(r) = \frac{1}{4\pi}\int_{4\pi} d\hat{r}\; V_{i\,\text{mol}}(\mathbf{r}). \tag{15}$$

This averaging procedure is illustrated through Fig. 6, where the effective charges $rU_{i\,\text{mol}}(r)$ and $rV_{i\,\text{mol}}(\mathbf{r})$ for two set of angles (θ,ϕ) are compared in the case of CH_4. The effective charge goes from -6 at $r=0$ and to -1 asymptotically. The minimum is located at $r \approx 2.08$ a.u., i.e., at the equilibrium position of each H atom; its depth and sharpness depend on the orientation and whether the angular average has been performed or not.

The model potential (12) proposed in this work can certainly be improved in many respects, some of which are under current investigation. One of them is inclusion of exchange. Also, as we use an independent particle approximation, some many-body aspects (i.e., correlation) are only included indirectly through the use of MO in Eq. (12) but not explicitly.

5.2.2 Driven Equation and Cross Section

To introduce our Sturmian approach, we start with the use of an arbitrary potential $U(\mathbf{r})$, such as the one given by Eq. (12). We describe the PI process using first-order perturbation theory for a molecule that interacts with a radiation field. The Hamiltonian can be written as

$$\widehat{\mathcal{H}} = \widehat{H}_0 + \widehat{W}(t), \tag{16}$$

where $\widehat{H}_0 = \widehat{T} + U(\mathbf{r}, \hat{\mathfrak{R}})$ is the field-free Hamiltonian of the target with $\hat{\mathfrak{R}} = (\alpha, \beta, \gamma)$ the set of Euler angles that specify the spatial orientation of the molecule; $\widehat{T}$ is the kinetic energy operator, and

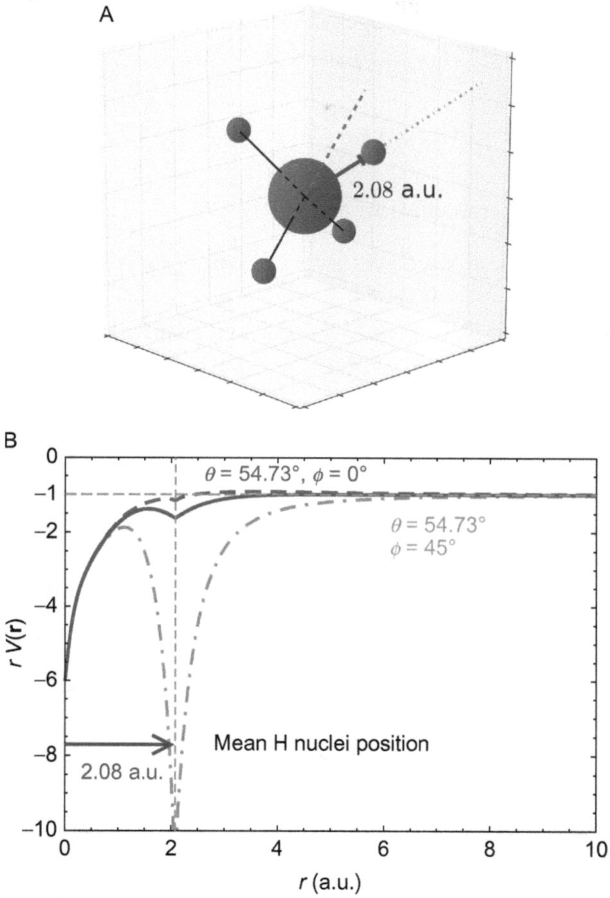

Figure 6 *Upper panel*: Representation of the spatial localization of the individual atoms on CH_4; the mean equilibrium distance is indicated. *Bottom panel*: Molecular model (12) (red (dark gray in the print version) dash and green (gray in the print version) dash-dot) and angular averaged (15) (blue (dark gray in the print version), solid) potentials for CH_4, at indicated angles. The potential in green (gray in the print version) (dash-dot) corresponds to the variation of the potential on the green (gray in the print version) (dot) path on the figure of the upper panel.

$$\widehat{W}(t) = \begin{cases} -F^{(L)}(t) \;\; \hat{\varepsilon}\cdot\mathbf{r} = F(t)\widehat{D}^{(L)}, & \text{length gauge} \\ -F^{(V)}(t) \;\; \hat{\varepsilon}\cdot\mathbf{p} = F(t)\widehat{D}^{(V)}, & \text{velocity gauge} \end{cases} \tag{17}$$

where $F^{(g)}(t)$ is the electric field in the length gauge or the vector potential in the velocity gauge, $\hat{\boldsymbol{\varepsilon}}$ gives the polarization of the field and $\widehat{D}$ are the dipolar operators (2); $F(t)$ contains the time-dependent profile of the radiation field.

Dropping the explicit $\hat{\mathfrak{R}}$-dependence for the moment, we begin with the TDSE for the total Hamiltonian (16)

$$\left(i\frac{\partial}{\partial t}-\widehat{\mathcal{H}}\right)\Psi(\mathbf{r},t)=\left(i\frac{\partial}{\partial t}-\widehat{H}_0-\widehat{W}(t)\right)\Psi(\mathbf{r},t)=0, \tag{18}$$

and propose the general solution to be

$$\Psi(\mathbf{r},t)=\mathrm{e}^{-i\omega_0 t}\left[\Phi^{(0)}(\mathbf{r})+\Psi_{\text{scatt}}(\mathbf{r},t)\right], \tag{19}$$

where $\Phi^{(0)}(\mathbf{r})$ is the wavefunction of the initial ground state of the molecule, usually the active MO to ionize, with energy $E_0=\omega_0$, and $\Psi_{\text{scatt}}(\mathbf{r},t)$ is the wavefunction of the photoelectron, with energy $\omega=E$ (in atomic units). Replacing (19) in (18), we obtain

$$\left[i\frac{\partial}{\partial t}-\omega_0-\widehat{H}_0-\widehat{W}(t)\right]\Psi_{\text{scatt}}(\mathbf{r},t)=\widehat{W}(t)\Phi^{(0)}(\mathbf{r}). \tag{20}$$

Now, if we apply a Fourier transform to (20), we obtain the TISE

$$\begin{aligned}\left(\omega-\omega_0-\widehat{H}_0\right)\Psi_{\text{scatt}}(\mathbf{r},\omega)-\frac{1}{\sqrt{2\pi}}\int_{-\infty}^{\infty}\mathrm{d}\omega'\,\widehat{\mathcal{W}}(\omega')\Psi_{\text{scatt}}(\mathbf{r},\omega-\omega')\\=\widehat{\mathcal{W}}(\omega)\Phi^{(0)}(\mathbf{r}),\end{aligned} \tag{21}$$

where $\widehat{\mathcal{W}}(\omega)$ is the Fourier transform of $\widehat{W}(t)$.

Equation (21) contains the interaction with the field to all orders, and therefore $\Psi_{\text{scatt}}(\mathbf{r},\omega)$ contains information over all possible processes. Neglecting the integral term of (21), we can introduce a perturbation expansion on the scattering wavefunction.[26] Since we are interested here only in single PI processes, we retain the first order, and then Eq. (21) results in the driven equation for the final state wavefunction

$$\left(\omega-\omega_0-\widehat{H}_0\right)\Psi^{(1)}(\mathbf{r},\omega)=\widehat{\mathcal{W}}(\omega)\Phi^{(0)}(\mathbf{r}). \tag{22}$$

This is the equation that we want to solve; the scattering wavefunction at first order, $\Psi^{(1)}(\mathbf{r},\omega)$, will provide the PI information.

To solve Eq. (22), we separate first the scattering wavefunction in its radial and angular parts

$$\Psi^{(1)}(\mathbf{r},\omega)=\frac{1}{r}\sum_{\ell m}\varphi_\ell(r,\omega)Y_\ell^m(\hat{r}). \tag{23}$$

Usually, the radial wavefunction $\varphi_\ell(r,\omega)$ is expanded in some radial basis set. Within our Sturmian approach it is expanded in a GSF set (see Section 5.1)

$$\varphi_\ell(r,\omega) = \sum_j a_j^{(\ell,\,E)}(\omega)\mathcal{S}_j^{(\ell,\,E)}(r). \tag{24}$$

Performing an angular projection, Eq. (22) is converted into a set of angular-coupled differential equations

$$\sum_{\ell m}\left[\left(\omega - \omega_0 + \frac{1}{2}\frac{\mathrm{d}^2}{\mathrm{d}r^2} - \frac{\ell(\ell+1)}{2r^2}\right)\delta_{\ell'\ell}\delta_{m'm} - U_{\ell'\ell}^{m'm}(r)\right]\varphi_\ell(r,\omega) = \varrho_{\ell'}^{m'}(r,\omega), \tag{25}$$

where $U_{\ell'\ell}^{m'm}(r) = \langle \ell' m' | U(\mathbf{r}) | \ell m \rangle$ and $\varrho_{\ell'}^{m'}(r,\omega) = r\left\langle \ell' m' \middle| \widehat{\mathcal{W}}(\omega) \middle| \Phi^{(0)} \right\rangle$. As mentioned in Section 1, the use of a noncentral potential to describe the molecular target couples directly the different angular momenta of the initial state. For atoms or angular averaged molecular potentials, on the other hand, there is no coupling, $U_{\ell'\ell}^{m'm}(r)$ is diagonal and we have a single radial equation

$$\left(\omega - \omega_0 + \frac{1}{2}\frac{\mathrm{d}^2}{\mathrm{d}r^2} - \frac{\ell(\ell+1)}{2r^2} - U(r)\right)\varphi_\ell(r,\omega) = \varrho_\ell^{m}(r,\omega). \tag{26}$$

Recall now that the potential $U(\mathbf{r},\hat{\mathfrak{R}})$ of the field-free Hamiltonian $\widehat{H}_0$ contains the orientation $\hat{\mathfrak{R}}$ of the molecule. This orientation dependence is to be accounted for by $\varphi_\ell(r,\omega)$ and finally by the coefficients $a_j^{(\ell,\,E)}(\omega)$, so that we actually have $a_j^{(\ell,\,E)}(\omega,\hat{\mathfrak{R}})$.

Now, to solve the coupled system of Eq. (25), we use the GSF expansion (24) and obtain

$$\begin{aligned}\sum_{\ell m}\sum_j a_j^{(\ell,\,E)}(\omega,\hat{\mathfrak{R}})\left[\left(\omega - \omega_0 + \frac{1}{2}\frac{\mathrm{d}^2}{\mathrm{d}r^2} - \frac{\ell(\ell+1)}{2r^2}\right)\delta_{\ell'\ell}\delta_{m'm} - U_{\ell'\ell}^{m'm}(r)\right] \\ \times\, \mathcal{S}_j^{(\ell,\,E)}(r) = \varrho_{\ell'}^{m'}(r,\omega).\end{aligned} \tag{27}$$

The final step consists in projecting (27) on $\mathcal{S}_i^{(\ell,\,E)}(r)$ (note that it is not the complex conjugate, see Eq. (11)): then, all the resulting matrices are calculated as indicated in Refs. 22 and 23. Solving the matrix problem with standard numerical methods provides the coefficients $a_j^{(\ell,\,E)}(\omega,\hat{\mathfrak{R}})$.

All GSFs of the basis set have the same and correct asymptotic behavior, in this case the behavior dictated by the Coulomb potential (see Section 5.1). This means that our basis functions possess, by construction, important physical information and need to expand essentially the inner region, whose size will be determined by the range of the driven term. This makes the basis set adequate and finally computationally efficient. From the asymptotic property of the GSFs, we obtain the transition amplitude directly from the expansion coefficients of the scattering wavefunction in (24)[270]

$$\mathcal{T}\left(\omega,\hat{r},\hat{\mathfrak{R}}\right) = -\sqrt{2\pi}\left\langle \Psi_{\mathbf{k}}^{(-)} \middle| \widehat{\mathcal{W}}(\omega) \middle| \Phi^{(0)} \right\rangle = \sum_{\ell m j} a_j^{(\ell,E)}\left(\omega,\hat{\mathfrak{R}}\right) \mathrm{e}^{-i\left(\delta_\ell - \ell\frac{\pi}{2}\right)} Y_\ell^m(\hat{r}). \tag{28}$$

After an angular projection, we finally have the PI cross section as a function of the photon energy[270]

$$\frac{\mathrm{d}\sigma^{(\ell)}\left(\hat{\mathfrak{R}}\right)}{\mathrm{d}E} = \frac{4\pi^2\, \omega_{ki}^{(g)}}{c}\, k\, \frac{1}{2\pi} \frac{\left| \sum_j a_j^{(\ell,\ E)}\left(\omega,\hat{\mathfrak{R}}\right) \right|^2}{|\mathscr{F}(\omega)|^2}. \tag{29}$$

where $\omega^{(\mathrm{L})} = E - E_0$ or $\omega^{(\mathrm{V})} = (E - E_0)^{-1}$ is the difference between final and initial energies in either length or velocity gauges, and $\mathscr{F}(\omega)$ is the Fourier transform of the radiation field profile $F(t)$.

5.2.3 Example: Hydrogen Atom

The coupled system of Eq. (27) allows us to study PI processes for any potential. For systems that are described with a central potential, we are left with a single differential equation i.e. Eq. (26).[270–271] Applications to a set of different molecules will be given in Section 5.3 and is illustrated here for the hydrogen atom. For this atomic target, we solved the TISE (27) in both length and velocity gauges for electron energies in the range $[0.00, 3.00]$. Each one of these energies was used as the fixed energy E to calculate our GSFs basis through Eq. (9), where a Coulomb potential with charge -1 was taken as auxiliary potential and a Yukawa potential with an energy-dependent parameter as generating potential. For the initial state, we used the exact ground state wavefunction of the atom. Our calculated PI cross section (29) is shown in Fig. 7, and is compared with the analytical formula given by Harriman.[272] Agreement between the cross sections in both gauges is perfect. Comparing with the analytical

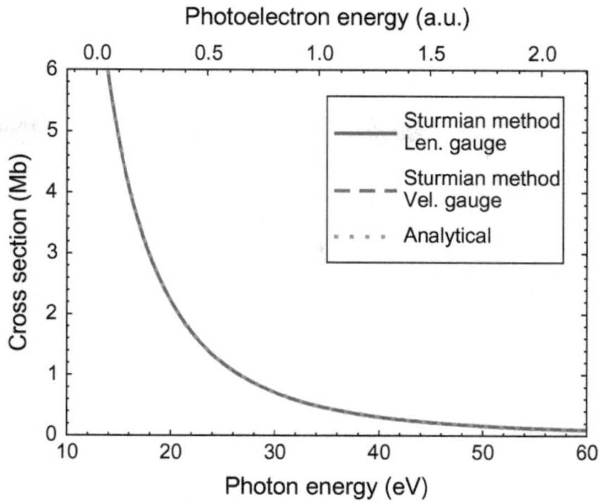

Figure 7 PI cross section of H atom from the ground state 1*s*, in Mb versus photon energy in eV. Our results in length (blue (dark gray in the print version), solid) and velocity (red (dark gray in the print version), dash) gauges are compared with the exact analytical formula by Harriman[272] (green (gray in the print version), dots).

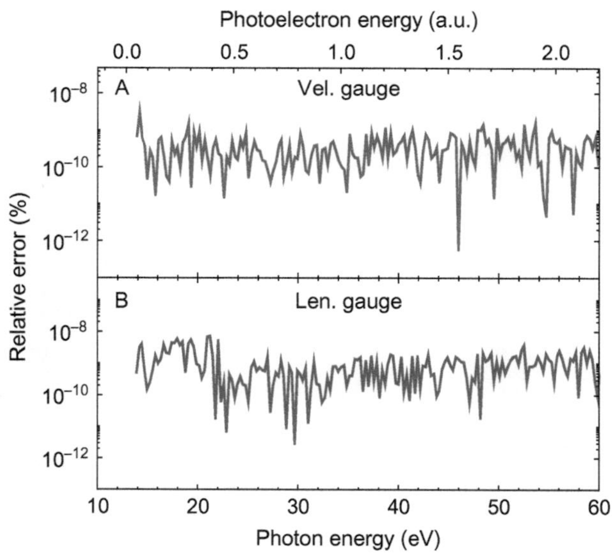

Figure 8 Relative errors for the calculated PI cross sections presented in Fig. 7: (A) velocity gauge and (B) length gauge.

formula, we obtain errors of the order of $10^{-8} \sim 10^{-11}$ % over all the energy range (see Fig. 8), showing that our results with the selected GSF parameters gives very stable and "numerically exact" solutions to the TISE (22).

5.3 Results for Molecules

In this section, we report some results obtained by applying our Sturmian approach for molecular single PI, first solving Eq. (27) for the angular averaged potential (15), and then using the noncentral potential (12). Some results have been partially published before[271,273] for CH_4 and H_2O. The treatment of molecular systems with an averaged (central) potential is similar to that of atomic systems. For all cases, we used 60 GSFs for each final energy and final (ℓ, m) set; the basis functions are defined in a box of 50 a.u., using an auxiliary Coulomb potential with charge -1 and a generating Yukawa potential with an energy-dependent parameter. We have verified that in all cases the cross sections are converged in terms of number of GSFs. The initial MO were taken from Moccia's publications, specifically for H_2O from Ref. 274, for NH_3 from Ref. 275 and for CH_4 from Ref. 276. For the calculation with the noncentral potential, with the fixed spatial molecular orientation given by Moccia, we use exactly the same GSF basis and initial MOs. The respective PI cross sections were calculated using Eq. (29), and we shall present our results in both length and velocity gauges. It is worth emphasizing here that the majority of theoretical publications on molecular PI present results obtained with the length gauge but do not provide a detailed analysis of gauge agreement, as often done for atomic systems.

5.3.1 H_2O

First, we start to study PI from the valence orbitals of H_2O whose electronic ground state configuration is $1a_1^2\, 2a_1^2\, 1b_2^2\, 3a_1^2\, 1b_1^2\ {}^1A_1$. We study here only the two valence MOs. For the inner valence orbital $3a_1$ ($E_0 = -15.1323$ eV), the calculated PI cross sections are shown in Fig. 9, and for the outer orbital $1b_1$ ($E_0 = -13.4805$ eV) in Fig. 10. Both are compared with TD-DFT calculations by Stener et al.,[92] GIPM/D by Kilcoyne et al.,[156], STT by Cacelli et al.,[200] and ISM by Machado et al.[235]; the experimental data were reported by Banna et al.[277]

For the MO $3a_1$, we observe a good agreement between our results in velocity gauge and other theoretical calculations, in particular for photon energies beyond 30 eV, where our results are close to the TD-DFT and GIPM/D; on the other hand, the length gauge results considerably overestimate the cross sections for all calculated energies. In general, the cross sections for inner valence orbitals are difficult to calculate accurately, due to the presence of different many-body effects, such as relaxation of the core.

For the MO $1b_1$, the gauge discrepancy is of the same order as for the $3a_1$ case. Our cross sections compare fairly with other theoretical results, ours

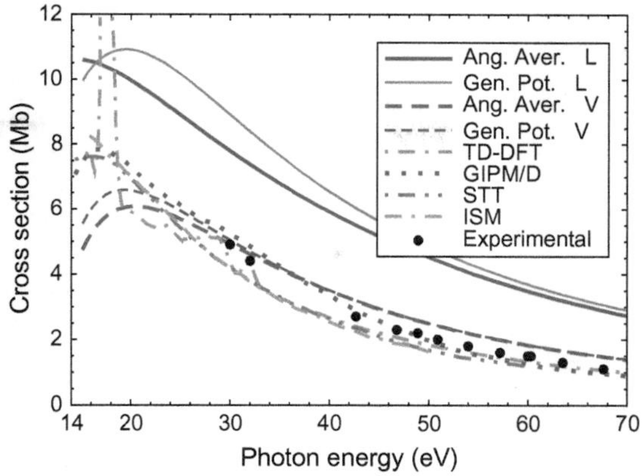

Figure 9 Partial PI cross section in Mb versus photon energy in eV from the MO $3a_1$ of H_2O. Our results using the angular averaged molecular potential (15) for length (blue (dark gray in the print version), solid) and velocity (red (dark gray in the print version), dash) gauges, and using the noncentral potential (12) in length (light blue (gray in the print version), thin solid) and velocity (purple (dark gray in the print version), thin dash) gauges are compared with results for TD-DFT[92] (green (gray in the print version), dash-dot); GIPM/D[156] (brown (dark gray in the print version), dots); STT[200] (gray, dash-dot-dot); ISM[235] (orange (gray in the print version), dash-dash-dot) and with experimental data[277] (black dots).

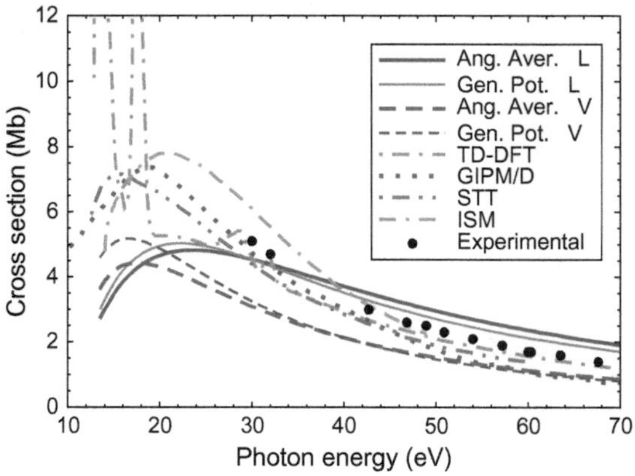

Figure 10 Same as Fig. 9 for MO $1b_1$ of H_2O.

being seemingly too low in the threshold region where unfortunately no experimental data are available.

The results obtained using the noncentral potential (12), are only slightly better, indicating therefore that the central potential (15) is good enough to study this particular molecule.

5.3.2 NH_3

Next we study PI for both valence orbitals of NH_3 whose ground state electronic structure is $1a_1^2\,2a_1^2\,1e^4\,3a_1^2\,{}^1A_1$. For the inner valence MO $1e$ ($E_0 = -16.2071$ eV), the cross section is shown in Fig. 11, and for the outer valence MO $3a_1$ ($E_0 = -11.2819$ eV) in Fig. 12. Our results are compared

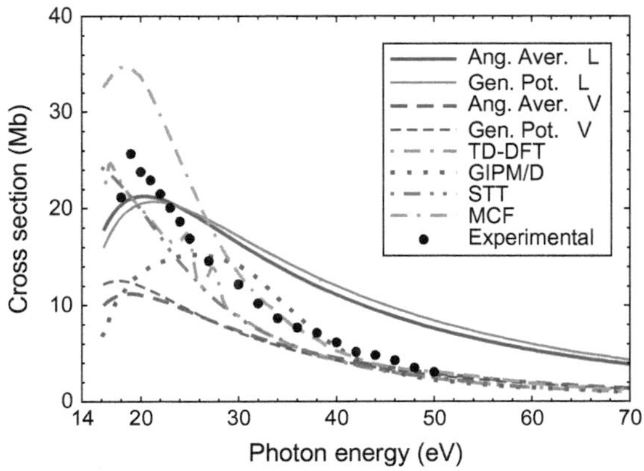

Figure 11 Partial PI cross section in Mb versus photon energy in eV from the MO 1*e* of NH_3. Our results using the angular averaged potential (15) for length (blue (dark gray in the print version), solid) and velocity (red (dark gray in the print version), dash) gauges, and using the noncentral (12) in length (light blue (gray in the print version), thin solid) and velocity (purple (dark gray in the print version), thin dash) gauges, are compared with results for TD-DFT[92] (green (gray in the print version), dash-dot); GIPM/D[156] (brown (dark gray in the print version), dots); STT[200] (gray, dash-dot-dot); MCF[249] (orange (gray in the print version), dash-dash-dot) and with experimental data[278] (black dots).

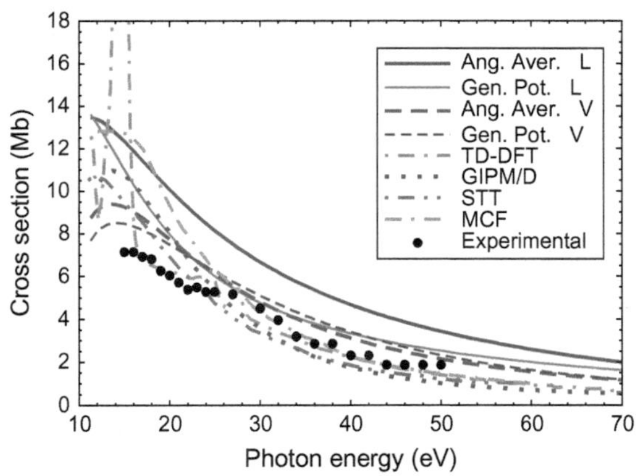

Figure 12 Same as Fig. 11 for the MO $3a_1$ of NH_3.

with the TD-DFT results by Stener et al.,[92] GIPM/D by Kilcoyne et al.,[156] STT by Cacelli et al.,[200] and with calculations using MCF by Nascimento et al.[249]; the experimental data were reported by Brion et al.[278]

For the orbital $1e$, our results in velocity gauge show only a fair agreement with all reported data, in particular at high photon energies. Gauge discrepancy is again important and, fortuitously, the length gauge results reproduce the experimental magnitude around 22 eV.

For the orbital $3a_1$, our results show a slightly better gauge agreement; the length gauge cross section presenting the same shape but with a larger magnitude. The results in velocity gauge are in acceptable agreement with the experimental data over the whole energy range.

As for H_2O, the use of the noncentral potential (12) has a small effect, except in length gauge for the $3a_1$ orbital.

5.3.3 CH_4

Finally, we show our results for CH_4 whose ground state electronic structure is $1a_1^2\,2a_1^2\,1t_2^6\;{}^1A_1$. The calculated PI cross sections in both length and velocity gauges for the inner valence MO $2a_1$ ($E_0 = -25.0454$ eV) are shown in Fig. 13, and for the outer valence MO $1t_2$ ($E_0 = -13.7199$ eV) in Fig. 14. They are compared with TD-DFT calculations by Stener et al.,[92] GIPM/D by Kilcoyne et al.,[156] MS Xα by Rosi et al.,[133] and with STT by Cacelli et al.;[201] the experimental data are taken from Backx and van der Wiel.[279]

For the inner valence orbital $2a_1$, the length gauge calculation shows no agreement with any other calculations. For higher energies, say beyond 40 eV, we have a good agreement between our velocity results and experimental and other theoretical data.

For outer valence orbital $1t_2$, results obtained in velocity gauge show a fair agreement with experimental data, at least for photon energies higher than 30 eV; near threshold the position of the experimental peak is rather well reproduced but not its magnitude. Length gauge results are about a factor two too large.

For this molecule, the effect of using the noncentral potential (12) is almost unnoticeable for the $1t_2$ orbital but improves the velocity gauge result for the $2a_1$ orbital at lower energies.

As can be observed from Figs. 9 to 14, the Sturmian approach can give reasonable PI cross sections, in particular for ionization from the outer MOs. Some general features are: (i) little difference is seen between the use of the angular averaged (central) potential (15) and the noncentral potential (12)

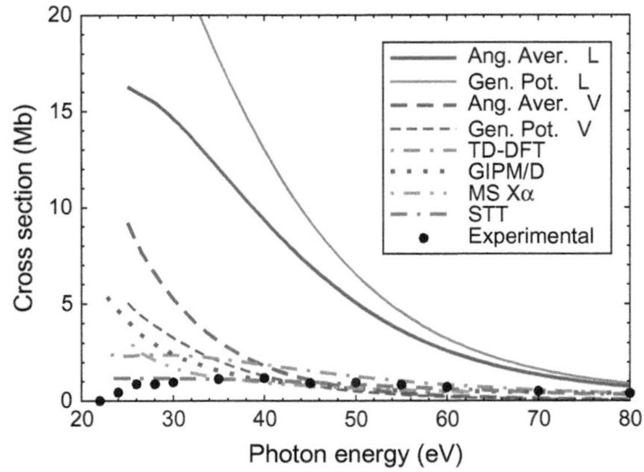

Figure 13 Partial PI cross section in Mb versus photon energy in eV from the MO $2a_1$ of CH_4. Our results using the central potential (15) for length (blue (dark gray in the print version), solid) and velocity (red (dark gray in the print version), dash) gauges, and using the noncentral (12) for length (light blue (gray in the print version), thin solid) and velocity (purple (dark gray in the print version), thin dash) gauges, are compared with TD-DFT[92] (green (gray in the print version), dash-dot); GIPM/D[156] (brown (dark gray in the print version), dots); MSM[133] (orange (gray in the print version), dash-dot-dot); STT[201] (gray, dash-dash-dot) and with experimental data[279] (black dots).

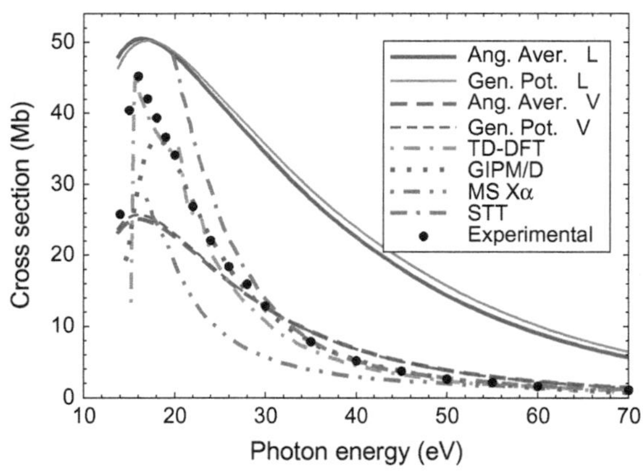

Figure 14 Same as Fig. 13 for the MO $1t_2$ of CH_4.

(the latter improves marginally the cross sections); (ii) the length gauge results are systematically much larger over the whole energy range than those obtained with the velocity gauge, and are generally not in agreement with other data (experimental or theoretical). This discrepancy indicates that the initial state description needs to be improved; (iii) our velocity gauge results are in overall fair agreement with other theoretical cross sections, in particular for energies above, say, 15–20 eV over the ionization threshold;

(iv) exactly as illustrated in Section 3 with other molecules, the agreement between theoretical results (including ours) and experimental data is not uniform. For energies below 15 eV above ionization threshold agreement is generally poor.

6. CONCLUSIONS

In this contribution, we explored different theoretical aspects of PI of molecules. The description of this process requires solving quantum mechanically a very difficult many-body and multicenter problem. Contrary to molecular electronic structure calculations, one needs to evaluate a continuum state with appropriate asymptotic conditions. To make calculations feasible, a number of approximations must be made. Moreover, over the years, a range of theoretical methods and numerical techniques have been proposed. Among them, one finds those familiar in quantum chemistry such as HF, CI, and DFT, but also others, such as the RMM, CS or RPA, which encountered great success in atomic scattering calculations and were extended to molecular targets. Since in most experiments the molecules are randomly oriented, an average of the cross sections must be performed over all spatial orientations. This is an extra computational challenge that does not appear when studying collisions with atomic targets.

We began (Section 3) by describing the degree of theoretical–experimental agreement one may find in the literature. The PI of the valence orbitals of H_2, N_2, CO_2, and C_6H_6, are taken as an illustration and pinpoint some of the challenges one is confronted with. Except for H_2, a nonuniform picture arises. Agreement between theoretical results and experimental data is not always satisfactory; moreover, depending on the molecular orbital which is ionized and the energy range, severe discrepancies are often observed between different calculations. This is due to both the approximations made and the technique adopted. For this reason, we provided, in Section 4, a brief description of each method, together with the list of molecules to which they are applied.

We then introduced (Section 5) our Sturmian approach for PI. Essentially, the method provides one-electron basis functions (named Generalized Sturmian Functions) with appropriately chosen asymptotic boundary conditions. As they intrinsically contain this property, the basis functions are particularly adequate in describing the ionized electron continuum state. We briefly described how the Sturmian method, developed originally for atoms, may be implemented for molecules with a noncentral molecular potential.

As indicated above, in order to reduce the complexity of the problem, different approximations must be considered to make it tractable. In this very first implementation of the Sturmian approach, we started with the OCE and the SAE approximations. These reduce considerably the dimension of the problem and allow one to deal with one-electron wavefunctions, an ideal starting point to test the versatility of our method for molecules. It is well known that these approximations are good enough to study symmetric molecules, and in particular the ones with a heavy nucleus in its center of mass; this is the case for the selected molecules in this work (H_2O, NH_3, and CH_4) for which we reported results for PI from their valence orbitals. We stress here that the computational procedure to obtain such results with an angular averaged molecular model potential (15) is exactly the same as that used to study PI in a hydrogen atom. The same GSF basis and initial state wavefunctions are used for calculations with the noncentral potential (12). It turns out that the use of the latter only slightly improves the calculated cross sections. A systematic gauge comparison clearly showed that the length gauge largely overestimates the spectra at all energies. Considering some of the crude approximations, we may state that the implemented technique yields velocity gauge cross sections in reasonable agreement with experimental data, in particular at higher photoelectron energies. Although clearly perfectible (see below), these results are promising since they demonstrate that we have a working computational tool to study the electronic spectra of different molecules.

Let us add a short comment on computational efficiency. As illustrated by several bound-state calculations reported in our review[22] (and references therein), the GSF method is able to deliver results with high accuracy and low computational cost. For scattering calculations, although there is not a rigorous way to perform efficiency comparisons, some estimations were given in Ref. 280 with studies of three-body atomic breakup problems. Comparisons between the GSF and state-of-the-art methods showed that our methodology improves the numerical efficiency by at least an order of magnitude. Recently, in a study of DPI of He,[25] the GSF method reproduced very precisely ECS differential cross sections with a substantial gain (more than 50%) in memory storage of Hamiltonian matrix. In the present molecular applications, the tool is similar. The built in properties make the GSF set very adequate (and, to our mind, efficient) to deal with scattering problems, here PI.

The use of the angular averaged molecular model potential, which is equivalent to include the random orientation of the molecule before the scattering calculations, gave us good results for high energies; for lower energies a better description of the target is clearly necessary. Besides the molecular potential itself, in that regime all the many-body effects are important, and the wavefunction for the initial state should include all active electrons. In that respect, we tried to include exchange terms in different manners. In our preliminary attempts, such terms did not remove the observed gauge discrepancies; we are currently investigating other ways to include in our model both exchange and correlations effects. Furthermore, the interactions with all nuclei becoming important, a many-center wavefunction should be employed; this may lead in a very expensive description of the system from the computational point of view, particularly for polyatomic molecules.

The use of a noncentral molecular potential (12) gave slightly better results in the low photon energy regime, since it is a more realistic description of the molecule. However, in this case the cross sections must be calculated for any Euler angles set. The final angular average over all possible spatial orientations of the molecule in the laboratory frame (defined by the polarization of the radiation field) is then computationally much more expensive.

In all investigated systems, in particular for inner valence orbitals, we saw that our length gauge results are always overestimating experimental data. The gauge discrepancies are known (see, for instance, Refs. 90, 200, 201, 203, 271 or 281) but are rarely discussed in the literature for molecular cases. They can be related to the quality of the wavefunctions for the initial state and also by the absence of different many-body effects. In order to obtain a better gauge agreement, one would need to use more sophisticated wave functions and avoid the FC and SAE approximations, since in some cases the relaxation effects are important, as shown, e.g., by TD-DFT results (see Sections 3 and 4.3.2 and Figs. 10 and 14).

As a summary, we have presented here some results of a first implementation of our Sturmian approach to study PI of molecules. Improvements of calculated cross sections may be obtained using more realistic molecular potentials and initial wavefunctions. Investigations in this direction are under way and will be presented elsewhere. GSFs offer also a promising tool to study other ionization processes in molecular systems such as the single ionization by electron impact, the so-called $(e, 2e)$ process.

ACKNOWLEDGMENTS

We acknowledge the CNRS (PICS project No. 06304) and CONICET (project No. Dl 158114) for funding our French-Argentinian collaboration, and CONICET (PIP 201301/607). G.G. also thanks the support by PGI (24/F059) of the Universidad Nacional del Sur.

APPENDIX. LIST OF PHOTOIONIZATION CALCULATIONS FOR DIFFERENT MOLECULES

We list here, molecule by molecule, the references of all applications of theoretical methods mentioned in Section 4.

Homonuclear diatomic

1. H_2^+ [109,110,115,116,119,212,214,252–254,255]
2. H_2 [28,33,34,36,37,58,60,61,76,77,111–113,117,119,122,129,146–148,151,154,155,165,166, 168,169,177–179,196,218,237,253]
3. D_2 [28,166]
4. Li_2 [69]
5. C_2 [1]
6. N_2 [39–43,52,68,90,102,118,122,129,130,132,146–150,152,154,155,171,172,175,176,180–182, 186,194,218]
7. O_2 [66]
8. F_2 [205]
9. Cl_2 [85]

Heteronuclear diatomic

10. HeH^{2+} [255]
11. HeH^+ [59,170]
12. LiH [184]
13. CH [195]
14. OH [238]
15. HF [78,83,92,149,154,189,201]
16. CN^- [172]
17. CO [35,41,45,52,68,72,93,94,102,103,129,132,147,149–152,172,204,220]
18. NO [122,171]
19. HCl [55,83,202]

Polyatomic

20. BH_3 [83]
21. H_2O [83,92,147,149–152,155,156,172,189,197–200,235,242,271]
22. NH_3 [83,92,150,156,172,189,200,221,249]
23. $LiCN$ [63]
24. AlH_3 [83]
25. H_2S [83,147,150,183,203,242]
26. PH_3 [83,90,150]
27. SiH_4 [83,135,236]

28. CO_2[42,45–47,54,70,122,132,172]
29. NO_2[42,45,71,172]
30. NO_2^-[172]
31. O_3[56]
32. C_2N_2[42,149]
33. CaHN[149]
34. C_2O_2[85]
35. COS[132]
36. NF_3[139]
37. CS_2[50,132]
38. CF_4[96,97,133,234]
39. PF_3[136,139]
40. CF_3Cl[138]
41. SiF_4[135]
42. SF_6[42,222,233]
43. CCl_4[234]
44. $SiCl_4$[134,135]
45. $TiCl_4$[42]

Organic molecules
46. CH_3[242]
47. CH_4[67,83,92,133,146–148,150–152,156,172,189,201,234,239,271,273]
48. CH_3I[137]
49. H_2CO[242]
50. C_2H_2[40,95,99,152,172,182,185,232,240]
51. C_2H_4[51,99,131,146,147,151,172]
52. C_2H_6[172]
53. $C_2(CN)_2$[149]
54. C_3H_6O (methyl-oxirane)[87]
55. C_3H_8[172]
56. $(CH_3)_2S$[150]
57. C_4H_6[64]
58. C_4H_4O (furan)[149]
59. C_4H_5N (pyrrole)[149]
60. $C_4H_4N_2$ (pyrimidine)[98]
61. $C_4H_4N_2$ (pyrazine)[98]
62. $C_4H_4N_2O_2$ (uracil)[88]
63. C_6H_6 (benzene)[49,51,52,86,206]
64. C_6F_6[150]

65. $C_4F_4N_2$[149]
66. $Cr(CO)_6$[42,85]
67. $C_{10}H_8$ (naphthalene)[86]
68. $C_{14}H_{10}$ (anthracene)[86]
69. $C_{16}H_{10}$ (pyrene)[86]

Fullerenes
70. C_{20}[188]
71. C_{60}^{+}[187]
72. C_{60}[100,188,241]

REFERENCES

1. Padial, N. T.; Collins, L. A.; Schneider, B. I. Photoionization of Ground-State Molecular Carbon C_2. *Astrophys. J.* **1985**, *298*, 369.
2. Liedahl, D. A.; Paerels, F. Photoionization-Driven X-Ray Line Emission in Cygnus X-3. *Astrophys. J.* **1996**, *468*, L33.
3. Bautista, M. A.; Romano, P.; Pradhan, A. K. Resonance-Averaged Photoionization Cross Sections for Astrophysical Models. *Astrophys. J. Suppl. Ser.* **1998**, *118*, 259.
4. Dopita, M. A.; Meatheringham, S. J. Photoionization Modeling of Magellanic Cloud Planetary Nebulae. I. *Astrophys. J.* **1991**, *367*, 115.
5. García-Segura, G.; Langer, N.; Rożyczka, M.; Franco, J. Shaping Bipolar and Elliptical Planetary Nebulae: Effects of Stellar Rotation, Photoionization Heating, and Magnetic Fields. *Astrophys. J.* **1999**, *517*, 767.
6. Monteiro, H.; Schwarz, H. E.; Gruenwald, R.; Heathcote, S. Three-Dimensional Photoionization Structure and Distances of Planetary Nebulae. I. NGC 6369. *Astrophys. J.* **2004**, *609*, 194.
7. Mallard, G.; Miller, J. H.; Smyth, K. C. Resonantly Enhanced Two-Photon Photoionization of NO in an Atmospheric Flame. *J. Chem. Phys.* **1982**, *76*, 3483.
8. Robb, D. B.; Blades, M. W. State-of-the-Art in Atmospheric Pressure Photoionization for LC/MS. *Anal. Chim. Acta* **2008**, *627*, 34.
9. Levine, J. S.; Javan, A. Observation of Laser Oscillation in a 1-atm CO_2-N_2-He Laser Pumped by an Electrically Heated Plasma Generated Via Photoionization. *Appl. Phys. Lett.* **1973**, *22*, 55.
10. Killian, T. C.; Kulin, S.; Bergeson, S. D.; Orozco, L. A.; Orzel, C.; Rolston, S. L. Creation of an Ultracold Neutral Plasma. *Phys. Rev. Lett.* **1999**, *83*, 4776.
11. Amusia, M. Y.; Baltenkov, A. S. Effect of Plasma Oscillations of C_{60} Collectivized Electrons on Photoionization of Endohedral Noble-Gas Atoms. *Phys. Rev. A.* **2006**, *73*, 062723.
12. Hubbell, J. H. Review of Photon Interaction Cross Section Data in the Medical and Biological Context. *Phys. Med. Biol.* **1999**, *44*, R1.
13. Stepanek, J.; Blattmann, H.; Laissue, J. A.; Lyubimova, N.; Di Michiel, M.; Slatkin, D. N. Physics Study of Microbeam Radiation Therapy with PSI-Version of Monte Carlo Code GEANT as a New Computational Tool. *Med. Phys.* **2000**, *27*, 1664.
14. Horsley, J. A.; Stöhr, J.; Hitchcock, A. P.; Newbury, D. C.; Johnson, A. L.; Sette, F. Resonances in the K Shell Excitation Spectra of Benzene and Pyridine: Gas Phase, Solid, and Chemisorbed States. *J. Chem. Phys.* **1985**, *83*, 6099.
15. Piancastelli, M. N.; Lindle, D. W.; Ferrett, T. A.; Shirley, D. A. The Relationship Between Shape Resonances and Bond Lengths. *J. Chem. Phys.* **1987**, *86*, 2765.

16. Sheehy, J. A.; Gil, T. J.; Winstead, C. L.; Farren, R. E.; Langhoff, P. W. Correlation of Molecular Valence- and K-shell Photoionization Resonances with Bond Lengths. *J. Chem. Phys.* **1989**, *91*, 1796.
17. Cassuto, A.; Mane, M.; Jupille, J. Ethylene Monolayer and Multilayer on Pt(111) Below 52 K: Determination of Bond Lengths by Near-Edge X-ray Fine Structure. *Surf. Sci. Lett.* **1991**, *249*, 8.
18. Tonner, B. P.; Kao, C. M.; Plummer, E. W.; Caves, T. C.; Messmer, R. P.; Salaneck, W. R. Intermolecular Screening in Core-Level Photoemission from the Nitric-Oxide Dimer. *Phys. Rev. Lett.* **1983**, *51*, 1378.
19. Stöhr, J.; Outka, D. A.; Baberschke, K.; Arvanitis, D.; Horsley, J. A. Identification of CH Resonances in the K-shell Excitation Spectra of Gas-Phase, Chemisorbed, and Polymeric Hydrocarbons. *Phys. Rev. B* **1987**, *36*, 2976.
20. Solomon, J. L.; Madix, R. J.; Stöhr, J. π Bonded Intermediates in Alcohol Oxidation: Orientations of Allyloxy and Propargyloxy on Ag(110) by Near Edge X-Ray Absorption Fine Structure. *J. Chem. Phys.* **1988**, *89*, 5316.
21. Liu, A. C.; Stöhr, J.; Friend, C. M.; Madix, R. J. A Critical Interpretation of the Near-Edge X-ray Absorption Fine Structure of Chemisorbed Benzene. *Surf. Sci.* **1990**, *235*, 107.
22. Gasaneo, G.; Ancarani, L. U.; Mitnik, D. M.; Randazzo, J. M.; Frapiccini, A. L.; Colavecchia, F. D. Three-Body Coulomb Problems with Generalized Sturmian Functions. *Adv. Quantum Chem.* **2013**, *67*, 153.
23. Mitnik, D. M.; Colavecchia, F. D.; Gasaneo, G.; Randazzo, J. M. Computational Methods for Generalized Sturmians basis. *Comp. Phys. Comm.* **2011**, *182*(5), 1145.
24. Ambrosio, M. J.; Colavecchia, F. D.; Gasaneo, G.; Mitnik, D. M.; Ancarani, L. U. Double Ionization of Helium by Fast Electrons with the Generalized Sturmian Functions Method. *J. Phys. B: At. Mol. Opt. Phys.* **2015**, *48*, 055204.
25. Randazzo, J. M.; Mitnik, D.; Gasaneo, G.; Ancarani, L. U.; Colavecchia, F. D. Double photoionization of helium: a generalized Sturmian approach. *Eur. J. Phys. D* **2015**, *69*, 189.
26. Messiah, A. *Quantum Mechanics*. North-Holland: Amsterdam, 1972.
27. Sanz-Vicario, J.; Bachau, H.; Martín, F. Time-Dependent Theoretical Description of Molecular Autoionization Produced by Femtosecond xuv Laser Pulses. *Phys. Rev. A* **2006**, *73*, 033410.
28. Sansone, G.; Kelkensberg, F.; Pérez-Torres, J. F.; Morales, F.; Kling, M. F.; Siu, W.; Ghafur, O.; Johnsson, P.; Swoboda, M.; Benedetti, E.; Ferrari, F.; Lépine, F.; Sanz-Vicario, J. L.; Zherebtsov, S.; Znakovskaya, I.; L'huillier, A.; Ivanov, M. Y.; Nisoli, M.; Martín, F.; Vrakking, M. J. J. Electron Localization Following Attosecond Molecular Photoionization. *Nature* **2010**, *465*, 763.
29. Fernández, J.; Martín, F. Electron and Ion Angular Distributions in Resonant Dissociative Photoionization of H_2 and D_2 Using Linearly Polarized Light. *New J. Phys.* **2009**, *11*, 043020.
30. Chandra, N. Photoelectron Spectroscopic Studies of Polyatomic Molecules: I. Theory. *J. Phys. B: At. Mol. Phys.* **1987**, *20*, 3405.
31. Edmonds, A. R. *Angular Momentum in Quantum Mechanics*; Princeton University Press: Princeton, NJ, 1957.
32. Chung, Y. M.; Lee, E. M.; Masuoka, T.; Samson, J. A. R. Dissociative Photoionization of H_2 from 18 to 124 eV. *J. Chem. Phys.* **1993**, *99*, 885.
33. Kelly, H. P. The Photoionization Cross Section for H_2 from Threshold to 30 eV. *Chem. Phys. Lett.* **1973**, *20*, 547.
34. Sanz-Vicario, J.; Palacios, A.; Cardona, J.; Bachau, H.; Martín, F. Ab Initio Time-Dependent Method to Study the Hydrogen Molecule Exposed to Intense Ultrashort Laser Pulses. *J. Electron Spectros. Relat. Phenom.* **2007**, *161*, 182.

35. Hilton, P. R.; Nordholm, S.; Hush, N. S. Ground-State Inversion Method Applied to Calculation of Molecular Photoionization Cross-Sections by Atomic Extrapolation: Interference Effects at Low Energies. *J. Electron Spectros. Relat. Phenom.* **1980**, *18*, 101.
36. Martín, P. H. S.; Rescigno, T. N.; McKoy, V.; Henneker, W. H. Photoionization Cross Sections for H_2 in the Random Phase Approximation with a Square-Integrable Basis. *Chem. Phys. Lett.* **1974**, *29*, 496.
37. Raşeev, G. Variational Calculation of the Logarithmic Derivative of the Wavefunction: The Electronic Autoionisation Region in Photoionisation of H_2. *J. Phys. B: At. Mol. Phys.* **1985**, *18*, 423.
38. Plummer, E. W.; Gustafsson, T.; Gudat, W.; Eastman, D. E. Partial Photoionization Cross Sections of N_2 and CO Using Synchrotron Radiation. *Phys. Rev. A* **1977**, *15*, 2339.
39. Stratmann, R. E.; Bandarage, G.; Lucchese, R. R. Electron-Correlation Effects in the Photoionization of N_2. *Phys. Rev. A* **1995**, *51*, 3756.
40. Levine, Z. H.; Soven, P. Time-Dependent Local-Density Theory of Dielectric Effects in Small Molecules. *Phys. Rev. A* **1984**, *29*, 625.
41. Davenport, J. Ultraviolet Photoionization Cross Sections for N_2 and CO. *Phys. Rev. Lett.* **1976**, *36*, 945.
42. Stener, M.; Decleva, P.; Lisini, A. Molecular Photoionization Cross Sections by the Local Density LCAO Stieltjes Imaging Approach. *J. Electron Spectrosc. Relat. Phenom.* **1995**, *74*, 29.
43. Lucchese, R.; Raseev, G.; McKoy, V. Studies of Differential and Total Photoionization Cross Sections of Molecular Nitrogen. *Phys. Rev. A* **1982**, *25*, 2572.
44. Brion, C.; Tan, K. Partial Oscillator Strengths for the Photoionization of N_2O and CO_2 (20-60 eV). *Chem. Phys.* **1978**, *34*, 141.
45. Kilcoyne, D. A. L.; Nordholm, S.; Hush, N. S. An Analysis of Photoionisation Cross Sections for Carbon Monoxide and Dioxide and Nitrous Oxide by Diffraction Theory. *Chem. Phys.* **1986**, *107*, 225.
46. Lucchese, R.; McKoy, V. Studies of Differential and Total Photoionization Cross Sections of Carbon Dioxide. *Phys. Rev. A* **1982**, *26*, 1406.
47. Harvey, A. G.; Brambila, D. S.; Morales, F.; Smirnova, O. An R-Matrix Approach to Electron-Photon-Molecule Collisions: Photoelectron Angular Distributions from Aligned Molecules. *J. Phys. B: At. Mol. Opt. Phys.* **2014**, *47*, 215005.
48. Carlson, T. A.; Gerard, P.; Krause, M. O.; Grimm, F. A.; Pullen, B. P. Photoelectron Dynamics of the Valence Shells of Benzene as a Function of Photon Energy. *J. Chem. Phys.* **1987**, *86*, 6918.
49. Venuti, M.; Stener, M.; Decleva, P. Valence Photoionization of C_6H_6 by the B-Spline One-Centre Expansion Density Functional Method. *Chem. Phys.* **1998**, *234*, 95.
50. Stener, M.; Fronzoni, G.; Decleva, P. Time-Dependent Density-Functional Theory for Molecular Photoionization with Noniterative Algorithm and Multicenter B-Spline Basis Set: CS_2 and C_6H_6 Case Studies. *J. Chem. Phys.* **2005**, *122*, 234301.
51. Kilcoyne, D. A. L.; Nordholm, S.; Hush, N. S. Photoionisation of Ethylene and Benzene: A Theoretical Analysis of Multicentre Diffraction Effects. *Chem. Phys.* **1986**, *107*, 255.
52. Wilhelmy, I.; Ackermann, L.; Görling, A.; Roösch, N. Molecular Photoionization Cross Sections by the Lobatto Technique. I. Valence Photoionization. *J. Chem. Phys.* **1994**, *100*, 2808.
53. Langhoff, P. W. Aspects of Electronic Configuration Interaction in Molecular Photoionization. In: Hinze, J. Ed.; Electron-Atom and Electron-Molecule Collisions; Springer: New York, NY, 1983.
54. Daasch, W. R.; Davidson, E. R.; Hazi, A. U. Oxygen K Hole Photoionization Cross Section of CO_2. *J. Chem. Phys.* **1982**, *76*, 6031.

55. van Dishoeck, E. F.; van Hemert, M. C.; Dalgarno, A. Photodissociation Processes in the HCl Molecule. *J. Chem. Phys.* **1982**, *77*, 3693.
56. Decleva, P.; De Alti, G.; Lisini, A. Theoretical Study of the Valence Photoelectron Spectrum of Ozone: An Analysis of Correlation Effects and Configuration Interaction (CI) Model Spaces. *J. Chem. Phys.* **1988**, *89*, 367.
57. Bachau, H.; Cormier, E.; Decleva, P.; Hansen, J. E.; Martín, F. Applications of *B*-splines in Atomic and Molecular Physics. *Rep. Prog. Phys.* **2001**, *64*, 1815.
58. Apalategui, A.; Saenz, A. Multiphoton Ionization of the Hydrogen Molecule H_2. *J. Phys. B: At. Mol. Opt. Phys.* **2002**, *35*, 1909.
59. Vanne, Y. V.; Saenz, A. Numerical Treatment of Diatomic Two-Electron Molecules Using a B-Spline Based CI Method. *J. Phys. B: At. Mol. Opt. Phys.* **2004**, *37*, 4101.
60. Fojón, O. A.; Fernández, J.; Palacios, A.; Rivarola, R. D.; Martín, F. Interferences Effects in H_2 Photoionization at High Energies. *J. Phys. B: At. Mol. Opt. Phys.* **2004**, *37*, 3035.
61. Dowek, D.; Pérez-Torres, J. F.; Picard, Y. J.; Billaud, P.; Elkharrat, C.; Houver, J. C.; Sanz-Vicario, J. L.; Martín, F. Circular Dichroism in Photoionization of H_2. *Phys. Rev. Lett.* **2010**, *104*, 233003.
62. Klamroth, T. Laser-Driven Electron Transfer through Metal-Insulator-Metal Contacts: Time-Dependent Configuration Interaction Singles Calculations for a Jellium Model. *Phys. Rev. B* **2003**, *68*, 245421.
63. Klinkusch, S.; Saalfrank, P.; Klamroth, T. Laser-Induced Electron Dynamics Including Photoionization: A Heuristic Model within Time-Dependent Configuration Interaction Theory. *J. Chem. Phys.* **2009**, *131*, 114304.
64. Sonk, J. A.; Schlegel, H. B. TD-CI Simulation of the Electronic Optical Response of Molecules in Intense Fields II: Comparison of DFT Functionals and EOM-CCSD. *J. Phys. Chem. A* **2011**, *115*, 11832.
65. Nesbet, R. K. *Variational Methods in Electron-Atom Scattering Theory*; Plenum: New York, NY, 1979.
66. Stratmann, R. E.; Lucchese, R. R. A Graphical Unitary Group Approach to Study Multiplet Specific Multichannel Electron Correlation Effects in the Photoionization of O_2. *J. Chem. Phys.* **1995**, *102*, 8493.
67. Dalgarno, A. The Photo-Ionization Cross Section of Methane. *Proc. Phys. Soc. A* **1952**, *65*, 663.
68. Schirmer, J.; Cederbaum, L.; Domcke, W.; von Niessen, W. Strong Correlation Effects in Inner Valence Ionization of N_2 AND CO. *Chem. Phys.* **1977**, *26*, 149.
69. Larkins, F. P.; Richards, J. A. Photoionisation and Auger Electron Emission from the Lithium Molecule: Calculations Using Multicentre Numerical Continuum Functions. *Aust. J. Phys.* **1986**, *39*, 809.
70. Saito, N.; Fanis, A. D.; Kubozuka, K.; Machida, M.; Takahashi, M.; Yoshida, H.; Suzuki, I. H.; Cassimi, A.; Czasch, A.; Schmidt, L.; Dörner, R.; Wang, K.; Zimmermann, B.; McKoy, V.; Koyano, I.; Ueda, K. Carbon K-shell Photoelectron Angular Distribution from Fixed-In-Space CO2 Molecules. *J. Phys. B: At. Mol. Opt. Phys.* **2003**, *36*, L25.
71. Saito, N.; Toffoli, D.; Lucchese, R. R.; Nagoshi, M.; De Fanis, A.; Tamenori, Y.; Oura, M.; Yamaoka, H.; Kitajima, M.; Tanaka, H.; Hergenhahn, U.; Ueda, K. Symmetry- and Multiplet-Resolved N 1s Photoionization Cross Sections of the NO_2 Molecule. *Phys. Rev. A* **2004**, *70*, 062724.
72. Semenov, S. K.; Cherepkov, N. A.; Jahnke, T.; Dörner, R. Theoretical Study of Vibrationally Resolved Photoionization for the C K-Shell of the CO Molecule. *J. Phys. B: At. Mol. Opt. Phys.* **2004**, *37*, 1331.
73. Ågren, H.; Carravetta, V.; Vahtras, O.; Pettersson, L. G. M. Direct SCF Direct Static-Exchange Calculations of Electronic Spectra. *Theor. Chem. Acc.* **1997**, *97*, 14.

74. Nest, M.; Klamroth, T.; Saalfrank, P. The Multiconfiguration Time-Dependent Hartree-Fock Method for Quantum Chemical Calculations. *J. Chem. Phys.* **2005**, *122*, 124102.
75. Alon, O. E.; Streltsov, A. I.; Cederbaum, L. S. Many-Body Theory for Systems with Particle Conversion: Extending the Multiconfigurational Time-Dependent Hartree Method. *Phys. Rev. A* **2009**, *79*, 022503.
76. Kato, T.; Kono, H. Time-Dependent Multiconfiguration Theory for Ultrafast Electronic Dynamics of Molecules in an Intense Laser Field: A Description in Terms of Numerical Orbital Functions. *J. Chem. Phys.* **2008**, *128*, 184102.
77. Haxton, D. J.; Lawler, K. V.; McCurdy, C. W. Multiconfiguration Time-Dependent Hartree-Fock Treatment of Electronic and Nuclear Dynamics in Diatomic Molecules. *Phys. Rev. A* **2011**, *83*, 063416.
78. Haxton, D. J.; Lawler, K. V.; McCurdy, C. W. Single Photoionization of Be and HF Using the Multiconfiguration Time-Dependent Hartree-Fock Method. *Phys. Rev. A* **2012**, *86*, 013406.
79. Hohenberg, P.; Kohn, W. Inhomogeneous Electron Gas. *Phys. Rev.* **1964**, *136*, B864.
80. Kohn, W.; Sham, L. J. Self-Consistent Equations including Exchange and Correlation Effects. *Phys. Rev.* **1965**, *140*, A1133.
81. van Leeuwen, R.; Baerends, E. J. Exchange-Correlation Potential with Correct Asymptotic Behavior. *Phys. Rev. A* **1994**, *49*, 2421.
82. Görling, A. New KS Method for Molecules Based on an Exchange Charge Density Generating the Exact Local KS Exchange Potential. *Phys. Rev. Lett.* **1999**, *83*, 5459.
83. Stener, M.; Decleva, P. Photoionization of First and Second Row Hydrides by the B-Spline One-Centre Expansion Density Functional Method. *J. Electron Spectros. Relat. Phenom.* **1998**, *94*, 195.
84. Stener, M.; Decleva, P. Photoionization of CH_4, SiH_4, BH_3 and AlH_3 by the B-Spline One-Centre Expansion Density Functional Method. *J. Electron Spectros. Relat. Phenom.* **1999**, *104*, 135.
85. Toffoli, D.; Stener, M.; Fronzoni, G.; Decleva, P. Convergence of the Multicenter B-Spline DFT Approach for the Continuum. *Chem. Phys.* **2002**, *276*, 25.
86. Woon, D. E.; Park, J. Photoionization of Benzene and Small Polycyclic Aromatic Hydrocarbons in Ultravioletprocessed Astrophysical Ices: A Computational Study. *Astrophys. J.* **2004**, *607*, 342.
87. Stranges, S.; Turchini, S.; Alagia, M.; Alberti, G.; Contini, G.; Decleva, P.; Fronzoni, G.; Stener, M.; Zema, N.; Prosperi, T. Valence Photoionization Dynamics in Circular Dichroism of Chiral Free Molecules: The Methyl-Oxirane. *J. Chem. Phys.* **2005**, *122*, 244303.
88. Toffoli, D.; Decleva, P.; Gianturco, F. A.; Lucchese, R. R. Density Functional Theory for the Photoionization Dynamics of Uracil. *J. Chem. Phys.* **2007**, *127*, 234317.
89. Runge, E.; Gross, E. K. U. Density-Functional Theory for Time-Dependent Systems. *Phys. Rev. Lett.* **1984**, *52*, 997.
90. Stener, M.; Decleva, P. Time-Dependent Density Functional Calculations of Molecular Photoionization Cross Sections: N_2 and PH_3. *J. Chem. Phys.* **2000**, *112*, 10871.
91. Zangwill, A.; Soven, P. Density-Functional Approach to Local-Field Effects in Finite Systems: Photoabsorption in the Rare Gases. *Phys. Rev. A* **1980**, *21*, 1561.
92. Stener, M.; Fronzoni, G.; Toffoli, D.; Decleva, P. Time Dependent Density Functional Photoionization of CH_4, NH_3, H_2O and HF. *Chem. Phys.* **2002**, *282*, 337.
93. Stener, M.; Decleva, P.; Cacelli, I.; Moccia, R.; Montuoro, R. Response Function Study of CO Photoionization: Ab Initio SCF and Density Functional Results. *Chem. Phys.* **2001**, *272*, 15.

94. Stener, M. Photoionization of Oriented Molecules: A Time Dependent Density Functional Approach. *Chem. Phys. Lett.* **2002**, *356*, 153.
95. Fronzoni, G.; Stener, M.; Decleva, P. Valence and Core Photoionization Dynamics of Acetylene by TD-DFT Continuum Approach. *Chem. Phys.* **2004**, *298*, 141.
96. Toffoli, D.; Stener, M.; Fronzoni, G.; Decleva, P. Photoionization Cross Section and Angular Distribution Calculations of Carbon Tetrafluoride. *J. Chem. Phys.* **2006**, *124*, 214313.
97. Patanen, M.; Kooser, K.; Argenti, L.; Ayuso, D.; Kimura, M.; Mondal, S.; Plésiat, E.; Palacios, A.; Sakai, K.; Travnikova, O.; Decleva, P.; Kukk, E.; Miron, C.; Ueda, K.; Martín, F. Vibrationally Resolved C 1s Photoionization Cross Section of CF_4. *J. Phys. B: At. Mol. Opt. Phys.* **2014**, *47*, 124032.
98. Holland, D. M. P.; Potts, A. W.; Karlsson, L.; Stener, M.; Decleva, P. A Study of the Valence Shell Photoionisation Dynamics of Pyrimidine and Pyrazine. *Chem. Phys.* **2011**, *390*, 25.
99. Russakoff, A.; Bubin, S.; Xie, X.; Erattupuzha, S.; Kitzler, M. Time-Dependent Density-Functional Study of the Alignment-Dependent Ionization of Acetylene and Ethylene by Strong Laser Pulses. *Phys. Rev. A* **2015**, *91*, 023422.
100. Madjet, M. E.; Chakraborty, H. S.; Rost, J. M.; Manson, S. T. Photoionization of C_{60}: A Model Study. *J. Phys. B: At. Mol. Opt. Phys.* **2008**, *41*, 105101.
101. Stener, M.; Toffoli, D.; Fronzoni, G.; Decleva, P. Recent Advances in Molecular Photoionization by Density Functional Theory Based Approaches. *Theor. Chem. Acc.* **2007**, *117*, 943.
102. Plésiat, E.; Decleva, P.; Martín, F. Vibrationally Resolved Photoelectron Angular Distributions from Randomly Oriented and Fixed-in-Space N_2 and CO Molecules. *J. Phys. B: At. Mol. Opt. Phys.* **2012**, *45*, 194008.
103. Kukk, E.; Ayuso, D.; Thomas, T. D.; Decleva, P.; Patanen, M.; Argenti, L.; Plésiat, E.; Palacios, A.; Kooser, K.; Travnikova, O.; Mondal, S.; Kimura, M.; Sakai, K.; Miron, C.; Martín, F.; Ueda, K. Effects of Molecular Potential and Geometry on Atomic Core-Level Photoemission over an Extended Energy Range: The Case Study of the CO Molecule. *Phys. Rev. A* **2013**, *88*, 033412.
104. Reinhardt, W. P. Complex Coordinates in the Theory of Atomic and Molecular Structure and Dynamics. *Ann. Rev. Phys. Chem.* **1982**, *33*, 223.
105. Moiseyev, N. Quantum theory of Resonances: Calculating Energies, Widths and Cross-Sections by Complex Scaling. *Phys. Rep.* **1998**, *302*, 212.
106. Nicolaides, C. A.; Beck, D. R. The Variational Calculation of Energies and Widths of Resonances. *Phys. Lett. A* **1978**, *65*, 11.
107. Simon, B. The Definition of Molecular Resonance Curves by the Method of Exterior Complex Scaling. *Phys. Lett. A* **1979**, *71*, 211.
108. McCurdy, C. W.; Martín, F. Implementation of Exterior Complex Scaling in B-Splines to Solve Atomic and Molecular Collision Problems. *J. Phys. B: At. Mol. Opt. Phys.* **2004**, *37*, 917.
109. McCurdy, C. W.; Rescigno, T. N. Complex-Basis-Function Calculations of Resolvent Matrix Elements: Molecular Photoionization. *Phys. Rev. A* **1980**, *21*, 1499.
110. Rescigno, T. N.; McCurdy, C. W. Locally Complex Distortions of the Energy Spectrum in the Calculation of Scattering Amplitudes and Photoionization Cross Sections. *Phys. Rev. A* **1985**, *31*, 624.
111. Vanroose, W.; Martín, F.; Rescigno, T. N.; McCurdy, C. W. Nonperturbative Theory of Double Photoionization of the Hydrogen Molecule. *Phys. Rev. A* **2004**, *70*.
112. Vanroose, W.; Horner, D. A.; Martín, F.; Rescigno, T. N.; McCurdy, C. W. Double Photoionization of Aligned Molecular Hydrogen. *Phys. Rev. A* **2006**, *74*, 052702.

113. Rescigno, T. N.; Vanroose, W.; Horner, D. A.; Martín, F.; McCurdy, C. W. First Principles Study of Double Photoionization of H_2 Using Exterior Complex Scaling. *J. Electron Spectros. Relat. Phenom.* **2007**, *161*, 85.
114. Light, J. C.; Hamilton, I. P.; Lill, J. V. Generalized Discrete Variable Approximation in Quantum Mechanics. *J. Chem. Phys.* **1985**, *82*, 1400.
115. Tao, L.; McCurdy, C. W.; Rescigno, T. N. Grid-Based Methods for Diatomic Quantum Scattering Problems: A Finite-Element Discrete-Variable Representation in Prolate Spheroidal Coordinates. *Phys. Rev. A* **2009**, *79*, 012719.
116. Tao, L.; McCurdy, C. W.; Rescigno, T. N. Grid-Based Methods for Diatomic Quantum Scattering Problems. II. Time-Dependent Treatment of Single- and Two-Photon Ionization of H_2^+. *Phys. Rev. A* **2009**, *80*, 013402.
117. Tao, L.; McCurdy, C. W.; Rescigno, T. N. Grid-Based Methods for Diatomic Quantum Scattering Problems. III. Double Photoionization of Molecular Hydrogen in Prolate Spheroidal Coordinates. *Phys. Rev. A* **2010**, *82*, 023423.
118. Yu, C.-h.; Pitzer, R. M.; McCurdy, C. W. Molecular Photoionization Cross Sections by the Complex-Basis- Function Method. *Phys. Rev. A* **1985**, *32*, 2134.
119. Morita, M.; Yabushita, S. Photoionization Cross Sections of H_2^+ and H_2 with Complex Gaussian-Type Basis Functions Optimized for the Frequency-Dependent Polarizabilities. *J. Comput. Chem.* **2008**, *29*, 2471.
120. Collins, L. A.; Schneider, B. I. Linear-Algebraic Approach to Electron-Molecule Collisions: General Formulation. *Phys. Rev. A* **1981**, *24*, 2387.
121. Collins, L. A.; Schneider, B. I. 2. The Linear Algebraic Method for Electron-Molecule Collisions. In: Huo, M. W., Gianturco, F. A., Eds.; Computational Methods for Electron-Molecule Collisions; Springer: New York, NY, 1995.
122. Collins, L.; Schneider, B. Molecular Photoionization in the Linear Algebraic Approach: H_2, N_2, NO, and CO_2. *Phys. Rev. A* **1984**, *29*, 1695.
123. Schneider, B. I.; Collins, L. A. Ab Initio Optical Potentials Applied to Low-Energy e-H_2 and e-N_2 Collisions in the Linear-Algebraic Approach. *Phys. Rev. A* **1983**, *27*, 2847.
124. Agassi, D.; Gal, A. Scattering from Non-Overlapping Potentials. I. General Formulation. *Ann. Phys. (N.Y.)* **1973**, *75*, 56.
125. Korringa, J. On the Calculation of the Energy of a Bloch Wave in a Metal. *Physica* **1947**, *13*, 392.
126. Dill, D.; Dehmer, J. L. Electron-Molecule Scattering and Molecular Photoionization Using the Multiple-Scattering Method. *J. Chem. Phys.* **1974**, *61*, 692.
127. Johnson, K. H. Scattered-Wave Theory of the Chemical Bond. In: Lödwin, P. O. Ed.; Advances in Quantum Chemistry; 7. Academic: New York, NY, 1973.
128. Slater, J. C.; Johnson, K. H. Self-Consistent-Field Xα Cluster Method for Polyatomic Molecules and Solids. *Phys. Rev. B* **1972**, *5*, 844.
129. Davenport, J. W. Multiple Scattering Theory of Photoemission. *Int. J. Quantum Chem.* **1977**, *12*, 89.
130. Dehmer, J. L.; Dill, D. Molecular Effects on Inner-Shell Photoabsorption. K-Shell Spectrum of N_2. *J. Chem. Phys.* **1976**, *65*, 5327–5334.
131. Grimm, F. A. Calculations of the Partial Differential Photoionization Cross Sections for the Valence Bands of Ethylene. *Chem. Phys.* **1983**, *81*, 315.
132. Grimm, F. A.; Carlson, T. A.; Dress, W. B.; Agron, P.; Thomson, J. O.; Davenport, J. W. Use of the Multiple-Scattering Method for Calculating the Asymmetry Parameter in the Angle-Resolved Photoelectron Spectroscopy of N_2, CO, CO_2, COS, and CS_2. *J. Chem. Phys.* **1980**, *72*, 3041.
133. Rosi, M.; Sgamellotti, A.; Tarantelli, F.; Andreev, V. A.; Gofman, M.; Nefedov, V. Theoretical Investigation of the Energy Dependence of Photoionization Cross-Sections and Angular Distributions of Photoemission of CH_4 and CF_4. *J. Electron Spectros. Relat. Phenom.* **1986**, *41*, 439.

134. Tse, J. S.; Liu, Z. F.; Bozek, J. D.; Bancroft, G. M. Multiple-Scattering Xα Study of the Silicon and Chlorine Core-Level Photoabsorption Spectra of $SiCl_4$. *Phys. Rev. A* **1989**, *39*, 1791.
135. Ishikawa, H.; Fujima, K.; Adachi, H.; Miyauchi, E.; Fujii, T. Calculation of Electronic Structure and Photoabsorption Spectra of Monosilane Molecules SiH_4, SiF_4, and $SiCl_4$. *J. Chem. Phys.* **1991**, *94*, 6740.
136. Powis, I. Continuum MS-Xα Calculations for Core and Outer Valence Shell Photoionization of PF_3. *Chem. Phys. Lett.* **1993**, *215*, 269.
137. Powis, I. A Theoretical CMS-Xα Treatment of CH_3I Photoionization Dynamics: Outer Valence Shell and Iodine 4d Levels. *Chem. Phys.* **1995**, *201*, 189.
138. Powis, I. Oriented Molecule Photoelectron Angular Distributions in the Vicinity of a Photoionization Shape Resonance: Continuum Multiple Scattering- Xα Calculations for Valence Ionization of CF_3Cl. *J. Chem. Phys.* **1997**, *106*, 5013.
139. Jürgensen, A.; Cavell, R. G. Valence Shell Photoionization Energies and Cross-Sections of NF_3 and PF_3. *J. Electron Spectros. Relat. Phenom.* **2003**, *128*, 245.
140. Ellison, F. O. Theoretical Equations for Photoionization Cross Sections of Polyatomic Molecules in Plane-Wave and Orthogonalized Plane-Wave Approximations. *J. Chem. Phys.* **1974**, *61*, 507.
141. Kaplan, I. G.; Markin, A. P. Calculation of Photoionization Cross Sections of Molecular Systems. I. Equations for Photoionization Cross Sections in AMO LCAO Approximation. *Opt. Spectrosc.* **1968**, *24*, 475.
142. Kaplan, I. G.; Markin, A. P. Calculation of Photionization Cross Sections of Molecular Systems. II. Ethylene, Butadiene and Benzene. *Opt. Spectrosc.* **1968**, *25*, 275.
143. Lohr, L. L.; Robin, M. B. Theoretical Study of Photoionization Cross Sections for π-Electron Systems. *J. Am. Chem. Soc.* **1970**, *92*, 7241.
144. Thiel, W.; Schweig, A. Photoionization Cross Sections in the Valence Electron Approximation. I. Linear Molecules. *Chem. Phys. Lett.* **1971**, *12*, 49.
145. Schweig, A.; Thiel, W. Photoionization Cross Sections: He I and He II Photoelectron Spectra of Saturated Three-Membered Rings. *Chem. Phys. Lett.* **1973**, *21*, 541.
146. Rabalais, J. W.; Debies, T. P.; Berkosky, J. L.; Huang, J. J.; Ellison, F. O. Calculated Photoionization Cross Sections and Relative Experimental Photoionization Intensities for a Selection of Small Molecules. *J. Chem. Phys.* **1974**, *61*, 516.
147. Dewar, M. J. S.; Komornicki, A.; Thiel, W.; Schweig, A. Calculation of Photoionization Cross Sections Using Ab-initio Wavefunctions and the Plane Wave Approximation. *Chem. Phys. Lett.* **1975**, *31*, 286.
148. Huang, J. T. J.; Ellison, F. O. Angular Asymmetry Parameters of Photoelectrons from H_2, N_2 and CH_4; an Extended Orthogonalized Plane-Wave Calculation. *Chem. Phys. Lett.* **1975**, *32*, 196.
149. Beerlage, M. J. M.; Feil, D. A Modified Plane Wave Model for Calculating UV Photo-Ionization Cross-Sections. *J. Electron Spectros. Relat. Phenom.* **1977**, *12*, 161.
150. Schweig, A.; Thiel, W. Photoionization Cross Sections: Interpretation of Band Intensities in He I and He II Photoelectron Spectra. *J. Electron Spectros. Relat. Phenom.* **1974**, *3*, 27.
151. Hilton, P. R.; Nordholm, S.; Hush, N. S. Molecular Photoionization Cross Sections Calculated by an Effective Plane Wave Method. *Chem. Phys.* **1976**, *15*, 345.
152. Deleuze, M.; Pickup, B. T.; Delhalle, J. Plane Wave and Orthogonalized Plane Wave Many-Body Green's Function Calculations of Photoionization Intensities. *Mol. Phys.* **1994**, *83*, 655.
153. Hilton, P. R.; Nordholm, S.; Hush, N. S. The Ground State Inversion Potential Method: Application to the Calculation of Photoionization Cross Sections. *J. Chem. Phys.* **1977**, *67*, 5213.
154. Kilcoyne, D. A. L.; McCarthy, C.; Nordholm, S.; Hush, N. S.; Hilton, P. R. An Atomic Diffraction Theory of Molecular Photoionization Cross Sections. *J. Electron Spectros. Relat. Phenom.* **1985**, *36*, 153–185.

155. Hilton, P. R.; Hordholm, S.; Hush, N. S. Photoionization Cross Section of Water by an Atomic Extrapolation Method. *Chem. Phys. Lett.* **1979**, *64*, 515.
156. Kilcoyne, D. A. L.; Nordholm, S.; Hush, N. S. Diffraction Analysis of the Photoionisation Cross Sections of Water, Ammonia and Methane. *Chem. Phys.* **1986**, *107*, 213.
157. In: *Atomic and Molecular Processes, an R-Matrix Approach*; Burke, P. G.; Berrington, K. A., Eds.; Institute of Physics Publishing: Bristol, 1993.
158. Bartschat, K. The R-matrix with Pseudo-States Method: Theory and Applications to Electron Scattering and Photoionization. *Comput. Phys. Commun.* **1998**, *114*, 168.
159. Schneider, B. I. 8. An *R*-Matrix Approach to Electron-Molecule Collisions. In: *Computational Methods for Electron-Molecule Collisions;* Huo, M. W., Gianturco, F. A., Eds.; Springer: New York, NY, 1995.
160. Noble, C. J. 14. *R*-Matrix for Intermediate Energy Scattering and Photoionization. In: *Computational Methods for Electron-Molecule Collisions;* Huo, M. W., Gianturco, F. A., Eds.; Springer: New York, NY, 1995.
161. Schneider, B. I.; LeDourneuf, M.; Burke, P. G. Theory of Vibrational Excitation and Dissociative Attachment: An R-matrix Approach. *J. Phys. B: At. Mol. Phys.* **1979**, *12*, L365.
162. Burke, P. G.; Seaton, M. J. The Vicinity of an R-Matrix Pole. *J. Phys. B: At. Mol. Phys.* **1984**, *17*, L683.
163. Seaton, M. J. Use of the R Matrix Method for Bound-State Calculations. I. General Theory. *J. Phys. B: At. Mol. Phys.* **1985**, *18*, 2111.
164. Seaton, M. J. Outer-Region Contributions to Radiative Transition Probabilities. *J. Phys. B: At. Mol. Phys.* **1986**, *19*, 2601.
165. Tennyson, J.; Noble, C.; Burke, P. Continuum States of the Hydrogen Molecule With the R-Matrix Method. *Int. J. Quantum Chem.* **1986**, *XXIX*, 1033.
166. Tennyson, J. Fully Vibrationally Resolved Photoionization of H_2 and D_2. *J. Phys. B: At. Mol. Phys.* **1987**, *20*, L375.
167. Joachain, C. J. R-matrix-Floquet Theory of Multiphoton Processes: Concepts, Results and Perspectives. *J. Mod. Opt.* **2007**, *54*, 1859.
168. Burke, P. G.; Colgan, J.; Glass, D. H.; Higgins, K. R-matrix-Floquet Theory of Molecular Multiphoton Processes. *J. Phys. B: At. Mol. Opt. Phys.* **2000**, *33*, 143.
169. Colgan, J.; Glass, D. H.; Higgins, K.; Burke, P. G. R-matrix Floquet Theory of Molecular Multiphoton Processes: II. Multiphoton Ionization of H_2. *J. Phys. B: At. Mol. Opt. Phys.* **2001**, *34*, 2089.
170. Saenz, A. Photoabsorption and Photoionization of HeH^+. *Phys. Rev. A* **2003**, *67*, 033409.
171. Tashiro, M. Application of the R-Matrix Method to Photoionization of Molecules. *J. Chem. Phys.* **2010**, *132*, 134306.
172. Amusia, M. Y.; Cherepkov, N. A. Many-Electron Correlations in the Scattering Processes. *Case Stud. At. Phys.* **1975**, *5*, 47.
173. Amusia, M. Y. Theory of Photoionization: VUV and Soft X-Ray Frequency Region. In: Becker, U., Shirley, D. A., Eds.; VUV and Soft X-Ray Photoionization; Plenum: New York, NY, 1996.
174. Rowe, D. Equations-of-Motion Method and the Extended Shell Model. *Rev. Mod. Phys.* **1968**, *40*, 153.
175. Yabushita, S.; McCurdy, C.; Rescigno, T. Complex-Basis-Function Treatment of Photoionization in the Random-Phase Approximation. *Phys. Rev. A* **1987**, *36*, 3146.
176. Semenov, S.; Cherepkov, N.; Fecher, G.; Schönhense, G. Generalization of the Atomic Random-Phase-Approximation Method for Diatomic Molecules: N_2 Photoionization Cross-Section Calculations. *Phys. Rev. A* **2000**, *61*, 032704.
177. Semenov, S. K.; Cherepkov, N. A. Photoionization of the H_2 Molecule in the Random Phase Approximation. *J. Phys. B: At. Mol. Opt. Phys.* **2003**, *36*, 1409.

178. Schirmer, J.; Mertins, F. A New Approach to the Random Phase Approximation. *J. Phys. B: At. Mol. Opt. Phys.* **1996**, *29*, 3559.
179. Semenov, S. K.; Cherepkov, N. A. Generalization of the Atomic RPA Method for Diatomic Molecules: H_2 Photoionization Cross-Section Calculation. *Chem. Phys. Lett.* **1998**, *291*, 375.
180. Lucchese, R.; Zurales, R. Comparison of the Random-Phase Approximation with the Multichannel Frozen-Core Hartree-Fock Approximation for the Photoionization of N_2. *Phys. Rev. A* **1991**, *44*, 291.
181. Semenov, S.; Cherepkov, N. Generalization of Atomic Random-Phase-Approximation Method for Diatomic Molecules. II. N_2 K-shell Photoionization. *Phys. Rev. A* **2002**, *66*, 022708.
182. Montuoro, R.; Moccia, R. Photoionization Cross Sections Calculation with Mixed L^2 Basis Set: STOs Plus B-Splines. Results for N_2 and C_2H_2 by KM-RPA Method. *Chem. Phys.* **2003**, *293*, 281.
183. Cacelli, I.; Carravetta, V.; Moccia, R. Differential Photoionization Cross Section Calculations for H_2S Using the Random Phase Approximation with L2 Basis Functions. *Chem. Phys.* **1994**, *184*, 213.
184. Carmona-Novillo, E.; Moccia, R.; Spizzo, P. Photoionization Cross Section and Asymmetry Parameter of LiH: A Mixed GTO/STOCOS L^2 Basis Set Calculation. *Chem. Phys.* **1996**, *210*, 435.
185. Yasuike, T.; Yabushita, S. Valence Photoionization and Autoionizing States of Acetylene Studied by the Complex Basis Function Method in the Random Phase Approximation. *Chem. Phys. Lett.* **2000**, *316*, 257.
186. Cherepkov, N.; Semenov, S.; Hikosaka, Y.; Ito, K.; Motoki, S.; Yagishita, A. Manifestation of Many-Electron Correlations in Photoionization of the K Shell of N_2. *Phys. Rev. Lett.* **2000**, *84*, 250.
187. Polozkov, R. G.; Ivanov, V. K.; Solov'yov, A. V. Photoionization of the Fullerene Ion C_{60}^{+}. *J. Phys. B: At. Mol. Opt. Phys.* **2005**, *38*, 4341.
188. Ivanov, V. K.; Kashenock, G. Y.; Polozkov, R. G.; Solov'yov, A. V. Photoionization Cross Sections of the Fullerenes C_{20} and C_{60} Calculated in a Simple Spherical Model. *J. Phys. B: At. Mol. Opt. Phys.* **2001**, *34*, L669.
189. Cacelli, I.; Carravetta, V.; Moccia, R.; Rizzo, A. Photoionization and Photoabsorption Cross Section Calculations in Methane, Ammonia, Water, and Hydrogen Fluoride Molecules. *J. Phys. Chem.* **1988**, *92*, 979.
190. Langhoff, P. W. Stieltjes Imaging of Atomic and Molecular Photoabsorption Profiles. *Chem. Phys. Lett.* **1973**, *22*, 60.
191. Langhoff, P. W. Stieltjes-Tchebycheff Moment-Tehory Approach to Molecular Photoionization Studies. In: *Electron-Molecule and Photon-Molecule Collisions* Rescigno, T., McKoy, V., Schneider, B., Eds.; Plenum: New York, NY, 1979.
192. Shohat, J. A.; Tamarkin, J. D. The Problem of Moments. *Mathematical Surveys,* 1. American Mathematical Society: Providence, RI, 1943.
193. Corcoran, C. T.; Langhoff, P. W. Moment-Theory Approximations for Nonnegative Spectral Densities. *J. Math. Phys.* **1977**, *18*, 651.
194. Rescigno, T. N.; Bender, C. F.; McKoy, B. V.; Langhoff, P. W. Photoabsorption in Molecular Nitrogen: A Moment Analysis of Discrete-Basis-Set Calculations in the Static-Exchange Approximation. *J. Chem. Phys.* **1978**, *68*, 970.
195. Barsuhn, J.; Nesbet, R. K. The Photoionization and Photodissociation of CH in the Vicinity of the Ionization Threshold. *J. Chem. Phys.* **1978**, *68*, 2783.
196. ONeil, S. V.; Reinhardt, W. P. Photoionization of Molecular Hydrogen. *J. Chem. Phys.* **1978**, *69*, 2126.
197. Williams, G. R. J.; Langhoff, P. W. Photoabsorption in H_2O: Stieltjes-Tchebycheff Calculations in the Time-Dependent Hartree-Fock Approximation. *Chem. Phys. Lett.* **1979**, *60*, 201.

198. Delaney, J. J.; Saunders, V. R.; Hillier, I. H. Stieltjes-Tchebycheff Calculations in the Static-Exchange Approximation of Photoexcitation and Ionisation in Water. *J. Phys. B: At. Mol. Phys.* **1981**, *14*, 819.
199. Diercksen, G. H. F.; Kraemer, W. P.; Rescigno, T. N.; Bender, C. F.; Mckoy, B. V.; Langhoff, S. R.; Langhoff, P. W. Theoretical Studies of Photoexcitation and Ionization in H_2O. *J. Chem. Phys.* **1982**, *76*, 1043.
200. Cacelli, I.; Moccia, R.; Carravetta, V. Photoionisation Cross Section Calculations for H_2O and NH_3 by One-Center Expansion and Sticltjies Technique. *Chem. Phys.* **1984**, *90*, 313.
201. Cacelli, I.; Carravetta, V.; Moccia, R. Transition Probability and Photoionisation Cross Section Calculations for CH_4 and HF by One-Centre Expansion and Stieltjes Technique. *J. Phys. B: At. Mol. Phys.* **1985**, *18*, 1375.
202. Cacelli, I.; Carravetta, V.; Moccia, R. Photoionization Cross Section Calculations of HCl by the Stieltjes Technique. *Mol. Phys.* **1986**, *59*, 385.
203. Cacelli, I.; Carravetta, V.; Moccia, R. H_2S Photoabsorption and Photoionization Cross Sections by Stieltjes Imaging. *Chem. Phys.* **1988**, *120*, 51.
204. Görling, A.; Rösch, N. Molecular Photo Cross Sections with the LCGTO-Xα Method Using Stieltjes Imaging. *J. Chem. Phys.* **1990**, *93*, 5563.
205. Orel, A. E.; Rescigno, T. N.; Mckoy, B. V.; Langhoff, P. W. Photoexcitation and Ionization in Molecular Fluorine: Stieltjes-Tchebycheff Calculations in the Static-Exchange Approximation. *J. Chem. Phys.* **1980**, *72*, 1265.
206. Gokhberg, K.; Vysotskiy, V.; Cederbaum, L. S.; Storchi, L.; Tarantelli, F.; Averbukh, V. Molecular Photoionization Cross Sections by Stieltjes-Chebyshev Moment Theory Applied to Lanczos Pseudospectra. *J. Chem. Phys.* **2009**, *130*, 064104.
207. Kohn, W. Variational Methods in Nuclear Collision Problems. *Phys. Rev.* **1948**, *74*, 1763.
208. Rescigno, T. N.; McCurdy, C. W.; Orel, A. E.; Lengsfield, B. H., III 1. The Complex Kohn Variational Method. In: Huo, M. W., Gianturco, F. A., Eds.; Computational Methods for Electron-Molecule Collisions; Springer: New York, NY, 1995.
209. McCurdy, C. W.; Rescigno, T. N. Collisions of Electrons with Polyatomic Molecules: Electron-Methane Scattering by the Complex Kohn Variational Method. *Phys. Rev. A* **1989**, *39*, 4487.
210. Manolopoulos, D.; Wyatt, R. Quantum Scattering Via the Log Derivative Version of the Kohn Variational Principle. *Chem. Phys. Lett.* **1988**, *152*, 23.
211. Manolopoulos, D. E.; Wyatt, R. E.; Clary, D. C. Iterative Solution in Quantum Scattering Theory. The Log Derivative Kohn Approach. *J. Chem. Soc. Faraday Trans.* **1990**, *86*, 1641.
212. Le Rouzo, H.; Raşeev, G. Finite-Volume Variational Method: First Application to Direct Molecular Photoionization. *Phys. Rev. A* **1984**, *29*, 1214.
213. Manolopoulos, D. E. Lobatto Shape Functions. In: Cerjan, C. Ed.; Numerical Grid Methods and Their Application to Schrödinger's Equation; Springer: The Netherlands, 1993.
214. Rösch, N.; Wilhelmy, I. Representation of Electronic Wavefunctions by Lobatto Shape Functions: Application to the Photoionization Cross Section of H_2^+. *Chem. Phys. Lett.* **1992**, *189*, 499.
215. Wilhelmy, I.; Rösch, N. Molecular Photoionization Cross Sections by the Lobatto Technique. II. Core Level Photionization. *Chem. Phys.* **1994**, *185*, 317–332.
216. Orel, A. E.; Rescigno, T. N. Variational Expressions for First-Order Properties involving Continuum Wave Functions. *Phys. Rev. A* **1990**, *41*, 1695.
217. Rescigno, T.; Lengsfield, B.; McCurdy, C. Electronic Excitation of Formaldehyde by Low-Energy Electrons: A Theoretical Study Using the Complex Kohn variational Method. *Phys. Rev. A* **1990**, *41*, 2462.
218. Lynch, D. L.; Schneider, B. I. Molecular Photoionization Using the Complex Kohn Variational Method. *Phys. Rev. A* **1992**, *45*, 4494.

219. Nesbet, R. K. Comparison of the R-matrix and Hulthén-Kohn methods for a model multichannel scattering problem. *Phys. Rev. A* **1981**, *24*, 2975.
220. Rescigno, T. N.; Lengsfield, B. H.; Orel, A. E. Interchannel Coupling and Ground State Correlation Effects in the Photoionization of CO. *J. Chem. Phys.* **1993**, *99*, 5097.
221. Orel, A. E.; Rescigno, T. N. Photoionization of Ammonia. *Chem. Phys. Lett.* **1997**, *269*, 222.
222. Jose, J.; Lucchese, R. R.; Rescigno, T. N. Interchannel Coupling Effects in the Valence Photoionization of SF_6. *J. Chem. Phys.* **2014**, *140*, 204305.
223. Schwinger, J. Minutes of the Meeting at Stanford University, California July 11-12, 1947 [14]. *Phys. Rev.* **1947**, *72*, 738, See p. 742.
224. Lippmann, B. A.; Schwinger, J. Variational Principles for Scattering Processes. I. *Phys. Rev.* **1950**, *79*, 469.
225. Huo, W. M. 15. The Schwinger Variational Method. In: Huo, M. W., Gianturco, F. A., Eds.; Computational Methods for Electron-Molecule Collisions; Springer: New York, NY, 1995.
226. Taylor, J. R. *Scattering Theory: The Quantum Theory of Nonrelativistic Collisions*. Wiley: New York, 1972.
227. Lucchese, R. R.; McKoy, V. Application of the Schwinger variational principle to electron scattering. *J. Phys. B: At. Mol. Phys.* **1979**, *12*, L421.
228. Takatsuka, K.; McKoy, V. Extension of the Schwinger variational principle beyond the Static-Exchange Approximation. *Phys. Rev. A* **1981**, *24*, 2473.
229. Watson, D. K.; McKoy, V. Discrete-Basis-Function Approach to Electron-Molecule Scattering. *Phys. Rev. A* **1979**, *20*, 1474.
230. Lucchese, R. R.; Watson, D. K.; McKoy, V. Iterative Approach to the Schwinger Variational Principle for Electron-Molecule Collisions. *Phys. Rev. A* **1980**, *22*, 421.
231. Lucchese, R.; McKoy, V. Iterative Approach to the Schwinger Variational Principle Applied to ElectronMolecular-Ion Collisions. *Phys. Rev. A* **1981**, *24*, 770.
232. Lynch, D.; Lee, M. T.; Lucchese, R. R.; McKoy, V. Studies of the Photoionization Cross Sections of Acetylene. *J. Chem. Phys.* **1984**, *80*, 1907.
233. Natalense, A. P. P.; Lucchese, R. R. Cross Section and Asymmetry Parameter Calculation for Sulfur 1s Photoionization of SF_6. *J. Chem. Phys.* **1999**, *111*, 5344.
234. Natalense, A.; Brescansin, L.; Lucchese, R. Cross Section and Asymmetry Parameter Calculations for the C 1s Photoionization of CH_4, CF_4, and CCl_4. *Phys. Rev. A* **2003**, *68*, 032701.
235. Machado, L. E.; Brescansin, L. M.; Lima, M. A. P.; Braunstein, M.; McKoy, V. Cross Sections and Photoelectron Asymmetry Parameters for Photoionization of H_2O. *J. Chem. Phys.* **1990**, *92*, 2362.
236. Machado, L. E.; Lee, M. T.; Brescansin, L. M. Photoionization Cross Sections and Asymmetry Parameters for Silane. *J. Chem. Phys.* **1999**, *110*, 7228.
237. Machado, A. M.; Masili, M. Variationally Stable Calculations for Molecular Systems: Polarizabilities and Two-Photon Ionization Cross Section for the Hydrogen Molecule. *J. Chem. Phys.* **2004**, *120*, 7505.
238. Stephens, J. A.; McKoy, V. Photoionization of the Valence Orbitals of OH. *J. Chem. Phys.* **1988**, *88*, 1737.
239. Braunstein, M.; McKoy, V.; Machado, L. E.; Brescansin, L. M.; Lima, M. A. P. Studies of the Photoionization Cross Sections of CH_4. *J. Chem. Phys.* **1988**, *89*, 2998.
240. Wells, M.; Lucchese, R. R. The Inner Valence Photoionization of Acetylene. *J. Chem. Phys.* **1999**, *110*, 6365.
241. Gianturco, F.; Lucchese, R. Cross Sections and Asymmetry Parameters in Gas-Phase Photoionization of C_60. *Phys. Rev. A* **2001**, *64*, 032706.
242. Wiedmann, R. T.; White, M. G.; Wang, K.; McKoy, V. Rotationally Resolved Photoionization of Polyatomic Hydrides: CH_3, H_2O, H_2S, H_2CO. *J. Chem. Phys.* **1994**, *100*, 4738.

243. Horáček, J.; Sasakawa, T. Method of Continued Fractions with Application to Atomic Physics. *Phys. Rev. A* **1983**, *28*, 2151.
244. Horáček, J.; Sasakawa, T. Method of Continued Fractions with Application to Atomic Physics. II. *Phys. Rev. A* **1984**, *30*, 2274.
245. Lee, M. T.; Iga, I.; Fujimoto, M. M.; Lara, O. The Method of Continued Fractions for Electron (Positron)-Atom Scattering. *J. Phys. B: At. Mol. Opt. Phys.* **1995**, *28*, L299.
246. Lee, M. T.; Iga, I.; Fujimoto, M. M.; Lara, O. Application of the Method of Continued Fractions for Electron Scattering by Linear Molecules. *J. Phys. B: At. Mol. Opt. Phys.* **1995**, *28*, 3325.
247. Ribeiro, E. M. S.; Machado, L. E.; Lee, M. T.; Brescansin, L. M. Application of the Method of Continued Fractions to Electron Scattering by Polyatomic Molecules. *Comput. Phys. Commun.* **2001**, *136*, 117.
248. Machado, A.; Fujimoto, M.; Taveira, A.; Brescansin, L.; Lee, M. T. Application of the Method of Continued Fractions to Multichannel Studies on Electronic Excitation of H_2 by Electron Impact. *Phys. Rev. A* **2001**, *63*, 032707.
249. Nascimento, E. M.; Ribeiro, E. M. S.; Brescansin, L. M.; Lee, M. T.; Machado, L. E. Extension of the Method of Continued Fractions to Molecular Photoionization: An Application to Ammonia. *J. Phys. B: At. Mol. Opt. Phys.* **2003**, *36*, 3621.
250. Crank, J.; Nicolson, P. A practical Method for Numerical Evaluation of Solutions of Partial Differential Equations of the Heat-Conduction Type. *Proc. Cambridge Philos. Soc.* **1947**, *43*, 50.
251. Goldberg, A.; Shore, B. W. Modelling Laser Ionisation. *J. Phys. B: At. Mol. Phys.* **1978**, *11*, 3339.
252. Picón, A.; Bahabad, A.; Kapteyn, H. C.; Murnane, M. M.; Becker, A. Two-Center Interferences in Photoionization of a Dissociating H_2^+ Molecule. *Phys. Rev. A* **2011**, *83*, 013414.
253. Yuan, K. J.; Lu, H.; Bandrauk, A. Linear- and Circular-Polarization Photoionization Angular Distributions in H_2 and H_2^+ by Attosecond Xuv Laser Pulses. *Phys. Rev. A* **2011**, *83*, 043418.
254. Silva, R. E. F.; Catoire, F.; Rivière, P.; Bachau, H.; Martín, F. Correlated Electron and Nuclear Dynamics in Strong Field Photoionization of H_2^+. *Phys. Rev. Lett.* **2013**, *110*, 113001.
255. Bian, X. B. Photoionization of Atoms and Molecules Studied by the Crank-Nicolson Method. *Phys. Rev. A* **2014**, *90*, 033403.
256. Shull, H.; Loöwdin, P. O. Superposition of Configurations and Natural Spin Orbitals. Applications to the He Problem. *J. Chem. Phys.* **1959**, *30*, 617.
257. Goscinski, O. Conjugate Eigenvalue Problems and Generalized Sturmians. *Adv. Quantum Chem.* **2002**, *41*, 51. Preliminary research unpublished. Included as an appendix.
258. Aquilanti, V.; Cavalli, S.; Coletti, C.; Grossi, G. Alternative Sturmian Bases and Momentum Space Orbitals: An Application to the Hydrogen Molecular Ion. *Chem. Phys.* **1996**, *209*, 405.
259. Aquilanti, V.; Cavalli, S.; De Fazio, D. Hyperquantization Algorithm. I. Theory for Triatomic Systems. *J. Chem. Phys.* **1998**, *109*(10), 3792.
260. Avery, J.; Shim, R. Molecular Sturmians. Part 1. *Int. J. Quantum Chem.* **2001**, *83*, 1.
261. Avery, J.; Avery, J. The Generalized Sturmian Method for Calculating Spectra of Atoms and Ions. *J. Math. Chem.* **2003**, *33*, 145.
262. Rawitscher, G. Positive Energy Weinberg States for the Solution of Scattering Problems. *Phys. Rev. C* **1982**, *25*, 2196.
263. Rawitscher, G. Iterative Solution of Integral Equations on a Basis of Positive Energy Sturmian Functions. *Phys. Rev. E* **2012**, *85*, 026701. http://link.aps.org/doi/10.1103/PhysRevE.85.026701.

264. Ovchinnikov, S. Y.; Macek, J. H. Positive Energy Sturmian States for Two-Coulomb-Center Problems. *Phys. Rev. A* **1997**, *55*, 3605.
265. Macek, J. H.; Yu Ovchinnikov, S.; Gasaneo, G. Exact Solution for Three Particles Interacting Via Zero-Range Potentials. *Phys. Rev. A* **2006**, *73*, 032704.
266. Rotenberg, M. Application of Sturmian Functions to the Schroedinger Three-Body Problem: Elastic e+-H Scattering. *Ann. Phys. (N.Y.)* **1962**, *19*, 262.
267. Rotenberg, M. Theory and Application of Sturmian Functions. *Adv. At. Mol. Phys.* **1970**, *6*, 233.
268. Fano, U.; Cooper, J. Spectral Distribution of Atomic Oscillator Strengths. *Rev. Mod. Phys.* **1968**, *40*, 441.
269. Fernández-Menchero, L.; Otranto, S. Single ionization of CH_4 by Bare Ions: Fully Differential Cross Sections. *Phys. Rev. A* **2010**, *82*, 022712.
270. Granados-Castro, C. M.; Gómez, I. A.; Ancarani, L. U.; Gasaneo, G.; Mitnik, D. M. Perturbative-Generalized Sturmian Method for the Study of Photoionization in Atoms 2015. Submitted for publication.
271. Granados-Castro, C. M.; Ancarani, L. U.; Gasaneo, G.; Mitnik, D. M. Sturmian Approach to Single Photoionization of Many Electron Atoms and Molecules. *J. Phys.: Conf. Ser.* **2015**, *601*, 012009.
272. Harriman, J. Numerical Values for Hydrogen Fine Structure. *Phys. Rev.* **1956**, *101*, 594.
273. Granados-Castro, C. M.; Ancarani, L. U.; Gasaneo, G.; Mitnik, D. M. Sturmian Approach to Single Photoionization of CH_4. *Few-Body Syst.* **2014**, *55*, 1029.
274. Moccia, R. One-Center Basis Set SCF MO's. III. H_2O, H_2S, and HCl. *J. Chem. Phys.* **1964**, *40*, 2186.
275. Moccia, R. One-Center Basis Set SCF MO's. II. NH_3, NH_4^+, PH_3, PH_4^+. *J. Chem. Phys.* **1964**, *40*, 2176.
276. Moccia, R. One-Center Basis Set SCF MO's. I. HF, CH_4, and SiH_4. *J. Chem. Phys.* **1964**, *40*, 2164.
277. Banna, M. S.; McQuaide, B. H.; Malutzki, R.; Schmidt, V. The Photoelectron Spectrum of Water in the 30 to 140 eV Photon Energy Range. *J. Chem. Phys.* **1986**, *84*, 4739.
278. Brion, C.; Hamnett, A.; Wight, G.; Van der Wiel, M. Branching Ratios and Partial Oscillator Strengths for the Photoionization of NH_3 in the 15-50 eV Region. *J. Electron Spectros. Relat. Phenom.* **1977**, *12*, 323.
279. Backx, C.; der Wiel, M. J. V. Electron-ion Coincidence Measurements of CH_4. *J. Phys. B: At. Mol. Phys.* **1975**, *8*, 3020.
280. Randazzo, J. M.; Buezas, F.; Frapiccini, A. L.; Colavecchia, F. D.; Gasaneo, G. Solving Three-Body-Breakup Problems with Outgoing-Flux Asymptotic Conditions. *Phys. Rev. A* **2011**, *84*, 052715.
281. Cacelli, I.; Moccia, R.; Rizzo, A. Gaussian Type Orbital Basis Sets for the Calculation of Continuum Properties in Molecules: The Photoionization Cross Section of H_2. *J. Chem. Phys.* **1993**, *98*, 8742.

CHAPTER TWO

General Coalescence Conditions for the Exact Wave Functions: Higher-Order Relations for Coulombic and Non-Coulombic Systems

Yusaku I. Kurokawa[1], Hiroyuki Nakashima, Hiroshi Nakatsuji
Quantum Chemistry Research Institute, Kyoto, Japan
[1]Corresponding author: e-mail address: y_kurokawa@qcri.or.jp

Contents

Abstract

We have extended and generalized Kato's cusp condition (CC) and Rassolov and Chipman's CC, which are necessary conditions for the exact solution to the Schrödinger equation. We named them general coalescence conditions (GCCs). The GCCs can be applied not only to Coulombic systems. In this article, we review the GCCs for two- and many-particle systems. Then, we numerically verify that the GCC equations are satisfied by the free complement (FC) wave function of a hydrogen atom in the 1*s* and 2*p* states, a harmonic oscillator, $V=r$ system, and a helium atom. Based on the GCCs, a new

Advances in Quantum Chemistry, Volume 73
ISSN 0065-3276
http://dx.doi.org/10.1016/bs.aiq.2015.09.001

method is proposed to determine parameters in a trial wave function. We named them GCC constrained (GCCC) method. Using the GCCC method combined with the FC wave function, we could obtain the exact wave function that behaves correctly in the coalescence point.

1. INTRODUCTION

A coalescence region is important and interesting in quantum chemistry. When two charged particles coalesce with each other $\left(r_{ij} \to 0\right)$, the Coulomb potential between them $\left(\propto 1/r_{ij}\right)$ diverges to plus or minus infinity. Even so, the local energy $E_{\mathrm{L}} \equiv \hat{V} + \hat{K}\psi/\psi$ must be everywhere a constant E, where $\hat{V}$ and $\hat{K}$ are the potential and kinetic operators, respectively, ψ is the wave function of the system, and E is the solution of the Schrödinger equation (SE) $\hat{H}\psi = E\psi$. This means that the divergence of the Coulomb potential $\hat{V}$ must be exactly canceled by the "kinetic local energy" $\hat{K}\psi/\psi$.

This cancelation occurs when a wave function satisfies a special condition called "Kato's cusp condition (CC)."[1]

If an approximate wave function satisfies the Kato CC, the local energy does not diverge at the coalescence points.

There are two types of Kato's CCs: one is the electron–electron (e–e) CC,

$$\left.\frac{\partial \overline{\psi}}{\partial r_{ij}}\right|_{r_{ij}=0+} = \frac{1}{2}\overline{\psi}\Big|_{r_{ij}=0+}, \tag{1}$$

where particles i and j are both electrons, and the other is the nucleus–electron (n–e) CC,

$$\left.\frac{\partial \overline{\psi}}{\partial r_{ij}}\right|_{r_{ij}=0+} = -Z_i\overline{\psi}\Big|_{r_{ij}=0+}, \tag{2}$$

where particles i and j are a nucleus and an electron, respectively, r_{ij} is the distance between particles i and j, Z_i is the charge on particle i, and $\overline{\psi}$ represents the wave function spherically averaged over a small sphere centered at the point $r_{ij} = 0$. The Born–Oppenheimer (BO) approximation is assumed in (2). The Kato CC states the relation between the zeroth and first-order derivatives of the wave function with respect to r_{ij}, and their ratio (first/zeroth derivatives) is 1/2 for e–e coalescence and minus nucleus charge for n–e coalescence.

Kato's CC plays important roles in many theories, such as the Monte Carlo method,[2] the R12 method,[3,4] the F12 method,[5,6] etc.

Pack and Byers-Brown (PB) unified Kato's e–e and n–e CCs into one form as[7]

$$(l+1)f_{lm}^{(1)}=\zeta f_{lm}^{(0)} \quad (\text{for } l=0,1,2,\cdots \text{ and } m=-l,\cdots,l), \tag{3}$$

where

$$f_{lm}^{(k)}\equiv\frac{1}{(k+l)!}\lim_{r_{ij}\to 0+}\frac{\partial^{k+l}}{\partial {r_{ij}}^{k+l}}\iint \mathrm{d}\Omega Y_{lm}^{*}(\theta,\phi)\psi, \tag{4}$$

$\zeta\equiv Z_iZ_j\mu_{ij}$, μ_{ij} is the reduced mass of particles i and j, (r,θ,ϕ) are the polar coordinates of the vector $\mathbf{r}_{ij}$, Y_{lm}^{*} is the complex conjugate of the spherical harmonics, and $\iint \mathrm{d}\Omega=\iint \sin\theta \mathrm{d}\theta \mathrm{d}\phi$ represents the integration over the angles. $f_{lm}^{(k)}$ is the kth-order coefficient of the radial part of the Y_{lm} component of the wave function.

The PB CC is applicable to both the BO and non-BO wave functions: If particles i and j are the nucleus and electron, respectively, and μ_{ij} is unity instead of the true reduced mass, (3) holds for the BO wave function. The PB CC states the relation between the zeroth and first-order derivatives of the radial part of the Y_{lm} component of the wave function. The Kato's CCs are a special case of the PB's CC: $l=m=0$ and the BO approximation is assumed for the n–e case.

Rassolov and Chipman (RC) derived a higher-order relation of a wave function as[8]

$$f_{00}^{(3)}=\frac{2\zeta}{3}f_{00}^{(2)}-\frac{\zeta^3}{6}f_{00}^{(0)}, \tag{5}$$

for the Y_{00} component, and

$$f_{10}^{(3)}-\frac{7}{36}f_{10}^{(2)}+\frac{1}{288}f_{10}^{(0)}=0 \tag{6}$$

for a triplet pair of electrons.

These equations state the relation between the zeroth, second, and third derivatives of the wave function. Tew derived similar equations and examined different kinds of wave functions.[9]

Several authors investigated various aspects of these Coulombic CCs; Gaussian functions with linear r term,[10] unnatural-parity state,[11] combination with the boundary conditions,[12,13] three particle coalescence,[14,15] the density on nucleus,[16] mathematical rigorous-related theories[17,18], etc.

Loos and Gill studied the spherium "atom,"[19,20] where two electrons move on the surface of the sphere. It is an interesting system but we do not investigate it here since the potential in this "atom" is completely different from those in ordinary atoms and molecules.

Silanes, Ugalde, and Boyd (SUB) extended Kato's CCs to a non-Coulombic system[21]: They extended the potential operator to any one written in general as

$$\hat{V} \equiv \sum_{i=1}^{N-1} \sum_{j>i}^{N} \sum_{a=-1}^{A} C_{ij}^{(a)} r_{ij}^{a}, \tag{7}$$

where N is the number of particles of the system, A is a finite (given) integer, and $C_{ij}^{(a)}$ are linear combination coefficients. Various types of potential were represented in this form; for example, the Coulomb potential is represented by $C_{ij}^{(a)} = Z_i Z_j$ and $A = -1$, and the harmonic potential $\left(\hat{V} = \frac{1}{2} \sum_{i>j}^{N-1,N} \omega_{ij} r_{ij}^2 \right)$ is represented by $C_{ij}^{(-1)} = C_{ij}^{(0)} = C_{ij}^{(1)} = 0$, $C_{ij}^{(2)} = \frac{1}{2}\omega_{ij}$, and $A = 2$. The SUB CCs are limited to the relations between the zeroth and first derivatives of a wave function. Note that we call these conditions a "coalescence condition" instead of a "cusp condition" because there is no "cusp" in the exact wave function of the harmonic oscillator, for example.

If a wave function satisfies the CCs mentioned above, its local energy will not diverge at the coalescence point. However *not diverging* is insufficient: the local energy must be everywhere a constant E. Other conditions need to be satisfied by the exact wave function at the coalescence point. Recently, we derived relations among the higher-order derivatives of wave functions at the coalescence point with a general form of the potential.[22,23] In this article, we review them and some applications. Throughout this paper, we mainly focus on a situation where two particles coalesce.

2. BASIC FORMULATION

We consider the non-relativistic time-independent SE of an N-particle system, $(\hat{H} - E)\psi = 0$. The Hamiltonian is written as

$$\hat{H} \equiv -\sum_{i=1}^{N} \frac{1}{2m_i} \nabla_i^2 + \hat{V}, \tag{8}$$

where m_i is the mass of particle i, and $\hat{V}$ is the potential written generally as

$$\hat{V} \equiv \sum_{i=1}^{N-1} \sum_{j>i}^{N} \sum_{a=-1}^{A_{ij}} C_{ij}^{(a)} r_{ij}^{a}, \tag{9}$$

with integer A_{ij}. Equation (9) expresses any potential that is written in a power series of the inter-particle distance r_{ij}. A combination of different kinds of potential in a system is allowed.

We investigate the situation in which particles 1 and 2 come close with each other ($r_{12} \to 0$) and coalesce, while the other particles i ($i \geq 3$) are well separated from the coalescence region and fixed at given positions. Using the relative coordinate, $\mathbf{r}_{12}$, and the center of mass coordinate, $\mathbf{G}$, the SE can be written as

$$\left[-\frac{1}{2\mu_{12}} \nabla_{r_{12}}^2 + V_{12} + W(1,2:3,4,\ldots,N) + \hat{O} - E \right] \psi = 0, \tag{10}$$

where V_{12} is the potential acting between particles 1 and 2, W is the remaining part of the potentials acting on particles 1 and 2, E is the solution to the SE, and the fourth term $\hat{O}$ is the remaining part of the Hamiltonian. The operator $\hat{O}$ includes all the variables other than $\mathbf{r}_{12}$. The total energy E is generally unknown.

The most general bounded solution of (10) can be expressed as

$$\psi \equiv \sum_{l=0}^{\infty} r_{12}{}^{l} \sum_{m=-l}^{l} \left(\sum_{k=0}^{\infty} f_{lm}^{(k)} r_{12}{}^{k} \right) Y_{lm}(\theta_{12}, \phi_{12}), \tag{11}$$

where $f_{lm}^{(k)}$ is the expansion coefficient depending on the coordinate of **G**, **3**, **4**, ..., **N**. The $f_{lm}^{(k)}$ is associated with the wave functions by (4). If the **G** and the position of the non-coalescence particles are fixed, the $f_{lm}^{(k)}$ is just a number.

3. PRIMITIVE GENERAL COALESCENCE CONDITIONS

Using the quantities defined in Section 2, the SE can be written as

$$\left[-\nabla_r^2+2\mu\sum_{a=-1}^{A}C^{(a)}r^a+2\mu\sum_{q=0}^{\infty}r^q\sum_{q'=-q}^{q}\left(\sum_{q''=0}^{\infty}R_{qq'}^{(q'')}r^{q''}\right)Y_{qq'}(\theta,\phi)\right.$$
$$\left.+2\mu\hat{O}-2\mu E\right]\left[\sum_{l=0}^{\infty}r^l\sum_{m=-l}^{l}\sum_{k=0}^{\infty}f_{lm}^{(k)}r^kY_{lm}(\theta,\phi)\right]=0 \quad . \tag{12}$$

For simplicity in the following discussion and (12), we remove the subscript "12" from A_{12}, C_{12}, μ_{12}, θ_{12}, ϕ_{12}, and r_{12} if it is not specifically mentioned.

If we integrate out the angular parts in (12) by operating $Y_{LM}^*(\theta,\phi)\mathrm{d}\Omega$ with arbitrary non-negative integers L and M ($L=0,1,2,\ldots$ and $-L\leq M\leq L$), the radial equation remains as

$$\sum_{n=-1}^{\infty}\left[-(n+2)(n+3+2L)f_{LM}^{(n+2)}+2\mu\sum_{a=-1}^{A}C^{(a)}f_{LM}^{(n-a)}+2\mu(\hat{O}-E)f_{LM}^{(n)}\right.$$
$$\left.+2\mu\sum_{q=0}^{n+L}\sum_{q''=0}^{n+L-q}\sum_{l=0}^{n+L-q-q''}\sum_{q'=-q}^{q}R_{qq'}^{(q'')}I_{qq'l,M-q'}^{LM}f_{l,M-q'}^{(n+L-q-q''-l)}\right]r^n=0. \tag{13}$$

Here, we define $f_{lm}^{(p)}\equiv 0$ for all l and m when $p\leq -1$.

Because the part in [ ] of (13) is independent of small r, it must be zero for all n, leading to the following equation:

$$\begin{aligned}&-(n+2)(n+3+2L)f_{LM}^{(n+2)}+2\mu\sum_{a=-1}^{A}C^{(a)}f_{LM}^{(n-a)}\\&+2\mu\sum_{q=0}^{n+L}\sum_{q''=0}^{n+L-q}\sum_{l=0}^{n+L-q-q''}\sum_{q'=-q}^{q}R_{qq'}^{(q'')}I_{qq'l,M-q'}^{LM}f_{l,M-q'}^{(n+L-q-q''-l)}\\&+2\mu\hat{O}f_{LM}^{(n)}=2\mu Ef_{LM}^{(n)}\end{aligned} \tag{14}$$

for arbitrary $n\geq -1$. In (14), the $f_{lm}^{(k)}$s are associated to the wave function by (4), $R_{qq'}^{(q'')}$s are associated to potentials W by

$$R_{qq'}^{(q'')}\equiv\lim_{r_{12}\to 0}\frac{1}{(q+q'')}\left(\frac{\partial}{\partial r_{12}}\right)^{q+q''}\int Y_{qq'}^*(\theta_{12},\phi_{12})W\mathrm{d}\Omega \tag{15}$$

and I is a constant defined by

$$I^{LM}_{l_1 m_1 l_2 m_2} = (-1)^M \sqrt{\frac{(2l_1+1)(2l_2+1)(2L+1)}{4\pi}} \begin{pmatrix} l_1 & l_2 & L \\ 0 & 0 & 0 \end{pmatrix} \begin{pmatrix} l_1 & l_2 & L \\ m_1 & m_2 & M \end{pmatrix}. \tag{16}$$

Equation (14) states the relation between $\{f^{(k)}_{lm}\}$, but they are essentially the same as the SE. We call (14) "primitive general coalescence conditions (pGCCs)." Equation (14) must hold; otherwise the SE does not hold at the coalescence point.

In the special case of $n=-1$, (14) is simplified to

$$-(1+L)f^{(1)}_{LM} + \mu \sum_{a=-1}^{A} C^{(a)} f^{(-1-a)}_{LM} = 0. \tag{17}$$

This equation represents Kato's CCs (3) when the potential is Coulombic ($A=-1$, $C^{(-1)}=Z_1 Z_2$). When $L=1$, (17) represents the CC for p-wave function. Similarly, $L=2, 3, 4, \ldots$ correspond to that for the d-, f-, g-, ... wave function, respectively.

4. GENERAL COALESCENCE CONDITIONS

The pGCCs (14) are series of necessary conditions which must be satisfied by the exact wave function at the coalescence point. However, the pGCCs for $n \geq 0$ are not useful since they contain the unknown term. (The "primitive" in pGCCs implies the presence of an unknown term.)

4.1 Two-Particle Systems

In two-particle systems, the $\hat{O}$ and W in (10) vanish since particles 3, 4,..., do not exist, and we can regard $R^{(q'')}_{qq'}=0$ for all q, q', and q''. After eliminating the unknown terms, we can derive the nth-order general coalescence conditions (GCCs) for two-particle systems as $F^{(n)}_{LM}=0$, where

$$F^{(n)}_{LM} \equiv \frac{2\mu \sum_{a=-1}^{A} C^{(a)} f^{(n-a-1)}_{LM} - (n+1)(n+2+2L) f^{(n+1)}_{LM}}{2\mu \sum_{a=-1}^{A} C^{(a)} f^{(n-a)}_{LM} - (n+2)(n+3+2L) f^{(n+2)}_{LM}} - \frac{f^{(n-1)}_{LM}}{f^{(n)}_{LM}} \tag{18}$$

$$(n=0,1,2,\ldots)$$

when $f^{(n)}_{LM} \neq 0$, and

$$F_{LM}^{(n)} \equiv 2\mu \sum_{a=-1}^{A} C^{(a)} f_{LM}^{(n-a)} - (n+2)(n+3+2L) f_{LM}^{(n+2)} \tag{19}$$

when $f_{LM}^{(n)} = 0$.[22] Each variable in (18) and (19) is given and fs are evaluated from a wave function. The GCC $F_{LM}^{(n)} = 0$ with (18) and (19) are equivalent to (17) when $n=0$. Any wave function of two-particle systems must satisfy the GCCs, $F_{LM}^{(n)} = 0$. We call Fs as the nth-order GCC values for the (L, M) component. All of these GCC values should be zero if they are calculated with the exact wave function. So the GCC values can be used as a measure to indicate the accuracy of a wave function at the coalescence point. In Section 5, we will show some GCC values of some systems.

4.2 Many-Particle Systems

In many-particle systems, we eliminate the unknown term inductively as follows. First, we confirm that (17) holds with arbitrary L and M ($L=0,1,2,\ldots$ and $-L \leq M \leq L$), it is the zeroth-order GCC. It is written in a linear form with respect to $\{f_{lm}^{(n)}\}$ as

$$F_{LM}^{(n)} = \sum_{s=0}^{S} \sum_{t=0}^{T} \sum_{u=-t}^{t} a_{tu}^{(s)} f_{tu}^{(s)} = 0, \tag{20}$$

where $\{a_{tu}^{(s)}\}$ are the linear combination coefficients or operators acting on $f_{lm}^{(n)}$ and all the terms are known. The form of (20) is the same as the nth-order GCC.

Then, the $(n+1)$th-order GCC is derived as $F_{LM}^{(n+1)} = 0$,[23] where

$$F_{LM}^{(n+1)} = \sum_{s=0}^{S} \sum_{t=0}^{T} \sum_{u=-t}^{t} a_{tu}^{(s)} \left[-(s+2)(s+3+2t) f_{tu}^{(s+2)} + 2\mu \sum_{a=-1}^{A} C^{(a)} f_{tu}^{(s-a)} \right. \\ \left. + 2\mu \sum_{q=0}^{s+t} \sum_{q''=0}^{s+t-q} \sum_{l=0}^{s+t-q-q''} \sum_{q'=-q}^{q} R_{qq'}^{(q'')} I_{qq'l,u-q'}^{tu} f_{l,u-q'}^{(s+t-q-q''-l)} + 2\mu \hat{O} f_{tu}^{(s)} \right]. \tag{21}$$

We can rewrite (21) to be in the same form as (20). Thus we can generate new GCCs inductively. Consequently, we can obtain an infinite number of GCCs without unknown terms.

The GCCs (21) state the relation between $\{f_{lm}^{(k)}\}$. In the zeroth-order GCC (17), the angular index (L) is only one kind. But as the order of

GCC increases, the number of the angular indices, represented by "t," "l," and "q" in (21), increases. So even though we investigate the two-particle coalescence, coupling terms between the p- (d-, f-...) component of the wave function and the p- (d-, f-...) component of the distinct particles appear in the GCC for S component ($L=0$).

4.3 Nucleus–Electron GCC Values of Helium Atom

In this section, we show the GCCs for the n–e coalescence of helium as an example. Let us assume that the positions of nucleus 1 and electron 3 are fixed at $(r_1, \theta_1, \phi_1) = (0, 0, 0)$ and $(r_3, \theta_3, \phi_3) = (1, 0, 0)$, respectively, and that electron 2 approaches the nucleus, as shown in Fig. 1. The parameters are set as $\mu = 1$, $A = -1$, $C^{(-1)} = -2$, $R_{q0}^{(0)} = \sqrt{4\pi/(2q+1)}(2\delta_{q0} - 1)$, $R_{qq'}^{(q'')} = 0$ (for $q' \neq 0$ or $q'' = 0$), and $\zeta = -2$.

The zeroth-order GCC for the (L, M) component is (17). The zeroth-order GCC values $F_{00}^{(0)}$ is written as

$$F_{00}^{(0)} \equiv f_{00}^{(1)} + 2f_{00}^{(0)}, \tag{22}$$

and $F_{00}^{(0)} = 0$ is the zeroth-order GCC. The low-order n–e GCC values of the helium atom for the (L, M) = (0, 0) components are as follows:

$$F_{00}^{(1)} \equiv f_{00}^{(3)} + \frac{4}{3}f_{00}^{(2)} - \frac{4}{3}f_{00}^{(0)} \tag{23}$$

$$F_{00}^{(2)} \equiv f_{00}^{(5)} + \frac{46}{45}f_{00}^{(4)} - \frac{68}{135}f_{00}^{(2)} + \frac{16}{27}f_{00}^{(0)} - \frac{\sqrt{3}}{45}f_{10}^{(1)} - \frac{4\sqrt{3}}{135}f_{10}^{(0)}, \text{ and} \tag{24}$$

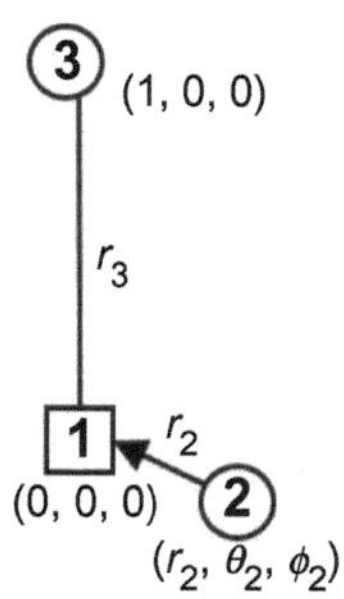

Figure 1 Coalescence of a nucleus and electron in a helium atom. The particle 1 is a nucleus and the particles 2 and 3 are electrons.

$$F_{00}^{(3)} \equiv f_{00}^{(7)} + \tfrac{88}{105} f_{00}^{(6)} + \tfrac{137}{1260} f_{00}^{(5)} - \tfrac{271}{1890} f_{00}^{(4)} - \tfrac{34}{945} f_{00}^{(3)} + \tfrac{43}{945} f_{00}^{(2)} - \tfrac{121}{1890} f_{00}^{(0)}$$
$$- \tfrac{1}{28} \hat{O} f_{00}^{(5)} - \tfrac{23}{630} \hat{O} f_{00}^{(4)} + \tfrac{17}{945} \hat{O} f_{00}^{(2)} - \tfrac{4}{189} \hat{O} f_{00}^{(0)} - \tfrac{2\sqrt{3}}{105} f_{10}^{(3)} - \tfrac{2\sqrt{3}}{105} f_{10}^{(2)} \tag{25}$$
$$- \tfrac{11\sqrt{3}}{3780} f_{10}^{(1)} - \tfrac{2\sqrt{3}}{405} f_{10}^{(0)} + \tfrac{\sqrt{3}}{1260} \hat{O} f_{10}^{(1)} + \tfrac{\sqrt{3}}{945} \hat{O} f_{10}^{(0)} - \tfrac{23\sqrt{5}}{3150} f_{20}^{(0)} - \tfrac{\sqrt{5}}{140} f_{20}^{(1)} = 0.$$

The first-order GCC, $F_{00}^{(1)} = 0$, is the same as the RC's CC (5).[8] The fifth and sixth terms in (24) are the coupling terms between the p-components of both the wave function and geometry of distinct particles. The last two terms on the left-hand side of (25) are the coupling terms between the d-components of both the wave function and the geometry of distinct particles. Thus, the second- and higher-order GCCs depend on the geometry information for distinct particles, while the zeroth- and first-order GCCs do not.

We can obtain the GCCs for the $(L, M) = (1, 0)$ component, if we start the zeroth-order GCC, $F_{10}^{(0)} = 0$ where

$$F_{10}^{(0)} \equiv f_{10}^{(1)} + f_{10}^{(0)}. \tag{26}$$

The higher-order GCC values for the (1, 0) component are as follows:

$$F_{10}^{(1)} \equiv f_{10}^{(3)} + \frac{7}{9} f_{10}^{(2)} - \frac{2}{9} f_{10}^{(0)} + \frac{\sqrt{3}}{27} f_{00}^{(0)}, \tag{27}$$

$$F_{10}^{(2)} \equiv f_{10}^{(5)} + \tfrac{29}{45} f_{10}^{(4)} + \tfrac{23}{180} f_{10}^{(3)} - \tfrac{1}{60} f_{10}^{(2)}$$
$$+ \tfrac{7}{450} f_{10}^{(0)} - \tfrac{1}{20} \hat{O} f_{10}^{(3)} - \tfrac{7}{180} \hat{O} f_{10}^{(2)} + \tfrac{1}{90} \hat{O} f_{10}^{(0)} - \tfrac{\sqrt{3}}{540} \hat{O} f_{00}^{(0)} \tag{28}$$
$$- \tfrac{\sqrt{3}}{60} f_{00}^{(3)} - \tfrac{\sqrt{3}}{135} f_{00}^{(2)} - \tfrac{\sqrt{15}}{150} f_{20}^{(1)} - \tfrac{7\sqrt{15}}{1350} f_{20}^{(0)} - \tfrac{\sqrt{3}}{540} f_{00}^{(0)}$$

and so on. All of these GCC values must be zero if they are calculated with the exact wave function.

4.4 Electron–Electron GCC Values of Helium Atom

Next, we show the GCCs for the e–e coalescence of helium. Let us assume that the position of nucleus 3 is fixed at $(r_3, \theta_3, \phi_3) = (1, 0, 0)$, and that electrons 1 and 2 come close to each other and coalesce at $\mathbf{g} = (0, 0, 0)$, as shown in Fig. 2. The parameters are set as $\mu = 1/2$, $A = -1$, $C^{(-1)} = 1$, $R_{q0}^{(0)} = -4\sqrt{\pi/(2q+1)}[(-1)^q + 1/2^q]$, and $R_{qq'}^{(q'')} = 0$ (for $q' \neq 0$ or $q'' = 0$).

Similar to the n–e case, we can obtain the e–e GCC values of the helium atom for the $(L, M) = (0, 0)$ components are as follows:

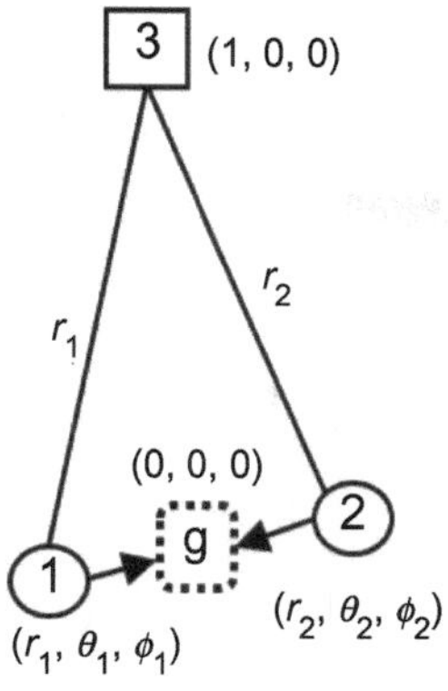

Figure 2 Coalescence of two electrons in a helium atom. The particle 3 is a nucleus and the particles 1 and 2 are electrons. The electrons coalesce at their center of mass coordinate g, and now we assume $g=(0, 0, 0)$.

$$F_{00}^{(0)} \equiv f_{00}^{(1)} - \frac{1}{2} f_{00}^{(0)}, \tag{29}$$

$$F_{00}^{(1)} \equiv f_{00}^{(3)} - \frac{1}{3} f_{00}^{(2)} + \frac{3}{8} f_{00}^{(1)} - \frac{1}{6} f_{00}^{(0)}, \tag{30}$$

$$F_{00}^{(2)} \equiv f_{00}^{(5)} - \frac{23}{90} f_{00}^{(4)} + \frac{53}{180} f_{00}^{(3)} - \frac{13}{144} f_{00}^{(2)} + \frac{1}{18} f_{00}^{(1)} - \frac{1}{45} f_{00}^{(0)}, \tag{31}$$

$$\begin{aligned} F_{00}^{(3)} \equiv & f_{00}^{(7)} - \tfrac{22}{105} f_{00}^{(6)} + \tfrac{589}{2520} f_{00}^{(5)} - \tfrac{281}{5040} f_{00}^{(4)} + \tfrac{199}{5760} f_{00}^{(3)} - \tfrac{11}{1120} f_{00}^{(2)} + \tfrac{11}{2520} f_{00}^{(1)} \\ & - \tfrac{1}{630} f_{00}^{(0)} + \tfrac{\sqrt{5}}{280} f_{20}^{(1)} - \tfrac{23\sqrt{5}}{25200} f_{20}^{(0)}. \end{aligned} \tag{32}$$

and so on. All of the GCC values must be zero if they are calculated with the exact wave function. The zeroth-order GCC, $F_{00}^{(0)} = 0$ is the Kato's CC. The $f_{00}^{(1)}$ term in (30) vanishes using Kato's CC, giving the RC's CC (5). Thus the first-order GCC is essentially the same as the RC CC.

The second order GCC, $F_{00}^{(2)} = 0$, does not have the p-components unlike the n–e GCCs. Generally in the singlet pair of e–e GCCs, no $f_{lm}^{(n)}$ (l: odd) terms appear. It is because the wave function is symmetric with respect to the interchange of the singlet pair of electrons (since the spin part is anti-symmetric) but $f_{lm}^{(n)}$ (l: odd) terms are always anti-symmetric (see (4)), i.e., $f_{lm}^{(n)} = 0$ (l: odd) for the singlet pair of electrons.

5. VERIFICATION OF GCCs

In this section, we apply the free complement (FC) wave functions to the GCCs derived above to see how accurately those equations hold. The

FC wave functions were obtained in our group,[24–27] and they are guaranteed to be exact at convergence. The FC wave function used in the present study is briefly explained in Section 5.1. More details are provided in the references.[24–27]

5.1 Free Complement Wave Function

The FC wave function of order n is a linear combination of complement functions (cfs) of order n, represented as

$$\psi^{(n)} = \sum_{i=1}^{M^{(n)}} C_i^{(n)} \phi_i^{(n)}. \tag{33}$$

One method to determine the coefficients $\{C^{(n)}\}$ is FC-variation principle (FC-VP) method, where the coefficients are determined by the variation principle, i.e., by solving the secular equation, $HC = ESC$, where H and S are the Hamiltonian and overlap matrices, respectively. In this method, we can obtain very accurate wave functions and energies.[15,28,29] Another way to determine the $\{C^{(n)}\}$ is the local SE method.[26,27] In the following discussion, we used the FC-VP wave functions. The cfs $\{\phi^{(n)}\}$ are generated from the initial function $\phi^{(0)}$ by operating the Hamiltonian and g function n times as

$$\left\{\phi^{(n)}\right\} = \left\{[1 + g(H - E)]^n \phi^{(0)}\right\}, \tag{34}$$

where E is a constant. The g function is defined as a sum of inter-particle distances plus unity, $g = 1 + \sum_{i,j} r_{ij}$. The only linearly independent and non-divergent functions on the right-hand side of (34) are selected in $\{\phi^{(n)}\}$. The number of the selected functions is $M^{(n)}$. It is proven that the $\psi^{(n)}$ determined by (33) converges to the exact solution of the SE as the order n increases.[24–27]

5.2 Two-Particle Systems

We apply the FC-VP method to the $1s$ and $2p$ states of a hydrogen atom, harmonic oscillator, and $V = r$ system. The initial choices of the FC wave functions in this study are $\psi_{1s}^{0} \equiv \exp(-1.1r)$ and $\psi_{2p}^{0} \equiv \exp(-0.6r) Y_{10}(\theta, \phi)$ for the $1s$ and $2p$ states of a hydrogen atom, respectively, $\psi_{\text{Harmonic}}^{0} \equiv \exp\left(-0.6r^2\right)$ for the harmonic oscillator and $\psi_{V=r}^{0} \equiv \exp(-3.5r)$ for the $V = r$ system.

For the harmonic oscillator, we use the potential $V=r^2/2$, which is expressed by $C^{(-1)}=C^{(0)}=C^{(1)}=0$, $C^{(2)}=1/2$, $\mu=1$, and $A=2$. The GCC values for the harmonic oscillator are given by (18) for even n and (19) for odd n since $f_{LM}^{(n)}=0$ for odd n.

A system with a linear potential of $V=r$ ($C^{(-1)}=C^{(0)}=0$, $C^{(1)}=1$, $\mu=1$, and $A=1$), namely a constant force system, is not real but we can derive its GCCs because the SE of such a system exists. In this system, the pGCC equations become

$$2f_{LM}^{(n-1)}-(n+2)(n+3+2L)f_{LM}^{(n+2)}=-2\hat{O}f_{LM}^{(n)}\quad (n=\;-1,0,1,2,\ldots). \tag{35}$$

We use (19) as the first-order GCC value since $f_{LM}^{(1)}=0$, and (18) as the other order GCC values.

The exact energies of these systems are, in atomic units, $E(H(1s))=$-0.5, $E(H(2p))=$-0.125, $E(\text{harmonic})=1.5$, and $E(V=r\,\text{system})=$-1.855757081 489238478416,.... The calculated FC energies of these molecules have more than 20-digit accuracy.

Figure 3 shows the GCC values calculated with the FC wave functions for the hydrogen atom in the (A) 1s ($L=0$) and (B) 2p ($L=1$) states, where the x-axis represents the order of the FC wave function; i.e., the accuracy of the wave function. Note that only the $L=0$ components of the 1s state and only the $L=1$ components of the 2p state, have non-zero GCC values, respectively.

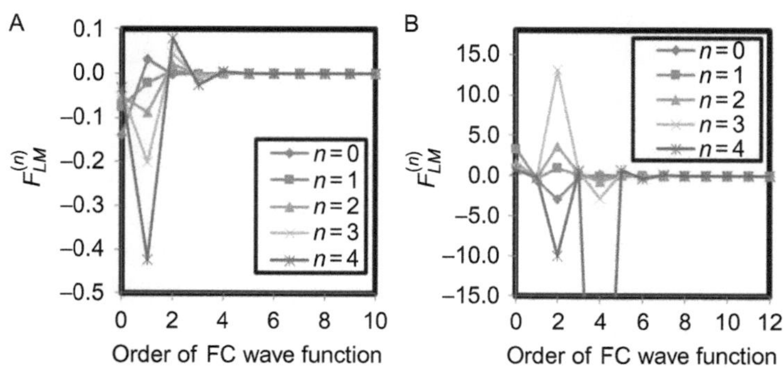

Figure 3 Convergence of the GCC values of the hydrogen atom in the (A) 1s ($L=0$) and (B) 2p ($L=1$) states by the FC wave functions. "n" represents the order of the GCC values. In the 2p state, the zeroth-order GCC value of the zeroth-order FC wave function is minus infinity, and that of the fourth-order FC wave function is −58.8.

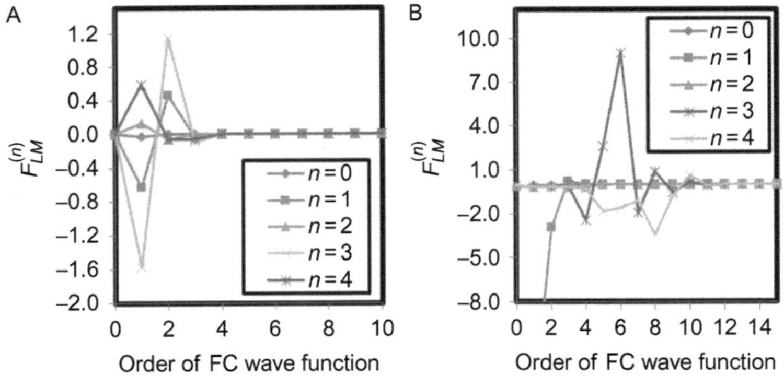

Figure 4 Convergence of the GCC values of (A) the harmonic oscillator in the ground state ($L=0$) and (B) the $V=r$ system in the ground state ($L=0$) by the FC wave functions. "n" represents the order of the GCC values. In (B), the first-order GCC values of the zeroth- and first-order FC wave function are -45.8 and -15.2, respectively.

Figure 4 shows the GCC values of (A) the harmonic oscillator and (B) the $V=r$ system calculated with the FC wave functions for the $L=0$ component.

5.3 GCC Values of Helium Atom

We apply the FC-VP wave functions of a helium atom to the GCCs. The wave functions employed here are reported in literature,[28] and it has an accuracy of more than 42 digits in energy.

Figure 5 shows the n–e GCC values of the FC wave functions of the helium atom in the ground state for the (A) $L=0$ and (B) $L=1$ components, respectively, where the x-axis represents the order of the FC wave function and the y-axis represents $\ln|F_{LM}^{(n)}|$. Figure 6 shows the e–e GCC values of the FC wave functions of the helium atom in the ground state for the $L=0$ component. The wave functions in the singlet spin state have no $L=1$ component.

It is observed that as the wave function becomes more accurate, the n–e GCC values of $n=0$, 1, 2, and 3 for the $(L, M)=(0, 0)$ component and those of $n=0$, 1, and 2 for the $(L, M)=(1, 0)$ component converge to zero. It is also observed that the e–e GCC values of $n=0$, 1, 2, and 3 converge to zero as the wave function becomes more accurate.

5.4 Discussion

In all cases in Figs. 3–6, the GCC values of any order converge to zero as the wave functions converge to the exact function. This means that the exact wave functions satisfy the GCCs equations for two- and many-particle cases.

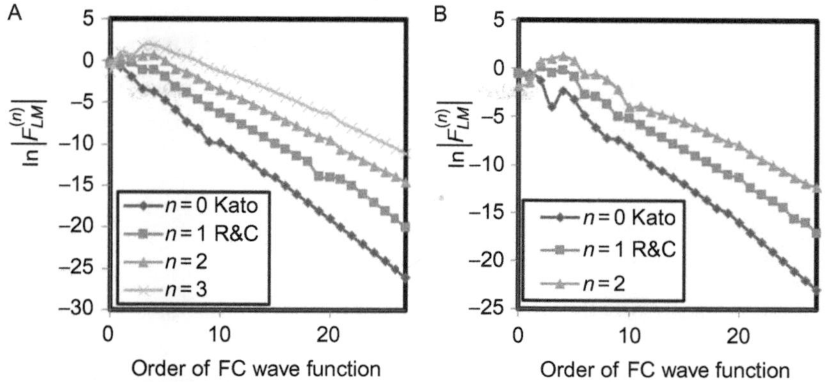

Figure 5 Logarithmic plots of the nucleus–electron GCC values of the helium atom in the ground state for the (A) $(L, M)=(0, 0)$ and (B) $(L, M)=(1, 0)$ components by the FC wave functions. "n" represents the order of the GCC values. As going right on the x-axis, the FC wave functions become more accurate.

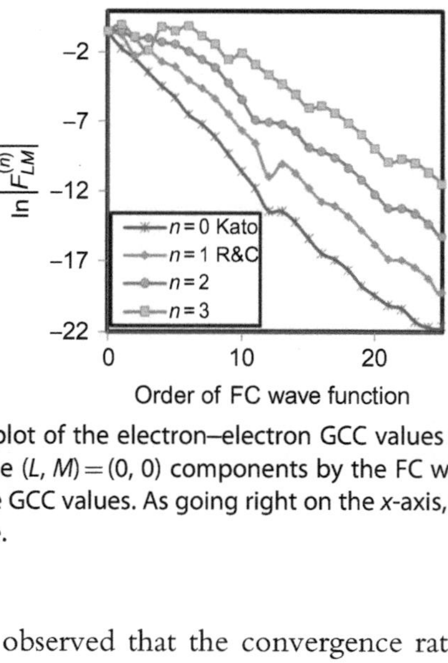

Figure 6 Logarithmic plot of the electron–electron GCC values of the helium atom in the ground state for the $(L, M)=(0, 0)$ components by the FC wave functions. "n" represents the order of the GCC values. As going right on the x-axis, the FC wave functions become more accurate.

It is commonly observed that the convergence rate of the low-order (small n) GCC values is faster than that of the high-order (large n) GCC values. Generally speaking, in the variational method, the energy error is roughly the square of the wave function error, i.e., the convergence rate of the total energy is faster than that of the wave function itself. The errors in the wave function directly influence the errors in the GCC values. Thus, even a good wave function in the variational sense is not good enough from the viewpoint of the GCCs.[22,23] Therefore, we believe that the GCCs can be used as a strict measure of accuracy of a given wave function.

It is interesting that the converged FC-VP wave functions automatically satisfy the GCC equations, even though the parameters in the FC-VP wave functions are determined using completely different conditions, i.e., the variation principle. This is one of the numeric proofs showing that the FC wave function becomes exact at convergence.

6. SOLVING THE SE WITH GCCs

In our current study for solving the SE, we use the local Schrödinger equation (LSE) as the necessary conditions to determine the variables included in the cfs.[26,27,30] This is an integration-free method and can be a general methodology for atoms and molecules.

The GCCs are the necessary conditions for wave functions to be exact, too. In this section, we use the GCCs to determine parameters in a trial wave function for a two-particle system.

The exact wave function can be expressed as a linear combination of cfs as shown in (33), which are *sufficient* to represent the exact wave function. The remaining task is to determine the coefficients $\{C\}$. In our previous method, the variation principle or LSE method were used to determine the coefficients.

Here we propose a new simple method, GCC constrained (GCCC) method. In this method, the parameters are determined to satisfy from 0th to $(M-2)$th GCC equations (M is the number of the cfs). The rest parameter is determined by the normalization condition. This method is based on the idea that if the number of unknown parameters equals the number of necessary conditions, all parameters are determined.

We applied the GCCC method for a hydrogen atom in the ground state. The order m cfs are:

$$\left\{\phi^{(m)}\right\} = \left\{\exp(-\alpha r) r^k | k = 0, \ldots, m\right\} \tag{36}$$

when we use the initial wave function $\phi^{(0)} = \exp(-\alpha r)$ ($\alpha = 1.1$, 0.6 for the 1s and 2s states, respectively), and $M^{(m)} = m + 1$. Substituting (33) with (36) for (4) and (18) ($n = 0,1,2,\ldots m^{-1}$) yields m coupling equations with respect to $\{C^{(k)} | (k = 0,1,2,\ldots m)\}$. These coupling equations and the normalization condition can be solved simultaneously to get $\{C^{(k)}\}$.

Table 1 shows the energies of the GCCC and FC-VP wave functions for the hydrogen atom in the 1s state. Both the expectation energy of the GCCC wave function and the eigenvalue of the FC-VP wave function

Table 1 The Energies of the GCC Constrained and FC-VP Wave Functions for the Hydrogen Atom in the 1*s* State[a]

		GCC Constrained Wave Function		FC-VP Wave Function	
Order	***M***	**Expectation Energy**	**Local Energy at Cusp**	**Eigenvalue**	**Local Energy at Cusp**
0	1	−0.4950000000000000000	∞	−0.4950000000000000000000000	∞
1	2	−0.4999044585987261146	−0.485000000000	−0.499969386607534940955427895	∞
2	3	−0.4999967035489199131	−0.498846153846	−0.499999861035142876742602560	∞
3	4	−0.4999999329823483432	−0.499909632650	−0.499999999474413686337008197	∞
4	5	−0.4999999990392850219	−0.499993506429	−0.499999999998211610465558019	∞
5	6	−0.4999999999886663374	−0.499999564865	−0.499999999999994321277856924	∞
6	7	−0.4999999999998796033	−0.499999972275	−0.499999999999999982828163725	∞
7	8	−0.4999999999999987918	−0.499999998298	−0.499999999999999999949930907	∞
Exact		−0.5000000000000000000	−0.50000000000	−0.5000000000000000000000000	−0.50

[a]The exponent $\alpha = 1.1$ is used.

converge to the exact energy (-0.5 a.u.) as the order of the FC wave function increases. It seems that the FC-VP method gives better energy than the GCCC method by a few digits at the same order. This is not surprising, because the FC-VP method is variational. However, the local energy at the cusp point ($r = 0$) diverges in the FC-VP method. In the GCCC method, on the other hand, the local energies at the cusp point are calculated, finite and converge to the exact value. Table 2 shows the energies of the hydrogen atom in the $2s$ excited state. The tendencies of the energies are similar to the $1s$ ground state mentioned above.

Consequently, the GCCC method can be used to determine parameters in a trial wave function. We believe that the combination of GCCs and other necessary conditions, for example, the LSE method, are possible way to determine the parameters in a trial wave function.

7. CONCLUSIONS

In this paper, we have reviewed the general coalescence conditions (GCCs) for the exact wave functions focusing on the higher-order relations for two- and many-particle systems. We first represented the SE in the $\mathbf{r}_{12}$ coordinate. Then, we derived the pGCCs. The pGCCs include an unknown term such as the total energy E of the system. After eliminating the unknown terms, we derived the general coalescence conditions (GCCs) without any unknown terms for two- and many-particle systems. The GCC equations state the relations between the $\{f_{lm}^{(n)}\}$, which are the coefficients of the power expansion of a wave function around the coalescence point. The zeroth- and first-order GCCs for the Coulombic system are the same as Kato's and Rassolov and Chipman's CCs, respectively. The second- and higher-order GCCs were not known. The GCCs can be applied not only to Coulombic systems but also to any systems with potentials between the coalescencing particles that are written as a power series of inter-particle distances. We show the GCC equations are satisfied numerically by the FC wave function of a hydrogen atom in the $1s$ and $2p$ states, a harmonic oscillator, $V = r$ system, and a helium atom in the n–e and e–e coalescence situations.

In this study, we proposed a new method to determine parameters in a trial wave function; GCCC method. Using the GCCC method with the FC wave function, we could obtain the exact wave function of a hydrogen atom that behaves correctly in the coalescence point.

Table 2 The Energies of the GCC Constrained and FC-VP Wave Functions for the Hydrogen Atom in the 2*s* State[a]

		GCC Constrained Wave Function		FC-VP Wave Function	
Order	***M***	**Expectation Energy**	**Local Energy at Cusp**	**Eigenvalue**	**Local Energy at Cusp**
1	2	−0.17999999999999999	−0.260000000	−0.11669697220176639	∞
2	3	−0.12666666666666666	−0.020000000	−0.12471045771736545	∞
3	4	−0.12490142343976020	−0.116851966	−0.12499204267240784	∞
4	5	−0.12498517444631722	−0.123504966	−0.12499983416693426	∞
5	6	−0.12499855036823718	−0.124720421	−0.12499999710790568	∞
6	7	−0.12499989771690573	−0.124951533	−0.12499999995525163	∞
7	8	−0.12499999432766827	−0.124992264	−0.12499999999936475	∞
Exact		−0.12500000000000000	−0.125000000	−0.12500000000000000	−0.1250

[a]The exponent $\alpha=0.6$ is used.

ACKNOWLEDGMENT

Y.I.K. acknowledges JSPS KAKENHI grant number 26810014 for its financial support.

REFERENCES

1. Kato, T. On the Eigenfunctions of Many-Particle Systems in Quantum Mechanics. *Commun. Pure Appl. Math.* **1957**, *10*, 151.
2. Hammond, B. L.; Lester, W. A., Jr.; Reynolds, P. J. *Monte Carlo Methods in Ab Initio Quantum Chemistry*; World Scientific: Singapore, 1994.
3. Kutzelnigg, W.; Klopper, W. Wave Functions with Terms Linear in the Interelectronic Coordinates to Take Care of the Correlation Cusp. I. General Theory. *J. Chem. Phys.* **1991**, *94*, 1985.
4. Klopper, W.; Manby, F. R.; Ten-No, S.; Valeev, E. F. R12 Methods in Explicitly Correlated Molecular Electronic Structure Theory. *Int. Rev. Phys. Chem.* **2006**, *25*, 427.
5. Ten-no, S.; Noga, J. Explicitly Correlated Electronic Structure Theory from R12/F12 Ansätze. *WIREs Comput. Mol. Sci.* **2012**, *2*, 114.
6. Ten-no, S. Initiation of Explicitly Correlated Slater-Type Geminal Theory. *Chem. Phys. Lett.* **2004**, *398*, 56.
7. Pack, R. T.; Byers-Brown, W. Cusp Conditions for Molecular Wavefunctions. *J. Chem. Phys.* **1966**, *45*, 556.
8. Rassolov, V. A.; Chipman, D. M. Behavior of Electronic Wave Functions Near Cusps. *J. Chem. Phys.* **1996**, *104*, 9908.
9. Tew, D. P. Second Order Coalescence Conditions of Molecular Wave Functions. *J. Chem. Phys.* **2008**, *129*, 014104.
10. Pachucki, K.; Komasa, J. Gaussian Basis Sets with the Cusp Condition. *Chem. Phys. Lett.* **2004**, *389*, 209.
11. Kutzelnigg, W.; Morgan, J. D., III. Rates of Convergence of the Partial—Wave Expansions of Atomic Correlation Energies. *J. Chem. Phys.* **1992**, *96*, 4484.
12. Liverts, E. Z.; Amusia, M. Y.; Krivec, R.; Mandelzweig, V. B. Boundary Solutions of the Two-Electron Schrödinger Equation at Two-Particle Coalescences of the Atomic Systems. *Phys. Rev. A* **2006**, *73*, 012514.
13. Toth, P. V. *Boundary Conditions for Many-Electron Systems.* 2011, arXiv:1010.2700v4.
14. Myers, C. R.; Umrigar, C. J.; Sethna, J. P.; Morgan, J. D., III. Fock's Expansion, Kato's Cusp Conditions, and the Exponential Ansatz. *Phys. Rev. A* **1991**, *44*, 5537.
15. Kurokawa, Y. I.; Nakashima, H.; Nakatsuji, H. Solving the Schrödinger Equation of Helium and Its Isoelectronic Ions with the Exponential Integral (EI) Function in the Free Iterative Complement Interaction Method. *Phys. Chem. Chem. Phys.* **2008**, *10*, 4486.
16. Pan, X.-Y.; Sahni, V. Criticality of the Electron-Nucleus Cusp Condition to Local Effective Potential-Energy Theories. *Phys. Rev. A* **2003**, *67*, 012501.
17. Fournais, S.; Hoffmann-Ostenhof, M.; Hoffmann-Ostenhof, T.; Østergaard Sørensen, T. Sharp Regularity Results for Coulombic Many-Electron Wave Functions. *Commun. Math. Phys.* **2005**, *255*, 183.
18. Fournais, S.; Hoffmann-Ostenhof, M.; Hoffmann-Ostenhof, T.; Østergaard Sørensen, T. Analytic Structure of Many-Body Coulombic Wave Functions. *Comm. Math. Phys.* **2008**, *289*, 291.
19. Loos, P.-F.; Gill, P. M. W. Two Electrons on a Hypersphere: A Quasiexactly Solvable Model. *Phys. Rev. Lett.* **2009**, *103*, 123008.
20. Loos, P.-F.; Gill, P. M. W. Ground State of two Electrons on a Sphere. *Phys. Rev. A* **2009**, *79*, 062517.
21. Silanes, I.; Ugalde, J. M.; Boyd, R. J. Cusp Conditions for Non-Coulombic Interactions. *J. Mol. Struct. (THEOCHEM)* **2000**, *527*, 27.

22. Kurokawa, Y. I.; Nakashima, H.; Nakatsuji, H. General Coalescence Conditions for the Exact Wave Functions: Higher-Order Relations for Two-Particle Systems. *J. Chem. Phys.* **2013**, *139. 044114-1.*
23. Kurokawa, Y. I.; Nakashima, H.; Nakatsuji, H. General Coalescence Conditions for the Exact Wave Functions. II. Higher-Order Relations for Many-Particle Systems. *J. Chem. Phys.* **2014**, *140. 214103-1.*
24. Nakatsuji, H. Scaled Schrodinger Equation and the Exact Wave Function. *Phys. Rev. Lett.* **2004**, *93*, 030403.
25. Nakatsuji, H. General Method of Solving the Schrodinger Equation of Atoms and Molecules. *Phys. Rev. A* **2005**, *72*, 062110.
26. Nakatsuji, H. Discovery of a General Method of Solving the Schrödinger and Dirac Equations that Opens a Way to Accurately Predictive Quantum Chemistry. *Acc. Chem. Res.* **2012**, *45*, 1480.
27. Nakatsuji, H.; Nakashima, H.; Kurokawa, Y.; Ishikawa, A. Solving the Schrodinger Equation of Atoms and Molecules Without Analytical Integration Based on the Free Iterative-Complement-Interaction Wave Function. *Phys. Rev. Lett.* **2007**, *99*, 240402.
28. Nakashima, H.; Nakatsuji, H. Solving the Schrodinger Equation for Helium Atom and Its Isoelectronic Ions with the Free Iterative Complement Interaction (ICI) Method. *J. Chem. Phys.* **2007**, *127*, 224104.
29. Kurokawa, Y.; Nakashima, H.; Nakatsuji, H. Free ICI (Iterative Complement Interaction) Calculations of Hydrogen Molecule. *Phys. Rev. A* **2005**, *72*, 062502.
30. Nakatsuji, H.; Nakashima, H. Free-Complement Local-Schrödinger-Equation Method for Solving the Schrödinger Equation of Atoms and Molecules. Basic Theories and Features. *J. Chem. Phys.* **2015**, *142*, 084117.

CHAPTER THREE

Exponentially Correlated Wave Functions for Four-Body Systems

Frank E. Harris[1]
Department of Physics, University of Utah, Salt Lake City, USA
Quantum Theory Project, University of Florida, Gainesville, Florida, USA
[1]Corresponding author: e-mail address: harris@qtp.ufl.edu

Contents

Abstract

This contribution surveys the current status of analytical methods based on the use of exponentially correlated wave functions for atom-like four-body problems. The topics reviewed include the history of the research area, Hylleraas wave functions, basic notions of correlated-orbital formulations, separation of external and internal (shape) coordinates, basic matrix-element formulas and recursive methods for their evaluation, and expressions for the kinetic and potential energies. Included is an overview of the results that have been obtained to date.

1. INTRODUCTION AND HISTORY

As early as 1929, Hylleraas[1] presented computations for the Helium atom showing that the use of the interelectron distance as an explicit coordinate yielded simple wave functions that were significantly superior to those of comparable complexity but were based only on products of atomic

Advances in Quantum Chemistry, Volume 73
ISSN 0065-3276
http://dx.doi.org/10.1016/bs.aiq.2015.09.005

orbitals. To keep the mathematics relatively simple, the wave functions used by Hylleraas had spatial parts of the generic form

$$\psi = r_1^{n_1} r_2^{n_2} r_{12}^{p} e^{\alpha(r_1 + r_2)}, \tag{1}$$

where r_1 and r_2 are the distances between the He nucleus and electron 1 or 2, r_{12} is the interelectron distance $|\mathbf{r}_2 - \mathbf{r}_1|$, n_1, n_2, p are nonnegative integers, and α is a parameter that controls the overall scale of the wave function.

The success of the above wave function caused many investigators to include the interelectron distance (or, in larger problems, distances) as powers multiplying an exponential that could be thought of as an orbital product; wave functions of this type became generally known as "Hylleraas functions." Among the earliest applications of Hylleraas functions was a ground-breaking study of the hydrogen molecule by James and Coolidge.[2] One of the first such computations of extremely high accuracy was that of Pekeris on the low-lying states of the Helium atom.[3,4]

Though the early studies of two-electron atomic systems used only powers of the interelectron distance, it was technically feasible to place this parameter in the exponent, as was first done in 1968 by Delves and Kalotas[5] and later exploited by Thakkar and Smith.[6] The resulting wave functions, which we characterize as *fully exponentially correlated*, proved very effective for the description not only of Helium and other members of its isoelectronic series (see, for example, work by Frolov and Smith[7] and by the present author,[8–10] but also for more exotic three-body systems of pseudoatomic character (those with no more than one "heavy" particle—e.g, muonic atoms or electron–positron systems).

There was clear motivation to extend the use of fully exponentially correlated wave functions to systems containing more than three particles, but attempts to do so encountered a stubborn technical problem, namely that the four-body (or sometimes called three-electron) integrals proved extremely resistant to analytical evaluation. For this reason, for many years the best approach for the Li atom and similar problems was to use Hylleraas functions. Even for the Hylleraas functions there were technical difficulties; the necessary integrals were usually evaluated either by purely numerical methods or by relatively cumbersome expansions.

The situation for four-body problems changed dramatically in 1987 when Fromm and Hill[11] published a closed analytical formula for a basic fully exponentially correlated integral, together with recipes for generating all the other integrals needed for computation of the electronic energy. It should be noted that Fromm and Hill's analysis represented a remarkable feat of complex integration, involving both ingenuity and evaluations by what appeared to be obscurely motivated inspection.

Despite its cleverness, Fromm and Hill's approach presented a number of practical problems:

- Individual terms in the formulas had singularities at physically realizable argument values, and there was no analysis showing how the individual-term singularities could be combined and canceled.
- The formulas involve multiple-valued functions (dilogarithms) that can assume complex values for physically relevant parameter values, and (unlike the logarithm) the multiplicity of values affects both the real and imaginary parts of the result. It is therefore necessary to have a procedure for selecting the function value (i.e., the *branch* of the function in a complex-variable-theory sense). The only solution offered was to track the branches of the various terms, starting from some limiting situation where the branches (choices of the multiple values) could be determined *a priori*.
- The process of generating the integrals actually needed in computations involved repeated differentiations with respect to various parameters, and this task was close to the limit of what was practical, even with the use of symbolic computing systems.
- The formula for the kinetic energy associated with wave functions written in terms of interparticle coordinates remained fairly cumbersome even for the simplest four-body problems; a single kinetic-energy matrix element involved as many as 43 different four-body integrals.
- States of nonzero angular momentum can be reduced to problems using the same types of integrals that are needed for S states, but there remained a need to develop formulas that were convenient when the wave function was written in terms of the interparticle coordinates.

Despite these issues, some computations were actually carried out, particularly in the group of T. K. Rebane in St. Petersburg.[12–14]

The integrals needed for four-body Hylleraas computations are a special case of the Fromm–Hill formulas, and a closed form for those integrals was reported in 1991 by Remiddi.[15] Remiddi's formulas, though obtained in a fairly complicated way, were far simpler than those of Fromm and Hill; the formulas still contained dilogarithm functions, but with arguments that were simpler, and with evaluations of the dilogarithm that did not involve singularities and that could be restricted to its principal branch. It was not then clear how to reconcile the Fromm–Hill and Remiddi formulas, but a strong indication that both were correct was that they gave identical numerical results for parameter values such that both were applicable.

Since the initial contributions of Fromm, Hill, and Remiddi, there have appeared a number of refinements that improve the technology of both Hylleraas and fully exponentially correlated computation on four-body systems. In particular,

- Rebane[16] showed how the 43 integrals needed for a basic kinetic-energy computation could be expressed entirely in terms of potential-energy integrals and an overlap integral.
- The present author resolved the singularity structure of the Fromm–Hill formula,[17] thereby reconciling the Fromm–Hill and Remiddi formulas, providing numerical stability at the termwise singularities and removing the need for branch tracking.
- Yan and Drake[18] presented formulas for calculation of the kinetic energy matrix elements for states of nonzero angular momentum written as an angular eigenstate multiplied by an expression involving only the interparticle coordinates; the present author showed in detail[19] how the kinetic-energy formulas could be framed in terms of rotational invariants in the interparticle-coordinate space, with the integrals involved thereby reduced to forms in which exponentials were multiplied by known polynomials in the interparticle coordinates.
- Zotev and Rebane[20] explicitly pointed out that the formal development for four-body integrals could be employed for complex values of the nonlinear parameters, thereby becoming applicable to wave functions in which the exponential variation could be supplemented by oscillatory factors. This observation generalized earlier work by Rebane and Yusupov[21] applicable to three-body wave functions.
- Pachucki, Puchalski, and Remiddi[22] presented recursion formulas for generating a set of four-body Hylleraas integrals of the form

$$\int r_1^{m_1} r_2^{m_2} r_3^{m_3} r_{23}^{n_1} r_{13}^{n_2} r_{12}^{n_3} e^{-u_1 r_1 - u_2 r_2 - u_3 r_3} d\tau, \tag{2}$$

  where the integration is over the complete three-particle space. The recurrence formulas relate integrals with $n_1 + n_2 + n_3 + m_1 + m_2 + m_3 = N$ to previously computed integrals whose exponent sums are less than N. These formulas simplify greatly the use of a large basis of Hylleraas functions for the Li atom and similar problems, and permitted Puchalski and Pachucki[23] to make the most accurate computation of the Li ground state that has thus far been reported.

- Motivated by the recursive procedure of Pachucki *et al*, the present author then developed recurrence formulas for integrals involving fully exponentially correlated wave functions.[24] This approach eliminated the need for cumbersome repeated differentiations and was the most recent step in the evolution and simplification of the use of fully exponentially correlated four-body wave functions.
- The present author showed how a procedure analogous to that which expressed all integrals needed for the kinetic energy entirely in terms of overlap and potential-energy integrals could be developed to treat wave functions containing preexponential powers of interelectron coordinates.[25] The resulting formulas share with those developed earlier by Rebane for the basic kinetic-energy integrals[16] the property that negligible computation beyond that for the potential energy is needed for the kinetic energy.

The foregoing historical summary highlights more than 25 years of effort toward the development of methods that go beyond that of Hylleraas in treating electron correlation through the explicit use of interparticle coordinates.. It is important to note that much additional effort, aimed at systems larger than the four-body problems discussed here, is represented by the activity on what are often called "r_{12} methods." That work, which does not rely on closed analytical expressions, is outside the scope of the present exposition; its current status has been the subject of a recent comprehensive review.[26]

The development of the "fully exponentially correlated" and "r_{12} methods" has apparently escaped the notice of some workers in theoretical atomic physics. For example, the following appeared in a recent paper by Frolov[27]: "It is interesting to note that for more than 40 years of extensive use of the Hylleraas variational expansion nobody has tried to improve its convergence rate. This work (i.e., that of Frolov) is the first attempt to replace the traditional Hylleraas basis set by another variational expansion which is more appropriate and rapidly convergent for three-electron atomic problems." That quotation is particularly difficult to understand in view of the fact that Frolov himself coauthored (as a junior colleague of the present author) a paper published in 2003 on fully exponentially correlated wave functions for four-body systems.[28]

The remainder of the present paper reviews the essential features of computational methods that involve the use of exponentially correlated interparticle coordinates, including

- Separation of external and internal (shape) coordinates

- Wave functions
- Potential and kinetic energies
- Matrix element evaluation
- Finite versus infinite nuclear mass
- Parameter optimization

The paper concludes with a survey of recent applications.

2. RELATIVE COORDINATES

For two-body systems in three-dimensional space, three of the six coordinates can locate the center of mass and two are needed to describe orientation; there is one remaining (internal) coordinate, usually taken to be the interparticle distance (r_{12}). A coordinate system specified in this way has a singular point at $r_{12} = 0$, but in practice this does not present problems.

For three-body systems, three of the nine coordinates can locate the center of mass, and except for the special (singular) case of linear geometry, three coordinates can specify overall orientation, leaving three internal coordinates (often called "shape coordinates") that can be the interparticle distances (r_{12},r_{13}, r_{23}), restricted to choices permitting these coordinates to form a triangle.

For four-body systems (our main interest here), except for the singular case of linear geometry, six of the 12 overall coordinates are needed to locate the center of mass and specify the overall orientation, leaving the possibility of using the six interparticle distances (r_{ij}) as shape coordinates. The values of the r_{ij} must then be restricted to sets permitting the formation of a tetrahedron. Except in the degenerate (planar) case two mirror-image tetrahedra are possible, so a complete description will require in addition a discrete two-valued index. The existence of this index can however be ignored unless discussing phenomena that involve chirality.

Systems with more than four particles have more interparticle distances than independent shape coordinates, making it unclear how the methods described here can be extended to larger systems.

It is useful to compare the above discussion with a purely orbital description of a four-body system. Letting (r_i,θ_i,φ_i) ($i = 1,2,3$) be coordinates of three particles relative to a fourth (typically a nucleus), we have nine internal coordinates, with the result that the shape and the orientation of the system have not been completely separated. The implications of this observation are far from trivial, and have for many years been the subject of intensive investigation. A review that reflects the current status of this topic has been provided by Littlejohn and Reinsch[29]; older work on three-body problems includes contributions by Breit,[30] Bhatia and Temkin,[31] and Schwartz.[32] The shape coordinates found optimum in the recent work seem too complicated for use in matrix-element computations, suggesting a continuation of the present practice involving the direct use of the interparticle distances as coordinates.

3. FOUR-BODY WAVE FUNCTIONS

In the approximation of an isolated nonrelativistic system, the stationary states will be angular-momentum eigenfunctions with angular quantum numbers L and M, and a convenient way[33] to create a basis set of four-body spatial wavefunctions Ψ_n in terms of the interparticle distances and their orientations is

$$\Psi_n = \mathcal{Y}_{LM}^{l_1 l_2 (l') l_3}(\mathbf{r}_{14}, \mathbf{r}_{24}, \mathbf{r}_{34}) \Phi_n(r_{12}, r_{13}, r_{14}, r_{23}, r_{24}, r_{34}), \tag{3}$$

where $\mathcal{Y}_{LM}^{l_1 l_2 (l') l_3}$ is a solid spherical harmonic of quantum numbers L,M constructed from spherical harmonics of respective angular momenta l_1, l_2, l_3, with the l_1 and l_2 harmonics coupled to yield an angular momentum l'. The solid spherical harmonics have definition

$$\mathcal{Y}_{LM}^{l_1 l_2 (l') l_3}(\mathbf{r}_{14}, \mathbf{r}_{24}, \mathbf{r}_{34}) = (-1)^{-l_1 + l_2 + l_3 - l'} r_{14}^{l_1} r_{24}^{l_2} r_{34}^{l_3} \sqrt{(2l' + 1)(2L + 1)}$$
$$\times \sum_{(m)} (-1)^{-m' - M} \begin{pmatrix} l_1 & l_2 & l' \\ m_1 & m_2 & -m' \end{pmatrix} \begin{pmatrix} l' & l_3 & L \\ m' & m_3 & -M \end{pmatrix} Y_{l_1 m_1}(\hat{\mathbf{r}}_{14}) Y_{l_2 m_2}(\hat{\mathbf{r}}_{24}) Y_{l_3 m_3}(\hat{\mathbf{r}}_{34}). \tag{4}$$

Here the parenthesized sets of six indices are Wigner 3-j symbols,[34] and the Y_{lm} are conventional spherical harmonics. The notation (m) attached to the summation indicates that it is to include all values of m_1, m_2, m_3, and m' for which the 3-j symbols are nonzero.

As discussed in some detail in Ref. 33, Eq. (3) is more restrictive than it appears; the most general angular-momentum function of this type is obtained with l_1, l_2, and l_3 such that $l_1 + l_2 + l_3$ is either L or $L + 1$, with none of the individual l values exceeding L. Moreover, for $l_1 + l_2 + l_3 = L$, the only possible value of l' is $l_1 + l_2$ and the coupling is unique. For $l_1 + l_2 + l_3 = L + 1$, the coupling is also unique unless all the individual l values are nonzero; in that case two couplings are possible: $l' = l_1 + l_2$ and

$l' = l_1 + l_2 - 1$. In addition, angular functions with $l_1 + l_2 + l_3 = L$ have different symmetry than those with $l_1 + l_2 + l_3 = L + 1$ and these two types of angular states are therefore orthogonal to each other.

The number of different angular functions to be considered can be minimized by limiting discussion to $\mathcal{Y}_{LL}$ and by defining the two functions with $l_1 + l_2 + l_3 = L + 1$ as

$$\mathcal{Y}_{LL}^{l_1 l_2 (l_1+l_2-1) l_3}(\mathbf{r}_{14}, \mathbf{r}_{24}, \mathbf{r}_{34}) \quad \text{and} \quad \mathcal{Y}_{LL}^{l_1 l_3 (l_1+l_3-1) l_2}(\mathbf{r}_{14}, \mathbf{r}_{34}, \mathbf{r}_{24}). \tag{5}$$

The shape function Φ_n is taken to be of the form

$$\Phi_n = \prod_{i<j} \left(r_{ij}^{\mu_{ij}} \mathrm{e}^{-u_{ij} r_{ij}} \right). \tag{6}$$

In Eqs. (3) and (6) the integer parameters l_i, l', μ_{ij}, and the continuous parameters u_{ij} can in principle have different values for different Ψ_n.

A spatial function built from the basis defined in Eqs. (3) and (6) will usually need to be used in conjunction with a spin function and (anti)symmetrized with respect to identical bosons(fermions). For example, to treat the Lithium ^2S ground state, a complete basis function with $L = M = 0$ can be formed by appending to Eq. (3) one of the two three-electron spin doublets (for particles 1, 2, and 3), then antisymmetrizing with respect to those three particles.

4. POTENTIAL AND KINETIC ENERGY OPERATORS

Because we are restricting consideration to Coulombic potentials, the potential energy operator V for a four-body system is a simple function of the shape coordinates only. If particles i have charges q_i,

$$V = \sum_{i<j} \frac{q_i q_j}{r_{ij}}. \tag{7}$$

The kinetic energy operator when written using the interparticle coordinates is more interesting and deserves some comment. For spherically symmetric (S) states, its form (for an arbitrary number of particles of masses m_i) was presented many years ago by Frost[35]:

$$T_{\mathrm{sph}} = -\sum_{ij,\; i<j} \frac{1}{2\mu_{ij}} \left(\frac{\partial^2}{\partial r_{ij}^2} + \frac{2}{r_{ij}} \frac{\partial}{\partial r_{ij}} \right) - \sum_{i=1}^{4} \frac{1}{m_i} \sum_{jk \neq i,\, j<k} (\hat{\mathbf{r}}_{ij} \cdot \hat{\mathbf{r}}_{ik}) \frac{\partial^2}{\partial r_{ij} r_{ik}}, \tag{8}$$

where $\mu_{ij}^{-1} = m_i^{-1} + m_j^{-1}$ and $\hat{\mathbf{r}}_{ij} \cdot \hat{\mathbf{r}}_{ik}$ (also sometimes identified as $\cos\theta_{ijk}$) can be expressed entirely in terms of interparticle distances:

$$\hat{\mathbf{r}}_{ij} \cdot \hat{\mathbf{r}}_{ik} = \cos\theta_{ijk} = \frac{r_{ij}^2 + r_{ik}^2 - r_{jk}^2}{2r_{ij}r_{ik}}. \tag{9}$$

A notable feature of Eq. (8) is that it does not assume any of the particles to be of infinite mass and therefore fixed in position. For this reason its use includes the consequences of finite nuclear mass and avoids the necessity of estimating the quantity usually referred to as "mass polarization."

For application of the kinetic-energy operator to a four-body wave function of the form given in Eq. (3) (in which particle 4 has been given a special role in the angular wave function), the relevant formula is[33]:

$$\begin{aligned} T\Psi &= \mathcal{Y}_{LM}^{l_1 l_2 (l') l_3} T_{\text{sph}} \Phi - \frac{1}{m_4} \sum_{jk \neq 4} \left(\mathbf{r}_{j4} \cdot \nabla_{k4} \mathcal{Y}_{LM}^{l_1 l_2 (l') l_3} \right) \frac{1}{r_{j4}} \frac{\partial \Phi}{\partial r_{j4}} \\ &- \sum_{i=1}^{3} \frac{1}{m_i} \sum_{j \neq i} \left(\mathbf{r}_{ij} \cdot \nabla_{i4} \mathcal{Y}_{LM}^{l_1 l_2 (l') l_3} \right) \frac{1}{r_{ij}} \frac{\partial \Phi}{\partial r_{ij}}. \end{aligned} \tag{10}$$

As pointed out in Ref. 33, Eq. (10) differs in form from a corresponding formula of Yan and Drake,[18] but both are correct. The form given here seems more appropriate when the matrix elements involved are evaluated as closed analytical expressions (rather than by using an expansion such as that of Sack[36]).

Equation (10) can be reduced to a more useful form for computation by evaluating the quantities $\mathbf{r} \cdot \nabla \mathcal{Y}$ that appear therein. This reduction, reported in detail in Ref. 33, converts each of these quantities to one or a linear combination of $\mathcal{Y}$ with the same L and M but with differing individual l_i. That observation is of course not surprising, as the kinetic-energy operator commutes with the total angular momentum operators L^2 and L_z, but not with the individual l_i^2.

After both the potential and kinetic energies are applied to a wave function of the type presently under consideration, the result will be a single term or a linear combination of terms, each of which consists of a product of r_{ij} (to integer powers), multiplied by a $\mathcal{Y}_{LM}$ and if needed a spin function, then (anti)symmetrized as appropriate for a particular application. Since the details of this process depend on the system under study, we do not develop explicit formulas here, but note that each term of $T\Psi_n$ or $V\Psi_n$ will be of the generic form given in Eq. (3), but with differing parameter sets.

5. MATRIX ELEMENTS

The analysis of the preceding two sections indicates that all matrix elements needed for the evaluation of the total energy can be brought to one of the forms

$$\begin{aligned} I_1 &= \left\langle \mathcal{Y}_{LM}^{l_1 l_2 (l') l_3}(\mathbf{r}_{14},\mathbf{r}_{24},\mathbf{r}_{34})\Phi_n \middle| \mathcal{Y}_{LM}^{\lambda_1 \lambda_2 (\lambda') \lambda_3}(\mathbf{r}_{14},\mathbf{r}_{24},\mathbf{r}_{34})\Phi_{n'} \right\rangle,; \\ I_2 &= \left\langle \mathcal{Y}_{LM}^{l_1 l_2 (l') l_3}(\mathbf{r}_{14},\mathbf{r}_{24},\mathbf{r}_{34})\Phi_n \middle| \mathcal{Y}_{LM}^{\lambda_1 \lambda_2 (\lambda') \lambda_3}(\mathbf{r}_{14},\mathbf{r}_{34},\mathbf{r}_{24})\Phi_{n'} \right\rangle, \end{aligned} \tag{11}$$

with either both $l' = l_1 + l_2$ and $\lambda' = \lambda_1 + \lambda_2$ or both $l' = l_1 + l_2 - 1$ and $\lambda' = \lambda_1 + \lambda_2 - 1$. Both I_1 and I_2 occur because the angular momenta may be coupled in different orders in the two angular functions being combined.

These matrix elements can be written as angular and shape-space integrals. Note however that the angular integration is not over the angular coordinates of the individual displacement vectors $\mathbf{r}_{ij}$ but over the possible orientations of the shape tetrahedron. In practice one may describe these orientations using rotation matrices defined in terms of a set of Euler angles (here designated α, β, γ), and carry out the angular integration by integrating over the Euler angles. Note the difference between the process described here and the integration over the angular coordinates of individual particles that are encountered in totally orbital formulations.

Writing the overall integrations using the notation

$$\langle \cdots | \cdots \rangle = \int \mathrm{d}\tau_r \int \mathrm{d}\tau_{\alpha\beta\gamma} \quad \cdots,$$

we treat first the angular integration, obtaining thereby a result that depends upon the shape coordinates. After noting the form of this result, we then proceed to an analysis of the shape-space integrals.

6. ANGULAR INTEGRATION

The angular integrals have values that depend only on the relative directions of the $\mathbf{r}_{ij}$. In Ref. 19, it is shown that these integrals can be written entirely in terms of integer powers of the cosines of the interior angles of the tetrahedron formed by the distances r_{ij}. Since these cosines can be expressed entirely in terms of the r_{ij}, as shown in Eq. (9), the angular integrals have values that are (when multiplied by the powers of r_{ij} in the solid harmonics) polynomials in the r_{ij}. These polynomials, designated below as $\hat{K}$, are known as rotational invariants, and have received some study (see Biedenharn and Louck[37]).

After carrying out the angular integrations, a process in which we used the graphical methods described by Brink and Satchler[38] for the angular-momentum algebra, we obtained the following results:

$$I_1 = \frac{(-1)^{L+l_1+l_2+l'}}{4\pi} \sum_{(\Lambda)} (\Lambda_1,\Lambda_2,\Lambda_3)(l_1,l_2,l',l_3,\lambda_1,\lambda_2,\lambda',\lambda_3)^{1/2} \begin{pmatrix} l_1 & \lambda_1 & \Lambda_1 \\ 0 & 0 & 0 \end{pmatrix}$$
$$\times \begin{pmatrix} l_2 & \lambda_2 & \Lambda_2 \\ 0 & 0 & 0 \end{pmatrix} \begin{pmatrix} l_3 & \lambda_3 & \Lambda_3 \\ 0 & 0 & 0 \end{pmatrix} \begin{Bmatrix} l_3 & \lambda_3 & \Lambda_3 \\ \lambda' & l' & L \end{Bmatrix} \begin{Bmatrix} \lambda_1 & \lambda_2 & \lambda' \\ l_1 & l_2 & l' \\ \Lambda_1 & \Lambda_2 & \Lambda_3 \end{Bmatrix}$$
$$\times \int d\tau_R r_{14}^{l_1+\lambda_1} r_{24}^{l_2+\lambda_2} r_{34}^{l_3+\lambda_3} \hat{K}_{\Lambda_1\Lambda_2\Lambda_3} \Phi_n \Phi_{n'}. \tag{12}$$

$$I_2 = \frac{(-1)^{L+\lambda_1+\lambda_2+\lambda'}}{4\pi} \sum_{(\Lambda)} (\Lambda_1,\Lambda_2,\Lambda_3)(l_1,l_2,l',l_3,\lambda_1,\lambda_2,\lambda',\lambda_3)^{1/2} \begin{pmatrix} l_1 & \lambda_1 & \Lambda_1 \\ 0 & 0 & 0 \end{pmatrix}$$
$$\times \begin{pmatrix} l_2 & \lambda_3 & \Lambda_2 \\ 0 & 0 & 0 \end{pmatrix} \begin{pmatrix} l_3 & \lambda_2 & \Lambda_3 \\ 0 & 0 & 0 \end{pmatrix} \sum_{\tau} (2\tau+1) \begin{Bmatrix} l_3 & l' & L \\ \tau & \lambda_2 & \Lambda_3 \end{Bmatrix} \begin{Bmatrix} \lambda_1 & \lambda_2 & \lambda' \\ L & \lambda_3 & \tau \end{Bmatrix}$$
$$\times \begin{Bmatrix} \lambda_3 & \lambda_1 & \tau \\ l_2 & l_1 & l' \\ \Lambda_2 & \Lambda_1 & \Lambda_3 \end{Bmatrix} \int d\tau_R r_{14}^{l_1+\lambda_1} r_{24}^{l_2+\lambda_3} r_{34}^{l_3+\lambda_2} \hat{K}_{\Lambda_1\Lambda_2\Lambda_3} \Phi_n \Phi_{n'}. \tag{13}$$

Here the notation $(\Lambda_1,\Lambda_2,\Lambda_3)$ stands for $(2\Lambda_1 + 1)(2\Lambda_2 + 1)(2\Lambda_3 + 1)$ and $(l_1,l_2,\cdots,\lambda_3)$ denotes $(2l_1 + 1)(2l_2 + 1)\cdots(2\lambda_3 + 1)$. The quantities $(\cdots)$ and $\{\cdots\}$ containing six elements are respectively 3-j and 6-j symbols, while those containing nine elements are 9-j symbols. These quantities are defined (with methods for their computation) in Ref. 34. The summation is over all values of the Λ_i such that none of the 3-, 6-, or 9-j symbols vanish. Each of the symbols that occurs in Eqs. (12) or (13) vanishes if any of its rows violates a triangle condition, and in addition the 3-j symbols in these formulas will vanish if the sum of the upper-row elements is not even. These conditions limit severely the number of terms actually contributing to the summations.

Table 1 Rotational Invariants $\hat{K}$, Expressed in Terms of the Quantities $\cos\theta_{ijk}$ Defined in Eq. (9).

Λ_1	Λ_2	Λ_3	$\hat{K}_{\Lambda_1\Lambda_2\Lambda_3}$
0	0	0	1
1	1	0	$-\cos\theta_{412}/\sqrt{3}$
2	2	0	$(3\cos^2\theta_{412}-1)/2\sqrt{5}$
2	2	2	$(3\cos^2\theta_{412}+3\cos^2\theta_{413}+3\cos^2\theta_{423}-9\cos\theta_{412}\cos\theta_{413}\cos\theta_{423})$ $/5\sqrt{70}$

Table entries additional to those listed here can be reached by permuting the indices (1,2,3) in both Λ_i and $\hat{K}$.

The rotational invariants $\hat{K}_{\Lambda_1\Lambda_2\Lambda_3}$ are defined by

$$\hat{K}_{\Lambda_1\Lambda_2\Lambda_3}=\frac{(4\pi)^{3/2}}{(\Lambda_1,\Lambda_2,\Lambda_3)^{1/2}}\sum_{(\sigma)}\begin{pmatrix}\Lambda_1 & \Lambda_2 & \Lambda_3\\ \sigma_1 & \sigma_2 & \sigma_3\end{pmatrix}Y_{\Lambda_1\sigma_1}(\hat{\mathbf{r}}_{14})Y_{\Lambda_2\sigma_2}(\hat{\mathbf{r}}_{24})Y_{\Lambda_3\sigma_3}(\hat{\mathbf{r}}_{34}), \tag{14}$$

where the sum is over all σ_i with nonzero contributions. Equation (14), though apparently designating $\hat{K}$ as a function of the individual vectors $\mathbf{r}_{ij}$, is actually dependent only on the shape coordinates, i.e., the lengths r_{ij}. The evaluation of the $\hat{K}$ is discussed in Ref. 19, together with a table of their values for parameter values sufficient for S, P, D, and F states. Note, however, that the values tabulated in Ref. 19 assumed that the solid harmonics were written in terms of $\mathbf{r}_{12}$, $\mathbf{r}_{13}$, and $\mathbf{r}_{14}$ rather than (as here) $\mathbf{r}_{14}$, $\mathbf{r}_{24}$, and $\mathbf{r}_{34}$, so for use here the index numbering in that table must be transformed according to (1,2,3,4) $\rightarrow$ (4,1,2,3). A list of these invariants for $\Lambda_i \le 2$ (in the present notation) is presented here as Table 1. For compactness the table gives the $\hat{K}$ in terms of the cosines of the angles θ_{4ij}; these quantities can be converted to expressions in the r_{ij} using Eq. (9). The instances of r_{ij} occurring in the denominator of that equation are offset by the premultipliers $r_{i4}^{l_i+\lambda_i}$ in Eqs. (12) and (13).

7. SHAPE INTEGRATION

After evaluating the angular part of a matrix element, there remain shape-space integral(s) of the generic form

$$F(n_1,n_2,n_3,m_1,m_2,m_3;u_1,u_2,u_3,w_1,w_2,w_3) = \int d\tau_r\left(\prod_{i<j\le 3} r_{ij}^{n_k-1}\,e^{-u_k r_{ij}}\right)\left(\prod_{k=1}^{3} r_{k4}^{m_k-1}\,e^{-w_k r_{k4}}\right). \tag{15}$$

In the first product appearing here, the index k has the value from the set (1,2,3) other than i or j. The form given in Eq. (15) for describing the integrals F is consistent with notational choices made in earlier work both by the present author and by Pachucki *et al.* The parameters n_k, m_k, u_k and w_k have values as needed for the rotational invariants $\hat{K}$ and the spatial wave functions $\mathbf{\Phi}_n$ and $\mathbf{\Phi}_{n'}$. Note also that we have written the exponents of the r_{ij} (and r_{k4}) as $n_k - 1$ (and $m_k - 1$). Integrals with one or more zero n_k or m_k are convergent, and some such integrals arise, for example, in evaluations of potential-energy matrix elements. In fact, the integral with all $n_k = m_k = 0$ is not only convergent, but is a starting point for obtaining formulas for all other sets $\{n,m\}$ with the n_k and/or $m_k \geq 0$.

Before proceeding further it is useful to adopt some notational conventions:

(1) Unless needed to avoid ambiguity, we do not always show explicitly the parameters u_k or w_k.

(2) Once a set of indices $\{n,m\} = (n_1,n_2,n_3,m_1,m_2,m_3)$ has been specified (either explicitly or implicitly), we write as arguments of F only the indices that differ from that set. Thus, for example,

$$F(n_{12}+1, n_{13}, n_{14}, n_{23}-1, n_{24}, n_{34}) \rightarrow F(n_{12}+1, n_{23}-1).$$

(3) We write F_0 to denote the basic integral for which all the n_k and m_k are zero.

It appears to be infeasible to evaluate Eq. (15) directly in the shape coordinates, in part because the volume element $d\tau_r$ is an unwieldy and rather complicated expression[28] (despite an incorrect assertion by Frolov and Smith[39] that it had a specific simple form). However, as shown by Fromm and Hill,[11] it is possible, starting from a formulation in spherical polar coordinates, to obtain by Fourier representation methods and sophisticated contour-integration procedures a closed analytical expression for the integral F_0.

Then, by repeated differentiation with respect to the parameters w_k and/or u_k, it is possible to obtain closed forms for integrals of the type given in Eq. (15) with arbitrary nonnegative n_k and m_k.

As indicated in the Introduction of this paper, the initial formulation of F_0 by Fromm and Hill presented some practical difficulties, which can be circumvented by methods described in Ref. 17. A concise procedure for the evaluation of F_0 can be found in appendix C of Ref. 24.

Even when evaluated as efficiently as presently possible, the relatively complicated form of F_0 leads to extremely voluminous formulas for its derivatives, motivating the development of recursion formulas for more general $F\{n\}$.

Before proceeding to the recursion formulas we note that integrals F satisfy various symmetries that express the invariance of F under permutations of the numbering of the four particles involved. In particular, note that

$$\begin{aligned} F(n_1,n_2,n_3,m_1,m_2,m_3;\ u_1,u_2,u_3,w_1,w_2,w_3) \\ = F(n_2,n_1,n_3,m_2,m_1,m_3;\ u_2,u_1,u_3,w_2,w_1,w_3), \\ = F(n_3,n_2,n_1,m_3,m_2,m_1;\ u_3,u_2,u_1,w_3,w_2,w_1), \\ = F(m_1,n_3,n_2,n_1,m_3,m_2;\ u_1,w_3,w_2,w_1,u_3,u_2), \end{aligned} \tag{16}$$

and that there are additional relations obtained by successively applying two or more of these formulas. These features indicate that we only need one recurrence formula, which we choose to be that which increases the parameter n_1 from the values corresponding to a reference index set.

An essential feature of the recurrence formulas is that they enable a single F whose indices sum to $N+1$ to be expressed as a linear combination of other F whose indices sum to N or less, plus additional "boundary" terms that correspond to three-body integrals. Thus, taking an index reference set n_1, n_2, n_3, m_1, m_2, m_3, we consider the recurrence formula for $F(n_1+1, n_2, n_3, m_1, m_2, m_3)$, which we write as $F(n_1+1)$. It is shown in Ref. 24 that

$$F(n_1+1) = \frac{C_1X_1 + C_2X_2 + C_3X_3}{D}, \tag{17}$$

where

$$C_1 = u_1(\mu_{23}^2 - 4u_2^2u_3^2), \tag{18}$$

$$C_2 = u_2(2u_3^2\mu_{12} - \mu_{13}\mu_{23}), \tag{19}$$

$$C_3 = u_3(2u_2^2\mu_{13} - \mu_{12}\mu_{23}), \tag{20}$$

$$\begin{aligned} D = 2u_2u_3(u_1^2\mu_{23}^2 + u_2^2\mu_{13}^2 + u_3^2\mu_{12}^2 \\ - \mu_{12}\mu_{13}\mu_{23} - 4u_1^2u_2^2u_3^2), \end{aligned} \tag{21}$$

with (i,j,k) equal (in any order) to (1,2,3), and

$$\mu_{ij} = u_i^2 + u_j^2 - w_k^2. \tag{22}$$

The X_i are given by the following formula, in which the jk sum is over the two ordered pairs in which j and k are the members of (1,2,3) other than i, and δ_p is unity if $p=0$ and zero otherwise. The symbol F without arguments denotes the integral with all indices at their reference values.

$$
\begin{aligned}
X_i &= 2u_j u_k (n_i + n_j + n_k + 1) F \\
&\quad - n_j n_k [2u_i F(n_i + 1, n_j - 1, n_k - 1) - (2n_i + n_j + n_k) F(n_j - 1, n_k - 1)] \\
&\quad + \sum_{jk} \{ m_j (m_j - 1)[u_j F(n_k - 1, m_j - 2) - n_j F(n_j - 1, n_k + 1, m_j - 2)] \\
&\quad + n_j \mu_{ik} F(n_j - 1, n_k + 1) + 2n_i u_i [u_j F(n_i - 1, n_k + 1) - n_j F(n_i - 1, n_j - 1, n_k + 1)] \\
&\quad - n_i (n_i - 1)[u_j F(n_i - 2, n_k + 1) - n_j F(n_i - 2, n_j - 1, n_k + 1)] \\
&\quad - 2m_j w_j [u_j F(n_k + 1, m_j - 1) - n_j F(n_j - 1, n_k + 1, m_j - 1)] \\
&\quad + n_j u_k [2u_i F(n_i + 1, n_j - 1) - (2n_i + n_j + 2n_k + 1) F(n_j - 1)] \\
&\quad - \delta_{m_j} [u_j F(n_k + 1, m_j = *) - n_j F(n_j - 1, n_k + 1, m_j = *)] \\
&\quad + \delta_{n_i} [u_j F(n_i = *, n_k + 1) - n_j F(n_i = *, n_j - 1, n_k + 1)] \}.
\end{aligned}
\tag{23}
$$

The F functions containing asterisks are "boundary" three-body integrals whose evaluation is discussed in an Appendix hereto.

In addition to the relations between F integrals implied by the recurrence formula, it was also found, first by Rebane,[14] and later in our research group,[28] that the combinations of integrals arising in the evaluation of the kinetic energy for spherically symmetric states could be expressed entirely in terms of overlap and potential-energy integrals. Similar relationships can be developed for wave functions of arbitrary angular symmetry, but we do not do so in the present contribution.

8. APPLICATIONS

The computational scheme outlined in the previous sections of this paper can be applied to four-body atoms (i.e., the Li isoelectronic series) and to exotic systems such as Ps_2 (the positronium "molecule": two electrons and two positrons) or "mesomolecules" (systems containing mesons). Accurate wave functions for these systems have value for a variety of purposes, including the study of relativistic effects (see, for example, Ref. 40) or the estimation of e^+–e^- annihilation rates (by a method such as that in Ref. 41).

Work on four-body systems in this author's research group has focused on the Li atom and Ps_2, with the immediate objective of finding wave functions that are sufficiently compact to be visualized easily. We note that more extensive computations in a fully exponentially correlated basis have been carried out in the group of T. K. Rebane; that work (to be discussed below) included a detailed study of Ps_2 and the examination of a variety of other mesomolecules.

We look first at the work from our group. In contrast to orbital-based formulations, the exponentials in the interparticle distances do not have optimum values that can be estimated based on prior experience or by extrapolation from values corresponding to noninteracting electrons. As a preliminary example, an optimum initial approximate wave function for the ground-state Li atom (with particle 4 the nucleus) would have spatial part (before considering spin and antisymmetry)

$$\Phi_1 = r_{34} \prod_{i<j} \mathrm{e}^{-a_{ij} r_{ij}}. \tag{24}$$

The factor r_{34} occurs, as expected, to correspond to the orbital structure $1s^2 2s$. The optimum set of the six a_{ij} for this wave function, which can be found by a fairly aggressive conjugate-gradient search, is,[42,43] in bohr^{-1},

$$\begin{aligned} a_{14} &= 3.3004778, \quad & a_{24} &= 2.353955, \quad & a_{34} &= 0.702684, \\ a_{12} &= -0.217022, \quad & a_{13} &= -0.022450, \quad & a_{23} &= -0.038799. \end{aligned}$$

Note that parameters a_{14} and a_{24} differ significantly, corresponding to what is often called a split-shell description of the $1s^2$ electron pair. With the parameters listed above we obtain the nonrelativistic energy (for an infinite-mass nucleus) − 7.471305 hartree, an accuracy intermediate between those of the four-term and five-term Hylleraas wave functions of Larsson,[44] respectively − 7.444731 and − 7.472405 hartree.

Continuing to multiple-term (multiconfiguration) wave functions,[43] several practical issues become apparent, of which the most important is the difficulty of finding optimum sets of the nonlinear parameters (of which there are, e.g., 12 for a two-term wave function and 24 for a four-term function). The optimization problem in those situations turns out to be extremely ill-conditioned, and is further complicated by the fact that there are multiple local energy minima. Despite these issues, we obtained with some effort well-converged wave functions, finding that the optimum second configuration had a structure corresponding mainly to correcting the form of the 2*s*-like function to make it more similar to a hydrogenic than to a Slater-type orbital (i.e., it was a term of the form $1s1s'1s''$), with overall two-term energy − 7.474682 hartree. We cannot compare the convergence

Table 2 Ground-State Energy of Li atom

Cfgs	E_H	$E_H - E_0$	E_E	$E_E - E_0$
1	−7.417 907	60,153	−7.471 305	6,755
2			−7.474 682	3,378
3	−7.430 046	48,014	−7.476 987	2,073
4	−7.444 731	33,329	−7.477 723	357
6	−7.474 021	4,039	−7.477 936	124
30	−7.477 935	125		

Energies in hartrees, energy differences in microhartrees. Hylleraas calculations E_H by Larsson[44]; exponentially correlated calculations E_E are the author's. The "exact" values E_0 are from work of Puchalski and Pachucki.[23]

of the Hylleraas and exponential two-term bases because of a misprint in the published Hylleraas two-term energy, but we did make these comparisons for other multiconfiguration wave functions.

Table 2 shows the calculated energies of the Li ground state (and energy errors relative to the extremely precise result of Puchalski and Pachucki[23]). The high rate of convergence of the exponential basis has led to an error of only 124 microhartrees at the six-configuration level, about 3% of the error of the corresponding-length Hylleraas expansion, which is itself more rapidly convergent than a conventional orbital approach.

Another example of the effective use of the correlated-exponential formulation is provided by the positronium "molecule" Ps_2 (the system $e^+e^-e^+e^-$), which illustrates another feature of the formulation, namely its applicability when none of the particles in a system under study can be approximated as having infinite mass. Older work in the Rebane group[14] showed how this feature enabled good computations of the energies of a variety of four-body mesomolecules involving protons and their isotopes d^+, t^+ and μ or K mesons without using approximations to account for "mass polarization."

Rather detailed studies of Ps_2 antedating the present research have been reported by Rebane's group[20–21]; they demonstrate the relatively rapid convergence of an exponentially correlated basis set. Our study of Ps_2[28] was focused on the added benefits made possible by computing the "molecular" integrals by methods that enabled close approach to values of the nonlinear parameters that exhibited spurious (i.e., termwise canceling) singularities. Use of these methods permitted an even more rapidly convergent computation than that exhibited in the work of the Rebane group, and illustrated the fact that parameter values near these termwise singularities arise in various practical computations.

ACKNOWLEDGMENTS

The early stages of this work were supported by the U.S. National Science Foundation, Grants NSF 0303412 and 0601758; more recent support was provided by U.S. Department of Energy Grant DE-SC 0002139. Some of the work reported here was carried out at the University of Florida by Victor Albert (then an undergraduate in Physics, now a graduate student at Yale University). The author thanks the University of Florida and the organizers of the Molecular Electronic Structure (MEST) Workshop in Amasya, Turkey, Philip Hoggan and Telhat Özdoğan, for financial support of his participation.

APPENDIX. BOUNDARY INTEGRALS

The functions F with index sets containing asterisks are integrals involving only three of the four particles; they then reduce to three-body integrals of the general form

$$\Gamma_{n_1,n_2,n_3}(\alpha,\beta,\gamma) = \int \frac{d\mathbf{r}_1 d\mathbf{r}_2}{16\pi^2} r_1^{n_1-1} r_2^{n_2-1} r_{12}^{n_3-1} e^{-\alpha r_1 - \beta r_2 - \gamma r_{12}}. \tag{A.1}$$

It is useful to note that these integrals are invariant with respect to the simultaneous permutation of (n_1, n_2, n_3) and (α, β, γ), as can be seen by writing the integral in orientation and shape coordinates, i.e., as integrations over Euler angles and (r_1, r_2, r_{12}). We now observe (cf. Ref. 24) that

$$F(*,n_2,n_3;m_1,m_2,m_3) = \Gamma_{m_1,m_2+m_3-1,n_2+n_3-1}(w_1,w_2+w_3,u_2+u_3), \tag{A.2}$$

$$F(n_1,n_2,n_3;m_1,m_2,*) = \Gamma_{m_1+n_2-1,m_2+n_1-1,n_3}(w_1+u_2,u_1+w_2,u_3). \tag{A.3}$$

The parameters u_k and w_k entering these equations are those introduced in Eq. (15).

The method for evaluating an integral Γ depends upon whether all its indices are nonnegative. Making use of the permutational invariance identified after Eq. (A.1), we note that all the Γ required here fall into two cases: (1) Γ_{n_1,n_2,n_3} with all n_i nonnegative, and (2) Equivalent to a Γ_{-1,n_2,n_3} with n_2 and n_3 nonnegative.

The first case (all indices nonnegative) can be treated by a procedure introduced by Sack, Roothaan, and Kolos,[45] which leads[24] to the recurrence formulas

$$\Gamma_{n_1 n_2 n_3} = \frac{1}{\alpha+\beta}\left[n_1 \Gamma_{n_1-1,n_2,n_3} + n_2 \Gamma_{n_1,n_2-1,n_3} + B_{n_1 n_2 n_3}\right], \quad \text{(A.4)}$$

$$B_{n_1 n_2 n_3} = \frac{1}{\alpha+\gamma}\left[n_1 B_{n_1-1,n_2,n_3} + n_3 B_{n_1,n_2,n_3-1} + A_{n_1 n_2 n_3}\right], \quad \text{(A.5)}$$

$$A_{n_1 n_2 n_3} = \frac{\delta_{n_1}(n_2+n_3)!}{(\beta+\gamma)^{n_1+n_3+1}}. \quad \text{(A.6)}$$

Here the notation δ_p (also used below) has the value unity if $p = 0$ and zero otherwise. It is computationally stable to evaluate first the array A, then B, and finally Γ.

For the integrals Γ_{-1,n_2,n_3}, the optimum procedure depends upon the magnitude of $\beta - \gamma$. If that quantity is not too small we can develop, analogously to Eqs. (A.4)–(A.6),

$$\Gamma_{-1,n_2,n_3} = \frac{1}{\beta+\gamma}\left[n_2 \Gamma_{-1,n_2-1,n_3} + n_3 \Gamma_{-1,n_2,n_3-1} + G_{n_2 n_3}\right], \quad \text{(A.7)}$$

$$G_{n_2 n_3} = \frac{1}{\beta-\gamma}\left[n_2 G_{n_2-1,n_3} - n_3 G_{n_2,n_3-1} + K_{n_2 n_3}\right], \quad \text{(A.8)}$$

$$K_{n_2 n_3} = \delta_{n_2}\delta_{n_3} \ln\left(\frac{\alpha+\beta}{\alpha+\gamma}\right) - \frac{\delta_{n_3}(1-\delta_{n_2})(n_2-1)!}{(\alpha+\beta)^{n_2}} + \frac{\delta_{n_2}(1-\delta_{n_3})(n_3-1)!}{(\alpha+\gamma)^{n_3}}, \quad \text{(A.9)}$$

and we can evaluate first the array K, then G, and finally Γ.

For $\beta - \gamma$ small, a more satisfactory approach is to develop a formula for $G_{n_2 n_3}$ as an expansion in powers of $y = (\beta - \gamma)/(\alpha + \beta)$. As discussed in detail in Ref. 46, one can start from the power-series expansion

$$G_{00} = \sum_{k=0}^{\infty} \frac{y^k}{(k+1)(\alpha+\beta)}, \quad \text{(A.10)}$$

and differentiate with respect to β and γ to obtain $G_{n_2 n_3}$ of more general index values. The result, which requires a significant amount of manipulation, can be written

$$G_{n_2,n_3} = \frac{1}{(\alpha+\beta)^{n_2+n_3+1}} \sum_{k=0}^{\infty} \frac{n_2!(n_3+k)!y^k}{k!(n_2+n_3+k+1)}. \quad \text{(A.11)}$$

This result is equivalent to one obtained by Korobov.[47] However, note that some of Korobov's formulas are in error by a factor of 2.

The convergence rate of the expansion in Eq. (A.11) can be improved by using a more symmetric expansion variable. Defining $x=\frac{1}{2}(\beta+\gamma)$ and $u=\beta-\gamma$, we now expand in powers of $w=u/(\alpha+x)$. The result is

$$G_{n_2n_3}=\frac{n_2!n_3!}{(\alpha+u)^{n_2+n_3+1}}\sum_{k=0}^{\infty}F(n_2+1,-k;n_2+n_3-2;2)\frac{(n_2+n_3+k)!}{(n_2+n_3+1)!}w^k. \tag{A.12}$$

Here F is a hypergeometric function,[48] sometimes written in the more explicit notation ${}_2F_1$. Since its last argument is a constant, 2, it is actually part of the coefficient of w^k, but its recognition as a hypergeometric function identifies methods for its efficient evaluation. Writing Eq. (A.12) in the form

$$G_{n_2n_3}=\frac{n_2!n_3!}{(\alpha+u)^{n_2+n_3+1}}\sum_{k=0}^{\infty}\frac{(-1)^kC_kw^k}{k!(n_2+n_3+k+1)}, \tag{A.13}$$

the coefficients C_k are tabulated in Ref. 46.

REFERENCES

1. Hylleraas, E. Neue Berechnung der Energie des Heliums im Grundzustande, sowie des Tiefsten Terms von Ortho-Helium. *Z. Phys.* **1929**, *54*, 347–366.
2. James, H. M.; Coolidge, A. S. The Ground State of the Hydrogen Molecule. *J. Chem. Phys.* **1933**, *1*, 825–835.
3. Pekeris, C. L. Ground State of Two-Electron Atoms. *Phys. Rev.* **1958**, *112*, 1649–1658.
4. Pekeris, C. L. $1\,^1S$ and $2\,^3S$ States of Helium. *Phys. Rev.* **1959**, *115*, 1216–1221.
5. Delves, L. M.; Kalotas, T. Variational Calculations for Three-Body S-States with Coulomb Interactions. *Aust. J. Phys.* **1968**, *21*, 431–438.
6. Thakkar, A. J.; Smith, V. H., Jr. Compact and Accurate Integral-Transform Wave Functions. I. The 1^1S State of the Helium-Like Ions from H^- through Mg^{10} +. *Phys. Rev. A* **1977**, *15*, 1–15.
7. Frolov, A. M.; Smith, V. H. J. Exponential Representation in the Coulomb Three-Body Problem. *J. Phys. B* **2004**, *37*, 2917–2932. Erratum 2004, 37, 4037.
8. Harris, F. E.; Smith, V. H. J. Highly Compact Wavefunctions for Two-Electron Systems. *Adv. Quantum. Chem.* **2005**, *48*, 407–419.
9. Harris, F. E.; Smith, V. H. J. Highly Compact Wave Functions for He-Like Systems. *J. Phys. Chem. A* **2005**, *109*, 11413–11416.
10. Harris, F. E.; Smith, V. H., Jr. Systematics of the Ground States of the He Isoelectronic Series. In: *Symmetry, Spectroscopy, and SHUR, Proceedings of the Prof. Brian G. Wybourne Commemorative Meeting*; King, R. C., Bylicki, M., Karwowski, J., Eds.; Nicolaus Copernicus University Press: Torun, Poland, 2006; pp 127–137.
11. Fromm, D. M.; Hill, R. N. Evaluation of Three-Electron Integrals. *Phys. Rev. A* **1987**, *36*, 1013–1044.
12. Rebane, T. K.; Yusupov, O. N. Application of Exponential Wave-Functions to Energy Calculations for Four-Body Systems. *Opt. Spektrosk.* **1992**, *73*, 875–879 [*Opt. Spectrosc.* **1992,** *73*, 523].

13. Rebane, T. K.; Zotev, V. S.; Yusupov, O. N. Variational Calculation of the Positronium Molecule in the Basis of Exponential Functions of Interparticle Separations. *Zh. Eksp. Teor. Fiz.* **1996**, *110*, 55–62 [*JETP* **1996,** *83*, 28].
14. Zotev, V. S.; Rebane, T. K. Calculation of Symmetric Mesomolecules in the Exponential Basis Set. *Opt. Spectrosc.* **1998**, *85*, 935–941 [*Opt. Spectrosc.* **1998,** *85*, 856].
15. Remiddi, E. Analytic Value of the Atomic Three-Electron Correlation Integral with Slater Wave Functions. *Phys. Rev. A* **1991**, *44*, 5492–5502.
16. Rebane, T. K. To the Energy Calculation of Coulomb System. *Opt. Spektrosk.* **1993**, *75*, 945–953 [*Opt. Spectrosc.* **1993,** *75*, 557].
17. Harris, F. E. Analytic Evaluation of Three-Electron Integrals with Slater Wave Functions. *Phys. Rev. A* **1997**, *55*, 1820–1831.
18. Yan, Z. C.; Drake, G. W. F. Computational Methods for Three-Electron Atomic Systems in Hylleraas Coordinates. *J. Phys. B* **1997**, *30*, 4723–4750.
19. Harris, F. E. Integrals for Exponentially Correlated Four-Body Systems of General Angular Symmetry. In *Fundamental World of Quantum Chemistry: A Tribute to Per-Olov Löwdin*; Vol. 3; Brändas, E. J., Kryachko, E. S., Eds.; Kluwer: Dordrecht, 2004; pp. 115–128.
20. Zotev, V. S.; Rebane, T. K. Analytic Evaluation of Four-Particle Integrals with Complex Parameters. *Phys. Rev. A* **2002**, *65*(06), 2501.
21. Rebane, T. K.; Yusupov, O. N. Complex Exponential Basis Wave Functions in the Coulomb Three-Body Problem. *Zh. Eksp. Teor. Fiz.* **1990**, *98*(1870) [*JETP* **1990,** *71*, 1050].
22. Pachucki, K.; Puchalski, M.; Remiddi, E. Recursion Relations for the Generic Hylleraas Three-Electron Integral. *Phys. Rev. A* **2004**, *70*(03), 2502.
23. Puchalski, M.; Pachucki, K. Ground-State Wave Function and Energy of the Lithium Atom. *Phys. Rev. A* **2006**, *73*(02), 2503.
24. Harris, F. E. Recurrence Formulas for Fully Exponentially Correlated Four-Body Wavefunctions. *Phys. Rev. A* **2009**, *79*(03), 2517.
25. Harris, F. E. Energy Computation for Exponentially Correlated Four-Body Wavefunctions. In *Advances in the Theory of Atomic and Molecular Systems*; Piecuch, P., Maruani, J., Delgado-Barrio, G., Wilson, S., Eds.; Springer: Dordrecht, 2009; pp 61–70.
26. Kong, L.; Bischoff, F. A.; Valeev, E. F. Explicitly Correlated R12/F12 Methods for Electronic Structure. *Chem. Rev.* **2012**, *112*, 75–107.
27. Frolov, A. M. Semi-Exponential Basis for Highly Accurate Computations of Three-Electron Atomic Systems. *Phys. Lett. A* **2010**, *374*, 2361–2366.
28. Harris, F. E.; Frolov, A. M.; Smith, V. H. J. Exponential Variational Expansion in Relative Coordinates for Highly Accurate Bound State Calculations in Four-Body Systems. *J. Chem. Phys.* **2003**, *119*, 8833–8841.
29. Littlejohn, R. G.; Reinsch, M. Internal Or Shape Coordinates in the N-Body Problem. *Phys. Rev. A* **1995**, *52*, 2035–2051.
30. Breit, G. Separation of Angles in the Two-Electron Problem. *Phys. Rev.* **1930**, *35*, 569–578.
31. Bhatia, A. K.; Temkin, A. Symmetric Euler-Angle Decomposition of the Two-Electron Fixed-Nucleus Problem. *Rev. Mod. Phys.* **1964**, *36*, 1050–1064.
32. Schwartz, C. L. Lamb Shift in the Helium Atom. *Phys. Rev.* **1961**, *123*, 1700–1705.
33. Harris, F. E. Angular Symmetry and Hylleraas Coordinates in Four-Body Problems. *Adv. Quantum. Chem.* **2005**, *50*, 61–75.
34. Edmonds, A. R. *Angular Momentum in Quantum Mechanics*. Princeton Univ. Press: Princeton, NJ, 1960.
35. Frost, A. A. The Use of Intcrparticle Coordinates in Electronic Energy Calculations for Atoms and Molecules. *Theor. Chim. Acta* **1962**, *1*, 36–41.
36. Sack, R. A. Generalization of Laplace's Expansion to Arbitrary Powers and Functions of the Distance between Two Points. *J. Math. Phys.* **1964**, *5*, 245–251.

37. Biedenharn, L. C.; Louck, J. D. *Angular Momentum in Quantum Physics: Theory and Applications*. Addison-Wesley: Reading, MA, 1981.
38. Brink, D. M.; Satchler, G. R. *Angular Momentum*, 3rd; Clarendon Press: Oxford, 1993.
39. Frolov, A. M.; Smith, V. H. J. Generalization of the Exponential Variational Ansatz in Relative Coordinates for Bound State Calculations in Four-Body Systems. *J. Chem. Phys.* **2001**, *115*, 1187–1196.
40. Puchalski, M.; Pachucki, K. Applications of Four-Body Exponentially Correlated Functions. *Phys. Rev. A* **2010**, *81*(05), 2505.
41. Ho, Y. K. Positron Annihilation in Positronium Negative Ions. *Phys. Lett. A* **1990**, *144*, 237–240.
42. Albert, V. V. *Few-Parameter Exponentially Correlated Wave Functions for the Ground State of Lithium; Undergraduate thesis (unpublished)*. Department of Physics University of Florida: Gainesville, 2010.
43. Harris, F. E.; Albert, V. V. Fully Correlated Wavefunctions for Three- and Four-Body Systems. *Adv. Quantum. Chem.* **2013**, *67*, 3–18.
44. Larsson, S. Calculations on the ^{2}S Ground State of the Lithium Atom Using Wawe Functions of Hylleraas Type. *Phys. Rev.* **1968**, *169*, 49–54.
45. Sack, R. A.; Roothaan, C. C. J.; Kolos, W. Recursive Evaluation of Some Atomic Integrals. *J. Math. Phys.* **1967**, *8*, 1093–1094.
46. Harris, F. E. Atomic Three- and Four-Body Recurrence Formulas and Related Summations. *Theor. Chem. Acc.* **2014**, *133*, 1475.
47. Korobov, V. I. Regular and Singular Integrals for Relativistic and QED Matrix Elements of the Coulomb Three-Body Problem, for an Exponential Basis Set. *J. Phys. B* **2002**, *35*, 1959–1966.
48. Abramowitz, M.; Stegun, I. A. *Handbook of Mathematical Functions*. Dover: New York, 1972.

CHAPTER FOUR

Analytic Formulas for Two-Center Two-Electron Integrals with Exponential Functions

Magdalena Zientkiewicz*, Krzysztof Pachucki*,[1]
*Faculty of Physics, University of Warsaw, Warsaw, Poland
[1]Corresponding author: e-mail address: krp@fuw.edu.pl

Contents

Abstract

We propose the use of exponential functions multiplied by the product of spherical harmonics at two centers in high precision calculations for diatomic molecules and demonstrate that all integrals can be obtained using analytic methods.

1. INTRODUCTION

In spite of many advantages of exponential bases for molecular systems, they are very rarely employed due to difficulties in performing two-electron two-center integrals.[1,2] The best-known accurate calculations with exponential functions have been performed by Kolos and coworkers for a hydrogen molecule, starting with the seminal work in Ref. 3. They used the Neumann expansion of $1/r_{12}$ in spheroidal coordinates and its slow convergence was the main factor limiting the numerical accuracy of two-center two-electron integrals. This expansion has recently been improved

Advances in Quantum Chemistry, Volume 73
ISSN 0065-3276
http://dx.doi.org/10.1016/bs.aiq.2015.04.002

in Ref. 4 with calculations for the Be_2 molecule. In this work, we overcome the problem of slow convergence of this expansion by direct calculation of all integrals with electronic functions of the type

$$\begin{aligned}\phi_A &= e^{-\alpha r_A} r_A^{n_A} Y_{lm}(\hat{r}_A) r_B^{n_B} Y_{l'm'}(\hat{r}_B),\\ \phi_B &= e^{-\beta r_B} r_B^{n_B} Y_{l'm'}(\hat{r}_B) r_A^{n_A} Y_{lm}(\hat{r}_A),\end{aligned} \tag{1}$$

where A,B are indices of the nuclei, r_A,r_B the electron distance from the corresponding nucleus, α and β are fixed nonlinear parameters, and $n_A \geq l$,$n_B \geq l'$ are arbitrary nonnegative integers. Matrix elements of nonrelativistic Hamiltonian in this basis can all be expressed in terms of the following integrals over powers of interparticle distances

$$f(n_0,n_1,n_2,n_3,n_4,r) = \int \frac{d^3 r_1}{4\pi} \int \frac{d^3 r_2}{4\pi} e^{-\alpha_1 r_{1A} - \beta_1 r_{1B} - \alpha_2 r_{2A} - \beta_2 r_{2B}} \; r_{12}^{n_0-1} r_{1A}^{n_1-1} r_{1B}^{n_2-1} r_{2A}^{n_3-1} r_{2B}^{n_4-1} \tag{2}$$

where r is the distance between nuclei. Note that products of spherical harmonics can be expressed as a polynomial in electron distances as in the above equation. Due to the choice of basis functions in Eq. (1), the nonlinear parameters may take the following values (up to $1 \Leftrightarrow 2$ and $A \Leftrightarrow B$ symmetries)

$$(\alpha_1, \beta_1, \alpha_2, \beta_2) = \begin{cases} (2\alpha, 0, 0, 2\beta) & \text{-direct} \\ (\alpha, \beta, \alpha, \beta) & \text{-exchange} \\ (2\alpha, 0, 2\alpha, 0) & \text{-ionic} \\ (2\alpha, 0, \alpha, \beta) & \text{-mixed} \end{cases}. \tag{3}$$

We propose here to generate an explicit analytic expression for all integrals up to some $\Omega = n_0 + n_1 + n_2 + n_3 + n_4$ with the help of a computer symbolic algebra program and the rest of this work explains how to achieve this. Due to presence of only two independent nonlinear parameters, the analytical expressions are not excessively large and can be stored in a few GB volume for Ω as large as 37.

Let us start from the integral with all $n_i = 0$; it will be called the master integral $f(r)$

$$f(r) = r \int \frac{d^3 r_1}{4\pi} \int \frac{d^3 r_2}{4\pi} \frac{1}{r_{12}} \frac{e^{-u_3 r_{1A}}}{r_{1A}} \frac{e^{-u_2 r_{1B}}}{r_{1B}} \frac{e^{-w_2 r_{2A}}}{r_{2A}} \frac{e^{-w_3 r_{2B}}}{r_{2B}} \tag{4}$$

where we changed the notation for nonlinear parameter to synchronize it with our former papers and introduced for convenience an extra factor of r. The function $f(r)$ satisfies the following differential equation[5]:

$$\left[\sigma_{20}\,\frac{d}{d\,r}\,r\,\frac{d}{d\,r}+\sigma_{00}\,r\right]f(r)=F(r), \tag{5}$$

where σ coefficients are

$$\sigma_{20}=-(u_2^2-u_3^2)\,(w_2^2-w_3^2)=16\,u\,w\,x\,y \tag{6}$$

$$\begin{aligned}\sigma_{00}&=(u_2^2-u_3^2+w_2^2-w_3^2)\,(u_2^2\,w_2^2-u_3^2\,w_3^2)\\&=16\,(w\,x-u\,y)\,(u\,x-w\,y)\,(u\,w-x\,y)\end{aligned} \tag{7}$$

and notation for subscripts is specified below. The new parameters y,x,u,w

$$w_2=w+x,\quad w_3=w-x,\quad u_2=u-y,\quad u_3=u+y, \tag{8}$$

are adapted to the symmetry of the master integral, namely

$$f(y,x,u,w)=f(x,y,u,w)=f(y,x,w,u)=f(x,y,w,u) \tag{9}$$

The inhomogeneous term in the differential equation (5) is the following

$$\begin{aligned}F(r)=&\,2\,(u\,w\,x-u\,w\,y-u\,x\,y+w\,x\,y)\,F_{1-}+2\,(u\,w\,x-u\,w\,y+u\,x\,y-w\,x\,y)\,F_{2-}\\&-2(u\,w\,x+u\,w\,y+u\,x\,y+w\,x\,y)\,F_{3-}+2\,(u\,w\,x+u\,w\,y-u\,x\,y-w\,x\,y)\,F_{4-},\end{aligned} \tag{10}$$

where

$$\begin{aligned}F_{1\pm}=&\,\mathrm{Ei}(-2\,r\,u)\,\mathrm{e}^{r\,(u-w-x+y)}\pm\mathrm{Ei}(-2\,r\,w)\,\mathrm{e}^{r\,(-u+w+x-y)},\\F_{2\pm}=&\,\mathrm{Ei}(-2\,r\,w)\,\mathrm{e}^{r\,(-u+w-x+y)}\pm\mathrm{Ei}(-2\,r\,u)\,\mathrm{e}^{r\,(u-w+x-y)},\\F_{3\pm}=&\left[\mathrm{Ei}(2\,r\,x)+\mathrm{Ei}(2\,r\,y)-\mathrm{Ei}(2\,r\,(x+y))+\ln\left(\frac{u\,w\,(x+y)}{x\,y\,(u+w)}\right)\right]\\&\mathrm{e}^{-r\,(u+w+x+y)}\pm\mathrm{Ei}(-2\,r\,(u+w))\,\mathrm{e}^{r\,(u+w+x+y)},\\F_{4\pm}=&\left[\mathrm{Ei}(-2\,r\,x)+\mathrm{Ei}(-2\,r\,y)-\mathrm{Ei}(-2\,r\,(x+y))+\ln\left(\frac{u\,w\,(x+y)}{x\,y\,(u+w)}\right)\right]\\&\mathrm{e}^{-r\,(u+w-x-y)}\pm\mathrm{Ei}(-2\,r\,(u+w))\,\mathrm{e}^{r\,(u+w-x-y)},\end{aligned} \tag{11}$$

and Ei is the exponential integral function. This differential equation is supplemented by the boundary conditions, namely $f(r)$ vanishes at small and large r.

2. INTEGRAL REPRESENTATION FOR THE MASTER INTEGRAL

Since the homogeneous differential equation (5) has solutions I_0 and K_0, modified Bessel functions, the inhomogeneous equation can be expressed as[5]

$$f(r) = -\frac{1}{\sigma_{20}}\left[I_0(p\,r)\int_r^{\infty} dr'\,F(r')\,K_0(p\,r') + K_0(p\,r)\int_0^{r} dr'\,F(r')\,I_0(p\,r')\right], \tag{12}$$

where $p = \sqrt{-\sigma_{00}/\sigma_{20}}$. However, a more elegant integral representation for $f(r)$ is the following[6]

$$\begin{aligned} f(r) = \int_0^{-\infty} dt\, e^{t\,r}\,\frac{1}{2\sqrt{\sigma}}[&\theta(t_1-t)\,\ln\beta_1 + \theta(t_2-t)\,\ln\beta_2 \\ &-\theta(t_3-t)\,\ln\beta_3 - \theta(t_4-t)\,\ln\beta_4]. \end{aligned} \tag{13}$$

where $\sigma = \sigma_{00} + t^2\,\sigma_{20}$ and

$$\begin{aligned} -t_1 &= u_3 + w_2 = u + y + w + x, \\ -t_2 &= u_2 + w_3 = u - y + w - x, \\ -t_3 &= u_3 + w_3 = u + y + w - x, \\ -t_4 &= u_2 + w_2 = u - y + w + x, \end{aligned} \tag{14}$$

and where

$$\beta_i = \frac{\sqrt{\sigma} - \gamma_i}{\sqrt{\sigma} + \gamma_i}, \tag{15}$$

with the following γ_i coefficients

$$\begin{aligned} \gamma_1 &= -4\,(u\,w\,x + u\,w\,y - u\,x\,y - w\,x\,y), \\ \gamma_2 &= 4\,(u\,w\,x + u\,w\,y + u\,x\,y + w\,x\,y), \\ \gamma_3 &= -16\,u\,w\,x\,y/(t + u + w - x + y) + 4\,(u\,w\,x - u\,w\,y + u\,x\,y + w\,x\,y), \\ \gamma_4 &= -16\,u\,w\,x\,y/(t + u + w + x - y) - 4\,(u\,w\,x - u\,w\,y - u\,x\,y - w\,x\,y), \end{aligned} \tag{16}$$

The evaluation of the logarithms in Eq. (13) proceeds as follows[7]

$$\mathrm{Re}\left[\frac{1}{2\sqrt{\sigma}}\ln\left(\frac{\sqrt{\sigma}-\gamma}{\sqrt{\sigma}+\gamma}\right)\right]=\begin{cases}-\frac{1}{\gamma}, & \text{for } \sigma=0,\\ -\frac{1}{\sqrt{\sigma}}\operatorname{arctanh}\left(\frac{\gamma}{\sqrt{\sigma}}\right), & \text{for } \sigma>0, |\gamma|<\sqrt{\sigma},\\ -\frac{1}{\sqrt{\sigma}}\operatorname{arctanh}\left(\frac{\sqrt{\sigma}}{\gamma}\right), & \text{for } \sigma>0, |\gamma|>\sqrt{\sigma},\\ -\frac{1}{\sqrt{-\sigma}}\left[\arctan\left(\frac{\sqrt{-\sigma}}{\gamma}\right)+k\pi\right], & \text{for } \sigma<0.\end{cases} \tag{17}$$

In the last line, the phase prefactor $k = 0,\pm1$ is introduced, which is determined as follows. On an integration path where σ is negative, the parameter γ may change a sign. At this point, a phase factor $\pm\,\pi$ is introduced so as to make the integrand continuous. Since γ changes sign twice, the correction term vanishes at the integration point where $\sigma = 0$.

In the special case of the direct integral

$$\begin{aligned} u_3 &= \alpha \\ w_3 &= \beta \\ u_2 &= w_2 = 0 \end{aligned} \tag{18}$$

the integral representation becomes

$$f(r)=\int_0^{-\infty} dt\,\frac{e^{t\,r}}{2\sqrt{\sigma}} \left[\theta(-\alpha-t)\,\ln\beta_1+\theta(-\beta-t)\,\ln\beta_2-\theta(-\alpha-\beta-t)\,\ln\beta_3-\ln\beta_4\right] \tag{19}$$

with

$$\begin{aligned} \sigma &= \alpha^2\beta^2\left(\alpha^2+\beta^2-t^2\right) \\ \gamma_1 &= -\alpha^2\beta \\ \gamma_2 &= -\alpha\beta^2 \\ \gamma_3 &= -\alpha\beta\left(\alpha+\beta\right)+\frac{\alpha^2\beta^2}{t+\alpha+\beta} \\ \gamma_4 &= \frac{\alpha^2\beta^2}{t} \end{aligned} \tag{20}$$

In the case of an exchange integral $w = u$, $x = y$, σ_{00} vanishes and the master integral becomes

$$
\begin{aligned}
f(r) = \frac{1}{8\,w\,x}\Bigg(&\int_{-2\,(w+x)}^{-\infty} dt\,\ln\left|\frac{t+2\,(w-x)}{t-2\,(w-x)}\right|\frac{e^{t\,r}}{t}\\
&+\int_{-2\,(w-x)}^{-\infty} dt\,\ln\left|\frac{t-2\,(w+x)}{t+2\,(w+x)}\right|\frac{e^{t\,r}}{t}\\
&-2\int_{-2\,w}^{-\infty} dt\,\ln\left|\frac{t+2\,(w-x)}{t+2\,(w+x)}\right|\frac{e^{t\,r}}{t}\Bigg).
\end{aligned}
\tag{21}
$$

This integral cannot be expressed in terms of standard functions but its derivative can, namely

$$
\begin{aligned}
f'(r) = \frac{1}{2\,r\,(u^2-w^2)}\,\Big[&\mathrm{Ei}[-2\,r\,(u+w)]\,[\exp(2\,r\,u)-\exp(2\,r\,w)]\\
&+\left(\mathrm{Ei}[2\,r\,(w-u)]-2\,\mathrm{Ei}[r\,(w-u)]+\ln\left|\frac{u-w}{u+w}\right|\right)\exp(-2\,r\,w)\\
&-\left(\mathrm{Ei}[2\,r\,(u-w)]-2\,\mathrm{Ei}[r\,(u-w)]+\ln\left|\frac{u-w}{u+w}\right|\right)\exp(-2\,r\,u)\Big]
\end{aligned}
\tag{22}
$$

Two other cases of Eq. (3) are similar, so the master integral can be calculated as accurately as the one-dimensional integral and we found that Gaussian integrations adapted to the logarithmic singularity at the end point[8] converge very quickly.

3. RECURRENCE RELATIONS FOR THE GENERAL INTEGRAL WITH $\sigma_{00} \neq 0$

In order to obtain the molecular integrals with higher powers of r_{12},

$$
f(r,n) = r\int\frac{d^3r_1}{4\pi}\int\frac{d^3r_2}{4\pi}\,r_{12}^{n-1}\,\frac{e^{-u_3\,r_{1A}}}{r_{1A}}\,\frac{e^{-u_2\,r_{1B}}}{r_{1B}}\,\frac{e^{-w_2\,r_{2A}}}{r_{2A}}\,\frac{e^{-w_3\,r_{2B}}}{r_{2B}}
\tag{23}
$$

we consider a more general function $\tilde{f}(r)$, which includes $\exp(-w_1\,r_{12})$ (see Eq. A.1). Its properties are very similar to that of $f(r)$. In particular, it satisfies the master differential equation (A.2),[6,9] several other differential equations (A.10,A.11), and it can also be represented as a one-dimensional integral.[6] The recurrence relations for $f(r,n)$ have already been presented in Refs. 5,6. Here, we rederive them in a compact way, namely let us take Eqs. (A.2) and (A.10), and differentiate them with respect to w_1, n, and $n-1$ times, respectively, at $w_1 = 0$

$$r\,\sigma_{00} f(r,n) + \sigma_{20} f'(r,n) + r\,\sigma_{20} f''(r,n) + (n-1)\,n\,r\,\sigma_{02} f(r,n-2) \\ + (n-1)\,n\,\sigma_{22} f'(r,n-2) + (n-1)\,n\,r\,\sigma_{22} f''(r,n-2) + 2\,(n-1)\,n f^{(3)}(r,n-2) \\ + (n-1)\,n\,r f^{(4)}(r,n-2) + (n-3)\,(n-2)\,(n-1)\,n f'(r,n-4) \\ + (n-3)\,(n-2)\,(n-1)\,n\,r f''(r,n-4) = F_{u_1}(r,n), \tag{24}$$

$$\sigma_{00} f(r,n) + \sigma_{20} f''(r,n) + (n-1)^2\,\sigma_{02} f(r,n-2) + (n-1)^2\,\sigma_{22} f''(r,n-2) \\ + (n-1)^2 f^{(4)}(r,n-2) + (n-3)\,(n-2)^2\,(n-1) f''(r,n-4) = F_{w_1}(r,n-1). \tag{25}$$

These are two linear equations for three unknowns $f(r,n)$, $f'(r,n)$, and $f''(r,n)$. The third equation is obtained by elimination of $f''(r,n)$ and further differentiation with respect to r. The solution of these three equations for $f(r,n)$ is

$$\begin{aligned} f(r,n) = \frac{1}{\sigma_{00}} \Big[& -(n-2)\,(n-1)\,\sigma_{02} f(r,n-2) + (n-1)\,r\,\sigma_{02} f'(r,n-2) \\ & + 2\,(n-1)\,\sigma_{22} f''(r,n-2) + (n-1)\,r\,\sigma_{22} f^{(3)}(r,n-2) \\ & + (n-1)\,(n+2) f^{(4)}(r,n-2) + (n-1)\,r f^{(5)}(r,n-2) \\ & + 4\,(n-3)\,(n-2)\,(n-1) f''(r,n-4) \\ & + 2\,(n-3)\,(n-2)\,(n-1)\,r f^{(3)}(r,n-4) \\ & + 2\,F_{w_1}(r,n-1) + r\,F'_{w_1}(r,n-1) - F'_{u_1}(r,n) \Big], \end{aligned} \tag{26}$$

where

$$F_X(r,n) = (-1)^n \frac{\partial^n}{\partial w_1^n}\Big|_{w_1=0} \widetilde{F}_X(r), \tag{27}$$

for $X = w_1, u_1$. Equation (26) allows one to obtain $f(r,n)$ in terms of $f(r)$, exponential and exponential integral functions, for example

$$\begin{aligned} f(r,0) &= f(r), \\ f(r,1) &= \frac{r^3}{4} h_0(r\,u)\,h_0(r\,w)\,j_0(r\,x)\,j_0(r\,y), \\ f(r,2) &= \frac{1}{\sigma_{00}} \Big[r\,\sigma_{02} f'(r) + 2\,\sigma_{22} f''(r) + r\,\sigma_{22} f^{(3)}(r) + 4 f^{(4)}(r) + r f^{(5)}(r) \\ &\quad + 2\,F_{w_1}(r,1) + r\,F'_{w_1}(r,1) - F'_{u_1}(r,2) \Big], \\ f(r,3) &= \frac{r^5}{24} \big[-3\,h_1(r\,u)\,h_1(r\,w)\,j_1(r\,x)\,j_1(r\,y) - h_0(r\,u)\,h_0(r\,w)\,j_0(r\,x)\,j_0(r\,y) \\ &\quad + h_0(r\,w)\,h_2(r\,u)\,j_0(r\,x)\,j_0(r\,y) + h_0(r\,u)\,h_2(r\,w)\,j_0(r\,x)\,j_0(r\,y) \\ &\quad + h_0(r\,u)\,h_0(r\,w)\,j_2(r\,x)\,j_0(r\,y) + h_0(r\,u)\,h_0(r\,w)\,j_0(r\,x)\,j_2(r\,y) \big], \end{aligned} \tag{28}$$

where j_n and h_n are (up to the sign) the modified spherical Bessel functions[10]

$$\begin{aligned} j_n(x) &= x^n \left(\frac{1}{x}\frac{d}{dx}\right)^n \frac{\sinh(x)}{x}, \\ h_n(x) &= x^n \left(\frac{1}{x}\frac{d}{dx}\right)^n \frac{\exp(-x)}{x}. \end{aligned} \tag{29}$$

Note that $f(r,n)$ for odd n can be expressed in terms of j_n and h_n only, so their numerical evaluation is straightforward.

What remains are the derivatives of f with respect to $\alpha = u,w,x,y$ at $w_1 = 0$. We adapt here the derivation of the corresponding formulas from Ref. 5. One takes fundamental differential equations in u_1 and α

$$r\,\sigma_{00} f(r) + \sigma_{20} f'(r) + r\,\sigma_{20} f''(r) = F_{u_1}(r) \tag{30}$$

$$\frac{1}{2}\frac{\partial\sigma_{00}}{\partial\alpha} f(r) + \sigma_{00}\frac{\partial f(r)}{\partial\alpha} + \frac{1}{2}\frac{\partial\sigma_{20}}{\partial\alpha} f''(r) + \sigma_{20}\frac{\partial f''(r)}{\partial\alpha} = -F_\alpha(r), \tag{31}$$

differentiates the first equation with respect to α, eliminates $\partial f''(r)/\partial\alpha$, and differentiates resulting equation again with respect to r. The obtained equation for the derivative of f using $\partial\sigma_{20}/\partial\alpha = \sigma_{20}/\alpha$ is

$$\frac{\partial f}{\partial\alpha} = -\frac{1}{2\,\alpha} f(r) + \frac{r}{2}\left(\frac{1}{\sigma_{00}}\frac{\partial\sigma_{00}}{\partial\alpha} - \frac{1}{\alpha}\right) f'(r) + \frac{G_\alpha(r)}{\sigma_{00}}, \tag{32}$$

where

$$\begin{aligned} \frac{G_w}{\sigma_{00}} &= \frac{1}{16\,w}\left(\frac{H_1}{w\,x-u\,y} + \frac{H_2}{w\,y-u\,x} + \frac{H_3}{u\,w-x\,y}\right) \\ \frac{G_u}{\sigma_{00}} &= \frac{1}{16\,u}\left(\frac{H_1}{u\,y-w\,x} + \frac{H_2}{u\,x-w\,y} + \frac{H_3}{u\,w-x\,y}\right) \\ \frac{G_x}{\sigma_{00}} &= \frac{1}{16\,x}\left(\frac{H_1}{w\,x-u\,y} + \frac{H_2}{u\,x-w\,y} + \frac{H_3}{x\,y-u\,w}\right) \\ \frac{G_y}{\sigma_{00}} &= \frac{1}{16\,y}\left(\frac{H_1}{u\,y-w\,x} + \frac{H_2}{w\,y-u\,x} + \frac{H_3}{x\,y-u\,w}\right) \end{aligned} \tag{33}$$

and where

$$\begin{aligned} H_1 &= -F_{1+} + F_{2+} - F_{3+} + F_{4+} \\ H_2 &= F_{1+} - F_{2+} - F_{3+} + F_{4+} \\ H_3 &= F_{1+} + F_{2+} - F_{3+} - F_{4+} \end{aligned} \tag{34}$$

The formula (32) allows one to obtain integrals with arbitrary powers of electronic distances, but it requires however, that $\sigma_{00} \neq 0$

4. RECURRENCE RELATIONS AT $\sigma_{00} = 0$

When $\sigma_{00} = 0$, as it is for an exchange integral, one derives recurrence relations not for $f(r)$, but for $f'(r)$, and afterward integrates analytically the expression. This analytic integration is possible, since $rf'(r)$ is a simple combination of exponential and exponential integral functions (see Eq. 22).

The recurrence relations for $f'(r)$ in powers of r_{12} are obtained from Eqs. (24) and (25), in analogous way as for $f(r)$, and the result is

$$\begin{aligned} f'(r,n) = -\frac{1}{\sigma_{20}} & [(n-1)\, r\, \sigma_{02} f(r,n-2) + (n-3)\,(n-2)\,(n-1)\, n f'(r,n-4) \\ & + (n-1)\, n\, \sigma_{22} f'(r,n-2) + 2\,(n-3)\,(n-2)\,(n-1)\, r f''(r,n-4) \\ & + (n-1)\, r\, \sigma_{22} f''(r,n-2) + 2\,(n-1)\, n f^{(3)}(r,n-2) \\ & + (n-1)\, r f^{(4)}(r,n-2) + r\, F_{w_1}(r,n-1) - F_{u_1}(r,n)], \end{aligned} \tag{35}$$

The derivatives of $f'(r)$ with respect to nonlinear parameters can be obtained from Eq. (32), by taking the derivative over r and using the master equation (5). Alternatively, one can use general equations (A.12) and set $w_1 = 0$ there. The result is

$$\begin{aligned} 4\, w^2 \frac{\partial f'(r)}{\partial w} &= \frac{r f(r)}{2} \frac{\partial \sigma_{02}}{\partial w} + 2\, r\, w f''(r) - \frac{-F_{1-} + F_{2-} - F_{3-} - F_{4-}}{2}, \\ 4\, u^2 \frac{\partial f'(r)}{\partial u} &= \frac{r f(r)}{2} \frac{\partial \sigma_{02}}{\partial u} + 2\, r\, u f''(r) - \frac{F_{1-} - F_{2-} - F_{3-} - F_{4-}}{2}, \\ 4\, x^2 \frac{\partial f'(r)}{\partial x} &= \frac{r f(r)}{2} \frac{\partial \sigma_{02}}{\partial x} + 2\, r\, x f''(r) - \frac{F_{1-} + F_{2-} - F_{3-} + F_{4-}}{2}, \\ 4\, y^2 \frac{\partial f'(r)}{\partial y} &= \frac{r f(r)}{2} \frac{\partial \sigma_{02}}{\partial y} + 2\, r\, y f''(r) - \frac{-F_{1-} - F_{2-} - F_{3-} + F_{4-}}{2}, \end{aligned} \tag{36}$$

where σ_{02} is defined in Eq. (A.7) and F_{i-} in Eqs. (11).

5. JAMES-COOLIDGE INTEGRALS $\sigma_{20} = 0$

This is the simplest case where all f's can be obtained explicitly.[11,12] It is related to the fact that the Neumann expansion of $1/r_{12}$ in spheroidal coordinates is finite when $x = 0$ or $y = 0$, what corresponds to $\sigma_{20} = 0$.[3] Let us use one of Eqs. (A.12) from the Appendix

$$(w_1^2 - 4\,x^2)\,\frac{\partial \tilde{f}'(r)}{\partial x} = -\frac{r\,\tilde{f}(r)}{2}\,\frac{\partial \sigma_{02}}{\partial x} - 2\,r\,x\tilde{f}''(r) + \frac{\tilde{F}_{1-} + \tilde{F}_{2-} - \tilde{F}_{3-} + \tilde{F}_{4-}}{2}. \tag{37}$$

If we set $w_1 = x = 0$, it becomes an algebraic equation for $f(r)$, namely

$$f(r)|_{x=0} = \frac{\sinh(r\,y)}{4\,r\,y\,u\,w}\{-\mathrm{Ei}[-2\,r\,u]\exp[r\,(u-w)] - \mathrm{Ei}[-2\,r\,w]\exp[r\,(w-u)] + \mathrm{Ei}[-2\,r\,(u+w)]\exp[r\,(u+w)] + \left(\gamma + \ln\left[\frac{2\,r\,u\,w}{u+w}\right]\right)\exp[-r\,(u+w)]\} \tag{38}$$

To obtain $f(n_0,n_1,r)$ defined as

$$f(n_0, n_2, r) = \left(-\frac{\partial}{\partial w_1}\right)^{n_0}\left(-\frac{\partial}{\partial x}\right)^{n_2}\tilde{f}(r)|_{w_1=x=0} \tag{39}$$

one takes derivatives of the above differential equation, and remaining derivatives can be taken directly on $f(n_0,n_2,r)$.

6. CONCLUSION

We have presented an approach to perform calculations for diatomic systems in the exponential basis. Assuming the nonlinear parameter is fixed for any center, all integrals can be explicitly derived from differential equations using any computer symbolic program and stored on the hard disk. Such calculations have already been performed for H_2[11] and HeH^+ molecules,[13] demonstrating the computational applicability and high accuracy.

ACKNOWLEDGMENT

This work was supported by NCN grant 2012/04/A/ST2/00105.

APPENDIX A. MASTER INTEGRAL

The master integral $\tilde{f}$ is defined by

$$\tilde{f}(r) = r \int \frac{d^3 r_1}{4\pi} \int \frac{d^3 r_2}{4\pi} \frac{e^{-w_1 r_{12}}}{r_{12}} \frac{e^{-u_3 r_{1A}}}{r_{1A}} \frac{e^{-u_2 r_{1B}}}{r_{1B}} \frac{e^{-w_2 r_{2A}}}{r_{2A}} \frac{e^{-w_3 r_{2B}}}{r_{2B}}, \tag{A.1}$$

where $r = r_{AB}$ is the distance between the nuclei. $\tilde{f}(r)$ satisfies several differential equations. The master differential equation is[6,9]

$$\left[\sigma_4 \frac{d^2}{d r^2} r \frac{d^2}{d r^2} + \sigma_2 \frac{d}{d r} r \frac{d}{d r} + \sigma_0 r\right] \tilde{f}(r) = \tilde{F}(r), \tag{A.2}$$

where $\sigma = \sigma_0 + u_1^2 \sigma_2 + u_1^4 \sigma_4$ and σ being the sixth-order polynomial

$$\begin{aligned} \sigma = {} & u_1^2 u_2^2 w_3^2 + u_2^2 u_3^2 w_1^2 + u_1^2 u_3^2 w_2^2 + w_1^2 w_2^2 w_3^2 \\ & + u_1^2 w_1^2 (u_1^2 + w_1^2 - u_2^2 - u_3^2 - w_2^2 - w_3^2) \\ & + u_2^2 w_2^2 (u_2^2 + w_2^2 - u_1^2 - u_3^2 - w_1^2 - w_3^2) \\ & + u_3^2 w_3^2 (u_3^2 + w_3^2 - u_2^2 - u_1^2 - w_1^2 - w_2^2). \end{aligned} \tag{A.3}$$

Using the following new parameters which reflect symmetries of $\tilde{f}$

$$w_2 = w + x, \quad w_3 = w - x, \quad u_2 = u - y, \quad u_3 = u + y, \tag{A.4}$$

σ polynomials are

$$\sigma_4 = w_1^2, \tag{A.5}$$

$$\begin{aligned} \sigma_2 &= w_1^4 - 2 w_1^2 (u^2 + w^2 + x^2 + y^2) + 16 u w x y \\ &= w_1^4 + w_1^2 \sigma_{22} + \sigma_{20}, \end{aligned} \tag{A.6}$$

$$\begin{aligned} \sigma_0 &= w_1^2 (u + w - x - y)(u - w + x - y)(u - w - x + y)(u + w + x + y) \\ &\quad + 16 (w x - u y)(u x - w y)(u w - x y) \\ &= w_1^2 \sigma_{02} + \sigma_{00}, \end{aligned} \tag{A.7}$$

and the inhomogeneous term is

$$
\begin{aligned}
\widetilde{F}(r) = {} & w_1\left(\frac{1}{r^2}+\frac{2\,w_1+u+w+x-y}{r}\right)\mathrm{e}^{-r\,(u+w+w_1+x-y)} \\
& + w_1\left(\frac{1}{r^2}+\frac{2\,w_1+u+w-x+y}{r}\right)\mathrm{e}^{-r\,(u+w+w_1-x+y)} \\
& - w_1\left(\frac{1}{r^2}+\frac{u+w-x-y}{r}\right)\mathrm{e}^{-r\,(u+w-x-y)} \\
& - w_1\left(\frac{1}{r^2}+\frac{u+w+x+y}{r}\right)\mathrm{e}^{-r\,(u+w+x+y)} \\
& + \left[\frac{w_1^2}{2}\,(u-w+x-y)+2\,u\,w\,(y-x)+2\,x\,y\,(w-u)\right]\widetilde{F}_{1-} \\
& + \left[\frac{w_1^2}{2}\,(w-u+x-y)+2\,u\,w\,(y-x)+2\,x\,y\,(u-w)\right]\widetilde{F}_{2-} \\
& - \left[\frac{w_1^2}{2}\,(u+w+x+y)+2\,u\,w\,(x+y)+2\,x\,y\,(u+w)\right]\widetilde{F}_{3-} \\
& - \left[\frac{w_1^2}{2}\,(u+w-x-y)-2\,u\,w\,(x+y)+2\,x\,y\,(u+w)\right]\widetilde{F}_{4-},
\end{aligned}
\tag{A.8}
$$

where

$$
\begin{aligned}
\widetilde{F}_{1-} = {} & \mathrm{Ei}[-r\,(w_1+2\,u)]\exp[r\,(u-w+x-y)] \\
& -\mathrm{Ei}[-r\,(w_1+2\,w)]\exp[r\,(w-u-x+y)], \\
\widetilde{F}_{2-} = {} & \mathrm{Ei}[-r\,(w_1+2\,w)]\exp[r\,(w-u+x-y)] \\
& -\mathrm{Ei}[-r\,(w_1+2\,u)]\exp[r\,(u-w-x+y)], \\
\widetilde{F}_{3-} = {} & -\mathrm{Ei}[-2\,r\,(u+w)]\exp[r\,(u+w+x+y)] \\
& + \Big\{-\mathrm{Ei}[2\,r\,(x+y)]+\mathrm{Ei}[-r\,(w_1-2\,x)]+\mathrm{Ei}[-r\,(w_1-2\,y)] \\
& + \ln\left[\frac{(w_1+2\,u)\,(w_1+2\,w)\,(x+y)}{(u+w)\,(w_1-2\,x)\,(w_1-2\,y)}\right]\Big\}\exp[-r\,(u+w+x+y)], \\
\widetilde{F}_{4-} = {} & -\mathrm{Ei}[-2\,r\,(u+w)]\exp[r\,(u+w-x-y)] \\
& + \Big\{-\mathrm{Ei}[-2\,r\,(x+y)]+\mathrm{Ei}[-r\,(w_1+2\,x)]+\mathrm{Ei}[-r\,(w_1+2\,y)] \\
& + \ln\left[\frac{(w_1+2\,u)\,(w_1+2\,w)\,(x+y)}{(u+w)\,(w_1+2\,x)\,(w_1+2\,y)}\right]\Big\}\exp[-r\,(u+w-x-y)],
\end{aligned}
\tag{A.9}
$$

and Ei is the exponential integral function.

The complementary differential equations are[6]

$$\left(\frac{1}{2}\frac{\partial\sigma_0}{\partial w_1}+\sigma_0\frac{\partial}{\partial w_1}\right)f(r)+\left(\frac{1}{2}\frac{\partial\sigma_2}{\partial w_1}+\sigma_2\frac{\partial}{\partial w_1}\right)f''(r)$$
$$+\left(w_1+w_1^2\frac{\partial}{\partial w_1}\right)f^{(4)}(r)=-F_{w_1}(r), \tag{A.10}$$

where $F_{w_1}(r)$ is given in Appendix B and that for an arbitrary parameter $\alpha=u,w,x,y$

$$\left(\frac{1}{2}\frac{\partial\sigma_0}{\partial\alpha}+\sigma_0\frac{\partial}{\partial\alpha}\right)f(r)+\left(\frac{1}{2}\frac{\partial\sigma_2}{\partial\alpha}+\sigma_2\frac{\partial}{\partial\alpha}\right)f''(r)+w_1^2\frac{\partial}{\partial\alpha}f^{(4)}(r)=-F_\alpha(r). \tag{A.11}$$

From the master differential equation (A.2) and from Eq. (A.11), one derives[7] the following simple formulas for derivatives with respect to nonlinear parameters

$$\begin{aligned}
(w_1^2-4w^2)\frac{\partial f'(r)}{\partial w}&=-\frac{rf(r)}{2}\frac{\partial\sigma_{02}}{\partial w}-2\,r\,w f''(r)+\frac{-F_{1-}+F_{2-}-F_{3-}-F_{4-}}{2},\\
(w_1^2-4u^2)\frac{\partial f'(r)}{\partial u}&=-\frac{rf(r)}{2}\frac{\partial\sigma_{02}}{\partial u}-2\,r\,u f''(r)+\frac{F_{1-}-F_{2-}-F_{3-}-F_{4-}}{2},\\
(w_1^2-4x^2)\frac{\partial f'(r)}{\partial x}&=-\frac{rf(r)}{2}\frac{\partial\sigma_{02}}{\partial x}-2\,r\,x f''(r)+\frac{F_{1-}+F_{2-}-F_{3-}+F_{4-}}{2},\\
(w_1^2-4y^2)\frac{\partial f'(r)}{\partial y}&=-\frac{rf(r)}{2}\frac{\partial\sigma_{02}}{\partial y}-2\,r\,y f''(r)+\frac{-F_{1-}-F_{2-}-F_{3-}+F_{4-}}{2},
\end{aligned} \tag{A.12}$$

where σ_{02} is defined in Eq. (A.7), the inhomogeneous F_i terms are given by Eq. (A.9), and prime in $f'(r)$ means derivative with respect to r.

APPENDIX B. INHOMOGENEOUS TERMS

The inhomogeneous terms in the master differential equation (A.2) and in the supplementary differential equations (A.10,A.11) are the inverse Laplace transform of P_α for $\alpha=u_i,w_i$.

$$F_\alpha=\frac{1}{2\pi\mathrm{i}}\int_{-\mathrm{i}\,\infty+\epsilon}^{\mathrm{i}\,\infty+\epsilon}dt\,\mathrm{e}^{t\,r}\,P_\alpha\Big|_{u_1=t} \tag{B.1}$$

P_α are related to each other by

$$\begin{aligned} P_{w_1} &= P(w_1,u_1;w_2,u_2;w_3,u_3) \\ &= P(w_1,u_1;w_3,u_3;w_2,u_2) \\ P_{u_1} &= P(u_1,w_1;w_2,u_2;u_3,w_3) \\ P_{w_2} &= P(w_2,u_2;w_3,u_3;w_1,u_1) \\ P_{u_2} &= P(u_2,w_2;w_3,u_3;u_1,w_1) \\ P_{w_3} &= P(w_3,u_3;w_1,u_1;w_2,u_2) \\ P_{u_3} &= P(u_3,w_3;u_1,w_1;w_2,u_2) \end{aligned} \tag{B.2}$$

The expression for P was obtained in Ref. 5 and is the following

$$\begin{aligned} &P(w_1,u_1;w_2,u_2;w_3,u_3) \\ &= \frac{u_1\,w_1\left[(u_1+w_2)^2-u_3^2\right]}{(-u_1+u_3-w_2)\,(u_1+u_3+w_2)}\ln\left[\frac{u_2+u_3+w_1}{u_1+u_2+w_1+w_2}\right] \\ &+\frac{u_1\,w_1\left[(u_1+u_3)^2-w_2^2\right]}{(-u_1-u_3+w_2)\,(u_1+u_3+w_2)}\ln\left[\frac{w_1+w_2+w_3}{u_1+u_3+w_1+w_3}\right] \\ &-\frac{u_1^2\,w_1^2+u_2^2\,w_2^2-u_3^2\,w_3^2+w_1\,w_2\,(u_1^2+u_2^2-w_3^2)}{(-w_1-w_2+w_3)\,(w_1+w_2+w_3)}\ln\left[\frac{u_1+u_2+w_3}{u_1+u_2+w_1+w_2}\right] \\ &-\frac{u_1^2\,w_1^2-u_2^2\,w_2^2+u_3^2\,w_3^2+w_1\,w_3\,(u_1^2+u_3^2-w_2^2)}{(-w_1+w_2-w_3)\,(w_1+w_2+w_3)}\ln\left[\frac{u_1+u_3+w_2}{u_1+u_3+w_1+w_3}\right] \\ &+\frac{u_2\,(u_2+w_1)\,(u_1^2+u_3^2-w_2^2)-u_3^2\,(u_1^2+u_2^2-w_3^2)}{(-u_2+u_3-w_1)\,(u_2+u_3+w_1)}\ln\left[\frac{u_1+u_3+w_2}{u_1+u_2+w_1+w_2}\right] \\ &+\frac{u_3\,(u_3+w_1)\,(u_1^2+u_2^2-w_3^2)-u_2^2\,(u_1^2+u_3^2-w_2^2)}{(u_2-u_3-w_1)\,(u_2+u_3+w_1)}\ln\left[\frac{u_1+u_2+w_3}{u_1+u_3+w_1+w_3}\right] \\ &-\frac{w_1\left[w_2\,(u_1^2-u_2^2+w_3^2)+w_3\,(u_1^2-u_3^2+w_2^2)\right]}{(w_1-w_2-w_3)\,(w_1+w_2+w_3)}\ln\left[\frac{u_2+u_3+w_1}{u_2+u_3+w_2+w_3}\right] \\ &-\frac{w_1\left[u_2\,(u_1^2+u_3^2-w_2^2)+u_3\,(u_1^2+u_2^2-w_3^2)\right]}{(-u_2-u_3+w_1)\,(u_2+u_3+w_1)}\ln\left[\frac{w_1+w_2+w_3}{u_2+u_3+w_2+w_3}\right] \end{aligned} \tag{B.3}$$

In the case, the nonlinear parameter α is the combination of w_i,u_i as in Eq. (8), P_α is given by

$$\begin{aligned} P_w &= P_{w_2}+P_{w_3} \\ P_x &= P_{w_2}-P_{w_3} \\ P_u &= P_{u_2}+P_{u_3} \\ P_y &= P_{u_3}-P_{u_2} \end{aligned} \tag{B.4}$$

In particular, $F_{u_1}=F(r)$ is given by Eq. (A.8), and F_{w_1} is

$$
\begin{aligned}
F_{w_1}(r) = \Bigg\{ & \left[\frac{r\,(u+w+x+y)+1}{r^3} \left(2 - \frac{w_1}{2\,u+w_1} - \frac{w_1}{2\,w+w_1} - \frac{w_1}{w_1-2\,x} - \frac{w_1}{w_1-2\,y} \right) \right. \\
& + \frac{2}{r} \left(\frac{(u+w)\,(u+x)\,(u+y)}{2\,u+w_1} + \frac{(u+w)\,(w+x)\,(w+y)}{2\,w+w_1} \right. \\
& \left.\left. - \frac{(u+x)\,(w+x)\,(x+y)}{w_1-2\,x} - \frac{(u+y)\,(w+y)\,(x+y)}{w_1-2\,y} - \frac{(u+w+x+y)^2}{2} \right) \right] \\
& \times e^{-r\,(u+w+x+y)} \\
& + \left[\frac{r\,(u+w+w_1-x+y)+1}{r^3} \left(-2 + \frac{w_1}{2\,u+w_1} + \frac{w_1}{2\,w+w_1} + \frac{w_1}{w_1-2\,x} + \frac{w_1}{w_1+2\,y} \right) \right. \\
& + \frac{w_1}{r^2} + \frac{1}{r} \left(w_1^2 + 4\,(x\,y+u\,x+w\,x-u\,y-w\,y-u\,w) + (w_1+u+w-x+y)^2 \right. \\
& - \frac{2\,(u-w)\,(u+x)\,(u-y)}{2\,u+w_1} + \frac{2\,(u-w)\,(w+x)\,(w-y)}{2\,w+w_1} \\
& \left.\left. + \frac{2\,(u+x)\,(w+x)\,(x+y)}{w_1-2\,x} + \frac{2\,(u-y)\,(w-y)\,(-x-y)}{w_1+2\,y} \right) \right] e^{-r\,(u+w+w_1-x+y)} \Bigg\} \\
& + \{x \to -x, y \to -y\}
\end{aligned}
\tag{B.5}
$$

Note that F_{w_1} involves only the exponential and rational functions.

REFERENCES

1. Harris, F. Analytic Evaluation of Two-Center STO Electron Repulsion Integrals via Ellipsoidal Expansion. *Int. J. Quant. Chem* **2002**, *88*, 701.
2. Ozdogan, T.; Ruiz, M. B. *Recent Advances in Computational Chemistry: Molecular Integrals over Slater Orbitals*; Transworld Research Network: India, 2008.
3. Kolos, W.; Roothaan, C. C. J. Correlated Orbitals for the Ground State of the Hydrogen Molecule. *Rev. Mod. Phys.* **1960**, *32*, 205, Accurate Electronic Wave Functions for the H_2 Molecule. *Rev. Mod. Phys.* **1960**, *32*, 219.
4. Lesiuk, M.; Przybytek, M.; Musiał, M.; Jeziorski, B.; Moszyński, R. Reexamination of the Calculation of Two-Center, Two-Electron Integrals over Slater-Type Orbitals. III. Case Study of the Beryllium Dimer. *Phys. Rev. A* **2015**, *91*, 012510.
5. Pachucki, K. Two-Center Two-Electron Integrals with Exponential Functions. *Phys. Rev. A* **2009**, *80*, 032520.
6. Pachucki, K. Correlated Exponential Functions in High Precision Calculations for Diatomic Molecules. *Phys. Rev. A* **2012**, *86*, 052514.
7. Pachucki, K.; Yerokhin, V. A. Application of the Fully Correlated Basis of Exponential Functions for Molecular Hydrogen. *Phys. Rev. A* **2013**, *87*, 062508.
8. Pachucki, K.; Puchalski, M.; Yerokhin, V. A. Extended Gaussian Quadratures for Functions with an End-Point Singularity of Logarithmic Type. *Comp. Phys. Commun.* **2014**, *185*, 2913.
9. Lesiuk, M.; Moszyński, R. Analytical Two-Center Integrals over Slater Geminal Functions. *Phys. Rev. A* **2012**, *86*, 052513.

10. Olver, F.; Lozier, D.; Boivert, R.; Clark, C. *NIST Handbook of Mathematical Functions*; Cambridge University Press: New York, NY, USA, 2010.
11. Pachucki, K. Born-Oppenheimer Potential for H_2. *Phys. Rev. A* **2010**, *82*, 032509.
12. Ferron, A.; Serra, P. Evaluation of Two Center Two Electron Integrals. *J. Chem. Theor. Comp* **2006**, *2*, 306.
13. Pachucki, K. Born-Oppenheimer Potential for HeH^+. *Phys. Rev. A* **2012**, *85*, 042511.

CHAPTER FIVE

Singlet and Triplet Bound State Spectra in the Four-Electron Be-Like Atomic Systems

María Belén Ruiz*,[1], Federico Latorre*,†, Alexei M. Frolov‡
*Department of Theoretical Chemistry, Friedrich-Alexander University Erlangen–Nürnberg, Erlangen, Germany
†Institute of Theoretical Chemistry, University of Vienna, Vienna, Austria
‡Department of Applied Mathematics, University of Western Ontario, London, Ontario, Canada
[1]Corresponding author: e-mail address: maria.belen.ruiz@fau.de

Contents

Abstract

A large number of bound singlet and triplet states in the four-electron Be atom and Be-like ions B^{+}, C^{2+}, F^{5+}, and Mg^{8+} are determined to milli-Hartree numerical accuracy (1×10^{-3} a.u.). These states include the bound singlet and triplet S, P, D, F, G, H, I, and K states. Based on computational results, we analyze the singlet and triplet series of the Be atom bound state spectrum and some four-electron ions: B^{+}, C^{2+}, F^{5+}, and Mg^{8+} (Be-like ions). The analogous study of the Be-atom triplet states was the topic of our earlier study (Frolov and Ruiz[1]). The computational data allowed us to draw the spectral diagram of the bound state spectrum of the Be atom and other four-electron ions mentioned above. The Be-atom spectrum contains the two optical series of bound states: singlets and triplets.

1. INTRODUCTION

The electronic structure of the beryllium atom is of great interest in various problems arising in different areas of modern science, including

Advances in Quantum Chemistry, Volume 73
ISSN 0065-3276
http://dx.doi.org/10.1016/bs.aiq.2015.05.002

stellar astrophysics and plasmas, high-temperature physics, and applied nuclear physics. Beryllium and some of its compounds (BeO, Be_2C) are extensively used in the nuclear industry mainly as very effective (almost ideal) moderators of fast/slow neutrons. Nevertheless, currently there are many gaps in our understanding of the Be-atom optical spectrum. Total energies of all rotationally excited (bound) states with $L \geq 4$ in particular have not yet been evaluated, to our knowledge.

Another interesting problem is to describe the transitions from the spectra of the low-lying bound states in the Be atom to the weakly bound (or Rydberg) states.

Recently, we have studied the general structure of the triplet bound state spectrum in the four-electron Be atom[1] and accurately calculated a large number of low-lying (bound) S, P, D, F, G, H, I, and K states, i.e., bound states with $L \leq 7$. These computational results allowed us to determine a spectral diagram of the triplet states of the Be atom. The theoretical/computational spectral diagram of the Be atom agrees well with the known experimental data (see, e.g., Ref. 2 and references therein). *Note:* High angular momentum states, e.g., G($L = 4$), H($L = 5$), I($L = 6$), and K($L = 7$), are absent from atomic data bases.[2]

After the landmark paper Ref. 3, the bound states in the four-electron Be atom and Be-like ions have been considered by many authors, using various highly accurate methods specifically designed for the four-electron atomic systems. Such calculations were restricted to the ground 2^1S state and a very few excited states only. For instance, the ground-state energy for the Be atom was determined by applying the Configuration Interaction (CI) method with Slater-type orbitals (STOs),[4,5] the Hylleraas method (Hy),[6,7] the Hylleraas-Configuration Interaction method (Hy-CI),[3,8] and the Exponential Correlated Gaussian (ECG) method[9,10] (also called variational expansion in multidimensional gaussoids[11]). A few selected bound singlet S, P, and D states were calculated by Monte Carlo methods[12] and by the ECG method.[9,10,13,14] In contrast to singlet states, the Be-atom triplet states were investigated only in a very few earlier studies[12,15,16] (see also Ref. 1 and references therein).

To complete the analysis of the bound state spectrum of the Be atom,[1] we need to consider the singlet series of bound states in this system. In this study, we determine the total energies of the bound singlet S, P, D, F, G, H, I, and K states.

The method of calculation allows the total energies of these states to be obtained to very high numerical accuracy, ≈ few milli-Hartree

(1×10^{-3} a.u.), which is significantly better (for highly excited states) than accuracy which Hartree–Fock-based methods can provide. It is important to note that currently there is no reliable experimental information about rotationally excited states with $L \geq 4$ in the Be atom. Therefore, the main goal of this study is to determine the bound state spectrum of low-lying singlet states in the four-electron Be atom, including rotationally excited states with $L = 4, 5, 6$, and 7. The results of this study have been represented as a spectral diagram for the singlet and triplet bound states in the Be atom. For readers benefit and to complete our analysis, we also present an analogous spectral diagram for the bound triplet states in the Be atom.[1]

2. HAMILTONIAN AND BOUND STATE WAVE FUNCTIONS IN THE CI METHOD

The computational goal of this study is to determine the accurate numerical solutions of the five-body (or four-electron) Schrödinger equation $H\Psi = E\Psi$, where the Hamiltonian written in Hylleraas coordinates for a CI wave function is written in the form (see, e.g., Ref. 17)

$$\begin{aligned}\hat{H} = & -\frac{1}{2}\sum_{i=1}^{n}\frac{\partial^2}{\partial r_i^2} - \sum_{i=1}^{n}\frac{1}{r_i}\frac{\partial}{\partial r_i} - \sum_{i=1}^{n}\frac{Z}{r_i} + \sum_{i<j}^{n}\frac{1}{r_{ij}} \\ & -\frac{1}{2}\sum_{i=1}^{n}\frac{1}{r_i^2}\frac{\partial^2}{\partial\theta_i^2} - \frac{1}{2}\sum_{i=1}^{n}\frac{1}{r_i^2\sin^2\theta_i}\frac{\partial^2}{\partial\varphi_i^2} - \frac{1}{2}\sum_{i=1}^{n}\frac{\cot\theta_i}{r_i^2}\frac{\partial}{\partial\theta_i}.\end{aligned} \tag{1}$$

Note that when the Hamiltonian in Hylleraas coordinates is applied to the CI wave function (the CI wave function does not explicitly include r_{ij} coordinates), some terms of Ref. 17 vanish. In addition, the kinetic energy operator is represented as a sum of a few terms each of which has its own radial and angular parts. The operator $\hat{H}$ is diagonal in the basis of the spherical harmonics which are used below as angular parts of the orbital functions. Note also that the angular momentum operator can easily be extracted from Eq. (1):

$$\sum_{i=1}^{n}\frac{1}{r_i^2}\hat{L}_i^2 = -\frac{1}{2}\sum_{i=1}^{n}\frac{1}{r_i^2}\frac{\partial^2}{\partial\theta_i^2} - \frac{1}{2}\sum_{i=1}^{n}\frac{1}{r_i^2\sin^2\theta_i}\frac{\partial^2}{\partial\varphi_i^2} - \frac{1}{2}\sum_{i=1}^{n}\frac{\cot\theta_i}{r_i^2}\frac{\partial}{\partial\theta_i}. \tag{2}$$

For the orbital basis functions ϕ_i (or orbitals, for short), we can write

$$L_i^2\phi_i = l_i(l_i + 1)\phi_i, \tag{3}$$

with l_i the angular quantum number of the orbital ϕ_i. The Hamiltonian is reduced to the form

$$\hat{H} = -\frac{1}{2}\sum_{i=1}^{n}\frac{\partial^2}{\partial r_i^2} - \sum_{i=1}^{n}\frac{1}{r_i}\frac{\partial}{\partial r_i} - \sum_{i=1}^{n}\frac{Z}{r_i} + \sum_{i<j}^{n}\frac{1}{r_{ij}} + \sum_{i=1}^{n}\frac{1}{r_i^2}l_i(l_i+1). \quad (4)$$

Now, from the variational principle, one obtains the following eigenvalue problem:

$$(\mathbf{H} - E\mathbf{S})\mathbf{C} = \mathbf{0}, \quad (5)$$

where the matrix elements of the Hamiltonian matrix **H** and overlap matrix **S** are

$$H_{\mathrm{kl}} = \int \Phi_{\mathrm{k}} H \Phi_{\mathrm{l}} \mathrm{d}\tau, \quad S_{\mathrm{kl}} = \int \Phi_{\mathrm{k}} \Phi_{\mathrm{l}} \mathrm{d}\tau. \quad (6)$$

The integrals occurring in the CI calculations of an n-electron atom are one- and two-electron integrals. The two-electron integrals are of the type[18]

$$\left\langle \phi(\mathbf{r}_1)\phi(\mathbf{r}_2)\frac{1}{r_{12}}\phi(\mathbf{r}_1)\phi(\mathbf{r}_2) \right\rangle, \quad (7)$$

and they are expressed as sums of the auxiliary two-electron integrals $V(m,n;\ \alpha,\beta)$, defined as

$$V(m,n;\ \alpha,\beta) = \int_0^\infty r_1^m e^{-\alpha r_1} \mathrm{d}r_1 \int_{r_1}^\infty r_2^n e^{-\beta r_2} \mathrm{d}r_2. \quad (8)$$

The auxiliary integrals $V(m,n;\ \alpha,\beta)$ for positive indices m,n consist of a sum of $A(n,\alpha)$ auxiliary integrals (Ref. 19, Eq. 5):

$$V(m,n;\ \alpha,\beta) = \sum_{n'=0}^{n}\binom{n}{n'}A(n',\alpha)A(m+n-n',\alpha+\beta), \quad m,n \geq 0. \quad (9)$$

This formula, developed by Frolov and Smith, is very useful because it is numerically stable and provides very fast convergence. For negative n and positive m (but $m + n \geq -1$ always), the formula for the Sims and Hagstrom sum (Ref. 20, Eq. 33) must be computed. In quantum mechanical calculations of two-electron systems, this formula is employed to calculate the V-auxiliary integrals with the lowest index $n = -1$:

$$V(m,n;\ \alpha,\beta) = \sum_{q=1}^{\infty}\frac{\alpha^{q-1}m!}{(m+q)!}A(m+n+q;\ \alpha+\beta), \quad m+n \geq -1,\ m>0,\ n<0, \quad (10)$$

where the $A(n,\alpha)$ auxiliary integrals are

$$A(n,\alpha) = \frac{n!}{\alpha^{n+1}}. \tag{11}$$

Let us briefly discuss the explicit construction of the trial wave functions which are used to approximate the exact wave functions of bound states in the four-electron Be atom. In this work, we shall use the CI wave functions constructed from STO and LS eigenfunctions. These wave functions are represented in the form

$$\Psi = \sum_{p=1}^{N} C_p \Phi_p, \quad \Phi_p = \hat{O}(\hat{L}^2)\hat{A}\phi_p \chi, \tag{12}$$

i.e., it is a linear combination of N symmetry-adapted configurations Φ_p, where the coefficients C_p are determined variationally by solving the eigenvalue problem which follows from the Schrödinger equation.

In this work, the symmetry-adapted configurations are constructed *a priori* so that they are eigenfunctions of the angular momentum operator $\hat{L}^2$. Another possibility would be the posterior projection of the configurations over the proper spatial space, as indicated in Eq. (12) by the projection operator $\hat{O}(\hat{L}^2)$, where $\hat{A}$ is the antisymmetrization operator and χ is the spin eigenfunction for $S = 0$ and $M_S = 0$

$$\chi = [(\alpha\beta - \beta\alpha)(\alpha\beta - \beta\alpha)]. \tag{13}$$

As discussed in the case of the Li atom in Ref. 21 and calculations of the Be atom,[8] it is sufficient to consider only one spin function. Strictly, a linear combination of all possible spin eigenfunctions should be employed but that it has proved not to be necessary. Indeed, the Slater determinants produced by the antisymmetrization of further spin functions would be repeated when considering the spin eigenfunction (Eq. 13). The spatial part of the basis functions consists of Hartree products of STOs:

$$\phi_p = \prod_{k=1}^{n} \phi_k(r_k, \theta_k, \varphi_k). \tag{14}$$

The basis functions, ϕ_p, are products of s, p, d, f, g, h, i, and k STOs, defined as

$$\phi(\mathbf{r}) = r^{n-1} e^{-\alpha r} Y_l^m(\theta, \varphi), \tag{15}$$

where $Y_l^m(\theta,\varphi)$ are the spherical harmonics.[22]

We have written a four-electron CI computer program for four-electron atomic systems in Fortran 90. Numerical calculations have been conducted in double precision arithmetic. This program has been thoroughly checked by comparing results of our numerical calculations with the results by Sims and Hagstrom for the Be atom.[3,8] In these calculations, we obtain complete agreement.

The ground-state configuration of the Be atom is ssss (i.e., s(1)s(2)s(3)s(4)). The other configurations considered for *S*-symmetry states ($L = 0$) are, ordered by decreasing energy contribution, sspp, spps, ppss, pppp, ssdd, sdds, ddss, sppd, dpps, ..., ffdd. The energetically important configurations for $L = 1,\ldots,7$ are listed in Table 1. The quantum number $M_L = 0$ was chosen, because for this case a smaller number of Slater determinants is required. We systematically selected the CI configurations according to their energy contribution. This was done by calculations on blocks constructed for all possible configurations. The eigenvalue equation was diagonalized upon each addition of a configuration. In this manner, the contribution of every single configuration and of each block of a given type to the total energy was evaluated. Configurations with an overall energy contribution below 1×10^{-8} a.u. were neglected.

The procedure of selection of the configurations is similar to that described in our previous works.[1,21] In this work, we construct the full-CI (FCI)

Table 1 List of the Different L Configurations Used in the CI Calculations of the S, P, D, F, G, H, I, and K States

State	*L*	*M*	Configurations
S	0	0	ssss, sspp, spps, ppss, pppp, ssdd, sdds, ddss, sppd, dpps, sdpp, ppds, ppdd, dddd, ddpp, ssff, ppff, ddff, ffss, ffpp, ffdd
P	1	0	sssp, spss, sppp, ppsp, sddp, ddsp, sffp, ffsp, sspd, spds, pdss, pdpp, pddd, pdff, pppd, ddpd, ffpd, ssdf, ppdf, dddf, ffdf, pdpp
D	2	0	sssd, sdss, sppd, ppsd, sddp, ddsd, sffd, ffsd, sspp, spps, ppss, pppp, ssdd, sdds, ddss
F	3	0	sssf, sfss, sppf, ppsf, sddf, ddsf, ffsf, ggsf, hhsf
G	4	0	sssg, sgss, sppg, ppsg, sddg, ddsg, ffsg, ggsg, hhsg
H	5	0	sssh, shss, spph, ppsh, sddh, ddsh, ffsh, ggsh, hhsh
I	6	0	sssi, siss, sppi, ppsi, sddi, ddsi, ffsi, ggsi, hhsi, iisi
K	7	0	sssk, skss, sppk, ppsk, sddk, ddsk, ffsk, ggsk, hhsk, kksk

In all configurations, $M = 0$. The exponents within a shell have been kept equal.

wave function for every symmetry and basis set including the types of configurations which contribute most to the total energy of the lowest state of every symmetry. The larger the contribution of a configuration, the smaller the sum of the l quantum numbers of the employed orbitals $l_1 + l_2 + l_3 + l_4$ is, i.e., the contribution of the configuration sssp>sppp for a P state. In cases such as the P states sspd and ppsp, where the sum of l_i is equal, the two inner electrons in ppsp form an S configuration. The resulting four-electron configuration is (^{1}S)sp (a P configuration) and contributes more than the sspd configuration. This is especially important in the case of F, G, H, I, and K states. Among the many possibilities to construct configurations of these symmetries, the energetically most important configurations were proven to be those with an inner S shell and a single occupied orbital with the symmetry of the state under consideration, i.e., (^{1}S)sf, (^{1}S)sg, (^{1}S)sh, (^{1}S)si, and (^{1}S)sk. The inner shell is described with a sum of configurations (^{1}S)= ss+pp+dd+ff+gg+hh+ii+kk. In the CI calculations of S, P, and D states, we used s, p, d, and f orbitals (see Table 1). In those of the F, G, H, I, and K states, we have also used g, h, i, and k orbitals (Table 1).

More types of configurations than the ones discussed here can be constructed for a given L quantum number. For instance, configurations like pssp could be considered, if the exponents $\alpha_1 \neq \alpha_2$. However, we have kept the orbital exponents in the K shell and L shell equal.

Note that there are more possible "degenerate L-eigenfunction" solutions with a larger number of Slater determinants. Specifically, these are degenerate with respect to the quantum numbers L and M, but with possible different energy contribution, i.e., nondegenerate with respect to the energy.[4] Although the inclusion of various degenerate configurations has been shown to improve the energy of the state, such a contribution is very small. This is important for very accurate CI calculations, as reported, e.g., by Bunge.[4] In our work, we have concentrated on the energetically most significant CI configurations.

Another important aspect in CI and Hy-CI calculations is the symmetry adaptation of the configurations. As mentioned above, the configurations are constructed *a priori* to be eigenfunctions of the angular momentum operator $\hat{L}^2$. The configurations of Table 1 are constructed as sums of Slater determinants. The determinants are pairwise symmetric (i.e., ssp_1p_{-1} and $ssp_{-1}p_1$ in the sspp configuration) and lead to the same values of the electronic integrals. Therefore, it is possible and desirable to consider only one of the determinants and to deduce the second result. In other words, the solution of the eigenvalue problem obtained when using reduced

1×1 matrix elements (where the integrals are added, configuration $sp_1p_{-1} + sp_{-1}p_1$) or when using explicit 2×2 matrix elements of the Slater determinants is the same. The symmetry adaption is computationally favorable, since the number of Slater determinants in the input is smaller and the repeated computation of equal integrals is avoided. The explicit sums of symmetry-adapted configurations in the three-electron case are listed in Ref. 21.

In this work, we start with the FCI wave function constructed with configurations of the type of the ground configuration of a given state (see first configuration of every symmetry in Table 1) and we use the large basis set $n = 8$. The notation $n = 4$ stands for the basis set [4s3p2d1f]. The first step consists in an optimization of the orbital exponents for this truncated wave function. The optimization is carried out using a parabolic procedure, explained in Ref. 21. The orbital exponents are optimized for each atomic state of the Be atom. A set of two exponents is used (one for the K shell and the other for the electrons in the L shell), and kept equal for all configurations. This technique accelerates computations, while still producing sufficiently accurate wave functions to determine the bound state properties. We use the virial theorem:

$$\chi = -\frac{\langle V \rangle}{\langle T \rangle} \tag{16}$$

as a criterion to check the quality of the wave function and guide the numerical optimization of the exponents in the trial wave functions.

Using the appropriate exponents for every state, we filtered the configurations of the first configuration block of the FCI wave function calculating the total energy E_i each time a single configuration was added, and comparing it to the total energy without this configuration E_{i-1}. If the difference of the energy was smaller than the threshold $|E_{i-1} - E_i| < 1 \times 10^{-7}$ a.u, the new configuration was discarded. In this manner, all configurations were checked, leading to a relatively compact CI wave function.

The next step consists in adding a new block of configurations (FCI) of the following types given in Table 1. As the wave function may become very large, a new selection of the newly added configurations is carried out. The resulting compact wave function is optimized again. The procedure is repeated when each new block of configurations is added. In this work, we employ the basis set $n = 8$. Using this method, we obtain precise energy values, and the addition of configurations with higher l_i quantum numbers contributes to the convergence to the nonrelativistic energy. The final wave function is a compact wave function containing one to two thousand configurations which has milli-Hartree accuracy for the lowest states of every symmetry. This technique is a compromise between selection and optimization.

By using the CI method, we have calculated the bound S, P, D, F, G, H, I, and K states in the Be atom. Concretely, we have determined the energies of the four (lowest) S, three P, three D, two F, two G, one H, one I, and one K state. The total energies of the F, G, H, I, and K states in the Be atom are reported here for the first time. To our knowledge, they have not been determined in earlier studies neither computationally nor experimentally. Our results are summarized in Table 2. The overall accuracy of our calculations for the lowest states of every symmetry can be evaluated as $\approx \pm 1$–5×10^{-3} a.u. Higher excited states are not necessarily less accurate. The results have this limited accuracy due to the exponent restrictions of the method used here. In Table 3, the CI calculations of the triplets bound states of the Be isoelectronic ions B^{+}, C^{2+}, F^{5+}, and Mg^{8+} are summarized.

3. GENERAL STRUCTURE OF THE BOUND STATE SPECTRA

As mentioned above, the Be-atom bound state spectrum contains two series of bound states: singlet series and triplet series. To the lowest order non-relativistic dipole approximation, these two series are independent of each other, i.e., any dipole transition between the two states from different spectral series is strictly prohibited. In reality, transitions between singlet and triplet states of the Be atom and other Be-like ions are always possible due to nonelastic collisions of these atoms with electrons, ions, and other atoms. It is clear that the probabilities of such collisional transitions substantially depend upon the spatial densities of electrons, Be atoms, etc. In very good vacuum ($\approx 10^{-12}$ atm.) and at relatively large temperatures, one can easily see the two different optical series (singlet and triplet) in the gaseous mixture of the ^{9}Be atoms. Note also that very small relativistic components of the exact four-electron wave functions also make these singlet–triplet transitions possible. Rates of such transitions are very low for the neutral Be atom but they rapidly increase with the nuclear charge in the series: B^{+}, C^{2+}, F^{5+}, and Mg^{8+}. The electronic structure of the Be-atom ground singlet state is $1s^2 2s^2$, while all excited states have a similar structure where the two electrons occupy the $1s^2$-electron shell (its excitation energy is extremely large), while the third electron is mostly located in the 2s shell.

Table 2 The Total Energies in a.u. of Some Low-Lying Singlet States of the $^{\infty}$Be Atom Determined with the Use of the CI Method and Comparison with the Nonrelativistic Values of the Bibliography

State	*N*	This Work	*N*	*E*(CI,MC)	Ref.	*N*	*E*(nr)	Ref.	Diff.
2^1S	1137	−14.665 730	≈ 2m	−14.667 347 30	4	4096	−14.667 356 4949	9	1.63
3^1S	1300	−14.416 247	1038	−14.417 957 27	5	10,000	−14.418 240 328	10	2.00
4^1S	1466	−14.365 107				10,000	−14.370 087 876	10	4.98
5^1S	1242	−14.348 206				10,000	−14.351 511 654	10	3.30
6^1S	1358	−14.338 766				10,000	−14.342 403 552	10	3.64
2^1P	1307	−14.470 359	1038	−14.473 009 65	5	10,700	−14.473 451 378	13	3.09
3^1P	1398	−14.389 310	1038	−14.392 788 28	5	11,600	−14.393 143 528	13	3.83
4^1P	1944	−14.353 697				11,900	−14.361 938 388	13	8.24
5^1P	2159	−14.326 239				12,200	−14.347 876 275	13	21.64
3^1D	943	−14.401 671		−14.404 36(6)	12	4200	−14.408 237 03(40)	14	6.57
4^1D	1754	−14.367 016	1038	−14.373 442 41	5	4200	−14.373 824 38(30)	14	6.81
5^1D	1131	−14.343 314				4200	−14.353 982 65(50)	14	10.67
6^1D		−14.319 243				4200	−14.343 857 75(60)	14	24.61
4^1F	2210	−14.352 803							
5^1F	1625	−14.334 737							
5^1G	1676	−14.338 416							
6^1G	1844	−14.318 712							
6^1H	1443	−14.327 148							
7^1I	1242	−14.319 606							
8^1K	1540	−14.285 509							

The total energies of the bound states in the upper part of this table are lower than the ionization threshold of the $^{\infty}$Be atom ($E_{\infty Be^+} = -14.324\,763\,176\,790\,43(22)$a.u.[23]). *N* is the number of configurations used in calculations. Diff. are the energy differences between the present and reference energies in milli-Hartree (1×10^{-3} a.u.).

Table 3 CI Calculations on the Triplet Excited States of Be Atom and Isoelectronic Ions

State	*E*(Be)[a]	*E*(B^+)[b]	*E*(C^{2+})[c]	*E*(F^{5+})[d]	*E*(Mg^{8+})[e]
2^3S	−14.428 858	−23.756 406	−35.448 470	−84.697 926	−155.200 640
3^3S	−14.371 277	−23.589 166	−35.123 226	−84.002 052	−154.198 338
4^3S	−14.343 789	−23.517 737	−35.061 579	−83.959 916	−154.131 317
5^3S	−14.314 690	−23.473 813	−35.037 812	−83.592 771	−152.883 937
6^3S		−23.457 241	−34.978 838	−83.115 745	−152.029 948
7^3S		−23.437 292	−34.885 511	−82.994 777	−152.014 389
8^3S		−23.329 741	−34.775 418	−82.975 972	−151.847 275
2^3P	−14.565 432	−24.176 888	−36.294 390	−87.659 827	−161.532 045
3^3P	−14.392 598	−23.687 970	−35.347 035	−84.498 979	−154.905 474
4^3P	−14.357 020	−23.559 695	−35.127 064	−84.132 562	−154.393 494
5^3P	−14.333 088	−23.515 121	−35.081 574	−83.881 773	−154.017 858
6^3P		−23.502 078	−34.989 957	−83.709 305	−153.901 853
7^3P		−23.428 174	−34.948 115	−83.798 028	−153.881 813
8^3P			−34.936 695	−83.494 189	−152.728 404
3^3D	−14.381 020	−23.651 575	−35.290 221	−84.383 794	−154.733 038
4^3D	−14.350 318	−23.548 131	−35.060 046	−83.474 253	−152.702 810
5^3D	−14.343 322	−23.493 956	−34.948 953	−83.052 943	−151.756 880
6^3D		−23.445 075	−34.869 323	−82.802 203	−151.228 105
7^3D		−23.357 994	−34.754 413	−82.135 779	−150.526 015
8^3D		−23.141 951	−34.334 354	−79.533 960	−147.349 779
4^3F	−14.351 575	−23.546 249	−35.053 823	−83.454 221	−152.661 603
5^3F	−14.334 657	−23.490 973	−34.941 051	−83.036 317	−151.737 664
5^3G	−14.336 099	−23.489 411	−34.941 295	−83.043 091	−151.749 511
6^3G	−14.335 897	−23.387 245	−34.810 143	−82.746 257	−151.139 034
6^3H	−14.326 506	−23.464 042	−34.888 913	−82.817 790	−151.248 300
7^3I	−14.320 140	−23.440 907	−34.853 261	−82.670 858	−150.914 473
8^3K	−14.315 243	−23.420 425	−34.821 733	−82.560 657	−150.729 034

Energy in a.u.
[a]The ionization threshold of the $^{\infty}$Be atom ($E_{\infty Be^+}$ =−14.324 763 176 790 43(22) a.u.[23]).
[b]The ionization threshold of the $^{\infty}B^+$ ion ($E_{\infty B^{2+}}$ =−23.424 605 665 a.u.).
[c]The ionization threshold of the $^{\infty}C^{2+}$ ion ($E_{\infty C^{3+}}$ =−34.775 510 611 a.u.).
[d]The ionization threshold of the $^{\infty}F^{5+}$ ion ($E_{\infty F^{6+}}$ =−82.330 336 543 a.u.).
[e]The ionization threshold of the $^{\infty}Mg^{8+}$ ion ($E_{\infty Mg^{9+}}$ =−150.136 154 391 a.u.).

The fourth (and sometimes the third) electron(s) can occupy any free electron orbital in the atom. The occupation numbers of the (third, fourth) electron(s) determine the actual state (or configuration) of the Be atom. The pair of the third and fourth electron can be either in the singlet state or in the triplet state (as the whole Be atom). It follows from here that the bound state spectrum of the Be atom must be similar to the bound state spectra of the two-electron He atom. Indeed, such a similarity can be observed (the two series of bound states, the ground state is the singlet S state). However, the actual order of different bound states is different for the He and Be atoms. For instance, the lowest state in the triplet series is 2^3S state in the helium atom and 2^3P state in the beryllium atom. For the excited bound states in the He and Be atoms, one finds more differences than similarities, while for singlet states close to the ground state, similarities with between the spectra of these two elements can easily be seen.

Since the three-electron core of the Be atom has the $1s^22s$ electron configuration, then the dissociation threshold for neutral Be corresponds to formation of the three-electron Be^+ ion in its ground 2^2S state (doublet). The nonrelativistic energy of this state is $E_{\infty Be^+} \approx -14.324\ 763\ 176\ 790\ 43(22)$ a.u.[23] This dissociation threshold corresponds to the following ionization process of the Be atom

$$\mathrm{Be} = \mathrm{Be}^+(2^2\mathrm{S}) + \mathrm{e}^-, \tag{17}$$

where the symbol $Be^+(2^2S)$ means that the final three-electron Be^+ ion is in its ground 2^2S state. Now we can write the following expression for the total energies of the weakly bound states, i.e., for the states close to the dissociation threshold of the Be atom (in atomic units):

$$E(\mathrm{Be};\, nL) = E(\mathrm{Be}^+;\, 2^2\mathrm{S}) - \frac{m_e e^4}{2\hbar^2}\frac{1}{(n+\Delta_\ell)^2} = -14.32476317679043 - \frac{1}{2(n+\Delta_\ell)^2}, \tag{18}$$

where $L = \ell$ (in this case), n is the principal quantum number of the nL state (L is the angular quantum number) of the Be atom, and Δ_ℓ is the Rydberg correction which explicitly depends upon ℓ (angular momentum of the outer most electron) and the total electron spin of this atomic state. It can be shown that the Rydberg correction rapidly vanishes when ℓ increases (for given n and L). Moreover, the Δ_ℓ correction also decreases when the principal quantum number n grows. The formula, Eq. (18), can be used to approximate the total energies of weakly bound, Rydberg states in the Be atom. However, by following the original ideas of Heisenberg[24] and Bethe (see, e.g., Ref. 25 and references therein), we can write a significantly more accurate formula which can be used to approximate the same Rydberg states to very high numerical accuracy. This formula is written in the form

$$\begin{aligned} E(\mathrm{Be};\, nL) &= E(\mathrm{Be}^+;\, 2^2\mathrm{S}) - \frac{m_e e^4}{2\hbar^2} \frac{1-\epsilon_L}{(n+\Delta_\ell+(-1)^S\Delta_A)^2} \\ &\approx -14.32476317679043 - \frac{1-\epsilon_L}{2(n+\Delta_\ell+(-1)^S\Delta_A)^2} \end{aligned} \tag{19}$$

where S is the total electron spin, while ϵ_L,Δ_ℓ, and Δ_A are the three parameters which must be varied in each specific case to obtain better numerical approximations. All these parameters rapidly decrease when ℓ (and L) grows. In reality, to apply the formula, Eq. (19), one needs to know the accurate values of the total energies of at least three bound states in each spectral series, i.e., the total energies of three singlet and three triplet bound states with $n \geq 5$.

Based on Eq. (19), one can predict that the total energies of the singlet and triplet highly excited states (with the same n) are equal to each other to high accuracy (near degeneracy). In general, such near degeneracy of energy levels becomes almost exact when n grows. It is a well-known property of the Rydberg states, and it can be observed in any atomic system which has energy spectrum consisting of a few different spectral series. Formally, based on the formula, Eq. (19), we can classify all bound states in the Be atom as the Rydberg states, pre-Rydberg and non-Rydberg states. Each group of these states has its unique electron density distribution.

4. SPECTRAL DIAGRAM OF THE FOUR-ELECTRON BE-LIKE ATOMS

In this study, we have determined the total energies of a large number of bound singlet states in the Be atom. Our computational results can be used to draw the energy levels of all computed singlet (bound) states of the $^{\infty}$Be atom as functions of angular momentum L of these states. In classic books on atomic spectroscopy, such pictures (or diagrams) were called the "spectral diagrams." In general, the spectral diagrams are very useful tools to study various effects related to the electron density distribution in different bound LS states of the atomic systems which contain the same number of electrons.

For neutral atoms and ions with the same nuclear charge Q, measured spectral diagrams are often used to investigate effects related with the role of electron–electron correlations in different atomic states. For instance, from our spectral diagram one finds that the 3^1D state in the Be atom is less bound than the analogous 3^1S state, while for the bound 4^1D and 4^1S states such an order of bound states is reversed. The true theory of electron–electron correlation in atoms must explain the observed order of the bound states (or energy levels) in the spectrum and approximately predict the energy differences between them.

In general, by performing numerical calculations of a large number of bound states in atomic systems one always needs to answer the following two questions: (1) predict the correct order of low-lying bound states and (2) describe transitions between the low-lying bound states and weakly bound, or Rydberg states. To solve the first problem, we can compare our results with the known experimental data for Be atom.[2] For the singlet states in the Be atom, the agreement between our computational results, Fig. 1, and data for the beryllium atom presented in Ref. 2 can be considered as very good. Combining the theoretical and experimental data, we can predict the total order of states in the singlet and triplet series, as shown in Table 4. It is also clear we have calculated only the nonrelativistic (total) energies, i.e., all relativistic and lowest-order QED corrections were ignored. Note also that the CI method using STOs is substantially more accurate (in the order of few milli-Hartree) than various procedures based

Figure 1 The energy levels of the singlet states in the beryllium atom. The threshold energy (or ionization limit) $E_{\infty Be^+} = -14.324\ 763\ 176\ 790\ 43(22)$ a.u. coincides with the total energy of the ground 2^2S state of the three-electron Be^+ ion.

Table 4 Order of Bound States in the Be Atom from the Theoretical Calculations and Experimental Observations of Spectral Lines[2]

n	Order in Singlet Series	Order in Triplet Series
	$^2Be^+$ Ionization Limit	$^2Be^+$ Ionization Limit
13	$13^1D < 13^1P$	13^3D
12	$12^1D < 12^1P$	12^3D
11	$11^1D < 11^1S < 11^1P$	11^3D
10	$10^1D < 10^1S < 10^1P$	10^3D
9	$9^1D < 9^1S < 9^1P$	9^3D
8	$8^1D < 8^1S < 8^1P$	8^3P
7	$7^1D < 7^1S < 7^1P < 7^1F < 7^1G < 7^1H < 7^1I$	$7^3D < 7^3F < 7^3S$
6	$6^1D < 6^1S < 6^1P < 6^1F < 6^1G < 6^1H$	$6^3P < 6^3D < 6^3F < 6^3S$
5	$5^1D < 5^1S < 5^1P < 5^1F < 5^1G$	$5^3P < 5^3D < 5^3F < 5^3G < 5^3S$
4	$4^1D < 4^1S < 4^1P < 4^1F$	$4^3P < 4^3D < 4^3F < 4^3S$
3	$3^1S < 3^1D < 3^1P$	$3^3P, 3^3P'^a < 3^3D < 3^3S$
2	$2^1S < 2^1P$	$2^3P < 2^3S$

The symbol "<" denotes here lower energy.
[a]Here we denote the doublet of states with same *n* energy level.

on Hartree–Fock approximation, but it still provides a restricted description of the electron–electron correlation in specific atoms and ions. Nevertheless, the observed agreement with the measured bound state spectra of the triplet states in the Be atom (or ^9Be atom) is very good not only for low-lying bound states but also for Rydberg states.

Now consider the second problem. As follows from the results of our calculations, all bound singlet states with $n \geq 6$ in the beryllium atom are the weakly bound or Rydberg states. On the other hand, all bound triplet states in the Be atom with $n \geq 4$ can be considered as the pre-Rydberg states. It follows from comparison of the total energies of the triplet 4^3F, 5^3G states and singlet 4^1F, 5^1G states (for more details, see Ref. 1). On the other hand, it is clear that the "boundary" principal quantum number n_R from which the Rydberg states begin (for $n \geq n_R$) must be exactly the same for both spectral series in the four-electron atoms and ions.

In this study, our main focus was on the singlet bound states in the four-electron Be atom. The triplet states in the Be atom were considered in our paper[1] which also contains the spectral diagram of the bound triplet states in the Be atom. For maximal convenience of the reader, we also included the updated spectral diagram of the triplet states in our present analysis (see Fig. 2). In many cases, it is useful to observe spectral diagrams for the both singlet and triplet series together and compare these diagrams with the analogous spectral diagrams for atoms which have bound state spectra represented as a combination of the two separate spectral series (singlet and triplet). For instance, it is very interesting and informative to make such a comparison of the spectra of the beryllium and helium atoms (see, e.g., Ref. 25 and references therein).

We have also drawn spectral diagrams of the triplet states for various positively charged ions, e.g., for the Be atom and B^+, C^{2+}, F^{5+}, and Mg^{8+} (see Figs. 3–6 using the computational results of this work and the known

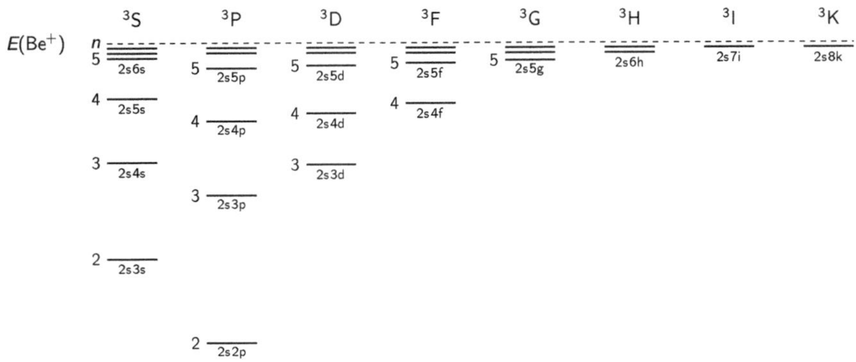

Figure 2 The energy levels of the triplet states in the beryllium atom. The threshold energy (or ionization limit) $E_{\infty Be^+} = -14.324\ 763\ 176\ 790\ 43(22)$ a.u. coincides with the total energy of the ground 2^2S state of the three-electron Be^+ ion.

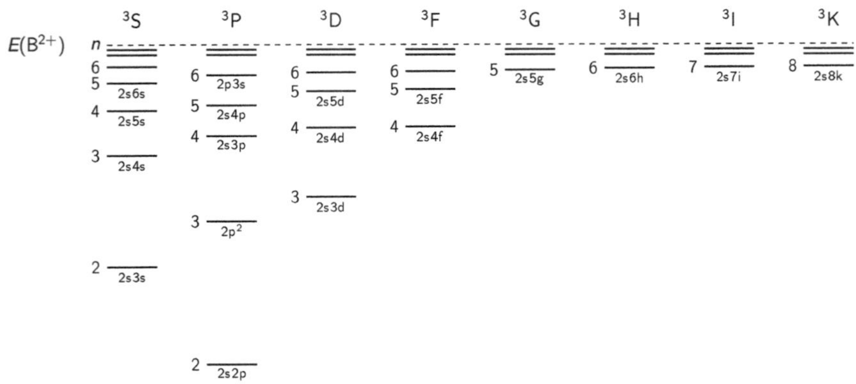

Figure 3 The energy levels of the triplet states in the B^+ ion. The threshold energy (or ionization limit) $E(B^{2+}) = -23.424\ 605\ 665$ a.u. coincides with the total energy of the ground 2^2S state of the three-electron B^{2+} ion.

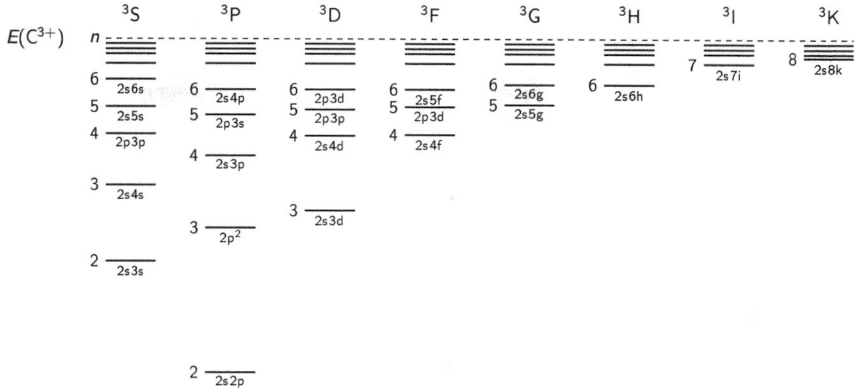

Figure 4 The energy levels of the triplet states in the C^{2+} ion. The threshold energy (or ionization limit) $E(C^{3+}) = -34.775\ 510\ 611$ a.u. coincides with the total energy of the ground 2^2S state of the three-electron C^{3+} ion.

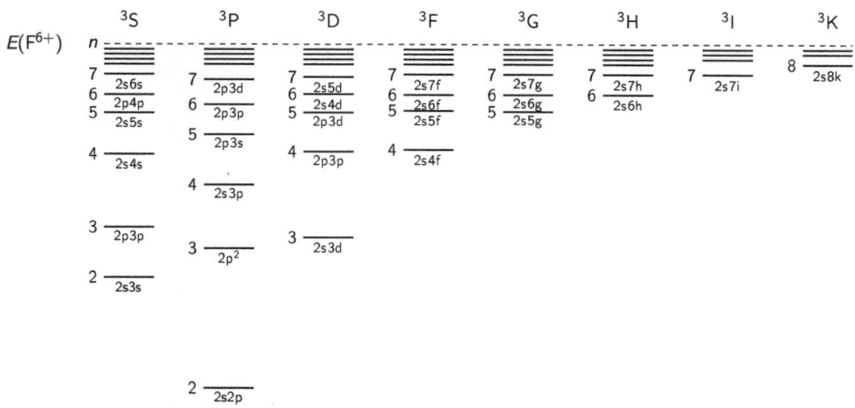

Figure 5 The energy levels of the triplet states in the F^{5+} ion. The threshold energy (or ionization limit) $E(F^{6+}) = -82.330\ 336\ 543$ a.u. coincides with the total energy of the ground 2^2S state of the three-electron F^{6+} ion.

experimental data from NIST Atomic Database and Refs. 26–29). The spectra of the negatively charged ions, e.g., the Li^- ion, contains only a very few bound states (usually one bound state[30]) and its spectral diagram is very simple and not informative. The corresponding spectral diagrams of the cations are very similar to each other, but there are few differences between them due to the Z-dependence. These differences may well improve our understanding of the electron–electron correlation in the four-electron atomic systems.

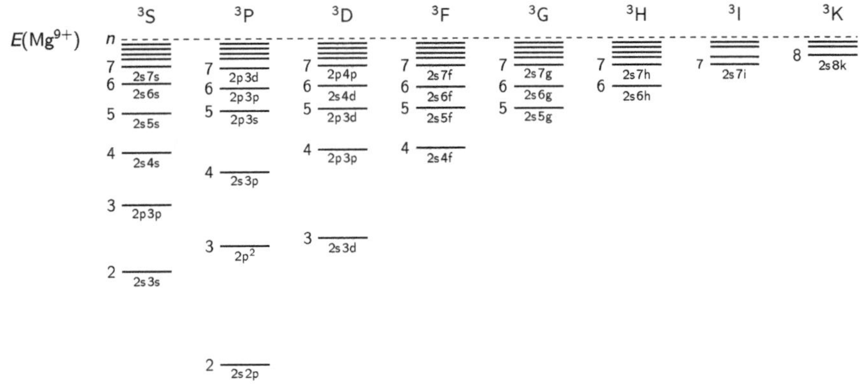

Figure 6 The energy levels of the triplet states in the Mg^{8+} ion. The threshold energy (or ionization limit) $E(Mg^{9+}) = -150.136\ 154\ 391$ a.u. coincides with the total energy of the ground 2^2S state of the three-electron Mg^{9+} ion.

5. CONCLUSION

We have considered the bound state spectrum of the singlet states in the four-electron Be atom and the spectra of the triplet states in the Be-like ions B^+, C^{2+}, F^{5+}, and Mg^{8+}. The analogous spectrum of the bound triplet states in the four-electron Be atom has been presented and discussed in our earlier study.[1] The results of both studies reproduce the complete "optical" spectrum of the four-electron Be atom. The agreement between our computational results and actual singlet/triplet spectrum of the Be atom[2] can be considered as very good. The quite complicated bound state spectrum of beryllium atom has been obtained and studied by using only computational methods. This study is based on a computational approach which has three following advantages: (1) it can be applied for accurate computations of all bound states in the spectrum, including rotationally excited states and weakly bound, Rydberg atomic states; (2) it provides overall accuracy in the order of few milli-Hartree which is beyond the level provided by various method based on the Hartree–Fock approximation; and (3) such an accuracy does not decrease for the excited LS states in the spectrum. Disadvantage of this approach is the slow convergence of the CI method, which requires selection of predominant configurations and successive optimization of the orbital exponents.

The results of this study allowed us to draw the spectral diagrams of the singlet and triplet spectra of the four-electron Be atom and Be-like ions. Such spectral diagrams for Be atom can now be compared with analogous spectral diagrams of other light atoms and ions. Theoretical comparison of the atomic spectra of the Be and He atoms seems to be very interesting, since each of these two spectra contains two independent series of bound states (singlet and triplet).

ACKNOWLEDGMENTS

We would like to thank James Sims and Stanley Hagstrom for providing us with numerical results from Hylleraas-CI calculations on the Be atom which have serve to check our Be computer program, and to Carlos Bunge for valuable advice about the construction of LS and degenerate configurations. One of us (F.L.) greatly acknowledges his PhD supervisor Leticia González at the University of Vienna for her scientific support. Finally, we are very grateful to Philip Hoggan and Telhat Özdogan for the invitation to contribute to this volume and for proofreading the manuscript.

REFERENCES

1. Frolov, A. M.; Ruiz, M. B. Bound State Spectrum of the Triplet States in the Be Atom. *Chem. Phys. Lett.* **2014**, *197*, 595–596.
2. Kramida, A.; Martin, W. C. A Compilation of Energy Levels and Wavelengths for the Spectrum of Neutral Beryllium (Be I). *J. Phys. Chem. Ref. Data* **1997**, *26*, 1185.
3. Sims, J. S.; Hagstrom, S. A. Combined Configuration-Interaction Hylleraas-Type Wave-Function Study of the Ground State of the Beryllium Atom. *Phys. Rev. A* **1971**, *4*, 908.
4. Bunge, C. F. Configuration Interaction Benchmark for Be Ground State. *Theor. Chem. Acc.* **2010**, *126*, 139.
5. Chung, K. T.; Zhu, X. W. Energies, Fine Structures, and Isotope Shifts of the $1s^2 2snl$ Excited States of the Beryllium Atom. *Phys. Rev. A* **1993**, *48*, 1944.
6. Büsse, G.; Kleindienst, H.; Lüchow, A. Nonrelativistic Energies for the Be Atom: Double-Linked Hylleraas-CI Calculation. *Int. J. Quantum Chem.* **1998**, *66*, 241.
7. King, F. W.; Quicker, D.; Langer, J. Compact Wave Functions for the Beryllium Isoelectronic Series, Li to Ne^6 +: A Standard Hylleraas Approach. *J. Chem. Phys.* **2011**, *134*, 124114.
8. Sims, J. S.; Hagstrom, S. A. Hylleraas-Configuration-Interaction Study of the ^1S Ground State of Neutral Beryllium. *Phys. Rev. A* **2011**, *83*, 032518.
9. Puchalski, M.; Komasa, J.; Pachucki, K. Testing Quantum Electrodynamics in the Lowest Singlet States of the Beryllium Atom. *Phys. Rev. A* **2013**, *87*, 030502(R).
10. Stanke, M.; Komasa, J.; Bubin, S.; Adamowicz, L. Five Lowest ^1s States of the Be Atom Calculated with a Finite-Nuclear-Mass Approach and with Relativistic and QED Corrections. *Phys. Rev. A* **2009**, *80*, 022514.
11. Kolesnikov, N. N.; Tarasov, V. I. Phenomenological N-N Potential from an Analysis of Three- and Four-Particle Nuclei. *Yad. Phys.* **1982**, *35*, 609, *Sov. Phys. Nucl. Phys.* **1982**, *35*, 354.
12. Gálvez, F. G.; Buendía, E.; Sarsa, A. Excited States of Beryllium Atom from Explicitly Correlated Wave Functions. *J. Chem. Phys.* **2002**, *117*, 6071.
13. Bubin, S.; Adamowicz, L. Prediction of ^1p Rydberg Energy Levels of Beryllium Based on Calculations with Explicitly Correlated Gaussians. *J. Chem. Phys.* **2014**, *140*, 024301.
14. Sharkey, K. L.; Bubin, S.; Adamowicz, L. ^1D States of the Beryllium Atom: Quantum Mechanical Nonrelativistic Calculations Employing Explicitly Correlated Gaussian Functions. *Phys. Rev. A* **2011**, *84*, 044503.

15. Bertini, L.; Mella, M.; Bressanini, D.; Morosi, G. Explicitly Correlated Trial Wavefunctions in Quantum Monte Carlo Calculations of Excited States of Be and Be^-. *J. Phys. B At. Mol. Opt. Phys.* **2001**, *34*, 257.
16. Frolov, A. M.; Wardlaw, D. M. Triplet 2^3s States in Four-Electron Be-Like Ions: Bound-State Properties and Hyperfine-Structure Splitting. *Phys. Rev. A* **2009**, *79*, 064501.
17. Ruiz, M. B. Hylleraas Method for Many-Electron Atoms. I. The Hamiltonian. *Int. J. Quantum Chem.* **2005**, *101*, 246.
18. Ruiz, M. B. Evaluation of Hylleraas-CI Atomic Integrals. III. Two-Electron Kinetic Energy Integrals. *J. Math. Chem.* **2011**, *49*, 2457.
19. Frolov, A. M.; Smith, V. H. Exact Finite Series for the Few-Body Auxiliary Functions. *Int. J. Quantum Chem.* **1997**, *63*, 269.
20. Sims, J. S.; Hagstrom, S. A. Mathematical and Computational Science Issues in High Precision Hylleraas-Configuration Interaction Variational Calculations: I. Three-Electron Integrals. *J. Phys. B At. Mol. Opt. Phys.* **2004**, *37*, 1519.
21. Ruiz, M. B.; Margraf, J. T.; Frolov, A. M. Hylleraas-Configuration-Interaction Analysis of the Low-Lying States in the Three-Electron Li Atom and Be^+ Ion. *Phys. Rev. A* **2013**, *88*, 012505.
22. Stevenson, R. *Multiplet Structure of Atoms and Molecules*; W. B. Saunders Company: Philadelphia and London, 1965.
23. Puchalski, M.; Kedziera, D.; Pachucki, K. Ground State of Li and Be^+ Using Explicitly Correlated Functions. *Phys. Rev. A* **2009**, *80*, 032521.
24. Heisenberg, W. Über die spektra von atomsystemen mit zwei elektronen. *Z. Phys.* **1926**, *39*, 499.
25. Bethe, H. A.; Salpeter, E. E. *Quantum Mechanics of One- and Two-Electron Atoms*; Plenum Publ. Corp.: New York, 1977.
26. Ryabtsev, A. N.; Kink, I.; Awaya, Y.; Ekberg, J. O.; Mannervik, S.; Ölme, A.; Martinson, I. Additions to the Spectrum and Energy Levels and a Critical Compilation of Singly-Ionized Boron, B II. *Phys. Scr.* **2005**, *71*, 489.
27. Moore, C. E.; Gallagher, J. W. *Tables of Spectra of Hydrogen, Carbon, Nitrogen, and Oxygen Atoms and Ions. CRC Series in Evaluated Data in Atomic Physics*, CRC Press: Boca Raton, FL, 1993, 339 pp.
28. Engström, L. The Spectrum of Five Times Ionized Fluorine. *Phys. Scr.* **1985**, *31*, 379.
29. Martin, W. C.; Zalubas, R. Energy Levels of Magnesium, Mg I Through Mg XII. *J. Phys. Chem. Ref. Data* **1980**, *9*, 1.
30. Frolov, A. M. Bound-State Calculations of Coulomb Three-Body Systems. *Phys. Rev. A* **1999**, *59*, 4270.

CHAPTER SIX

An Application of the Gaussian Transform for Approximating Some Bessel Functions and Multicenter Integrals Involving 1s Slater-Type Orbitals

Philip E. Hoggan*, Jorge E. Pérez[†,1], Ahmed Bouferguene[‡]
*Institut Pacal, UMR CNRS 6602, Campus Universitaire des Cézeaux, Aubiere Cedex, France
[†]Departamento de Física, Facultad de Ciencias Exactas Fco-Qcas y Naturales, Universidad Nacional de Río Cuarto, Río Cuarto, Argentina
[‡]Faculté Saint-Jean, University of Alberta, Alberta, Canada
[1]Corresponding author: e-mail address: eperez@exa.unrc.edu.ar

Contents

Abstract

New Gaussian expansions for molecular integrals over 1s Slater-type orbital are developed. Some well-documented integral representations for the modified Bessel of second kind are revisited by using Gaussian transforms. These integral representations are evaluated with numerical quadrature, which generates new series to be considered.

Simple examples of applications in the molecular context are presented.

1. INTRODUCTION

The eigen functions of hydrogen-like Hamiltonians decrease exponentially at long range and constitute appropriate sets to be used as bases in molecular electronic calculations. Unfortunately, the multicenter integrals are difficult and often time-consuming to evaluate. Therefore,

Advances in Quantum Chemistry, Volume 73
ISSN 0065-3276
http://dx.doi.org/10.1016/bs.aiq.2015.06.007

exponential basis sets are often represented in molecular calculations using the Slater-type orbital (STO)-NG approximation[1] (i.e., expanded in terms of N Gaussians). Within this framework, the STO-NG approach has been used for obtaining simple models[2a] and also for getting benchmark multicenter integral evaluations.[3] Normally, the linear combinations of Gaussian functions suitable for expanding exponential bases involve the determination of numerous parameters in very lengthy calculations which can be avoided by using numerical quadrature performed on some integral representations.[2b]

In this work, we also consider the Gaussian expansions for the function $k_\nu(z) = z^\nu K_\nu(z)$ where $K_\nu(z)$ is the modified Bessel function.[4a] This possibility is based on linear combinations of the original eigen functions, called *B* functions that have been used because they have convenient analytical properties. A recent paper[5] pointed out the importance of the Bessel function in relation to the exponential basis sets. For instance, they appear in the two-center product of two 1s STO, the multicenter overlap integrals, attraction Coulomb integrals, multicenter repulsion integrals, etc.

Here, we present two equivalent integral representations for the Bessel function $k_\nu(z)$ which are well suited for integrating numerically, i.e., these numerical treatments give the linear combinations of Gaussian sought.

2. METHODOLOGY

The following integral representation for the function $k_\nu(z) = z^\nu K_\nu(z)$[4b]

$$k_\nu(z) = 2^{\nu-1}\int_0^\infty \mathrm{d}t\, t^{\nu-1}\mathrm{e}^{-t}\mathrm{e}^{-(z^2/4t^2)} \tag{1}$$

can be obtained from two other integral representations,[4a] written as follows:

$$k_\nu(z) = z^\nu K_\nu(z) = \frac{2^\nu \Gamma(\nu+1/2)}{\pi^{1/2}}\int_0^\infty \mathrm{d}x \frac{\cos(zx)}{[1+x^2]^{\nu+1/2}} \tag{2}$$

and

$$[1+x^2]^{-(\nu+(1/2))} = \frac{1}{\Gamma(\nu+(1/2))}\int_0^\infty \mathrm{d}t\, t^{(\nu-(1/2))}\mathrm{e}^{-t}\mathrm{e}^{-tx^2} \tag{3}$$

where $\nu = 0, \frac{1}{2}, 1, \frac{3}{2}, 2, \frac{5}{2} \ldots$ Equation (1) is generated by introducing Eq. (3) into Eq. (2) and then using the property that the Fourier cosine transform of a Gaussian function is again a Gaussian function. Equations (1) and (3) are Gaussian transforms.[6]

In equations like those in (1) and (3), it has been useful[2a,7–10] to introduce the following change of variable

$$t = R(1+u)/(1-u) \tag{4}$$

where R is an appropriate scaling constant. After this change of variable and others, the authors of the cited work introduced numerical quadratures for generating suitable mathematical expressions and in particular, Gaussian expansions.

Following the same strategy, we will consider the Gauss–Chebyshev quadrature, with the weight factor $(1-u^2)^{-0.5}$,[8] for the particular values of $\nu =$ nonnegativeinteger; i.e., this corresponds to the Bessel functions which cannot be expressed in terms of elementary functions.

It is worth mentioning that the numerical approach applied to Eq. (1) is totally equivalent to that obtained from Eqs. (2) and (3). However, when ν is integer, it can be much better to work with an analytical expression given by Eq. (3), because it is possible to obtain very precise truncation of the infinity interval for further studies. See Appendix to this work.

The Gauss–Chebyshev numerical quadrature for Eq. (3) gives

$$\left[1 + x^2\right]^{-(\nu+1/2)} \cong \sum_{i=1}^{N} \sigma_i^{(\nu)} \mathrm{e}^{-\zeta_i x^2} \tag{5}$$

By using Eq. (5) in Eq. (2), we obtain

$$\begin{aligned} k_\nu(z) &\cong \frac{2^\nu \Gamma(\nu+1/2)}{\pi^{1/2}} \sum_{i=1}^{N} \sigma_i^{(\nu)} \int_0^\infty \mathrm{d}x \cos(zx) \mathrm{e}^{-\zeta_i x^2} \\ &= \frac{2^\nu \Gamma(\nu+1/2)}{\pi^{1/2}} \sum_{i=1}^{N} \sigma_i^{(\nu)} \frac{\pi^{1/2}}{2(\varsigma_i)^{1/2}} \mathrm{e}^{-(z^2/4\varsigma_i)} \end{aligned} \tag{6}$$

In the applications, the expansion given in Eq. (6) can be used for obtaining relevant functions in molecular physics.

For instance, the two-center product of two 1s STO[11] and the Coulomb potential generated by the product of two unnormalized 1s STO,[12] which, respectively, are

$$P_{\mu\nu}(\vec{r}) = \frac{\alpha_\mu \alpha_\nu}{\pi} \int_0^1 du \frac{k_1\left(\delta_{(\mu\nu)}(u)\left[D^2_{(\mu\nu)} + \frac{R^2_{(\mu\nu)}(u)}{u(1-u)}\right]^{1/2}\right)}{[u(1-u)]^{1/2}\delta^2_{(\mu\nu)}(u)} \tag{7}$$

$$\Phi_{\mu\nu}(\vec{r}) = 4\alpha_\mu\alpha_\nu \int_0^1 du \frac{[u(1-u)]^{1/2}}{\delta^4_{(\mu\nu)}(u)} \int_0^1 dt k_2\left(\delta_{(\mu\nu)}(u)\left[D^2_{(\mu\nu)} + \frac{R^2_{(\mu\nu)}(u)t^2}{u(1-u)}\right]^{1/2}\right) \tag{8}$$

where $\delta_{(\mu\nu)}(u) = \left[u\alpha_\mu^2 + (1-u)\alpha_\nu^2\right]^{1/2}$, $\vec{R}_{(\mu\nu)}(u) = u\vec{R}_\nu + (1-u)\vec{R}_\mu$, and $D_{(\mu\nu)} = |\vec{R}_\nu - \vec{R}_\mu|$, with α_λ and $\vec{R}_\lambda$ the orbital exponent and the localization of the λ 1s STO, respectively.

3. RESULTS FOR THE COULOMB POTENTIAL

Considering the Coulomb potential given by Eq. (8) and introducing Eq. (6), in it with $\nu = 2$, we obtain

$$\Phi_{\mu\nu}(\vec{r}) \cong 4\alpha_\mu\alpha_\nu \sum_{j=1}^{n} \frac{\pi^{0.5}\sigma_j^{(2)}}{2[\varsigma_J]^{0.5}} \int_0^1 du \frac{[u(1-u)]^{0.5}}{[\delta_{\mu\nu}(u)]^4} \exp\left[\frac{-[\delta_{\mu\nu}(u)]^2 D^2_{\mu\nu}}{4\varsigma_j}\right] \frac{\pi^{0.5}\operatorname{erf}\left[\frac{\delta_{\mu\nu}(u)\left|\vec{R}_{\mu\nu}(u)\right|}{2(\varsigma_j)^{0.5}}\right]}{2\frac{\delta_{\mu\nu}(u)\left|\vec{R}_{\mu\nu}(u)\right|}{2(\varsigma_j)^{0.5}}} \tag{9}$$

where $\frac{\pi^{0.5}\operatorname{erf}[z^{0.5}]}{2z^{0.5}} = \int_0^1 du \exp(-zu^2)$.

In Table 1, the convergence of the approximant, given by Eq. (9), for a chosen example is shown.

4. CONCLUDING REMARKS

The approach developed in this chapter gives an extremely simple way of obtaining Gaussian expansions of a large class of Bessel functions. These series are based on the Gauss–Chebyshev quadrature applied to particular

Table 1 The Example Corresponds to $\vec{R}_1 = (1,1,0)$, $\vec{R}_2 = (0,1,0)$, $\vec{r} = (2,2,0)$, $\alpha_1 = 1$, and $\alpha_2 = 1.5$

n	**Approx.**	**Relative Error**
10	0.6610778203	2.3 (−4)
20	0.6612426237	−1.9 (−5)
30	0.6612302492	1.2 (−7)
40	0.6612303313	−1.2 (−9)
50	0.6612303305	−1.3 (−11)

The reference value is 0.661230330525, evaluated with the algorithm of Bouferguene et al.[13] All the quantities are in a.u.

integral representations; i.e., they can be obtained to any desired order by increasing the number of integration points. This is an important property if we consider the Bessel functions with integer order; i.e., the Coulomb potential example is written as a very well-behaved series.

For each molecular integral, the convergence properties of the series will depend on the integral representation and numerical quadrature involved. The idea presented in Appendix makes possible future studies with other numerical schemes.

APPENDIX

Definite integral for $[1+x^2]^{-\mu}$

If we consider the partition

$$\begin{aligned}\left[1+x^2\right]^{-\mu} &= \frac{1}{\Gamma(\mu)}\int_0^{\infty} du\, u^{\mu-1}\exp\left[-\left(1+x^2\right)u\right] \\ &= \frac{1}{\Gamma(\mu)}\int_0^{L} du\, u^{\mu-1}\exp\left[-\left(1+x^2\right)u\right] + \frac{1}{\Gamma(\mu)}\int_L^{\infty} du\, u^{\mu-1}\exp\left[-\left(1+x^2\right)u\right]\end{aligned} \tag{A.1}$$

We can obtain an expression for the *relative error*, *rel*, arising in the truncation of the improper integral representation,

$$\begin{aligned} rel &\equiv 1 - \frac{(1+x^2)^{\mu}}{\Gamma(\mu)}\int_0^{L} du\, u^{\mu-1}\exp\left[-\left(1+x^2\right)u\right] \\ &= \frac{(1+x^2)^{\mu}}{\Gamma(\mu)}\int_L^{\infty} du\, u^{\mu-1}\exp\left[-\left(1+x^2\right)u\right]\end{aligned} \tag{A.2}$$

In Eq. (A.2), the change of variable $y = (1 + x^2)u$ generates the following bound:

$$rel = \frac{1}{\Gamma(\mu)} \int_{L(1+x^2)}^{\infty} dy\, y^{\mu-1} \exp[-y] \leq \frac{1}{\Gamma(\mu)} \int_{L}^{\infty} dy\, y^{\mu-1} \exp[-y] \qquad (A.3)$$

For the particular case for which $\mu = 2.5$, i.e., it corresponds to $k_2(z)$, for obtaining relative errors less than or equal to 10^{-8}, the appropriate L value is about 23.

REFERENCES

1. Boys, S. F. Electronic Wavefunctions. I. A General Method of Calculation for Stationary States of Any Molecular System. *Proc. R. Soc. Lond.* **1950**, *A200*, 542–554.
2. (a) Montagnani, R.; Salvetti, O. Computation of Two-, Three- and Four-Center Exchange Integrals over Slater Basis by Means of Optimized Gaussian Expansions. *Int. J. Quantum. Chem.* **1992**, *43*(2), 273–280; (b) Silver, D. M. Gaussian Expansions for Polyatomic Molecules. *Mol. Phys.* **1971**, *22*(6), 1069–1080.
3. Fernändez Rico, J.; López, R.; Ema, I.; Ramirez, G. Calculation of Many-Centre Two-Electron Molecular Integrals with STO. *Comput. Phys. Commun.* **1997**, *105*, 216–224.
4. (a) Abramowitz, M.; Stegun, I. A. *Handbook of Mathematical Functions.* Dover Publications: New York, 1972; (b) Gradshteyn, I. S.; Ryzhik, I. M. *Table of Integrals, Series and Products*, 6th ed.; Academic Press: San Diego, CA, 2000.
5. Pinchon, D.; Hoggan, P. E. Gaussian Approximation of Exponential Type Orbitals Based on B Functions. *Int. J. Quantum. Chem.* **2009**, *109*(2), 135–144.
6. McKemmish, L. K.; Gill, P. M. W. Gaussian Expansions of Orbitals. *J. Chem. Theory Comput.* **2012**, *8*(12), 4891–4898.
7. Matsuoka, O.; Aoyama, T. Molecular Integral of Diamagnetic Contribution to Nuclear Spin-Spin Coupling Constant. *J. Chem. Phys.* **1980**, *73*(11), 5718–5720.
8. Becke, A. D. A Multicenter Numerical Integration Scheme for Polyatomic Molecules. *J. Chem. Phys.* **1998**, *88*, 2547–2553.
9. Lindh, R.; Malmqvist, P. A.; Gargliardi, L. Molecular Integrals by Numerical Quadrature. I. Radial Integration. *Theor. Chem. Acc.* **2001**, *106*, 178–187.
10. Gill, P. M. W.; Chien, S. H. Radial Quadrature for Multiexponentials Integrals. *J. Comput. Chem.* **2002**, 732–740.
11. Fernandez Rico, J.; Lopez, R.; Ramirez, G.; Ema, I. Multiple One-Center Expansions of Charge Distributions Associated with Slater Orbitals. *J. Mol. Struct. (THEOCHEM)* **1998**, *433*, 7–18.
12. Shavitt, I.; Karplus, M. Gaussian-Transform Method for Molecular Integrals. I. Formulation for Energy Integrals. *J. Chem. Phys.* **1965**, *43*(2).
13. Bouferguene, A.; Fares, M.; Rinaldi, D. Integrals over **B** Functions Basis Sets. I. Three-Center Molecular Integrals, a Numerical Study. *J. Chem. Phys.* **1994**, *100*, 8156.

CHAPTER SEVEN

Size-Extensivity Corrections in Single- and Multireference Configuration Interaction Calculations

Murat Erturk[*,†,1], **Leszek Meissner**[†]

[*]Department of Physics, Faculty of arts and sciences, Onsekiz Mart University, Çanakkale, Turkey
[†]Institute of Physics, Nicholas Copernicus University, Torun, Poland
[1]Corresponding author: e-mail address: merturk@comu.edu.tr

Contents

Abstract

It is well known that size-inextensivity is a major drawback of truncated configuration interaction (CI) methods. In order to ease the problem, many attempts have been made to minimize size-extensivity error of the CI methods through *a posteriori* corrections. In this study, we investigate performance of various size-extensivity corrections proposed in the literature. While the corrections were basically formulated for the single-reference (SR) CI method with singles and doubles, very soon their straightforward generalizations to a multireference (MR) case started to be used without any deeper theoretical justification. The only correction that is universal enough to be applied in case of any kind of CI method is the coupled-cluster (CC) correction. For the purpose of this study, we have used the closed shell $\tilde{a}^1A_1$ singlet excited states of the CH_2 and NH_2^+ molecules. The states are of MR character but still can be quite well described by SR approaches. That allows us to investigate the performance of the corrections in both SR and MR cases. For the first time, the second-order configuration interaction method has been used as a kind of the MR-type CI method. The calculations show that the performance of the CC correction is significantly better than other corrections.

Advances in Quantum Chemistry, Volume 73
ISSN 0065-3276
http://dx.doi.org/10.1016/bs.aiq.2015.07.001

1. INTRODUCTION

The single-reference (SR) coupled-cluster (SR-CC) approaches have proven to be successful and very useful for an accurate description of ground states of atoms and molecules in their equilibrium geometry.[1,2] This mainly concerns SR situations when the wave function is dominated by a single determinant in the zeroth-order approximation. The CC method with singles and doubles (CCSD) together with its extended version that includes an approximate effect of triple excitations (CCSD(T))[1,2] has become the fundamental SR approaches to the problem of describing many-electron systems and is included in almost all packages of programs for quantum physical and chemical calculations. However, there are many situations when an important restriction for the SR-CC methods, that the wave function should be dominated by a single determinant, does not allow us to apply them successfully. This is when other determinants become as important as the reference one and the SR determinant must be replaced with a multidimensional reference space. Such a situation requires multireference (MR) formulations of the CC (MR-CC) methods. Unfortunately, the generalization of the SR-CC method to an MR case turned out to be nontrivial and it is not unique. Two basic MR-CC schemes exist which are rather rarely used in calculations. Both basic MR-CC formulations[3–8] are complicated and plagued with the so-called intruder state problem in their practical numerical applications.[8] Conversely, the configuration interaction (CI) method (that is, not as attractive as the CC one in the SR calculations) can easily be generalized to MR situations. The MR variant of the SR-CI approach retains formal and computational simplicity of its SR counterpart. Unlike the SR cases, the MR-CI methods have many advantages over the MR-CC ones and can be routinely used in calculation for systems requiring an MR treatment. Thus, different variants of the MR-CI method are available in many packages of quantum chemistry programs. Unfortunately, the MR-CI approaches also share the drawbacks of their SR counterparts. Firstly, they are neither size-extensive nor efficient in describing dynamic electron correlation. The reason is that they use linear expansion for both dynamic and nondynamic correlation effects, while it is well known that for the dynamic part an exponential expansion is much better.

The source of size-extensivity error in the truncated SR-CISD method has been analyzed by many authors and various *a posteriori* size-extensivity corrections have been suggested. Most of them are very simple and give a very rough estimate of the size-extensivity error. The first correction of this type was the Davidson correction (DC)[9] giving an estimate of the contribution from the so-called renormalization term in the CI equations to the correlation energy. This correction has been analyzed in more detail by Bartlett and Shavitt and it has been found that a more correct evaluation of the same contribution is given by the so-called renormalized DC (RDC)[10,11] that takes into account the proper normalization condition assumed for the wave function. Other corrections are mainly derived from considerations for an ensemble consisting of identical noninteracting two-electron systems. These are Davidson and Silver correction (DSC)[12] (also Siegbahn derived a similar formula in a different way[11]), the Pople correction (PC),[13] the Meissner correction (MC),[14] and the Duch and Diercksen correction (DDC).[15] It should be emphasized that the PC was introduced to make the size-extensivity correction vanish when applied to two-electron systems for which CISD is exact.[13] Some of the corrections (DC, RDC, DSC) do not have this property. Finally, a correction based on the idea of accounting for the full inclusion of the effect of nonlinear terms in the wave function was proposed. The correction is called the coupled-cluster correction (CCc)[16] and is aimed at reproducing the CCSD energy in an approximate way using the CISD results. This correction is more complicated than those above and potentially more precise and dependable.[16]

While the idea of correcting the SR-CISD method in an *a posteriori* way is not very attractive in the presence of the SR-CCSD method, that is only a bit more numerically demanding but much more precise and strictly size-extensive, the situation is completely different in the MR case. MR approaches must be used when we have to deal with two types of correlation effects: nondynamic and dynamic. As mentioned, the MR-CC methods in their basic formulations are not sufficiently dependable contrary to the MR-CI ones. Correcting the MR-CI methods to make them more accurate and less inextensive seems a valid idea. However, since the MR situation is much more complicated to analyze than the SR[17,18] one, then a simple generalization of the SR corrections has been proposed.[19] The prescription for obtaining MR analogs of the SR corrections is[19]: (i) replace the CISD squared coefficient corresponding to the reference determinant with the sum of MR-CISD squared coefficient corresponding to all reference determinants, (ii) replace the CISD correlation energy with the difference between MR-CISD energy and the reference energy obtained by diagonalization of the Hamiltionian in the reference space. Only the CCc requires no changes of this type. Here, the assumption is that only dynamic correlation is corrected and that can be done in an SR manner.[20–23]

The purpose of this study is to show that the CCc, because of its universal character, can be effectively applied to any kind of CI scheme. In SR situations, the correction tends to reproduce the CCSD results while in MR cases, when the CCSD method does not perform well, it improves the MR-type CI energies giving results better than both the SR-CC and MR-CI methods. This works on condition that the reference space is properly selected so the CI scheme provides a good description of nondynamic correlation effects.

In this chapter, we want to show that the CCc can be successfully applied to basically any kind of CI method. For the purpose of this study, we use the second-order CI (SOCI) method as an MR-type CI approach.[24] The calculations have been carried out for potential energy curves of the $\tilde{a}\,^1A_1$ closed shell singlet excited states of the CH_2 and $NH_2{}^+$ molecules. The GAMESS quantum chemistry program package has been used to perform SOCI calculations.[25]

2. THEORETICAL BACKGROUND

Davidson was the first who suggested that the SR-CISD correlation energy should be corrected to account for cancelation of some contributions to the energy that takes place when higher than double excitations are included in the wave function expansion. These reducible contributions are contained in the so-called renormalization term in the equations for CI coefficients. The fourth-order perturbative estimate of the renormalization term contribution to the CISD energy generalized to infinity gives

$$\mathrm{DC} = E_{\mathrm{CISD}}\left(1 - C_0^2\right), \tag{1}$$

here E_{CISD} stands for the CISD correlation energy and C_0 for the reference function coefficient.[9] It can be shown that a more correct formula is given by the renormalized DC

$$\mathrm{RDC} = E_{\mathrm{CISD}}\frac{1 - C_0^2}{C_0^2}. \tag{2}$$

This takes into account that the perturbation expansion uses intermediate normalization for the wave function while the CISD method eigenvector is assumed to be normalized to unity.[10] Also considerations performed for a supersystem comprising noninteracting identical two-electron subsystems can be used to suggest ways of modifying the CISD energy to minimize the size-extensivity error. There are several corrections that have been obtained in this way, for example, the Davidson–Silver correction[12]

$$\mathrm{DSC} = E_{\mathrm{CISD}} \frac{1 - C_0^2}{2C_0^2 - 1}, \tag{3}$$

and the electron-number-dependent PC[13]

$$\mathrm{PC} = \frac{\sqrt{N^2 + 2N\tan^2 2\theta} - N}{2(\sec 2\theta - 1)} E_{\mathrm{CISD}} - E_{\mathrm{CISD}}, \cos\theta = C_0, \tag{4}$$

where N is the number of electrons. It can immediately be seen that the PC vanishes for $N=2$, which is an exact limit of the method since CISD applied to any two-electron system gives the exact result so should not be corrected. Two further corrections are also dependent on electron number. They are the MC[14]

$$\mathrm{MC} = \mathrm{RDC} \frac{(N-2)(N-3)}{N(N-1)}, \tag{5}$$

and the DDC[15]

$$\mathrm{DDC} = E_{\mathrm{CISD}} \frac{1 - C_0^2}{2\frac{N-1}{N-2}C_0^2 - 1}. \tag{6}$$

They have the same property of vanishing for $N=2$. One observes that while the DC, or more precisely the RDC, can be treated as the leading term in expansions of the difference between correlation energies given by the coupled-electron pair approximation zero (CEPA(0)) and CISD,[11,18,26] the PC can be seen, at least in its simplified version,[13] as an approach that aims to reproduce the CEPA(2) result in which pair energies are averaged.[14] From this point of view, the MC can be treated as a similar treatment of the CEPA(3) method.[14] Of course, a reliable approximation of the CCSD energy is sought using the CISD results instead of having a rough estimate of its simplified versions of CEPA type. That requires, however, more substantial numerical effort than for calculation of the simple Davidson-type correction. As far as the CEPA-type methods are concerned, there are also more advanced CEPA schemes than CEPA(3) that are not only correct for two-electron systems but also invariant with respect to separate rotations of occupied and unoccupied orbitals.[27] One of the most successful MR analogs of CEPA schemes is the Averaged Quadratic CC method.[28]

The CISD and CCSD wave functions contain the same number of parameters. These are coefficients associated with single and double excitation operators with respect to the reference determinant Φ, usually the Hartree–Fock (HF) one. The difference is that these operators appear linearly in the CISD wave function while the CCSD method assumes exponential expansion. We have[22]

$$\Psi_{\text{CISD}} = (1 + C_1 + C_2)\Phi, \tag{7}$$

and

$$\Psi_{\text{CCSD}} = \exp(T_1 + T_2)\Phi. \tag{8}$$

It is seen immediately that the CCSD expansion contains higher than double excitations represented by products of single and double excitations. A product of two double excitations is a key term, giving the lowest order approximation of quadruple excitations. A relation between CI excitation operators (C_1 and C_2) and cluster operators (T_1 and T_2) is given by

$$\begin{aligned} C_1 &= T_1 \\ C_2 &= T_2 + \frac{1}{2}T_1^2 \end{aligned} \tag{9}$$

that holds exactly for the full expansion and has an approximate character for truncated schemes. Equations for the C coefficients are obtained from the variation principle applied to the energy expectation value

$$E = \frac{\langle\Phi|(1 + C_1^+ + C_2^+)H(1 + C_1 + C_2)|\Phi\rangle}{\langle\Phi|(1 + C_1^+ + C_2^+)(1 + C_1 + C_2)|\Phi\rangle}, \tag{10}$$

that can be also written as

$$\begin{aligned} E = {} & \langle\Phi|(1 + C_1^+ + C_2^+)H(1 + C_1 + C_2)|\Phi\rangle \\ & - E\langle\Phi|(C_1^+ + C_2^+)(C_1 + C_2)|\Phi\rangle. \end{aligned} \tag{11}$$

Variation with respect to C^+ and the energy leads to the following equations

$$\langle\Phi^1|H(1 + C_1 + C_2)|\Phi\rangle - E_{\text{CISD}}\langle\Phi^1|C_1|\Phi\rangle = 0, \tag{12}$$

$$\langle\Phi^2|H(1 + C_1 + C_2)|\Phi\rangle - E_{\text{CISD}}\langle\Phi^2|C_2|\Phi\rangle = 0, \tag{13}$$

$$E_{\text{CISD}} = \langle\Phi|H(1 + C_1 + C_2)|\Phi\rangle. \tag{14}$$

Here, Φ^1 and Φ^2 represent singly and doubly excited determinants with respect to the HF reference, respectively. The simple energy expression (14) is obtained from Eq. (11) if Eqs. (12) and (13) are used. Let us now assume that we want to improve the CISD wave function in a CC manner by adding higher than double excitations as products of T_1 and T_2 that can be obtained from Eq. (9). In such a case, we have

$$\Psi_{\mathrm{CISD}+\Delta\mathrm{CC}} = [1 + C_1 + C_2 + Q\exp(T_1 + T_2)]\Phi, \tag{15}$$

where Q is a projection operator on the subspace of higher than doubly excited functions (subspace orthogonal to the CISD variation space). That leads to modification of the right-hand side expressions in Eqs. (12) and (13)

$$\langle\Phi^1|H(1 + C_1 + C_2 + Q\exp(T_1 + T_2))|\Phi\rangle - E\langle\Phi^1|C_1|\Phi\rangle, \tag{16}$$

$$\langle\Phi^2|H(1 + C_1 + C_2 + Q\exp(T_1 + T_2))|\Phi\rangle - E\langle\Phi^2|C_2|\Phi\rangle. \tag{17}$$

Because of the modification, expressions (16) and (17) with CISD coefficients and cluster amplitudes obtained from them are generally no longer zero. Inserting relations (9) into Eqs. (16) and (17), one can easily see that both (16) and (17) are zero, if the T_1 and T_2 operators obtained from the CISD coefficients are replaced with those from the CCSD method. The reason is that for (16) and (17) to be zero we have (together with the energy expression (14)) a set of CCSD equations.

Assuming that we do not want to switch to the CCSD method giving zero for (16) and (17) but are interested in an *a posteriori* correction to the CISD energy instead, we can use the corrected CISD wave function (15) as the right-hand side wave function in the energy expectation value expression (10). As a result, we obtain an analog of Eq. (11) of the form

$$\begin{aligned} E_{\mathrm{corr}} &= \langle\Phi|(1 + C_1^+ + C_2^+)H[1 + C_1 + C_2 + Q\exp(T_1 + T_2)]|\Phi\rangle \\ &= E_{\mathrm{CISD}}\langle\Phi|(C_1^+ + C_2^+)[C_1 + C_2 + Q\exp(T_1 + T_2)]|\Phi\rangle. \end{aligned} \tag{18}$$

Using Eqs. (12)–(14) we obtain

$$\begin{aligned} E_{\mathrm{corr}} &= \langle\Phi|H(1 + C_1 + C_2)|\Phi\rangle + \langle\Phi|(C_1^+ + C_2^+)HQ\exp(T_1 + T_2)]|\Phi\rangle \\ &= E_{\mathrm{CISD}} + \mathrm{CCc}, \end{aligned} \tag{19}$$

where

$$\mathrm{CCc} = \langle\Phi|(C_1^+ + C_2^+)HQ\exp(T_1 + T_2)|\Phi\rangle. \tag{20}$$

One may see that the CCc is not so easy to calculate but should be quite precise in reproducing the CCSD results from the CISD ones.

Of course, the idea of reproducing the CCSD results or results of one of its simplified CEPA-type versions has very limited sense since the CCSD calculation is only slightly more computational demanding than CISD. The situation is completely different when we have to deal with non-dynamic correlation effects. That requires MR approaches. MR-CC methods, because of their formal complexity, their vulnerability to intruder states and convergence problems, are not dependable and do not have the status of a routine method. Conversely, the MR-CISD method preserves the formal and computational simplicity of the SR-CISD approach and is one of few methods that is employed on a regular basis in calculations requiring an MR description. Thus unlike the SR case, the idea of correcting the MR-CISD results looks very attractive. It is obvious that mainly the description of the dynamic part of correlation should be corrected. However, any theoretical analysis of the MR-CISD equations to find out how the MR correction should look like would be very difficult involving comparison with one of the MR-CC approaches.[17,18] Thus, instead of that, straightforward generalizations of the SR corrections were proposed in which the squared CISD coefficient of the HF determinant ($C_0{}^2$) was replaced with the sum of squared MR-CISD coefficients associated with all determinants spanning the reference space and the CISD correlation energy with the difference between the MR-CISD and the so-called reference energy that is obtained as an eigenvalue of the matrix representation of the Hamiltonian H within the reference space. Thus the MR versions of the SR corrections are given by the following replacements[19]

$$C_0^2 \rightarrow \sum_{i \in \mathrm{REF}} C_i^2, \quad E_{\mathrm{CISD}} \rightarrow E_{\mathrm{MR-CISD}} - E_{\mathrm{MR-REF}}. \tag{21}$$

Conversely, the CCc does not require any unjustified manipulations. Due to its conceptual simplicity, it can be directly adapted to the MR-CISD case. Note that the idea behind the CCc is simply to introduce a description of excitations absent from the CISD expansion in a coupled-cluster manner through products of T_1 and T_2. Following this strategy, we can do exactly the same for any CI scheme, in particular, in the MR-CISD case. The MR-CISD wave function can be written in an SR way as

$$\Psi_{\mathrm{MR-CISD}} = \left(1 + C_1 + C_2 + C_3' + C_4' + \cdots + C_k'\right)\Phi, \tag{22}$$

where Φ is one of the reference determinants. In our calculations, Φ is selected to be the HF determinant. The expansion contains higher than doubly excited determinants but only those which are singly or doubly excited with respect to at least one of the reference functions are included. The highest excitation level k, depends on the size of the reference space. Following the same strategy, we modify the equations for singles and doubles in the set of MR-CISD equations. With this, the correction looks exactly the same as before (Eq. 20) with the SR-CISD coefficients and cluster amplitudes obtained from them replaced with their MR counterparts. As before, the Q operator stands for projection out of the CI space. The MR-CISD space is now used. Thus the correction describes the effect of those triples and quadruples that are not included in the MR-CISD wave function.

The CCc in the MR case can be considered as a noniterative version of the CC methods that use an external source to obtain information about higher than double excitations. The reduced MRCC[29] and the tailored CC[30] approaches belong to this category of methods.

In the next section, we show the performance of the CI corrections in both SR and MR cases. The MR case is represented by the second-order CI method.

3. RESULTS AND DISCUSSION

Our calculations are performed for the first excited state of the CH_2 and NH_2^+ isoelectronic molecules. In both cases, the electronic configuration of the excited state is characterized by the closed shell configuration[31,32]

$$(1a_1)^2(2a_1)^2(1b_2)^2(3a_1)^2, \tilde{a}^1A_1 \tag{23}$$

This state may be more accurately presented by the two-configuration wave function as[31,32]

$$C_1(1a_1)^2(2a_1)^2(1b_2)^2(3a_1)^2 + C_2(1a_1)^2(2a_1)^2(1b_2)^2(1b_1)^2, \tilde{a}^1A_1. \tag{24}$$

For this reason, the state is suitable for testing both SR and MR approaches. Calculations for bent $\tilde{a}^1A_1$ closed shell singlet excited states were carried out for the equilibrium intermolecular separation R_e and then up to $2.5R_e$. The C_{2v} symmetry was used. For the CH_2 molecule, the experimental equilibrium bond length and bond angle are $R_e = 1.107$ Å and $\theta_e = 102.4°$[31,32] and they are used in the calculations. For the NH_2^+

molecule, the bond length and bond angle are taken from Ref. 33 as $R_e = 1.0523$ Å and $\theta_e = 107.15°$.

The double-zeta (DZ) basis sets have been chosen in the calculations since for such basis sets the exact Full CI (FCI) energies can be obtained (EMSL basis set library[34,35]) and used as reference points. The SR-CISD, SR-CCSD, SOCI, and FCI results have been calculated using the GAMESS program package. The HF method is used to generate molecular orbitals and the HF determinant is selected as the reference in all calculations. All energies given in tables are presented relative to the FCI energies (in mH). In all correlation calculations of two core orbitals $1a_1$ and $2a_1$ were frozen. In the SOCI calculation, active space is constructed from ($b_2a_1b_1a_1$) active orbitals. The reference determinants in MR calculations are obtained by allowing four active electrons to occupy the set of active orbital levels in all possible ways. The reference space in SOCI calculations is thus denoted by (4,4)R. The SOCI approach generates all possible singly and doubly substituted configurations from every reference space function but only excitations from active levels are permitted. In the SR-CISD approach, the restricted Hartree–Fock (RHF) function is used as the reference. It should be noted that since excitations from the core orbitals are absent from the SOCI approach, dynamic correlation effects are not fully taken into account. The reason is that the SOCI method was designed to describe valence electron effects in relatively large systems so the core correlation contribution was neglected. To be consistent with the SOCI methodology, also in the SR calculations, we keep the core orbitals frozen.

We start with presenting our results for the SR case. In Table 1 and 2, FCI energies for four intermolecular distances, R_e, $1.5R_e$, $2R_e$, $2.5R_e$, are shown in a.u. Then energies relative to the FCI ones are reported in mHartree (mH) for basic SR approaches like CISD and CCSD and are supplemented with the SR-CISD energies modified by *a posteriori* corrections. It can be seen that the SR-CISD method, which is relatively accurate for the equilibrium geometry, quite rapidly becomes less and less precise with increased intermolecular distance. The reason is the growing degree of quasi-degeneracy that calls for MR treatment. Conversely, the SR-CCSD method, up to a certain moment, is able to deal with increasing amount of nondynamic correlation quite efficiently. It can be seen that only for $2.5R_e$ the CCSD method breaks down. As expected, the CCc is quite precise in reproducing the CCSD energies up to $2R_e$ geometry giving for 1 and $1.5R_e$ practically the same result as CCSD. Other corrections can be compared with the FCI energies only. In spite of the fact that they try to evaluate the effect

Table 1 FCI Total Energies (in a.u.) and Energy Differences Relative to FCI (in mH) for the $\tilde{a}^1A_1$ Excited State of CH_2

Method	R_e	$1.5R_e$	$2R_e$	$2.5R_e$
FCI Total Energy -(E+38)	0.910512	0.802116	0.715414	0.686394
SR-CISD	1.874	5.879	25.024	58.972
SR-CCSD	0.900	1.836	1.897	−6.253
SR-CISD+CCc	0.900	1.821	0.917	−70.681
SR-CISD+PC	0.440	1.067	−0.820	483.248
SR-CISD+MC	1.396	4.269	16.144	10.663
SR-CISD+DC	−0.831	−2.785	−14.265	−64.166
SR-CISD+RDC	−0.996	−3.776	−28.253	−230.886
SR-CISD+DSC	−1.182	−5.023	−57.709	877.955
SR-CISD+DDC	0.394	0.758	−7.384	−389.674

Table 2 FCI Total Energies (in a.u.) and Energy Differences Relative to FCI (in mH) for the $\tilde{a}^1A_1$ Excited State of NH_2^+

Method	R_e	$1.5R_e$	$2R_e$	$2.5R_e$
FCI Total Energy -(E+54)	1.168772	1.043054	0.938168	0.890232
SR-CISD	2.333	6.171	19.289	37.633
SR-CCSD	1.124	2.392	4.479	9.609
SR-CISD+CCc	1.127	2.405	4.496	−7.035
SR-CISD+PC	0.559	1.354	−0.767	389.194
SR-CISD+MC	1.741	4.561	12.452	−1.090
SR-CISD+DC	−0.992	−2.535	−12.147	−62.251
SR-CISD+RDC	−1.219	−3.493	−21.736	−194.708
SR-CISD+DSC	−1.479	−4.689	−39.743	750.121
SR-CISD+DDC	0.494	1.058	−4.915	−307.137

of higher excitations represented in the coupled-cluster manner, it is hard to expect their results to be close to the CCSD ones. This is because of either additional approximations made while deriving some of them or the assumption that formulas derived for a specific system consisting from noninteracting identical two-electron subsystems can be applied in any atomic or molecular calculation. It can be seen that some of the corrections tend to underestimate the FCI energy while the other ones overshoot the exact result but none of them, however, is as close to the CCSD energy as that given by the CCc. Let us recall here that for corrections that are derived from the comparison of the FCI energy with the CISD one for an ensemble of noninteracting two-electron system, the limit of the corrected CISD energy should be the CCSD energy since in such a case the FCI result is equivalent to CCSD. The behavior of particular corrections can be analyzed in more detail, some of them perform better than others for given systems but CCc is the only one that seems reliable.

Of course, the SR case can only be of interest as a test case since correcting the CISD method is not competitive with the CCSD approach at our disposal that can be used directly. The situation is different in the MR case. Here, the MR generalization of the SR-CISD scheme is one of few methods that can be used in routine calculations requiring an MR description so the idea of correcting the MR-CISD description of dynamic correlation (for which the coupled-cluster-type expansion is much more effective than the linear one) seems very attractive. The results of the MR calculations are shown in Tables 3 and 4. It can be seen that even for the equilibrium geometry an MR approach is required. The (4,4)*R*-SOCI energy is better than both the SR-CISD and SR-CCSD ones. The quality of the MR-CISD description deteriorates for both systems with increased intermolecular distance. The CCc not only significantly reduces

Table 3 FCI Total Energies (in a.u.) and Energy Differences Relative to FCI (in mH) for the $\tilde{a}^1A_1$ Excited State of CH_2

Method	R_e	$1.5R_e$	$2R_e$	$2.5R_e$
FCI Total Energy -(E + 38)	0.910512	0.802116	0.715414	0.686394
(4,4)*R*-SOCI	0.746	1.810	3.234	3.727
(4,4)*R*-SOCI + CCc	0.332	0.437	0.090	−4.349
(4,4)*R*-SOCI + PC	0.145	−1.700	−18.319	422.969
(4,4)*R*-SOCI + MC	0.545	0.637	−4.155	−25.602
(4,4)*R*-SOCI + DC	−0.422	−4.608	−29.788	−83.917
(4,4)*R*-SOCI + RDC	−0.456	−5.226	−41.104	−172.245
(4,4)*R*-SOCI + DSC	−0.492	−5.976	−64.217	22572.917
(4,4)*R*-SOCI + DDC	0.136	−1.886	−23.518	−173.628

Table 4 FCI Total Energies (in a.u.) and Energy Differences Relative to FCI (in mH) for the $\tilde{a}^1A_1$ Excited State of NH_2^+

Method	R_e	$1.5R_e$	$2R_e$	$2.5R_e$
FCI Total Energy -(E + 54)	1.168772	1.043054	0.938168	0.890232
(4,4)*R*-SOCI	0.718	1.812	4.436	6.694
(4,4)*R*-SOCI + CCc	0.352	0.390	0.146	−11.354
(4,4)*R*-SOCI + PC	0.201	−0.984	−8.178	−33.396
(4,4)*R*-SOCI + MC	0.545	0.879	0.183	−7.650
(4,4)*R*-SOCI + DC	−0.292	−3.400	−16.558	−48.427
(4,4)*R*-SOCI + RDC	−0.316	−3.788	−21.082	−79.370
(4,4)*R*-SOCI + DSC	−0.341	−4.238	−28.090	−189.500
(4,4)*R*-SOCI + DDC	0.195	−1.096	−9.863	−53.129

the MR-CISD error relative to FCI but also improves description of the potential energy curves. In case of CH_2, for intermolecular distance up to and including $2R_e$, the nonparallelity error (NPE) for the (4,4)*R*-SOCI is 2.488 mH and the CCc lowers the error to 0.347 mH. For NH_2^+, the corresponding errors are 3.718 and 0.244 mH, respectively. For the $2.5R_e$ distance the (4,4)*R* reference is not sufficiently large to efficiently cope with the growing component of nondynamic correlation. That means that the part of the wave function that is mimicked by the cluster expansion must also handle the description of some nondynamic effects. That violates our related assumption and the quality of the CCc deteriorates. Other corrections do not seem reliable. While for the equilibrium geometry all of them reduce the MR-CISD gap relative to FCI (with some of them underestimating and some of them overshooting the FCI energies) the situation rapidly deteriorates with increased intermolecular separation. Note that the CCc is definitely superior to other corrections, especially when NPE is concerned.

Our results indicate that the CCc is not only theoretically well motivated but also behaves in a predictable way. The improvement of the CISD results is significant in both the SR and MR cases. Even though the CCc method is more complicated and time consuming, it is more efficient and reliable than the others. If the reference space is properly selected then the CCc method performs well. Therefore, this choice is essential for the quality of the results.

4. CONCLUSION

In this chapter, different size-extensivity corrections for truncated CI approaches are compared in calculation for the first excited state of CH_2 and $NH_2{}^+$. These states are interesting since they can be quite successfully described by both the SR and MR approaches. We employ both types of methods to investigate the performance of different types of *a posteriori* corrections to CISD. The SR CISD method with different size-extensivity corrections is confronted with the CCSD method. Assuming that the goal of the corrections is to include the effect of the terms in the wave function expansion that are responsible for cancelation of reducible contributions in the CISD equations, the corrections should be aimed at reproducing the CCSD results or results of one of the approximate versions of CCSD. Only the CCc seems capable of providing energies close to the CCSD ones in the region where the degree of quasi-degeneracy is not high.

The MR-CISD case is much more interesting from both theoretical and practical point of view. First of all there is no clear indication that the substitution given by Eq. (21) is fully justified and does not introduce additional errors. Generally, the MR versions of the SR corrections are not derived but simply postulated by analogy with the SR case. Contrary to that the CCc is sufficiently general to be applied in both SR and MR cases. The practical importance of the MR corrections rely on the fact that the MR-CISD scheme is one of few methods that are used in routine quantum chemical calculations when an MR approach is required. Thus the significance of correcting the MR-CISD energy to minimize size-extensivity error and increasing the accuracy of the calculation seems evident.

Our calculations show that in the MR case the CCc provides results that significantly reduce the MR-CISD gap but at the same time make the NPE much smaller. That gives better description of potential energy curves which may be important for calculation of some molecular properties. On the other hand, the CCc results are also much more accurate than those of the SR-CCSD method. Hence, we have an approach that combines efficient description of the nondynamic correlation given by MR-CISD with an adequate description of the dynamic correlation based on the SR-CCSD methodology. None of the two methods, neither MR-CISD nor SR-CCSD, can be separately fully successful in the quasi-degenerate situations. Our results also show that the CCc performs much better than other MR-CISD corrections.

Finally, it should be noted that, unlike other corrections, calculation using the CC one requires some numerical effort. In our opinion, however, very sophisticated MR-CISD programs deserve a correction that gives a reliable estimate of extensivity error.

ACKNOWLEDGMENTS

This work has been financially supported by the Scientific and Technological Research Council of Turkey-TÜBİTAK, Grant Number: TUBITAK-BIDEB-2219 International Postdoctoral Research Fellowship Programme (2012-1). Also a partial support by the National Science Center under Grant No. DEC-2013/11/B/ST4/00771 is gratefully acknowledged.

REFERENCES

1. Bartlett, R. J. Coupled-Cluster Theory: An overview of Recent Developments. In: *Modern Electronic Structure Theory, Part II*; Yarkony, D. R. Ed.; World Scientific: Singapore, 1995 pp 1047–1131.
2. Paldus, J.; Li, X. Critical Assessment of Coupled Cluster Method in Quantum Chemistry. *Adv. Chem. Phys.* **1999**, *110*, 1–175.
3. Mukherjee, D.; Moitra, R. K.; Mukhopadhyay, A. Correlation Problem in Open-Shell Atoms and Molecules. *Mol. Phys.* **1975**, *30–6*, 1861–1888.
4. Offerman, R.; Ey, W.; Kummel, H. Degenerate many Fermion Theory in exp*S* Form: (I). General Formalism. *Nucl. Phys. A* **1976**, *273–2*, 349–367.
5. Lindgren, I. A Coupled-Cluster Approach to the Many-Body Perturbation Theory for Open-Shell Systems. *Int. J. Quantum Chem. Symp.* **1978**, *12*, 33–58.
6. Jeziorski, B.; Monkhorst, H. J. Coupled-Cluster Method for Multideterminantal Reference States. *Phys. Rev. A* **1981**, *24–4*, 1668–1681.
7. Meissner, L.; Jankowski, K.; Wasilewski, J. A Coupled-Cluster Method for Quasidegenerate States. *Int. J. Quantum Chem.* **1988**, *34–6*, 535–557.
8. Paldus, J.; Pylypow, L.; Jeziorski, B. Spin-Adapted Multi-Reference Coupled Cluster Formalism Including Non-Linear Terms and its Application to the H4 Model System. *Many-Body Methods in Quantum Chemistry, Lecture Notes in Chemistry*, Vol. 52; Springer: Berlin, 1989; pp 151–170.
9. Davidson, E. R. Configuration Interaction Description of Electron Correlation. In: *The World of Quantum Chemistry*; Daudel, R., Pullman, B., Eds.; Reidel: Dodrecht, 1974; Langhoff, S. R.; Davidson, E. R. Configuration Interaction Calculations on the Nitrogen Molecule. *Int. J. Quantum Chem.* **1974**, *8*, 61–72.
10. Bartlett, R. J.; Shavitt, I. Determination of the Size-Consistency Error in the Single and Double Excitation Configuration Interaction Model. *Int. J. Quantum Chem. Symp.* **1977**, *12-S11*, 165–173.
11. Siegbahn, P. E. M. Multiple Substitution Effects in Configuration Interaction Calculations. *Chem. Phys. Lett.* **1978**, *55–2*, 386–394.
12. Davidson, E. R.; Silver, D. W. Size Consistency in the Dilute Helium Gas Electronic Structure. *Chem. Phys. Lett.* **1977**, *52–3*, 403–406.
13. Pople, J. A.; Seeger, R.; Krishnan, R. Variational Configuration Interaction Methods and Comparison with Perturbation Theory. *Int. J. Quantum Chem. Symp.* **1977**, *11*, 149–163.
14. Meissner, L. Size-Consistency Corrections for Configuration-Interaction Calculations. *Chem. Phys. Lett.* **1988**, *146*(3-4), 204–210.
15. Duch, W.; Diercksen, G. H. Size-Extensivity Corrections in Configuration Interaction Methods. *J. Chem. Phys.* **1994**, *101–4*, 3018–3030.

16. Meissner, L. A Posteriori Corrections to the Configuration Interaction Method with Singles and Doubles. *Chem. Phys. Lett.* **1996**, *263*(3-4), 351–359.
17. Paldus, J.; Wormer, P. E. S.; Visser, F.; van der Avoird, A. Relationship between Configuration Interaction and Coupled Cluster Approaches. In: Nieuwpoort, W. C. Ed.; *Proceedings of the Fifth Seminar on Computational Methods in Quantum Chemistry, University of Groningen, Holland*, Max Planck Institute: München, 1982; pp 13–30.
18. Jankowski, K.; Meissner, L.; Wasilewski, J. Davidson-Type Corrections for Quasidegenerate States. *Int. J. Quantum Chem.* **1985**, *28–6*, 931–942.
19. Bruna, P. J.; Peyerimhoff, S. D.; Buenker, R. J. The Ground State of the CN+ Ion: A Multi-Reference CI Study. *Chem. Phys. Lett.* **1980**, *72–2*, 278–284.
20. Meissner, L.; Grabowski, I. A Coupled-Cluster Correction to the Multi-Reference Configuration Interaction Method. *Chem. Phys. Lett.* **1999**, *300*(1-2), 53–60.
21. Meissner, L.; Nooijen, M. The Coupled-Cluster Correction to Multi-Reference Configuration Interaction Method: Application to Excited States. *Chem. Phys. Lett.* **2000**, *316*(5-6), 501–509.
22. Meissner, L.; Gryniaków, J.; Hubač, I. Breaking N(2) Triple Bond: The Coupled-Cluster Corrected Multi-Reference Configuration Interaction Description. *Chem. Phys. Lett.* **2004**, *397*(1-3), 34–39.
23. Meissner, L.; Gryniaków, J.; Hubač, I. The Coupled-Cluster Corrections to Multi-Reference Configuration Interaction Method: The HF and F(2) Bond Breaking Description. *Mol. Phys.* **2005**, *103*(15-16), 2173–2182.
24. Schaefer, H.F. Configuration interaction wave functions and the properties of atoms and diatomic molecules, Ph. D. Thesis, Stanford University: Stanford, CA, 1969.
25. Schmidt, M. W.; Baldridge, K. K.; Boatz, J. A.; Elbert, S. T.; Gordon, M. S.; Jensen, J. H.; Koseki, S.; Matsunaga, N.; Nguyen, K. A.; Su, S.; Windus, T. L.; Dupuis, M.; Montgomery, J. A. General Atomic and Molecular Electronic Structure System. *J. Comput. Chem.* **1993**, *14–11*, 1347–1363.
26. Paldus, J.; Wormer, P. E. S.; Visser, F.; van der Avoird, A. Relationship Between Configuration Interaction and Coupled Cluster Approaches. *J. Chem. Phys.* **1982**, *76–5*, 2458–2470.
27. Bartlett, R. J.; Musial, M. Addition by Subtraction in Coupled-Cluster Theory: A Reconsideration of the CC and CI Interface and the nCC Hierarchy. *J. Chem. Phys.* **2006**, *125–20*, 204105.
28. Szalay, P. G.; Bartlet, R. J. Multi-Reference Averaged Quadratic Coupled-Cluster Method: A Size-Extensive Modification of Multi-Reference CI. *Chem. Phys. Lett.* **1993**, *214–5*, 481–488.
29. Li, X.; Paldus, J. Reduced Multireference CCSD Method: An Effective Approach to Quasidegenerate States. *J. Chem. Phys.* **1997**, *107–16*, 6257–6269.
30. Kinoshita, T.; Hino, O.; Bartlett, R. J. Coupled-Cluster Method Tailored by Configuration Interaction. *J. Chem. Phys.* **2005**, *123*–7, 074106.
31. Petek, H.; Nesbitt, D. J.; Darwin, D. C.; Ogilby, P. R.; Moore, C.; Ramsay, D. A. Analysis of CH2 $\tilde{a}^1A_1$ (1,0,0) and (0,0,1) Coriolis-Coupled States, $\tilde{a}^1A_1$-$X\sim{}^3B_1$ spin–orbit coupling, and the equilibrium structure of CH2 $\tilde{a}^1A_1$ state. *J. Chem. Phys.* **1989**, *91–11*, 6566–6578.
32. Yamaguchi, Y.; Sherrill, C. D.; Schaefer, H. F., III. The $\widetilde{X}^3B_1$, $\tilde{a}^1A_1$, b˜B11, and c˜A11 Electronic States of CH2. *J. Phys. Chem.* **1996**, *100–19*, 7911–7918.
33. Stephens, J. C.; Yamaguchi, Y.; Sherrill, C. D.; Schaefer, H. F., III. $\widetilde{X}^3B_1$, $\tilde{a}^1A_1$, b˜B11, and c˜$\sum$g + 1 Electronic States of $NH_2{}^+$. *J. Phys. Chem. A* **1998**, *102*(22), 3999–4006.
34. Feller, D. J. The Role of Databases in Support of Computational, Chemistry Calculations. *J. Comput. Chem.* **1996**, *17–13*, 1571–1586.
35. Schuchardt, K. L.; Didier, B. T.; Elsethagen, T.; Sun, L.; Gurumoorthi, V.; Chase, J.; Li, J.; Windus, T. L. Basis Set Exchange: A Community Database for Computational Sciences. *J. Chem. Inf. Model.* **2007**, *47–3*, 1045–1052.

CHAPTER EIGHT

Introducing a Polynomial Expression of Molecular Integrals for Algebraic the Molecular Orbital (MO) Equation

Jun Yasui[*,†,1]
*RIKEN Innovation Center, Hirosawa, Wako, Saitama, Japan
†Frontier Research Center, Canon Inc., Ohta-ku, Tokyo, Japan
[1]Corresponding author: e-mail address: jun.yasui@riken.jp

Contents

Abstract

An algebraic molecular orbital (MO) equation is proposed, based on molecular integrals expressed as polynomials with respect to internal molecular coordinates and orbital exponents of basis functions. The algebraic MO equation deals with molecular coordinates as variables and orbital exponents as nonlinear variational parameters, which are not considered in the Hartree–Fock–Roothaan (HFR) equation. In this chapter, a method to express molecular integrals over Slater-type orbitals in polynomial form, polynomial function is demonstrated for atoms. Applying Taylor-series expansion to analytical molecular integrals, polynomial molecular integrals are obtained with accuracy being controlled by the order of Taylor-series and their coefficients expressed as rational numbers. The algebraic MO simultaneous equations is stem from first-order necessary conditions for local minima of total electronic energy, orthonormality of MOs and the constraints named the Schrödinger condition, such as Kato's cusp condition and the virial theorem. Solution of the algebraic MO equation is expected to include electron correlation because the solution of the Schrödinger equation should obey the Schrödinger condition. Polynomial molecular integrals with controlled accuracy define the algebraic MO equation with enough accuracy to solve multivariable problems in quantum chemistry.

Advances in Quantum Chemistry, Volume 73
ISSN 0065-3276
http://dx.doi.org/10.1016/bs.aiq.2015.05.003

1. INTRODUCTION

Dirac's insight[1] was that construction of quantum chemistry is difficult because mechanism in chemistry depends on many parameters and is complex. Multivariate problems in quantum chemistry were not treated as a theoretical subject for a long time, even after the HFR equation was discovered.[2] An algebraic molecular orbital (MO) equation of many variables is proposed,[3–5] where the total energy functional is defined as a function of number of atoms N, number of electrons n, atomic number Z, geometric molecular structure X, linear combination coefficient of basis function C, and nonlinear orbital exponent of basis function ζ as follows:

$$E(N, n, Z, X, C, \zeta) = T(N, n, X, C, \zeta) + V(N, n, Z, X, C, \zeta).$$

In the basis-set expansion method, an MO equation of many parametric variables is defined formally as a simultaneous equation minimizing the total energy subject to not only ortho-normality of MOs,

$$C(X)^{+} S(X, \zeta) C(X) - 1 = 0,$$

but also several constraints that exact solution of the Schrödinger equation should obey, such as Kato's nuclear cusp condition,[6] the virial theorem, and other necessary conditions $\boldsymbol{\Phi}$, as follows:

$$\frac{\mathrm{d}}{\mathrm{d}r} R(N, n, Z, X, C, \zeta, r \to 0) + ZR(N, n, Z, X, C, \zeta, r \to 0) = 0,$$

$$2T(N, n, X, C, \zeta) + V(N, n, Z, X, C, \zeta) + \sum_{\alpha}^{N} \vec{X}_{\alpha} \cdot \frac{\partial E(N, n, Z, X, C, \zeta)}{\partial \vec{X}_{\alpha}} = 0,$$

$$\Phi^{i}(N, n, Z, X, C, \zeta) = 0.$$

Name the above set of constraints the Schrödinger condition.

In the traditional MO theory,[2,7–9] the Schrödinger condition is used as an internal check to judge the accuracy of any approximation. In the algebraic MO theory, it is used as an attempt to define the MO equation. The MO should obey the Schrödinger condition because the solution of the Schrödinger equation should satisfy it.

In order to define the MO equation of many variables to be solvable, I proposed a method to define it using molecular integrals expressed as polynomials in X and ζ.[4,5,10] A set of polynomial molecular integrals with rational number coefficients have a mathematical structure of a polynomial ring. Thus, we can define algebraic MO equation as simultaneous polynomial equation in affine space in which many variable problems are defined. The role of molecular integrals as polynomial functions is essentially different from that of traditional molecular integral (a numerical element of the Fock matrix). To find a set of solutions of the algebraic MO equation is a new problem in quantum chemistry.

In the 1960s, Buchberger and Hironaka discovered new algorithms for manipulating systems of polynomial equations.[11] The development of computers and these algorithms enabled investigation of complicated multivariate problems that would be impossible to solve by hand. For a long time, quantum chemists except Minimair and Barnett[12] did not use algebra in which many variable problems in quantum chemistry can be discussed.

There is no self-consistent field method (SCF) procedure in solving polynomial equations in the new algorithm using Gröbner bases or standard bases. On the other hand, the SCF iterative procedure, used in the traditional MO calculations, is a technique to solve a nonlinear MO equation by regarding it as a linear equation.[2] The SCF procedure is time consuming and encounters problems for quasidegenerate electronic states.

From the stand-point that nonadiabatic coupling between electronic states and nuclear states is the most fundamental mechanism in chemistry, it is expected that molecular integrals as a function of nuclear coordinates enables nonadiabatic connection between the Schrödinger equation of electron and that of nucleii symbolically. In the following, a method to express molecular integrals over Slater-type orbitals in polynomial is described, and some demonstrations are given for STOs of the helium atom.[13]

2. METHOD

Slater-type orbital centered at A assigned by quantum numbers, $\{n_a, l_a, m_a\}$, is employed in this chapter.

$$\chi\left(\vec{r}, \zeta\right) = R_{\mathrm{n}}(r, \zeta) Y_l^m(\theta, \phi) \tag{1}$$

where R_{n} is normalized radial function and Y_l^m is real normalized spherical harmonic function.

$$R_{\mathrm{n}}(r, \zeta) = N_r(n, \zeta) r^{n-1} e^{-\zeta r}, \tag{2}$$

$$N_r(n,\zeta)=\frac{(2\zeta)^{n+1/2}}{\Gamma(2n+1)^{1/2}}, \tag{3}$$

$$Y_l^m(\theta_A,\phi_A)=p_l^m(\cos\theta_A)f_m(\phi). \tag{4}$$

Overlap integrals, kinetic energy integrals, nuclear attraction energy integrals, and electron repulsion energy integrals in analytical form are given as follows[4,5]:

$$\begin{aligned} S_{AA}^{ab} &= \int \mathrm{d}V\chi_A^a\left(\vec{r}\right)\chi_A^b\left(\vec{r}\right)=\int \mathrm{d}V\rho_{AA}^{ab}\left(\vec{r}\right) \\ &= N_r(n_a,\zeta_a)N_r(n_b,\zeta_b)(n_a+n_b)!(\zeta_a+\zeta_b)^{-(n_a+n_b+1)}\delta_{l_al_b}\delta_{m_am_b} \end{aligned} \tag{5}$$

$$\begin{aligned} T_{AA}^{ab} &= \int \mathrm{d}V\chi_A^a\left(\vec{r}\right)\left[-\frac{1}{2}\Delta\chi_A^b\left(\vec{r}\right)\right] \\ &= N_r(n_a,\zeta_a)N_r(n_b,\zeta_b) \\ &\quad \left\{\frac{1}{2}\zeta_b^2H[0,n_a+n_b,\zeta_a+\zeta_b]-n_b\zeta_bH[0,n_a+n_b-1,\zeta_a+\zeta_b]\right. \\ &\quad \left.+\frac{1}{2}[n_b(n_b+1)-l_b(l_b+1)]H[0,n_a+n_b-2,\zeta_a+\zeta_b]\right\}\delta_{l_al_b}\delta_{m_am_b} \end{aligned} \tag{6}$$

$$\begin{aligned} V_{AA/A}^{ab} &= \int \mathrm{d}V\frac{\chi_A^a\left(\vec{r}_A\right)\chi_A^b\left(\vec{r}_A\right)}{r_A}, \\ &= N_r(n_a,\zeta_a)N_r(n_b,\zeta_b)(n_a+n_b-1)!(\zeta_a+\zeta_b)^{-(n_a+n_b)}\delta_{l_al_b}\delta_{m_am_b} \end{aligned} \tag{7}$$

$$\begin{aligned} V_{AAAA}^{abcd} &= \int \mathrm{d}V_1\chi_A^a\left(\vec{r}_1\right)\chi_A^b\left(\vec{r}_1\right)\int \mathrm{d}V_2\frac{\chi_A^c\left(\vec{r}_2\right)\chi_A^d\left(\vec{r}_2\mathrm{\vec{r}}_2\right)}{r_{12}} \\ &= 4\pi N_r(n_a,\zeta_a)N_r(n_b,\zeta_b)N_r(n_c,\zeta_c)N_r(n_d,\zeta_d) \\ &\quad \sum_l\frac{1}{2l+1}\sum_{m=-l}^{l}\mathrm{G}\begin{pmatrix} m_a & m_b & m \\ l_a & l_b & l\end{pmatrix}\mathrm{G}\begin{pmatrix} m_c & m_d & m \\ l_c & l_d & l\end{pmatrix} \\ &\quad \{-Hh[0,n_a+n_b-(l+1),\zeta_a+\zeta_b,n_a+n_b+l,\zeta_c+\zeta_d] \\ &\quad +H[0,n_a+n_b-(l+1),\zeta_a+\zeta_b]H[0,n_a+n_b+l,\zeta_a+\zeta_b] \\ &\quad +Hh[0,n_a+n_b+l,\zeta_a+\zeta_b,n_a+n_b-(l+1),\zeta_c+\zeta_d]\} \end{aligned} \tag{8}$$

where

$$H[X,n,a|\ a\neq 0]\equiv\int^{X}x^{n}\exp(-\mathrm{ax})\mathrm{d}x, \tag{9}$$

$$H[0,n,a|\ n\geq 0,a\neq 0]=-n!a^{-(n+1)}. \tag{10}$$

$$\begin{aligned} Hh[X,n_1,\alpha_1,n_2,\alpha_2] &= \int^{X}\mathrm{d}r_1{r_1}^{n_1}\exp(-\alpha_1 r_1)\int \mathrm{d}r_2{r_2}^{n_2}\exp(-\alpha_2 r_2) \\ &= \int^{X}\mathrm{d}r_1{r_1}^{n_1}\exp(-\alpha_1 r_1)H[r_1,n_2,\alpha_2] \end{aligned} \tag{11}$$

$$\mathrm{G}\begin{pmatrix} m_a & m_b & m_c \\ l_a & l_b & l_c \end{pmatrix}\equiv\int \mathrm{d}\Omega Y_{l_a}^{m_a}(\theta,\phi)Y_{l_b}^{m_b}(\theta,\phi)Y_{l_c}^{m_c}(\theta,\phi). \tag{12}$$

are Gaunt coefficients. Applying the Taylor expansion method, a molecular integral is obtained as a polynomial. Let x_0 be the center of Taylor expansion, ε displacement from x_0, and $x=x_0+\varepsilon$, the Taylor expansion for single-variable functions is given by

$$f(x)=f(x_0)+\sum_{n=1}^{\infty}\frac{1}{n!}\left[\frac{\mathrm{d}^n}{\mathrm{d}x^n}f(x)\right]_{x=x_0}\epsilon^n. \tag{13}$$

Replacing ε by $x-x_0$, $f(x)$ is expressed in a series of polynomial of x,

$$\begin{aligned} f(x) &=f(x_0)+\sum_{n=1}^{\infty}\frac{1}{n!}\left[\frac{\mathrm{d}^n}{\mathrm{d}x^n}f(x)\right]_{x=x_0}\sum_{k}^{n}\binom{n}{k}x^k(-x_0)^{n-k} \\ &=f(x_0)+\sum_{k}^{n}\left[\frac{1}{n!}(-x_0)^{n-k}\binom{n}{k}\left[\frac{\mathrm{d}^n}{\mathrm{d}x^n}f(x)\right]_{x=x_0}\right]x^k \end{aligned} \tag{14}$$

The function of one variable can be approximated in a finite Taylor-series with fixed order by choosing the expansion center x_0 appropriately as follows:

$$f(x)\simeq f(x_0)+\sum_{i=1}^{N}A_i(x_0)x^i. \tag{15}$$

In the same way, the function of two independent variables can be approximated in a finite Taylor-series with fixed order by choosing expansion center x_0 and y_0 appropriately as follows,

$$f(x, y) \simeq f(x_0, y_0) + \sum_{i+j=1}^{N,N} B_{i,j}(x_0, y_0)x^i y^j. \tag{16}$$

Taylor-series coefficients are converted to rational numbers that lies within a given small difference. For example, 0.37293 85619 27467 73982 with 15 digit precision is converted to 113/303 within 10^{-5} and 35911/96292 within 10^{-10}, respectively. There is no numerical error contamination in operation of rational numbers. In algebraic MO theory, molecular integrals are expressed as polynomial with such coefficients.

3. RESULTS AND DISCUSSION

One center overlap integral between 1s STOs with orbital exponent a and b is given as,

$$S_{1s,1s} = \langle 1s(a)|\, 1s(b)\rangle = \frac{8a^{3/2}b^{3/2}}{(a+b)^3}. \tag{17}$$

Applying Taylor expansion in Eq. (16), $S_{1s,1s}$ is expanded at expansion center a_0 and b_0 as shown in the Appendix. The degree of approximation of $S_{1s,1s}$ expressed in Taylor-series can be controlled by sliding expansion center, $\{a_0, b_0\}$, appropriately. It is possible to define an approximate Hamiltonian using such molecular integrals of controlled precision. It is used not only without introducing its expression but also with minimal molecular integrals with just enough accuracy to solve multivariate problems.

The single-center overlap integral between 2s and1s and that between 2s and 2s are given by

$$S_{2s,1s} = \langle 2s(a)|\, 1s(b)\rangle = \frac{8\sqrt{3}a^{5/2}b^{3/2}}{(a+b)^4}, \tag{18}$$

$$S_{2s,2s} = \langle 2s(a)|\, 2s(b)\rangle = \frac{32a^{5/2}b^{5/2}}{(a+b)^5}, \tag{19}$$

where a and b are orbital exponents of STO. Overlap integral functions, $S_{1s,1s}$ in Eq. (17), $S_{2s,1s}$ in Eq. (18), and $S_{2s,2s}$ in Eq. (19), are shown from left to right in Fig. 1 as a function of two orbital exponents lying between 0.5 and 9.0. Point of $\{a_0, b_0\}$ plotted in green is used as an expansion center for Taylor expansion of overlap integrals.

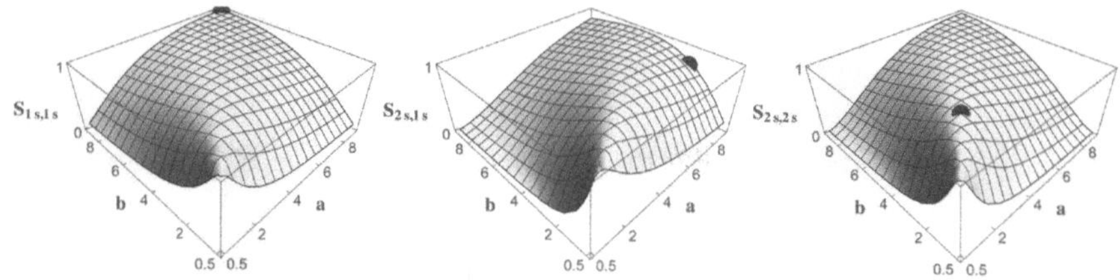

Figure 1 Overlap integral functions, $S_{1s,1s}$ in Eq. (17), $S_{2s,1s}$ in Eq. (18), and $S_{2s,2s}$ in Eq. (19), as a function of two orbital exponents between 0.5 and 9.0. Expansion centers, a_0 and b_0, are plotted in green (light gray in the print version).

Table 1 Accuracy of Fourth-Order Taylor-Series of Overlap Integrals, $S_{1s,1s}$, $S_{2s,1s}$, and $S_{2s,2s}$, where $\triangle x_{10}$ Means a Common Logarithm of the Numerical Difference $\triangle x_{10} = \log_{10}(|\triangle x|)$

δ	$S_{1s,1s}$,	$S_{2s,1s}$,	$S_{2s,1s}$
1	4.0×10^{-6}	3.5×10^{-5}	1.6×10^{-5}
1/2	1.5×10^{-7}	1.6×10^{-6}	2.0×10^{-5}
1/4	5.1×10^{-9}	5.9×10^{-8}	1.5×10^{-7}
1/8	1.7×10^{-10}	2.0×10^{-9}	2.1×10^{-8}
1/16	5.3×10^{-12}	6.6×10^{-11}	9.4×10^{-10}

The accuracy of Taylor-series depends on the domain curvature where molecular integral is defined. To keep the Taylor-series with a fixed order in high accuracy, increase the expansion center on surfaces with high curvature. To keep the Taylor-series with a fixed distribution of expansion center in high accuracy, increase the order of the Taylor-series according to the curvature of molecular integral surface.

To check the accuracy of Taylor-series of overlap integrals, fourth-order Taylor-series for $S_{1s,1s}$, $S_{2s,1s}$, and $S_{2s,2s}$ were calculated at five points displaced by δ from expansion points which are set to 8.7 for 1s and 2.6 for 2s. The results are compared with those calculated using analytical integrals of $S_{1s,1s}$, $S_{2s,1s}$, and $S_{2s,2s}$ in Eqs. (17)–(19), respectively.

Table 1 shows that differences between Taylor-series values and analytical integrals decrease with smaller displacement from the expansion center, except between $S_{2s,2s}$ at $\delta = 1$; $S_{2s,2s}$ at $\delta = 1/2$. Note as follows, a similar precision decreases in Tables 2 and 3 (overlap integrals), Tables 4 and 5 (kinetic energy integrals), and Tables 6 and 7 (nuclear attraction energy integrals): STOs.[13]

In the same way, the accuracy of Taylor-series of overlap integrals, kinetic energy integral, and nuclear attraction energy integral was checked for a set of 1s ($\zeta = 1.4595$), 2s ($\zeta = 1.7504$), and 2s ($\zeta = 2.6298$) given by Bunge.[13] The orbital exponent in parenthesis is used as an expansion center for the Taylor expansion. Radial parts of the set of STOs are shown in Fig. 2.

Accuracy of Taylor-series depends on their order. Table 2 shows the accuracy of fourth-order Taylor-series of overlap integrals between STOs given by Bunge. Table 3 shows that for sixth-order.

Comparison of Tables 2 and 3 shows that the precision of overlap integrals depends not only on the displacement of variables from the expansion center but also the number of terms in series. This means that there are two ways to define approximate molecular integrals to construct the algebraic MO equation. For a fixed expansion center, higher order Taylor-series is

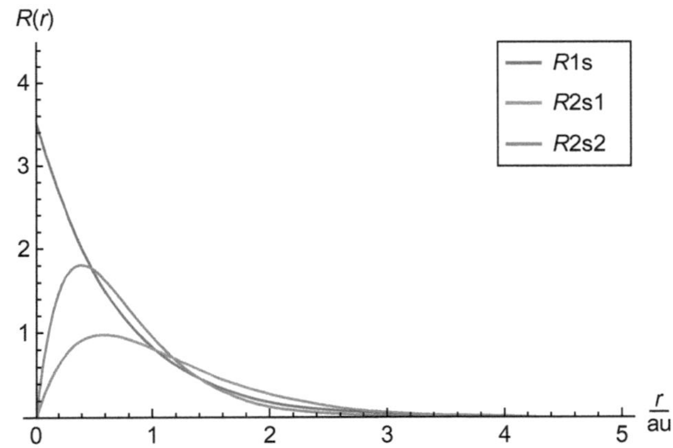

Figure 2 Radial part of STO; 1s ($\zeta = 1.4595$), 2s1 ($\zeta = 1.7504$), and 2s2 ($\zeta = 2.6298$).

Table 2 Accuracy of the Fourth-Order Taylor-Series in the Overlap Integral $\Delta x_{10} = \log_{10}(|\Delta x|)$

δ	(1s, 1s)	(2s1, 1s)	(2s1, 2s1)	(2s2, 1s)	(2s2, 2s1)	(2s2, 2s2)
1	3.4×10^{-3}	1.8×10^{-3}	1.7×10^{-2}	1.2×10^{-3}	5.6×10^{-3}	1.5×10^{-3}
1/2	4.3×10^{-4}	3.8×10^{-4}	2.6×10^{-4}	8.8×10^{-5}	1.9×10^{-5}	1.8×10^{-5}
1/4	2.4×10^{-5}	2.1×10^{-5}	1.9×10^{-6}	6.0×10^{-6}	5.9×10^{-6}	1.5×10^{-7}
1/8	1.0×10^{-6}	8.7×10^{-7}	9.1×10^{-8}	2.5×10^{-7}	1.8×10^{-7}	2.0×10^{-8}
1/16	3.6×10^{-8}	3.1×10^{-8}	5.7×10^{-9}	9.1×10^{-9}	5.5×10^{-9}	8.9×10^{-10}

Table 3 Accuracy of the Sixth-Order Taylor-Series in the Overlap Integrals $\Delta x_{10} = \log_{10}(|\Delta x|)$

δ	(1s, 1s)	(2s1, 1s)	(2s1, 2s1)	(2s2, 1s)	(2s2, 2s1)	(2s2, 2s2)
1	3.8×10^{-3}	2.9×10^{-3}	3.6×10^{-3}	1.7×10^{-3}	6.9×10^{-4}	2.5×10^{-4}
1/2	2.0×10^{-5}	1.7×10^{-5}	3.8×10^{-5}	1.7×10^{-5}	9.5×10^{-6}	2.5×10^{-6}
1/4	9.1×10^{-8}	8.4×10^{-8}	3.6×10^{-7}	1.5×10^{-7}	9.7×10^{-8}	2.2×10^{-8}
1/8	2.0×10^{-10}	2.6×10^{-10}	3.0×10^{-9}	1.2×10^{-9}	8.7×10^{-10}	1.8×10^{-10}
1/16	4.0×10^{-12}	3.5×10^{-14}	1.6×10^{-11}	4.0×10^{-12}	3.5×10^{-12}	1.6×10^{-12}

Table 4 Accuracy of the Fourth-Order Taylor-Series in the Kinetic Energy Integrals $\Delta x_{10} = \log_{10}(|\Delta x|)$

δ	(1s, 1s)	(2s1, 1s)	(2s1, 2s1)	(2s2, 1s)	(2s2, 2s1)	(2s2, 2s2)
1/4	3.8×10^{-5}	2.3×10^{-5}	3.3×10^{-5}	3.0×10^{-5}	8.3×10^{-6}	1.0×10^{-5}
1/8	1.2×10^{-6}	6.9×10^{-7}	1.1×10^{-6}	9.7×10^{-7}	3.0×10^{-7}	3.2×10^{-7}

Table 5 Accuracy of the Sixth-Order Taylor-Series in the Kinetic Energy Integrals $\Delta x_{10} = \log_{10}(|\Delta x|)$

δ	(1s, 1s)	(2s1, 1s)	(2s1, 2s1)	(2s2, 1s)	(2s2, 2s1)	(2s2, 2s2)
1/4	3.3×10^{-7}	2.6×10^{-7}	4.5×10^{-9}	1.5×10^{-7}	1.3×10^{-7}	5.0×10^{-9}
1/8	3.0×10^{-9}	2.3×10^{-9}	4.7×10^{-10}	1.4×10^{-9}	1.0×10^{-9}	8.6×10^{-11}

more accurate, and for a fixed Taylor-series order, lower displacement of variables from the center of expansion is more accurate.[14,16]

Table 4 shows the accuracy of fourth-order Taylor-series of kinetic energy integrals between STOs given by Bunge, and Table 5 shows that of sixth-order Taylor-series. Numerical value as a reference is calculated using analytical molecular integrals in Eq. (20),

$$K_{1s,1s} = \frac{4a^{5/2}b^{5/2}}{(a+b)^3}, K_{2s,1s} = \frac{4a^{5/2}b^{5/2}(2a-b)}{\sqrt{3}(a+b)^4}, K_{2s,2s} = \frac{8a^{5/2}b^{5/2}(a^2 - 4\mathrm{ab} + b^2)}{3(a+b)^5}. \tag{20}$$

Table 6 shows the accuracy of fourth-order Taylor-series of nuclear attraction potential energy integrals between STOs given by Bunge, and Table 7

Table 6 Accuracy of the Fourth-Order Taylor-Series in the Nuclear Attraction Potential Energy Integral

	$\Delta x_{10} = \log_{10}(\lvert\Delta x\rvert)$					
δ	(1s, 1s)	(2s1, 1s)	(2s1, 2s1)	(2s2, 1s)	(2s2, 2s1)	(2s2, 2s2)
1/4	1.6×10^{-6}	7.6×10^{-7}	1.3×10^{-5}	4.2×10^{-6}	8.0×10^{-6}	2.4×10^{-6}
1/8	4.3×10^{-8}	9.1×10^{-8}	3.7×10^{-7}	1.1×10^{-7}	2.6×10^{-7}	7.0×10^{-8}

Table 7 Accuracy of the Sixth-Order Taylor-Series of the Nuclear Attraction Potential Energy Integral

	$\Delta x_{10} = \log_{10}(\lvert\Delta x\rvert)$					
δ	(1s, 1s)	(2s1, 1s)	(2s1, 2s1)	(2s2, 1s)	(2s2, 2s1)	(2s2, 2s2)
1/4	2.7×10^{-7}	1.7×10^{-7}	2.0×10^{-7}	1.7×10^{-7}	5.2×10^{-8}	1.9×10^{-7}
1/8	2.1×10^{-9}	1.3×10^{-9}	1.8×10^{-9}	1.4×10^{-9}	4.7×10^{-10}	1.6×10^{-10}

shows that of sixth-order Taylor-series. Numerical value in high precision as a reference is calculated using analytical molecular integral in Eq. (21),

$$Vne_{1s,1s} = \frac{4a^{3/2}b^{3/2}}{(a+b)^2},\ Vne_{2s,1s} = \frac{8a^{5/2}b^{3/2}}{\sqrt{3}(a+b)^3},\ Vne_{2s,2s} = \frac{8a^{5/2}b^{5/2}}{(a+b)^4}. \tag{21}$$

4. CONCLUSION

In this chapter, a method to express molecular integrals over STO basis function in polynomial form of orbital exponent is presented with some examples for atomic systems. Molecular integrals over STO are approximated by Taylor-series. Accuracy of polynomial molecular integrals is controlled by the order of Taylor-series and their coefficients expressed in rational number with controlled accuracy.[15,17]

A set of polynomial functions of molecular integrals with rational number coefficients has a mathematical structure of a polynomial ring. Thus, we can define algebraic the MO equation as simultaneous polynomial equation in affine space based on the polynomial molecular integral. The algebraic MO equation can be constructed with just enough accuracy to solve many variable problems in quantum chemistry.

ACKNOWLEDGMENTS

I would like to express my heartfelt thanks to the following people: The late Professor A. Saika gave me the theme of molecular integrals over STFs to study the electron correlation effect in NMR chemical shifts by MBPT in my graduate school. He perceived the important nature of STF. For a long time from my graduate school, I have being inspired by Professor Shigeru Huzinaga. His book sparked my interest in molecular orbital theory.

APPENDIX. TAYLOR EXPANSION OF OVERLAP INTEGRAL OF ×16 ABOUT A_0 AND B_0

$$
\begin{aligned}
S_{1s,1s} \approx & -\frac{5775 b_0^{3/2} a_0^{19/2}}{2048(a_0+b_0)^{11}} - \frac{11319 b_0^{5/2} a_0^{17/2}}{256(a_0+b_0)^{11}} - \frac{201993 b_0^{7/2} a_0^{15/2}}{512(a_0+b_0)^{11}} - \frac{1102145 b_0^{9/2} a_0^{13/2}}{256(a_0+b_0)^{11}} + \frac{17710835 b_0^{11/2} a_0^{11/2}}{1024(a_0+b_0)^{11}} \\
& - \frac{1102145 b_0^{13/2} a_0^{9/2}}{256(a_0+b_0)^{11}} - \frac{201993 b_0^{15/2} a_0^{7/2}}{512(a_0+b_0)^{11}} - \frac{11319 b_0^{17/2} a_0^{5/2}}{256(a_0+b_0)^{11}} - \frac{5775 b_0^{19/2} a_0^{3/2}}{2048(a_0+b_0)^{11}} + \\
& a\left(\frac{17325\sqrt{a_0}\, b_0^{19/2}}{512(a_0+b_0)^{11}} + \frac{58905 a_0^{3/2} b_0^{17/2}}{128(a_0+b_0)^{11}} + \frac{107415 a_0^{5/2} b_0^{15/2}}{32(a_0+b_0)^{11}} + \frac{3444705 a_0^{7/2} b_0^{13/2}}{128(a_0+b_0)^{11}} - \frac{16532175 a_0^{9/2} b_0^{11/2}}{256(a_0+b_0)^{11}}\right. \\
& \left. + \frac{1691475 a_0^{11/2} b_0^{9/2}}{128(a_0+b_0)^{11}} + \frac{69135 a_0^{13/2} b_0^{7/2}}{64(a_0+b_0)^{11}} + \frac{14355 a_0^{15/2} b_0^{5/2}}{128(a_0+b_0)^{11}} + \frac{3465 a_0^{17/2} b_0^{3/2}}{512(a_0+b_0)^{11}}\right) + \\
& b\left(\frac{17325\sqrt{b_0}\, a_0^{19/2}}{512(a_0+b_0)^{11}} + \frac{58905 b_0^{3/2} a_0^{17/2}}{128(a_0+b_0)^{11}} + \frac{107415 b_0^{5/2} a_0^{15/2}}{32(a_0+b_0)^{11}} + \frac{3444705 b_0^{7/2} a_0^{13/2}}{128(a_0+b_0)^{11}} - \frac{16532175 b_0^{9/2} a_0^{11/2}}{256(a_0+b_0)^{11}}\right. \\
& \left. + \frac{1691475 b_0^{11/2} a_0^{9/2}}{128(a_0+b_0)^{11}} + \frac{69135 b_0^{13/2} a_0^{7/2}}{64(a_0+b_0)^{11}} + \frac{14355 b_0^{15/2} a_0^{5/2}}{128(a_0+b_0)^{11}} + \frac{3465 b_0^{17/2} a_0^{3/2}}{512(a_0+b_0)^{11}}\right) + \cdots \\
& a^4 b^4 \left(\frac{945 a_0^{11/2}}{2048 b_0^{5/2}(a_0+b_0)^{11}} + \frac{3465 a_0^{9/2}}{256 b_0^{3/2}(a_0+b_0)^{11}} + \frac{218295 a_0^{7/2}}{512\sqrt{b_0}(a_0+b_0)^{11}} - \frac{1528065\sqrt{b_0}\, a_0^{5/2}}{256(a_0+b_0)^{11}} + \frac{12733875 b_0^{3/2} a_0^{3/2}}{1024(a_0+b_0)^{11}}\right. \\
& \left. - \frac{1528065 b_0^{5/2}\sqrt{a_0}}{256(a_0+b_0)^{11}} + \frac{218295 b_0^{7/2}}{512(a_0+b_0)^{11}\sqrt{a_0}} + \frac{3465 b_0^{9/2}}{256(a_0+b_0)^{11} a_0^{3/2}} + \frac{945 b_0^{11/2}}{2048(a_0+b_0)^{11} a_0^{5/2}}\right)
\end{aligned}
$$

REFERENCES

1. Dirac, P. A. M. Quantum Mechanics of Many-Electron Systems. *Proc. R. Soc. Lond. A* **1929**, *123*, 714–733.
2. Roothaan, C. C. J. New Developments in Molecular Orbital Theory. *Rev. Mod. Phys.* **1951**, *23*, 69–89.
3. Yasui, J. Algebraic Molecular Orbital Equation. *Bull. Soc. DV-Xα* **2011**, *24*, 47–54.
4. Yasui, J. *Algebraic Molecular Orbital Theory. The DV-Xα Molecular-Orbital Calculation Method*, Springer: Cham, Heidelberg, New York, Dordrecht, London, 2014; pp 27–47.
5. Yasui, J. *Analytical Expression of Molecular Integrals over Slater-Type Functions for Generating Their Polynomial Expressions. The DV-Xα Molecular-Orbital Calculation Method*, Springer: Cham, Heidelberg, New York, Dordrecht, London, 2014; pp 49–106.
6. Kato, T. On the Eigenfunctions of Many-Particle Systems in Quantum Mechanics. *Commun. Pure Appl. Math.* **1957**, *10*, 151–177.
7. Huzinaga, S. *Molecular Orbital Method[Japanese]*. Iwanami Book Publisher/Cambridge University Press: Tokyo, 1980, 2000.
8. Kohn, W.; Sham, L. L. Self-Consistent Equations Including Exchange and Correlation Effects. *Phys. Rev. A* **1964**, *140*, 1133–1138.
9. Slater, J. C. *Quantum Theory of Molecules and Solids*; McGraw-Hill: New York, 1974.
10. Yasui, J. Polynomial Expressions of Molecular Integral Functionals over Slater-Type-Orbitals and Its Application to the Extension of Hartree-Fock-Roothaan Equation. *Bull. Soc. DV-Xα* **2010**, *23*, 54–59.
11. Cox, D.; Little, J.; O'Shea, D. *Ideals, Varieties, and Algorithms*, 3rd ed.; Springer: New York, 2006.
12. Minimair, M.; Barnett, M. P. Solving Polynomial Equations for Chemical Problems Using Gröbner Bases. *Mol. Phys.* **2004**, *102*, 2521–2535.
13. Bunge, C. F.; Barrientos, J. A. Roothaan-Hartree-Fock Ground-State Atomic Wave Functions: Slater-Type Orbital Expansions and Expectation Values for Z = 2–54. *Atomic Data and Nuclear Data Tables* **1993**, *53*, 113–162.
14. Barnett, M. P.; Coulson, C. A. The Evaluation of Integrals Occurring in the Theory of Molecular Structure. Parts I & II. *Philos. Trans. R. Soc. Lond.* **1951**, *A243*, 221–249.
15. Kotani, M.; Amemiya, A.; Ishiguro, A. E.; Kimura, T. *Table of Molecular Integrals*; Maruzen: Tokyo, 1955.
16. Sharma, R. R. Expansion of a Function About a Displaced Center for Multicenter Integrals: A General and Closed Expression of a Slater Orbital and for Overlap Integrals. *Phys. Rev. A* **1976**, *13*, 517–527.
17. Yasui, J.; Saika, A. Unified Analytical Evaluation of Two-Center, Two-Electron Integrals over Slater-Type Orbitals. *J. Chem. Phys.* **1982**, *76*, 468–472.

CHAPTER NINE

Analytic Calculation of Momentum Distribution and Compton Profiles of Atoms Using Hartree–Fock–Roothaan Method: Applications to Atoms $2 \leq Z \leq 10$

Telhat Ozdogan[1], Melek Eraslan
Department of Physics, Faculty of Arts and Sciences, Amasya University, Amasya, Turkey
[1]Corresponding author: e-mail address: telhat@gmail.com

Contents

Abstract

An analytical formula have been presented for momentum density and Compton profiles of atoms using Hartree–Fock–Roothaan method. The obtained formula includes the linear combination coefficients of molecular orbitals, auxiliary functions $B^{l}_{mn}(\alpha,\beta;q)$ and $K^{l}_{n}(\alpha,q)$, and Gaunt coefficients. Computer programs have been constructed for atomic Compton profiles and including functions. Utilizing these programs, Compton profiles of atoms $2 \leq Z \leq 10$ have been calculated for a wide range of incident photon energy. It is seen that the obtained results for Compton profiles of these atoms are in good agreement with the more recent theoretical and experimental works.

1. INTRODUCTION

Most of the experiments used in the determination of electronic structure of atoms, molecules, solids, gases, and crystals depend on photon–matter interactions and properties about these interactions. Some parameters

Advances in Quantum Chemistry, Volume 73
ISSN 0065-3276
http://dx.doi.org/10.1016/bs.aiq.2015.09.002

of the scattered photons enable us to get important properties about the studied matter.[1,2] The interaction of photons with atoms may be elastic or inelastic. Cross sections, intensity ratio, differential cross sections, double differential cross sections and intensity, etc., can be expressed in terms of elastic atomic scattering factors, inelastic atomic scattering functions, or atomic Compton profiles.[2]

All the works on the theoretical calculation of atomic Compton profiles originate from the earlier works of Duncanson and Coulson[3] that attributed many groups in 1940s. Then, Eisenberger,[4] Holm and Ribberfors,[5] and Biggs et al.[6] have paid special attention to the theoretical calculation of atomic Compton profiles. Nowadays, atomic Compton profiles are studied commonly due to their use especially in the determination of the electronic structure of matters.[7]

Following the work of Duncanson and Coulson,[3] atomic Compton profile is defined as

$$J(q) = \frac{1}{2}\int_q^{\infty} p^{-1} I(p)\,\mathrm{d}p \tag{1}$$

where

$$q = -13,7\frac{E_iE_f + E_iE_f((1-\cos\theta)/m_0c^2)}{\left(E_i^2 + E_f^2 - 2E_iE_f\cos\theta\right)^{1/2}} \tag{2}$$

and $I(p)$ is electron momentum distribution

$$I(p) = \int_{\Omega_p} \varphi*(\vec{p})\varphi(\vec{p})p^2\,\mathrm{d}\Omega_p \tag{3}$$

in which Ω_p solid angle and $\varphi(\vec{p})$ wave function in momentum space.

The aim of this study is to give an analytical formula for the calculation of atomic Compton profiles using HFR method and then apply the presented method to the calculation of Compton profiles of atoms $2 \leq Z \leq 10$ for testing the presented procedure.

2. ANALYTICAL EXPRESSION FOR MOMENTUM DENSITY AND ATOMIC COMPTON PROFILES

The momentum density of Z-electron atom can be expressed as

$$I(p) = 2\sum_{j=1}^{N} \int_{\Omega_p} \varphi_j^*(\vec{p})\varphi_j(\vec{p})p^2 \mathrm{d}p \tag{4}$$

where N is the number of molecular orbitals, $\varphi_j^*(\vec{p})$ is Fourier transformation of j-th molecular orbital wave function, and η_j is occupation number defined by

$$\eta_j = \begin{cases} 1/2; & \text{for semioccupied orbitals} \\ 1; & \text{for occupied orbitals} \end{cases} \tag{5}$$

Any j-th molecular orbital can be expanded through atomic orbitals by LCAO method[8] as in the following:

$$\varphi_j(\vec{p}) = \varphi_{n_j l_j m_j}(\vec{p}) = \sum_{k=1}^{M} c_{jk} U_{n_j l_j m_j}(\alpha_{n_j l_j}, \vec{p}) \tag{6}$$

Here, M is the size of j-th orbital basis set and linear combination coefficients c_{jk}, orbital exponents $a_{n_k l_j}$ and pseudo-quantum numbers n_j are variational parameters obtained from HFR method, and $U_{nlm}(\alpha, \vec{p})$ are Fourier transforms of STOs.

Substituting Eq. (6) into Eq. (4), we get the relation below for momentum density of Z-electron atom

$$I(p) = 2\sum_{j=1}^{N} \eta_j \sum_{k=1}^{M} c_{jk} \sum_{s=1}^{M} c_{js} I_{n_k l_j m_j, n_s l_j m_j}(\alpha_{n_k l_j}, \alpha_{n_s l_j}; p) \tag{7}$$

where $I_{nlm,n'lm}(\alpha, \beta)$ is electron momentum distribution integral defined by

$$I_{nlm,n'lm}(\alpha, \beta; p) = \int_{\Omega_p} U_{nlm}^*(\alpha, \vec{p}) U_{n'lm}(\beta, \vec{p}) p^2 \mathrm{d}\Omega_p \tag{8}$$

In Eqs. (6) and (8), $U_{nlm}(\alpha, \vec{p})$ is Fourier transform of Slater-type orbital $\chi_{nlm}(\alpha, \vec{r})$. Using definition of STO

$$\chi_{nlm}(\alpha, \vec{r}) = \frac{(2\alpha)^{n+(1/2)}}{\sqrt{(2n)!}} r^{n-1} \mathrm{e}^{-\alpha r} S_{lm}(\theta, \varphi) \tag{9}$$

Fourier transform of an STO is given by[9]

$$U_{nlm}(\alpha, \vec{p}) = (-i)^l f_{nl}(\alpha, p) S_{lm}(\theta, \phi) \tag{10}$$

Here $S_{lm}(\theta, \phi)$ is spherical harmonic and the function $f_{nl}(\alpha, p)$ is defined by

$$f_{nl}(\alpha, p) = \frac{1}{(p^2 + \alpha^2)^{n+1}} \sum_{s=0}^{E((n-1)/2)} B^s_{nl}(\alpha) p^{l+2s} \tag{11}$$

in which

$$E\left(\frac{n}{2}\right) = \frac{n}{2} - \frac{1}{4}[1 - (-1)^n] \text{ and}$$

$$A^s_{nl}(\alpha) = (-1)^s \frac{(2\alpha)^{2n-l-2s+(1/2)} F_{2l+2s+1}(n+l+1) F_s(l+s)}{\sqrt{F_l(2l) F_{n-l}(n+l) F_{n-l}(2n)}} \tag{12}$$

Using Eq. (10) in Eq. (8), we have

$$I_{nlm,n'lm}(\alpha, \beta; p) = f_{nl}(\alpha, p) f_{n'l}(\beta, p) p^2 \tag{13}$$

Substituting Eq. (11) in Eq. (13), we obtain the relation below for $I_{nlm,n'lm}(\alpha, \beta; p)$ as

$$I_{nlm,n'lm}(\alpha, \beta; p) = \frac{1}{(p^2 + \alpha^2)^{n+1} (p^2 + \beta^2)^{n'+1}} \sum_{s=0}^{E((n-l)/2)} A^s_{nl}(\alpha) \sum_{s'=0}^{E((n'-l)/2)} A^{s'}_{n'l}(\beta) p^{2l+2s+2s'+2} \tag{14}$$

Now, we can confine ourselves to the calculation of atomic Compton profiles. Using the expression (7) obtained for electron momentum distribution in Eq. (1), we have the formula below for atomic Compton profiles of Z-electron atom

$$I(q) = 2 \sum_{j=1}^{N} \eta_j \sum_{k=1}^{M} c_{jk} \sum_{s=1}^{M} c_{js} J_{n_k l_j m_j, n_s l_j m_j}\left(\alpha_{n_k l_j}, \alpha_{n_s l_j}; p\right) \tag{15}$$

where $J_{nlm,n'lm}(\alpha, \beta; q)$ is atomic Compton profile integral defined by

$$J_{nlm,n'lm}(\alpha, \beta; q) = \frac{1}{2} \int_q^{\infty} p^{-1} I_{nlm,n'lm}(\alpha, \beta; p)\, dp \tag{16}$$

Substituting Eq. (14) into Eq. (16), we derive the formula for atomic Compton profile integral as in the following

$$J_{nlm,n'lm}(\alpha,\beta;q)=\frac{1}{2}\sum_{s=0}^{e((n-l)/2)}B_{nl}^{s}(\alpha)\sum_{s'=0}^{E((n'-l)/2)}B_{n'l}^{s'}(\beta)B_{n+1,n'+1}^{2l+2s+2s'+1}(\alpha,\beta;q) \tag{17}$$

where $B_{mn}^{l}(\alpha,\beta;q)$ is an auxiliary function and investigated in next section.

3. AUXILIARY FUNCTIONS $B_{mn}^{l}(\alpha,\beta;q)$

The auxiliary function $B_{mn}^{l}(\alpha,\beta;q)$ is an integral of the form

$$B_{mn}^{l}(\alpha,\beta;q)=\int_{q}^{\infty}\frac{p^{l}}{(p^2+\alpha^2)^m(p^2+\beta^2)^n}\mathrm{d}p \tag{18}$$

For calculating this integral, we use the change of variables

$$p^2+\alpha^2=x,\quad \gamma=\beta^2-\alpha^2,\quad Q=q^2+\alpha^2 \tag{19}$$

and then get following compact formula

$$B_{mn}^{l}(\alpha,\beta;q)=\frac{1}{2}\sum_{k=0}^{((l-1)/2)}(-1)^kF_k\left(\frac{l-1}{2}\right)\alpha^{2k}b_{m-((l-1)/2)+k,n}(\beta^2-\alpha^2,q^2+\alpha^2). \tag{20}$$

Here, the function $b_{mn}(\gamma,Q)$ is defined as

$$b_{mn}(\gamma,Q)=\int_{Q}^{\infty}\frac{\mathrm{d}x}{x^m(x+\gamma)^n} \tag{21}$$

It is easy to derive the recurrence relation below for the function $b_{mn}(\gamma,Q)$

$$b_{mn}(\gamma,Q)=\frac{1}{\gamma}\left[b_{m,n-1}(\gamma,Q)-b_{m-1,n}(\gamma,Q)\right] \tag{22}$$

For the use of this recurrence formula, we need to be able to calculate the starting values $b_{1n}(\gamma,Q)$ and $b_{m1}(\gamma,Q)$. It is easy to have the following formulae with some basic mathematical tricks

$$\frac{1}{x(x+\gamma)^n}=\frac{1}{\gamma^n}\frac{1}{x}-\sum_{k=0}^{n-1}\frac{F_{k+1}(n)}{\gamma^{k+1}}\frac{x^k}{(x+\gamma)^n} \tag{23}$$

$$\frac{1}{x^m(x+\gamma)}=\frac{(-1)^m}{\gamma^m}\frac{1}{(x+\gamma)}-\sum_{k=0}^{m-1}\frac{(-1)^{m-k}}{\gamma^{m-k}}\frac{1}{x^{k+1}} \tag{24}$$

Substituting Eqs. (23) and (24) into the integral forms of $b_{1n}(\gamma, Q)$ and $b_{m1}(\gamma, Q)$, respectively, we get following analytical formulae

$$b_{1n}(\gamma, Q)=\frac{1}{\gamma^n}\ln\left(\frac{Q+\gamma}{Q}\right)-\frac{1}{\gamma^n}\sum_{s=0}^{n-1}(-1)^s F_s(n-1)\frac{\gamma^s}{(Q+\gamma)^s}$$
$$-\sum_{k=0}^{n-2}\frac{F_{k+1}(n)}{\gamma^{k+1}}\sum_{s=0}^{k}(-1)^s F_s(k)\frac{\gamma^s}{(Q+\gamma)^{n-k-s-1}} \tag{25}$$

$$b_{m1}(\gamma, Q)=\frac{(-1)^m}{\gamma^m}\ln\left(\frac{Q}{Q+\gamma}\right)-\sum_{k=1}^{m-1}\frac{(-1)^{m-k}}{\gamma^{m-k}}\frac{1}{Q^k} \tag{26}$$

In all formulae above, $F_{mn}(n)=n!/m!(n-m)!$ is a well-known binomial coefficient.

4. COMPTON PROFILE OF ATOMS $2 \leq Z \leq 10$

Using the formulae presented in this work, we have calculated Compton profile of atoms $2 \leq Z \leq 10$ and listed in Table 1 comparatively with the works of Mann et al.[6] In the calculations, Hartree–Fock–Roothaan wavefunctions of atoms have been taken from the work of Clementi and Roetti.[10]

5. RESULTS AND DISCUSSIONS

We have presented an analytical procedure for Compton profiles of atoms using Hartree–Fock–Roothaan method. Our formula contain the linear combination coefficients of molecular orbitals, auxiliary functions $B^l_{mn}(\alpha,\beta;q)$ and $K^l_n(\alpha,q)$, and Gaunt coefficients. Based on the derived expressions, we have constructed computer programs in Maple symbolic programming language.

Using the constructed computer programs, we have calculated Compton profiles of atoms $2 \leq Z \leq 10$. Calculated Compton profiles of atoms are listed in Table 1 comparatively with the work of Mann et al.[6] It is seen from Table 1 that there are small differences between our results and literature data. In the high momentum range ($q > 3$ a.u.), all the theoretical Compton

Table 1 Compton Profile of Atoms $2 \leq Z \leq 10$ Comparatively with Literature

	He		Li		Be	
q	**This Work**	**Ref. 6**	**This Work**	**Ref. 6**	**This Work**	**Ref. 6**
0.0	1.070303921E+00	1.07E+00	2.577377857E+00	2.59E+00	3.159105862E+00	3.16E+00
0.2	1.017092044E+00	1.02E+00	1.776939293E+00	1.78E+00	2.518512646E+00	2.52E+00
0.4	8.782623044E−01	8.78E−01	8.835537192E−01	8.84E−01	1.428738419E+00	1.43E+00
0.6	7.004988240E−01	7.00E−01	5.982577134E−01	5.98E−01	7.655144142E−01	7.66E−01
0.8	5.270995209E−01	5.27E−01	5.077588461E−01	5.08E−01	5.026400847E−01	5.03E−01
1.0	3.819851021E−01	3.82E−01	4.393605812E−01	4.39E−01	4.085998231E−01	4.09E−01
1.2	2.711746489E−01	2.71E−01	3.716221691E−01	3.72E−01	3.630277644E−01	3.63E−01
1.4	1.909418545E−01	1.91E−01	3.079294120E−01	3.08E−01	3.269754084E−01	3.27E−01
1.6	1.344706887E−01	1.34E−01	2.514870386E−01	2.51E−01	2.917378752E−01	2.92E−01
1.8	9.521337140E−02	9.52E−02	2.034139718E−01	2.03E−01	2.569606718E−01	2.57E−01
2.0	6.798904699E−02	6.8E−02	1.635300103E−01	1.64E−01	2.238092528E−01	2.24E−01
5.0	1.397237492E−03	1.4E−03	7.536352451E−03	7.53E−03	2.063955125E−02	2.06E−02
10.0	3.294514741E−05	3.3E−05	2.418175087E−04	2.4E−04	9.058232360E−04	9.1E−04
20.0	5.782140159E−07	5.8E−07	4.801521249E−06	4.8E−06	2.058827553E−05	2.1E−05
40.0	9.311966101E−09	9.3E−09	8.015090200E−08	7.9E−08	3.586947569E−07	3.6E−07
60.0	8.221891347E−10	8.6E−10	7.125511293E−09	7.1E−09	3.215974991E−08	3.2E−08
100.0	3.847291976E−11	7.8E−11	3.346062716E−10	3.6E−10	1.516864580E−09	1.5E−09

Continued

Table 1 Compton Profile of Atoms $2 \leq Z \leq 10$ Comparatively with Literature—cont'd

	B		C		N	
q	**This Work**	**Ref. 6**	**This Work**	**Ref. 6**	**This Work**	**Ref. 6**
0.0	2.989788762E+00	2.99E+00	2.877325914E+00	2.89E+00	2.798776238E+00	2.83E+00
0.2	2.690981318E+00	2.69E+00	2.713590736E+00	2.73E+00	2.699254290E+00	2.73E+00
0.4	1.999914912E+00	2.00E+00	2.280279904E+00	2.29E+00	2.416632540E+00	2.44E+00
0.6	1.312954707E+00	1.31E+00	1.737857961E+00	1.74E+00	2.013646005E+00	2.02E+00
0.8	8.431832033E−01	8.43E−01	1.249564431E+00	1.24E+00	1.586068820E+00	1.58E+00
1.0	5.784860874E−01	5.78E−01	8.878806790E−01	8.81E−01	1.209687429E+00	1.20E+00
1.2	4.387453461E−01	4.39E−01	6.474112510E−01	6.41E−01	9.153139818E−01	9.02E−01
1.4	3.617406845E−01	3.62E−01	4.946267428E−01	4.90E−01	7.004637228E−01	6.88E−01
1.6	3.134938164E−01	3.14E−01	3.973628859E−01	3.94E−01	5.488447771E−01	5.38E−01
1.8	2.780839344E−01	2.78E−01	3.330513218E−01	3.31E−01	4.426584098E−01	4.35E−01
2.0	2.486810493E−01	2.49E−01	2.878324421E−01	2.86E−01	3.674197422E−01	3.61E−01
5.0	3.830390779E−02	3.83E−02	5.686305738E−02	5.69E−02	7.416680809E−02	7.42E−02
10.0	2.310848457E−03	2.3E−03	4.643374544E−03	4.6E−03	7.921457924E−03	7.9E−03
20.0	6.091143778E−05	6.1E−05	1.434635751E−04	1.4E−04	2.886331139E−04	2.9E−04
40.0	1.115615229E−06	1.1E−06	2.782043882E−06	2.8E−06	5.964194389E−06	6.0E−06
60.0	1.010462233E−07	1.0E−07	2.550168919E−07	2.5E−07	5.542893693E−07	5.5E−07
100.0	4.791504100E−09	4.83−09	1.216948583E−08	1.2E−08	2.664641101E−08	2.7E−08

	O		F		Ne	
q	**This Work**	**Ref. 6**	**This Work**	**Ref. 6**	**This Work**	**Ref. 6**
0.0	2.776423681E+00	2.78E+00	2.751623947E+00	2.75E+00	2.727863126E+00	2.73E+00
0.2	2.711852535E+00	2.72E+00	2.707351889E+00	2.71E+00	2.696120969E+00	2.70E+00
0.4	2.515343504E+00	2.52E+00	2.567306525E+00	2.57E+00	2.593641590E+00	2.59E+00
0.6	2.205591951E+00	2.21E+00	2.331494027E+00	2.33E+00	2.413365884E+00	2.41E+00
0.8	1.839427538E+00	1.84E+00	2.029543883E+00	2.03E+00	2.168211876E+00	2.17E+00
1.0	1.480179304E+00	1.48E+00	1.707304232E+00	1.71E+00	1.888611415E+00	1.89E+00
1.2	1.168305143E+00	1.16E+00	1.403425576E+00	1.40E+00	1.606875332E+00	1.61E+00
1.4	9.178809102E−01	9.14E−01	1.139576624E+00	1.14E+00	1.346250429E+00	1.35E+00
1.6	7.260527112E−01	7.23E−01	9.225281737E−01	9.22E−01	1.118751103E+00	1.12E+00
1.8	5.827268562E−01	5.80E−01	7.499173780E−01	7.50E−01	9.277082589E−01	9.27E−01
2.0	4.765458227E−01	4.75E−01	6.152436085E−01	6.15E−01	7.712308412E−01	7.71E−01
5.0	9.025275342E−02	9.03E−02	1.064218964E−01	1.06E−01	1.245247453E−01	1.24E−01
10.0	1.203034776E−02	1.2E−02	1.674932913E−02	1.7E−02	2.188166684E−02	2.2E−02
20.0	5.173479515E−04	5.2E−04	8.486628675E−04	8.5E−04	1.297627219E−03	1.3E−03
40.0	1.145769776E−05	1.1E−05	2.020359516E−05	2.0E−05	3.329653267E−05	3.3E−05
60.0	1.081866754E−06	1.1E−06	1.939390007E−06	1.9E−06	3.251771330E−06	3.3E−06
100.0	5.246115324E−08	5.2E−08	9.489213459E−08	9.4E−08	1.606049983E−07	1.6E−07

profiles are effectively the same, which is expected because the contribution from inner electrons is identical in all the calculations. Due to different kinds of exchange and correlation approximations, Compton profiles differ on the low momentum side ($q \leq 2$ a.u.) and the Compton profile contribution of the valence electrons is insignificant in this region.

For testing the stability and the correctness of the procedure presented here, we have also look to the normalization condition

$$\int_0^\infty J(q)\,\mathrm{d}q = Z/2 \tag{27}$$

We have seen that this normalization condition is satisfied perfectly for all the atoms studied here.

As a result, based on the calculation results and comparison with the literature, the procedure and formulation for the calculation of atomic Compton profiles presented in this work can be used in the calculation of Compton profiles of all atoms and can be extended to molecules.

ACKNOWLEDGMENT

This work is supported by Scientific Research Foundation of Amasya University.

REFERENCES

1. (a) Cooper, M. J. Compton Scattering and Electron Momentum Determinations. *Rep. Prog. Phys.* **1985**, *48*, 415–481; (b) Cooper, M. J.; Mijnarends, P. E.; Shiotani, N.; Sakai, N.; Bansil, A. *X-Ray Compton Scattering*; Oxford Science Publications: New York, 2004.
2. Williams, B. G., Ed. *Compton Scattering. The Investigation of Electron Momentum Distributions*; McGraw-Hill: New York, 1977.
3. (a) Duncanson, W. E.; Coulson, C. A. Electron Momenta in Atoms. *Proc. Phys. Soc.* **1948**, *60*, 175–183; (b) Duncanson, W. E.; Coulson, C. A. Theoretical Shape of the Compton Profile for Atoms from H to Ne. *Proc. Phys. Soc.* **1945**, *57*, 190–198.
4. Eisenberger, P. Electron Momentum Density of He and H_2; Compton X-Ray Scattering. *Phys. Rev. A* **1970**, *2*, 1678–1686.
5. Holm, P.; Ribberfors, R. First Correction to the Nonrelativistic Compton Cross Section in the Impulse Approximation. *Phys. Rev. A* **1989**, *40*, 6251–6259.
6. Biggs, F.; Mendelsohn, L. B.; Mann, J. B. Hartree-Fock Compton Profiles for the Elements. *At. Data Nucl. Data Tables* **1975**, *16*, 201–309.
7. Schülke, W. *Electron Dynamics by Inelastic X-Ray Scattering*; Oxford University Press: Oxford, 2007.
8. Roothaan, C. C. J. New Developments in Molecular Orbital Theory. *Rev. Mod. Phys.* **1951**, *23*, 69–89.
9. Geller, M. Two-Center Integrals over Solid Spherical Harmonics. *J. Chem. Phys.* **1963**, *39*, 84–89.
10. Clementi, E.; Roetti, C. Roothaan-Hartree-Fock Atomic Wavefunctions. *At. Data Nucl. Data Tables* **1974**, *14*, 177–223.

CHAPTER TEN

Evaluation of One-Electron Basic Integrals of Irregular Solid Harmonics and Slater-Type Orbitals Using Fourier Transforms

Selda Akdemir*, Niyazi Yükçü†,[1], Emin Öztekin‡
*Department of Science Education, Faculty of Education, Sinop University, Sinop, Turkey
†Department of Energy Systems Engineering, Faculty of Technology, Adıyaman University, Adıyaman, Turkey
‡Department of Physics, Faculty of Science and Arts, Ondokuz Mayıs University, Samsun, Turkey
[1]Corresponding author: e-mail address: nyukcu@gmail.com

Contents

Abstract

In this paper, for one- and two-center one-electron integrals between irregular solid harmonics (ISHs) and Slater-type orbitals (STOs) with same screening parameters distinct analytical formulas are obtained. First, an expression is derived in terms of ISHs and STOs using a Fourier transform method and Taylor expansion. The second expression is written by taking the two-center overlap integrals over STOs with the different screening parameters to a limit. Numerical calculations of these two formulae have been performed for chosen quantum numbers. Finally, the two-center basic nuclear attraction integrals have been calculated as a special case of these integrals between ISHs and STOs. Results obtained have been compared with the literature. This comparison shows a perfect match.

1. INTRODUCTION

It is well known that the Schrödinger equation is soluble analytically only for some few quantum mechanical systems, among which is the

Advances in Quantum Chemistry, Volume 73
ISSN 0065-3276
http://dx.doi.org/10.1016/bs.aiq.2015.06.003

hydrogen atom. Although the hydrogen atom problem is old, it still leads to very interesting studies,[1] the mathematical results of which could finally be used in the treatment of more complicated systems, namely many electron atoms. Indeed, in the latter case, things become much more difficult, since it is impossible to construct the exact solutions of the Schrödinger equation. Approximation must be used to obtain solutions.

In the theoretical studies with help of approximation methods, a suitable basis function must be selected. Suitable basis functions for molecular calculations should satisfy two basic requirements[2]: (a) short expansions of the atomic wave-function in basis functions yield good accuracy and (b) the molecular multicenter integrals which occur inevitably and up to very large quantum numbers in that approach can be computed efficiently. For this purpose, some exponential-type functions are used as basis functions. Among the exponential-type functions, STOs, which were introduced by Slater in his article on shielding constants,[3] have the simplest analytical structure of all exponentially decreasing functions. Other commonly occurring functions of that class, for instance, hydrogen eigenfunctions, can normally be expressed quite easily as linear combinations of STOs. This implies that multicenter integrals of other exponentially decreasing functions can be expressed in terms of the basic multicenter integrals over STOs.

Currently, one promising approach for the evaluation of molecular integrals appears to be the so-called Fourier transform method where multicenter integrals are transformed into inverse Fourier integrals.[4–6] In this approach, it is not the analytical simplicity of basis function that matters but the analytical simplicity of its Fourier transform. In the case STOs, the Fourier transform is not simple but it is very useful for evaluation of multicenter molecular integrals, especially for overlap integrals. Therefore, calculations of this study have been done by using Fourier transform over STOs. The major emphasis of this chapter will be on the STOs, Fourier transform and overlap integrals which are not only needed in *ab initio* calculations but also in semiempirical calculations. Moreover, this study shows that overlap integrals are also an important intermediate step for the derivation of all other multicenter integrals. The evaluation of overlap and other many-center integrals over STOs is already today the bottleneck of any *ab initio* calculation in terms of accurate functions which have to satisfy the requirements of cusp condition and exponential decay. This explains the continued effort of theoretical investigation in this field, from the early study of Roothaan and Ruedenberg,[7–9] Coulson,[10] Löwdin,[11] up to more recent work by Silverstone,[12] Steinborn,[13] Jones,[14] Rinaldi's groups,[15] and other studies.[16–23]

In this study, we have calculated the one- and two-center one-electron basic integrals between irregular solid spherical harmonics (ISHs) and STOs analytically and numerically. Furthermore, the general expression of these integrals is arranged in terms of same screening parameters. Our general solutions for these integrals are given in terms of Gaunt coefficients, Gegenbauer polynomials, and linear combination of STOs. We present first definitions and some properties for regular and ISHs. Then, we sketch the calculation methods adapted to these expressions. In subsequent sections, the details of calculation over STOs are described, and for this purpose some tables of numerical results have been given.

2. PROPERTIES OF SOLID SPHERICAL HARMONICS

The solid spherical harmonics occur as solutions of the Laplace equation: $\nabla^2 V(\mathbf{r}) = 0$, in spherical coordinates r, θ, ϕ. These solutions take the form:

$$V(\mathbf{r}) = \sum_{l,m} \left(a_{lm} S_l^m(\mathbf{r}) + b_{lm} £_l^m(\mathbf{r}) \right) \tag{1}$$

The functions, $S_l^m(\mathbf{r})$ and $£_l^m(\mathbf{r})$, are regular solid spherical harmonics (RSHs):

$$S_l^m(\mathbf{r}) = S_l^m(r, \theta, \phi) = r^l Y_l^m(\theta, \phi) \tag{2}$$

and ISHs:

$$£_l^m(\mathbf{r}) = £_l^m(r, \theta, \phi) = \frac{1}{r^{l+1}} Y_l^m(\theta, \phi) = \frac{S_l^m(\theta, \phi)}{r^{2l+1}} \tag{3}$$

where $Y_l^m(\theta, \phi)$ are known as surface spherical harmonics.[24] They arise from solving the Laplace equation angular factor in spherical coordinates after separation of variables. They can be expressed as:

$$Y_l^m(\theta, \phi) = P_l^{|m|}(\cos\theta)\Phi_m(\phi) \tag{4}$$

where $P_l^{|m|}(\cos\theta)$ denotes a normalized associated Legendre polynomial[25] and for real surface spherical harmonics

$$\Phi_m(\phi) = \frac{1}{\sqrt{\pi\left(1+\delta_{m,0}\right)}}\begin{cases}\cos(m\phi), & m \geq 0\\ \sin(|m|\phi), & m < 0\end{cases} \tag{5}$$

for complex surface spherical harmonics:

$$\Phi_m(\phi) = \frac{1}{\sqrt{2\pi}} e^{im\phi} \tag{6}$$

For solid spherical harmonics, the following recurrence relations may be derived from the well-known relations of the surface spherical harmonics[25]:

$$S_l^m(\mathbf{r}) = \sqrt{\frac{(2l-1)(2l+1)}{(l-m)(l+m)}}\left\{r\cos\theta S_{l-1}^m(\mathbf{r}) - \sqrt{\frac{(l-m-1)(l+m+1)}{(2l-1)(2l-3)}} r^2 S_{l-2}^m(\mathbf{r})\right\} \tag{7}$$

$$£_l^m(\mathbf{r}) = \sqrt{\frac{(2l-1)(2l+1)}{(l-m)(l+m)}}\left\{\frac{\cos\theta}{r^2} £_{l-1}^m(\mathbf{r}) - \sqrt{\frac{(l-m-1)(l+m+1)}{(2l-1)(2l-3)}} \frac{£_{l-2}^m(\mathbf{r})}{r^2}\right\} \tag{8}$$

The solid spherical harmonics symmetry can be written related to ladder operators:

$$\begin{aligned} S_l^m(\mathbf{r}) &= \sqrt{\frac{(l+m)!}{(2l)!(l-m)!}}\left(\hat{L}_-\right)^{l-m} S_l^l(\mathbf{r}) \\ &= \sqrt{\frac{(l-m)!}{(2l)!(l+m)!}}\left(\hat{L}_+\right)^{l+m} S_l^{-l}(\mathbf{r}) \end{aligned} \tag{9}$$

Using the well-known addition theorem for spherical harmonics, the inverse distance between two distinct points, (r, θ_1, ϕ_1) and (a, θ_2, ϕ_2), is given by relations in terms of regular and ISHs:

$$\begin{aligned} \frac{1}{|\mathbf{r}-\mathbf{a}|} &= \sum_{l=0}^{\infty}\sum_{m=-l}^{l} \frac{4\pi}{2l+1} £_l^{m*}(\mathbf{r}) S_l^m(\mathbf{a}), \quad r > a \\ \frac{1}{|\mathbf{r}-\mathbf{a}|} &= \sum_{l=0}^{\infty}\sum_{m=-l}^{l} \frac{4\pi}{2l+1} S_l^{m*}(\mathbf{r}) £_l^m(\mathbf{a}), \quad r < a \end{aligned} \tag{10}$$

Solid spherical harmonics appear in many fields of physics and chemistry. Atomic orbitals located at distinct centers are products of radial functions and

RSHs. On the other hand, important quantum mechanical operators are given as a special case of ISHs, such as electric and magnetic potentials.[26]

Quantum mechanical calculations of molecular properties require the atomic orbitals which are centered on nuclei of the different atoms. Therefore, large numbers of molecular integrals which contain products of such atomic orbitals occur. Molecular integrals are particularly simple if all atomic orbitals are centered at the same nucleus. In order to facilitate the solution of multicenter integrals, atomic orbitals on different centers translated to a common center.[24,27–30]

3. ONE-ELECTRON INTEGRALS BETWEEN ISHs AND STOs WITH EQUAL SCREENING PARAMETERS

We will use the normalized STOs defined as below:

$$\chi_{n\,l}^{m}(\alpha,\mathbf{r}) = \frac{(2\alpha)^{n+1/2}}{\sqrt{(2n)!}} r^{n-1} e^{-\alpha r} Y_l^m(\theta,\phi) \tag{11}$$

where n, l, and m are quantum numbers and α is the screening parameter.

One-electron one-center integrals with ISHs and STOs are of the form:

$$\begin{aligned} Z_{l_1\,m_1}^{n_2\,l_2\,m_2}(\alpha,\alpha;0) &= \int \left(\pounds_{l_1}^{m_1}(\alpha\mathbf{r})\right)^{*} \chi_{n_2\,l_2}^{m_2}(\alpha,\mathbf{r})d\mathbf{r} \\ &= 2^{l_1+1}\sqrt{\frac{(2(n_2-l_1-1))!}{(2n_2)!}} \int Y_{l_1}^{m_1\,*}(\theta,\phi)\chi_{n_2-l_1-1\;l_2}^{m_2}(\alpha,\mathbf{r})d\mathbf{r} \end{aligned} \tag{12}$$

After the one-center integrals in Eq. (12) are reduced to a radial integral which consists of the definite integral form of the gamma function,[25] the following expression is readily obtained by the orthogonality relation for spherical harmonics:

$$Z_{l_1\,m_1}^{n_2\,l_2\,m_2}(\alpha,\alpha;0) = \delta_{l_1 l_2}\delta_{m_1 m_2}\frac{2^{n_2+1/2}(n_2-l_1)!}{\alpha^{3/2}\sqrt{(2n_2)!}} \tag{13}$$

One-electron two-center integrals between ISHs and STOs have the following compact form:

$$Z_{l_1\,m_1}^{n_2\,l_2\,m_2}(\alpha,\alpha;\mathbf{R}) = \int \left(\pounds_{l_1}^{m_1}(\alpha\mathbf{r})\right)^{*} \chi_{n_2\,l_2}^{m_2}(\alpha,\mathbf{r}-\mathbf{R})d\mathbf{r} \tag{14}$$

When we apply the Fourier transform method[31] to Eq. (14), the explicit integral representation in momentum space can be given as below:

$$Z_{l_1 m_1}^{n_2 l_2 m_2}(\alpha, \alpha; \mathbf{R}) = \int e^{-i\mathbf{p}.\mathbf{R}} \left(£_{l_1}^{m_1}(\alpha \mathbf{p}) \right)^* U_{n_2 l_2}^{m_2}(\alpha, \mathbf{p}) d\mathbf{p} \tag{15}$$

Here, $e^{i\mathbf{p}.\mathbf{R}}$ is the Rayleigh expansion of a plane wave and given by the well-known relation in terms of spherical Bessel functions $j_l(pR)$ and spherical harmonics

$$e^{\pm i\mathbf{p}.\mathbf{R}} = 4\pi \sum_{l=0}^{\infty} \sum_{m=-l}^{l} (\pm i)^l j_l(pR) \left(Y_l^m \left(\theta_p, \phi_p \right) \right)^* Y_l^m(\theta_R, \phi_R) \tag{16}$$

The Fourier transform of ISHs[6,32] and STOs are obtained from RSHs by:

$$£_l^m(\mathbf{p}) = \frac{\sqrt{2/\pi}}{(2l-1)!!} \frac{S_l^m(-i\mathbf{p})}{p^2} \tag{17}$$

$$U_{nl}^m(\alpha, \mathbf{p}) = \frac{2^{n+l+1} \alpha^{n+1/2} l!(n-l)!}{\sqrt{\pi (2n)! (\alpha^2 + p^2)^{n+l+2}}} C_{n-l}^{l+1} \left(\frac{\alpha}{\sqrt{\alpha^2 + p^2}} \right) S_l^m(-i\mathbf{p}) \tag{18}$$

where $C_n^l(x)$ is a Gegenbauer polynomial.[33]

Substituting Eq. (17) and Eq. (18) into Eq. (15), and using the properties of the product of two real spherical harmonics, we obtain:

$$Z_{l_1 m_1}^{n_2 l_2 m_2}(\alpha, \alpha; \mathbf{R}) = \frac{(-1)^{l_2} i^{l_1+l_2} 2^{2n_2+3/2} \alpha^{2n_2-l_1-l_2-1/2}}{\pi (2l_1-1)!! F_{l_2}(n_2) \sqrt{F_{n_2}(2n_2)}} \sum_{r=0}^{\left[\frac{n_2-l_2}{2}\right]} (-1)^r \frac{a_r(l_2+1, n_2-l_2)}{(2\alpha)^{2r}}$$
$$\sum_{l=|l_1-l_2|}^{l_1+l_2} {}^{2}\sum_{m=-l}^{l} \langle l_1 m_1 | l_2 m_2 | l m \rangle A_{m_1 m_2}^m \int e^{-i\mathbf{p}.\mathbf{R}} \frac{p^{l_1+l_2-l} S_l^m(\mathbf{p})}{p^2(\alpha^2+p^2)^{n_2-r+1}} d\mathbf{p} \tag{19}$$

Here, we use the relation for the product of two real spherical harmonics:

$$Y_{l_1}^{m_1}(\theta, \phi) Y_{l_2}^{m_2}(\theta, \phi) = \sum_{l=0}^{\infty} \sum_{m=-l}^{l} \langle l_1 m_1 | l_2 m_2 | l m \rangle A_{m_1 m_2}^m Y_l^m(\theta, \phi) \tag{20}$$

where $\langle l_1 m_1 | l_2 m_2 | l m \rangle$, so-called Gaunt coefficient, and $A_{m_1 m_2}^m$ are coefficients which are obtained with the integration of the product of three real spherical surface harmonics.[34,35]

$$\langle l_1 m_1 | l_2 m_2 | lm \rangle = \int_0^{\pi} P_{l_1}^{m_1}(\cos\theta) P_{l_2}^{m_2}(\cos\theta) P_l^m(\cos\theta) \sin\theta \, \mathrm{d}\theta \tag{21}$$

$$A_{m_1 m_2}^{m} = \frac{1}{\pi^{3/2}\sqrt{\left(1+\delta_{m,0}\right)\left(1+\delta_{m_1,0}\right)\left(1+\delta_{m_2,0}\right)}} \int_0^{2\pi} \begin{Bmatrix} \cos m\phi \\ \sin|m|\phi \end{Bmatrix} \begin{Bmatrix} \cos m_1\phi \\ \sin|m_1|\phi \end{Bmatrix} \begin{Bmatrix} \cos m_2\phi \\ \sin|m_2|\phi \end{Bmatrix} \mathrm{d}\phi \tag{22}$$

Using the Taylor expansion given by Eq. (4.2) of Ref. 6 and the Rayleigh expansion in Eq. (16), we find Eq. (19) as follows:

$$\begin{aligned} Z_{l_1 m_1}^{n_2 l_2 m_2}(\alpha, \alpha; \mathbf{R}) = & \frac{(-1)^{l_2} 2^{2n_2+7/2}}{(2l_1-1)!! F_{l_2}(n_2)\sqrt{F_{n_2}(2n_2)}} \sum_{r=0}^{\left[\frac{n_2-l_2}{2}\right]} (-1)^r \frac{a_r(l_2+1, n_2-l_2)}{2^{2r}} \\ & \sum_{l=|l_1-l_2|}^{l_1+l_2} (2) \sum_{m=-l}^{l} \langle l_1 m_1 | l_2 m_2 | lm \rangle A_{m_1 m_2}^{m} Y_l^m(\theta,\phi) \sum_{t=0}^{(l_1+l_2-l)/2} (-1)^t \frac{F_t((l_1+l_2-l)/2)}{\alpha^{l+5/2}} \\ & \left\{ \int_0^{\infty} p^l j_l(pR) \mathrm{d}p - \sum_{\nu=0}^{n_2-r-t} \alpha^{2\nu} \int_0^{\infty} \frac{p^{l+2} j_l(pR)}{(\alpha^2+p^2)^{\nu+1}} \mathrm{d}p \right\} \end{aligned} \tag{23}$$

The first integral in Eq. (23) can be obtained in terms of ISHs[32]

$$Y_l^m(\theta,\phi) \int_0^{\infty} j_l(pR) p^l \mathrm{d}p = \frac{\pi}{2}(2l-1)!! \pounds_l^m(\mathbf{R}) \tag{24}$$

The second integral is proved in terms of the modified Bessel function of second kind with the help of the integral tables of Ref. 33

$$\int_0^{\infty} \frac{p^{l+2} j_l(pR)}{(\alpha^2+p^2)^{\nu+1}} \mathrm{d}p = \sqrt{\pi} \frac{R^{\nu-1/2} \alpha^{l+1/2-\nu}}{2^{\nu+1/2} \nu!} K_{l-\nu+1/2}(\alpha R) \tag{25}$$

This integral can be written easily in terms of STOs when the series expansion of the modified Bessel function of second kind is used:

$$Y_l^m(\theta,\phi) \int_0^{\infty} \frac{p^{l+2} j_l(pR)}{(\alpha^2+p^2)^{\nu+1}} \mathrm{d}p = \pi \frac{\alpha^{l-2\nu-1/2}}{2^{2\nu+3/2}} \sum_{q=0}^{l-\nu} g_{\nu,q}^{l} \chi_{\nu-ql}^{m}(\alpha, \mathbf{R}) \tag{26}$$

where

$$g^{l}_{\nu,q} = \frac{(l-\nu+q)!\sqrt{(2(\nu-q))!}}{(l-\nu-q)!\nu!q!} \tag{27}$$

Finally, the final expression of the two-center integrals between ISHs and STOs with same screening parameters takes the following form:

$$Z^{n_2 l_2 m_2}_{l_1 m_1}(\alpha,\alpha;\mathbf{R}) = \frac{\pi(-1)^{l_2}2^{2n_2+5/2}\alpha^{-3/2}}{(2l_1-1)!!F_{l_2}(n_2)\sqrt{F_{n_2}(2n_2)}} \sum_{r=0}^{\left[\frac{n_2-l_2}{2}\right]} \frac{(-1)^r a_r(l_2+1,n_2-l_2)}{2^{2r}}$$
$$\sum_{l=|l_1-l_2|}^{l_1+l_2}(2)\sum_{m=-l}^{l}\langle l_1 m_1|l_2 m_2|lm\rangle A^{m}_{m_1 m_2} \sum_{t=0}^{(l_1+l_2-l)/2}(-1)^t F_t((l_1+l_2-l)/2)$$
$$\left\{(2l-1)!!\pounds^{m}_{l}(\alpha\mathbf{R}) - \sum_{\nu=0}^{n_2-r-t}\frac{\alpha^{-3/2}}{2^{2\nu+1/2}}\sum_{q=0}^{l-\nu} g^{l}_{\nu,q}\chi^{m}_{\nu-ql}(\alpha,\mathbf{R})\right\} \tag{28}$$

An alternative expression can be written for the one-center integrals ($\mathbf{R}=0$) in Eq. (12) and the two-center integrals in Eq. (14) in terms of a limit state of overlap integrals over STOs with different screening parameters:

$$Z^{n_2 l_2 m_2}_{l_1 m_1}(\alpha,\alpha;\mathbf{R}) = \frac{1}{\alpha^{l_1+1}}\lim_{\beta\to 0}\left\{\frac{\sqrt{(2n_1)!}}{(2\beta)^{n_1+1/2}} S^{n_2 l_2 m_2}_{n_1 l_1 m_1}(\beta,\alpha;\mathbf{R})\bigg|_{n_1=-l_1}\right\} \tag{29}$$

4. NUMERICAL RESULTS AND DISCUSSION

We have presented an analytical and numerical calculation for one-electron basic integrals between ISHs and STOs with the same screening parameters. First, for the case $\mathbf{R}=0$, the formula in Eq. (13) for the one-center basic integrals was obtained easily from the general formula Eq. (11) by using the orthogonality relation of the spherical harmonics. For the two-center integrals, we have applied the Fourier transform method to Eq. (14). Using some Taylor expansions in Ref. 6, an analytical expression has been obtained as given by Eq. (28) in terms of Gaunt coefficients, Gegenbauer polynomials, ISHs, and STOs. Another expression was then written as a limit case of the two-center overlap integrals over STOs with different screening parameters as given in Eq. (29). For these two different formulas, a computer program has been written in Mathematica 7.0 and run for physically significant values of atomic orbital parameters. The comparative results shown an exact match of the calculations are presented in Table 1. For the two-center overlap integrals with different screening parameters over STOs, we have used the formula in Eq. (25) of Ref. 36.

A special case ($l_1 = 0$, $m_1 = 0$) of the integrals between ISHs and STOs in Eq. (14) gives the two-center basic nuclear attraction integrals. Two-center basic nuclear attraction integral is of the form[32,37]:

$$A_{nl}^{m}(\alpha;\mathbf{R}) = \int \frac{\chi_{nl}^{m}(\alpha,\mathbf{r})}{|\mathbf{r}-\mathbf{R}|} d\mathbf{r} \tag{30}$$

In the literature, basic nuclear attraction integrals can be found with different symbol and a factor $1/\sqrt{4\pi}$.[38] In order to check the accuracy of our formulae, we calculated basic nuclear attraction integrals using the values of Eqs. (28) and (29) for $l_1 = 0$ and $m_1 = 0$. The numerical results given in Table 2 were compared with the literature. As seen from the tables, computed result accuracy is satisfactory.

Table 1 The Results of One- and Two-Center Integrals Between ISHs and STOs with Same Screening Parameters

l_1	m_1	n_2	l_2	m_2	α	R	θ	ϕ	Eq. (28) or Eq. (29)
1	0	2	1	0	2	0	0	0	$4.0824829046386301636 \times 10^{-1}$
1	1	2	1	1	3	5	0	0	$1.5799808616459912021 8\times 10^{-3}$
2	1	3	2	1	7	1.2	0	0	$-1.1010788826951090227 2\times 10^{-3}$
2	2	4	3	1	5	5	30	120	$4.2058853623026249730 9\times 10^{-6}$
3	−2	4	3	3	5	10	45	45	$1.2810682006609732618 8\times 10^{-9}$
4	1	5	3	1	0.5	25	120	30	$2.0454869408847230465 4\times 10^{-4}$
5	4	7	4	−2	3	20	60	30	$-4.8999758579187636376 8\times 10^{-13}$
7	3	15	7	1	4	10	45	210	$-1.0695603061319080506 2\times 10^{-12}$
9	8	14	13	7	3	5	300	18	$-1.3238984611010868134 2\times 10^{-11}$
10	10	15	10	10	4	10	45	210	$-1.5731483597280906161 0\times 10^{-18}$

Table 2 Comparison Results for Basic Nuclear Attraction Integrals in Eq. (30) as a Special Case of the Integrals Between ISHs and STOs

n	l	m	α	R	θ	ϕ	$A_{nl}^{m}(\alpha; R)$	Ref. 39
3	2	1	2.5	1.6	120	60	$1.21209793056389465092 \times 10^{-1}$	0.12120979305638946
4	2	0	0.5	0.6	150	60	$7.99934486690498648166 \times 10^{-3}$	0.007999344866904986
4	3	2	5.5	3.6	120	60	$1.72848868838081581979 \times 10^{-3}$	0.001728488688380815
5	4	3	5.5	3.6	120	60	$-2.72061600706482893649 \times 10^{-3}$	−0.002720616007064828
6	5	3	5.5	3.3	128	50	$3.61864429617501393268 \times 10^{-3}$	0.003618644296175013
10	9	8	3.5	2.3	115	75	$1.77592992099052455725 \times 10^{-2}$	0.01775929920990524 5
20	15	10	5.5	4.3	15	75	$7.49353677905022774528 \times 10^{-6}$	$7.4935367790502277 \times 10^{-6}$
30	20	10	5.5	4.4	150	60	$-4.63972547468125632240 \times 10^{-3}$	−0.004639725474681256 3

REFERENCES

1. (a) Swainson, R. A.; Drake, G. W. F. A Unified Treatment of the Nonrelativistic and Relativistic Hydrogen Atom: I. The Wavefunctions. *J. Phys. A* **1991**, *24*, 79–94; (b) Swainson, R. A.; Drake, G. W. F. A Unified Treatment of the Non-relativistic and Relativistic Hydrogen Atom: II. The Green Functions. *J. Phys. A* **1991**, *24*, 95–120; (c) Swainson, R. A.; Drake, G. W. F. A Unified Treatment of the Non-Relativistic and Relativistic Hydrogen Atom: III. The Reduced Green Functions. *J. Phys. A* **1991**, *24*, 1801–1824.
2. Weniger, E. J.; Steinborn, E. O. The Fourier Transforms of Some Exponential-Type Functions and Their Relevance to Multicenter Problems. *J. Chem. Phys.* **1983**, *78*, 6121–6132.
3. Slater, J. C. Atomic Shielding Constants. *Phys. Rev.* **1930**, *36*, 57–64.
4. Trivedi, H. P.; Steinborn, E. O. Fourier Transform of a Two-Center Product of Exponential-Type Orbitals. Application to One- and Two-Electron Multicenter Integrals. *Phys. Rev. A* **1983**, *27*, 670–679.
5. Grotendorst, J.; Steinborn, E. O. The Fourier Transform of a Two-Center Product of Exponential-Type Functions and its Efficient Evaluation. *J. Comput. Phys.* **1985**, *61*, 195–217.
6. Weniger, E. J.; Grotendorst, J.; Steinborn, E. O. Unified Analytical Treatment of Overlap, Two-Center Nuclear Attraction, and Coulomb Integrals of *B* Functions via the Fourier Transform Method. *Phys. Rev. A* **1986**, *33*, 3688–3705.
7. Roothaan, C. C. J. A Study of Two-Center Integrals Useful in Calculations on Molecular Structure. I. *J. Chem. Phys.* **1951**, *19*, 1445–1458.
8. Ruedenberg, K. A Study of Two-Center Integrals Useful in Calculations on Molecular Structure. II. The Two-Center Exchange Integrals. *J. Chem. Phys.* **1951**, *19*, 1459–1477.
9. Wahl, A. C.; Cade, P. E.; Roothaan, C. C. J. A Study of Two Center Integrals Useful in Calculations on Molecular Structure. *J. Chem. Phys.* **1964**, *41*, 2578–2599.
10. Barnett, M. P.; Coulson, C. A. Evaluation of Integrals Occurring in the Theory of Molecular Structure, Part I: Basic Functions. *Philos. Trans. R. Soc. Lond. A* **1951**, *243*, 221–233.
11. Löwdin, P. O. Quantum Theory of Cohesive Properties of Solids. *Adv. Phys.* **1956**, *5*, 1–171.
12. (a) Silverstone, H. J. Expansion about an Arbitrary Point of Three-Dimensional Functions Involving Spherical Harmonics by the Fourier-Transform Convolution Theorem. *J. Chem. Phys.* **1967**, *47*, 537–540; (b) Silverstone, H. J. On the Evaluation of Two-Center Overlap and Coulomb Integrals with Non-Integer-n Slater Type Orbitals. *J. Chem. Phys.* **1966**, *45*, 4337–4341.
13. (a) Steinborn, E. O.; Weniger, E. J. Advances of Reduced Bessel Functions as Atomic Orbitals: An Application to H_2^+. *Int. J. Quantum Chem. Symp.* **1977**, *11*, 509–516; (b) Filter, E.; Steinborn, E. O. Extremely Compact Formulas for Molecular 2-Center One-Electron Integrals and Coulomb Integrals Over Slater-Type Atomic Orbitals. *Phys. Rev. A* **1978**, *18*, 1–11.
14. (a) Jones, H. W.; Weatherford, C. A. A Modified form of Sharma's Formula for STO Löwdin Alpha Functions with Recurrence Relations for the Coefficient Matrix. *Int. J. Quantum Chem. Symp.* **1978**, *14*, 483–488; (b) Jones, H. W. Computer-Generated Formulas for Overlap Integrals of Slater-Type Orbitals. *Int. J. Quantum Chem.* **1980**, *18*, 709–713; (c) Jones, H. W. Computer-Generated Formulas for 3-Center Nuclear-Attraction Integrals (Electrostatic Potential) for Slater-Type Orbitals. *Phys. Rev. A* **1984**, *30*, 1–4; (d) Jones, H. W. Analytic Lowdin Alpha-Function Method for 2-Center Electron-Repulsion Integrals over Slater-Type Orbitals. *J. Comput. Chem.* **1991**, *12*, 1217–1222.

15. Bouferguene, A.; Fares, M.; Rinaldi, D. Integrals Over B-Functions Basis-Sets. 1. 3-Center Molecular Integrals, a Numerical Study. *J. Chem. Phys.* **1994**, *100*, 8156–8168.
16. Magnasco, V.; Casanova, M.; Rapallo, A. On the Evaluation of 2-Center Molecular Integrals over an STO Basis. *Chem. Phys. Lett.* **1998**, *289*, 81–89.
17. Magnusson, E. A.; Zauli, C. Evaluation of Molecular Integrals by a Numerical Method. *Proc. Phys. Soc. Lond.* **1961**, *78*, 53–64.
18. Magnasco, V.; Dellepiane, G. *Ric. Sci. Rome IIA* **1963**, *33*, 1173.
19. Musso, G. F.; Magnasco, V. Evaluation of Four-Centre Molecular Integrals. *J. Phys. B* **1971**, *4*, 1415–1421.
20. Harris, F. E.; Michels, H. H. Multicenter Integrals in Quantum Mechanics. I. Expansion of Slater-Type Orbitals about a New Origin. *J. Chem. Phys.* **1965**, *43*, 165–169.
21. Sharma, R. R. Expansion of a Function about a Displaced Center for Multicenter Integrals: A General and Closed Expression for the Coefficients in the Expansion of a Slater Orbital and for Overlap Integrals. *Phys. Rev. A* **1976**, *13*, 517–527.
22. Rico, J. F.; Lopez, R. Recurrence Relations for the Expansion of Slater-Type Orbitals about Displaced Centers. *J. Chem. Phys.* **1986**, *85*, 5890–5894.
23. Figari, G.; Costa, C.; Pratolongo, R.; Magnasco, V. Two-centre Coulomb Integrals over STOs from Analytical Evaluation of k-Integrals by Contour Integration in the Complex Plane. *Chem. Phys. Lett.* **1990**, *167*, 547–554.
24. Steinborn, E. O.; Ruedenberg, K. Rotation and Translation of Regular and Irregular Solid Spherical Harmonics. *Adv. Quantum Chem.* **1973**, 7, 1–81.
25. Arfken, G. B.; Weber, H. J. *Mathematical Methods for Physicists*; Academic Press: London, 2005.
26. Steinborn, O. Poly-Polar Expansions for Regular and Irregular Spherical Harmonics in Molecules. *Chem. Phys. Lett.* **1969**, *3*, 671–676.
27. Hobson, E. W. *The Theory of Spherical and Ellipsoidal Harmonics*; Chelsea Publishing Company: New York, 1955.
28. Guseinov, I. I. Spherically Symmetrical Properties of Expansion Coefficients for Translation of Spherical Harmonics. *J. Mol. Struct. (Theochem.)* **1996**, *367*, 83–86.
29. Barnett, M. P. Transformation of Harmonics for Molecular Calculations. *J. Chem. Inf. Comput. Sci.* **2003**, *43*, 1158–1165.
30. Rico, J. F.; Lopez, R.; Ema, I.; Ramirez, G. Translation of Real Solid Spherical Harmonics. *Int. J. Quantum Chem.* **2013**, *113*, 1544–1548.
31. (a) Prosser, F. P.; Blanchard, C. H. On the Evaluation of Two-Center Integrals. *J. Chem. Phys.* **1962**, *36*, 1112; (b) Geller, M. Two-Electron, One- and Two-Center Integrals. *J. Chem. Phys.* **1963**, *39*, 853–854.
32. (a) Öztekin, E.; Özcan, S. Overlap Integrals Between Irregular Solid Harmonics and STOs via the Fourier Transform Methods. *J. Math. Chem.* **2007**, *42*, 337–351; (b) Moharerrzadeh, M.; Chow Chiu, L. Y. Multicenter Molecular Integrals of Spherical Gaussian Functions by Fourier Transform Convolution Theorem. *J. Chem. Phys.* **1996**, *104*, 616–628.
33. Gradshteyn, I. S.; Ryzhik, I. M. *Tables of Integrals, Sums, Series and Products*; Academic Press: New York, 2000.
34. Gaunt, J. A. The Triplets of Helium. *Trans. R. Soc. Lond. Ser. A* **1929**, *228*, 151–196.
35. Guseinov, I. I. Analytical Evaluation of Two-Center Coulomb, Hybrid and One-Electron Integrals for Slater-Type Orbitals. *J. Phys. B* **1970**, *3*, 1399–1412.
36. Yavuz, M.; Yükçü, N.; Öztekin, E.; Yılmaz, H.; Döndür, S. On the Evaluation Overlap Integrals with the Same and Different Screening Parameters over Slater Type Orbitals Via the Fourier-Transform Method. *Commun. Theor. Phys.* **2005**, *43*, 151–158.
37. Özcan, S.; Öztekin, E. Analytical Evaluation for Two-Center Nuclear Attraction Integrals over Slater Type Orbitals by Using Fourier Transform Method. *J. Math. Chem.* **2009**, *45*, 1153–1165.

38. (a) Guseinov, I. I. Evaluation of Multielectron Molecular Integrals over Slater-Type Orbitals Using Binomial Coefficients. *J. Mol. Struct. (Theochem.)* **1997**, *417*, 117–121; (b) Guseinov, I. I.; Aydin, R.; Mamedov, B. A. Computation of Molecular Integrals over Slater-Type Orbitals. III. Calculation of Multicenter Nuclear-Attraction Integrals Using Recurrence Relations for Overlap Integrals. *J. Mol. Struct. (Theochem.)* **2000**, *503*, 173–177; (c) Guseinov, I. I.; Mamedov, B. A. Calculation of Multicenter Nuclear-Attraction and Electron Repulsion Integrals over Slater Orbitals by Fourier Transform Method Using Gegenbauer Polynomials. *J. Math. Chem.* **2002**, *32*, 309–322; (d) Öztekin, E.; Özcan, S.; Orbay, M.; Yavuz, M. Calculation of Nuclear-Attraction and Modified Overlap Integrals Using Gegenbauer Coefficients. *Int. J. Quantum Chem.* **2002**, *90*, 136–143; (e) Mamedov, B. A. Calculation of Two-Center Nuclear Attraction Integrals over Slater Type Orbitals in Molecular Coordinate System. *Chin. J. Chem.* **2004**, *22*, 545–548.
39. Mamedov, B. A.; Çopuroğlu, E. Use of Binomial Coefficients in Fast and Accurate Calculation of Löwdin-α Radial Functions. *J. Math. Chem.* **2011**, *49*, 201–207.

SECTION II

Electron Correlation in Molecules and Solids

CHAPTER ELEVEN

Excitation Energies of Molecules from Ensemble Density Functional Theory: Multiconfiguration Approaches

Katarzyna Pernal*, Nikitas I. Gidopoulos†, Ewa Pastorczak*,[1]
*Institute of Physics, Lodz University of Technology, Lodz, Poland
†Department of Physics, Durham University, Durham, United Kingdom
[1]Corresponding author: e-mail address: ewapastorczak@gmail.com

Contents

Abstract

Ensemble methods for excited states are based on the ensemble variation principle and in their simplest formulations can be either based on the wavefunction or the electron density. The latter group shares the favorable scaling of ground state density functional theory (DFT) and as such can be considered a computationally inexpensive alternative to time-dependent (TD)-DFT in cases where TD-DFT is not sufficiently accurate. The failures of TD-DFT most prominently include the poor description of conical intersections and excitations of multiple character, i.e., when multiconfigurational effects play a significant role. To deal with such issues, quite recently a number of multiconfiguration ensemble methods have been designed that combine a wavefunction-based formulation with ensemble density functional theory. This chapter discusses the merits and shortcomings of such approaches. It also attempts to elucidate some of the essential problems associated with the ensemble DFT methods and their variants to the computational chemistry community.

Advances in Quantum Chemistry, Volume 73
ISSN 0065-3276
http://dx.doi.org/10.1016/bs.aiq.2015.06.001

1. ENSEMBLE VARIATION PRINCIPLE AND ITS DESCENDANTS

In the 1970s, when the ensemble methods for describing the excited states of molecules first appeared, available methods for calculating the excitation energies for molecules were either very inaccurate or very expensive. The subspace density functional theory, introduced by Theophilou,[1] gave hope that it would be possible to describe the correlation in the excited states at modest cost. Then, many difficulties associated with using ensemble methods and the rapid development of time-dependent density functional theory (TD-DFT) impeded the progress of ensemble methods, as well as the other time-independent density functional methods for excited states. Currently, it seems that TD-DFT is reaching the limits of its capabilities and while it is of impressive accuracy in describing single valence excitations, it is still having problems with Rydberg[2] and charge-transfer states[3,4] and excitations of multiple character.[5] These observations prompted a renaissance of time-independent density functional methods for excited states[6–12]. Among them, the ensemble methods are prominent.

The foundation of all the ensemble methods is the ensemble minimum principle[1,13] which states that if $\psi_1,\psi_2,\ldots,\psi_m$ are orthonormal trial functions, then for m lowest eigenfunctions of N-electron Hamiltonian $\hat{H}$ with eigenvalues $E_1 \leq E_2 \leq \ldots \leq E_m$ the weighted sum of expectation values of energy obtained with the trial functions is bounded from below by a weighted sum of pertinent eigenvalues, i.e.,

$$\sum_{I=1}^{m} \omega_I \langle \psi_I | \hat{H} | \psi_I \rangle \geq \sum_{I=1}^{m} \omega_I E_I \tag{1}$$

provided that weights ω_I fulfil the condition

$$\omega_1 \geq \omega_2 \geq \cdots \geq \omega_m > 0. \tag{2}$$

Inequality (1) allows all excited states, or just a chosen number of the lowest states, of a system to be treated as one ensemble (mixed) state. The sum on the left side of the inequality (1) is called the ensemble energy, E_{ens}. The minimum of the ensemble energy uniquely determines the eigenstates of the system if all the weights $\omega_1,\ldots,\omega_m$ are different. If the weights are equal, the minimum of the ensemble energy determines only the subspaces in which the eigenstates lie, therefore the methods based on the simplified ensemble minimum principle are called subspace methods (i.e., subspace Hartree–Fock (HF),[14] subspace DFT[1]).

The simplest method based on inequality (1) is ensemble HF,[14–16] which consists in minimization of the ensemble energy within a set of single Slater determinants. While conceptually straightforward, this single-determinant wavefunction method is able to capture very little dynamic and no static correlation.

A more promising method is ensemble DFT (originally introduced under the name subspace DFT by Theophilou[1]). It is based on a Hohenberg–Kohn (HK) type theorem introduced by Theophilou and extended by Gross et al.[17] for ensembles of unequal weights (see also Refs. 18,19), chosen for an m-state ensemble as follows

$$\omega_1 = \omega_2 = \ldots = \omega_{m-g} \equiv \frac{1-\omega g}{m-g} \tag{3}$$

and

$$\omega_{m-g+1} = \omega_{m-g+2} = \ldots = \omega_m \equiv \omega \tag{4}$$

where g is an integer satisfying $1 \leq g \leq m-1$ and ω is a real parameter in the range $0 \leq \omega \leq 1/m$. The ensemble density matrix expressed in the basis of eigenstates $\{|j\rangle\}$ of the Hamiltonian $\hat{H} = \hat{T} + \hat{V}_{ee} + \hat{V}$ (with $\hat{T}$ denoting the kinetic energy, $\hat{V}_{ee}$ the electron–electron interaction and $\hat{V}$ the external potential) reads

$$\hat{\Gamma}^{m,g}(\omega) = (1-\omega_m)\left[\frac{1}{m-g}\sum_{j=1}^{m-g}|j\rangle\langle j|\right] + \omega_m\left[\frac{1}{m}\sum_{j=1}^{m}|j\rangle\langle j|\right]. \tag{5}$$

The theorem states that if for m lowest eigenstates $\{|m'\rangle\}$ of a Hamiltonian $\hat{H}' = \hat{T} + \hat{V}_{ee} + \hat{V}'$ another density matrix is constructed:

$$\hat{\Gamma}^{m,g}(\omega)' = (1-\omega_m)\left[\frac{1}{m-g}\sum_{j'=1}^{m-g}|j'\rangle\langle j'|\right] + \omega_m\left[\frac{1}{m}\sum_{j'=1}^{m}|j'\rangle\langle j'|\right] \tag{6}$$

then for fixed numbers m,g, ω the ensemble electron densities

$$\rho_{\text{ens}} = Tr\left\{\hat{\Gamma}^{m,g}(\omega)\hat{\rho}(r)\right\} \tag{7}$$

and

$$\rho'_{\text{ens}} = Tr\{\hat{\Gamma}^{m,g}(\omega)'\hat{\rho}(r)\} \tag{8}$$

are different, provided that potentials $\hat{V}$ and $\hat{V}'$ differ by more than a constant. The proof of the HK theorem for ensembles allowed[17] a universal HK functional to be defined

$$F^m[\omega,\rho_{\text{ens}}] = \text{Tr}\{\hat{\Gamma}^{m,g}(\omega)(\hat{T} + \hat{V}_{\text{ee}})\}, \tag{9}$$

where $\hat{\Gamma}^{m,g}(\omega)$ is a density matrix corresponding to a given density ρ. Note that there may be more than one density matrix generated by a unique potential and yielding ρ but each gives the same ensemble energy.[17] The functional (9) is defined for ensemble v-representable densities. As with a ground state theory, the extension of the universal functional to ensemble N-representable densities is possible if the constrained-search formalism construction is used. Namely, the universal functional for an ensemble density can be defined as

$$F_{m,\omega}[\rho_{\text{ens}}] = \min_{\{\Psi_I\}\to\rho_{\text{ens}}} \sum_{I=1}^{m} \omega_I \langle \Psi_I | \hat{T} + \hat{V}_{\text{ee}} | \Psi_I \rangle, \tag{10}$$

where the notation $\{\Psi_I\}\to\rho_{\text{ens}}$ indicates a set of m orthonormal wavefunctions forming an ensemble, which pertains to a given ensemble density by the relation

$$\rho_{\text{ens}}(\mathbf{r}) = \sum_{I=1}^{m} \omega_I \langle \Psi_I | \hat{\rho}(\mathbf{r}) | \Psi_I \rangle. \tag{11}$$

It is convenient to define an exchange-correlation (XC) ensemble density functional by introducing a Kohn–Sham (KS) system which shares an ensemble density with the interacting system (assuming that both densities are constructed from the m lowest states and with the same ensemble weights). The XC ensemble functional is then written by analogy to its ground state counterpart as

$$E^{\text{XC}}_{m,\omega}(\rho_{\text{ens}}) = F_{m,\omega}[\rho_{\text{ens}}] - T^s_{m,\omega}[\rho_{\text{ens}}] - \frac{1}{2}\int\int \frac{\rho_{\text{ens}}(\mathbf{r})\rho_{\text{ens}}(\mathbf{r}')}{|\mathbf{r}-\mathbf{r}'|} d\mathbf{r}d\mathbf{r}', \tag{12}$$

where $T^s_{m,\omega}$ stands for the kinetic energy of the KS system and the Coulomb term depends explicitly on the ensemble density ρ_{ens}.

The ensemble density functional theory formulated by Gross et al.[13,17] has become a standard formulation for ensemble DFT over many years and was used by Gross et al.[20] to calculate the spectra of a helium atom and a helium ion. In place of the ensemble XC energy, they used the ground-state local density approximation (LDA, Ref. 21) and also Kohn's quasi-local approximation (QLDA) for ensembles.[22] The spectra obtained were blurred, a result which the authors blamed on the self-interaction error. Subsequently, many approaches based on the approximate ground-state XC functionals were formulated and tested: the exact-exchange[23–27] and the local density approximation[28] or other approximations developed for the ground state DFT, reoptimized for ensemble calculations.[29,30]

Most of the aforementioned work on ensemble DFT used optimized effective potential approach, which employs an ensemble KS potential. There are two reasons why it is desirable to use a local, multiplicative ensemble exchange and correlation potential $v_{\mathrm{XC}}^{\mathrm{ens}} = v_{\mathrm{XC}}^{\mathrm{ens}}(\mathbf{r})$. The first reason is formal. In ensemble DFT, the ensemble density is in one-to-one correspondence with the external (local) potential of the system.[1,17] Furthermore, for the ensemble KS system, i.e., the virtual noninteracting system of electrons with the same ensemble density as the interacting system under study, the ensemble density is in one-to-one correspondence with the ensemble KS potential. The latter is the local effective potential that binds the noninteracting electrons in such a way that its ensemble density is the same as the ensemble density of the interacting system of interest. Therefore, determination of the ensemble KS potential is at the heart of ensemble DFT. The second reason is increased computational efficiency. The ensemble spin orbitals can be obtained by optimizing the ensemble energy (1). Like the HF equations, the direct minimization of the ensemble energy leads to single-particle eigenvalue-like equations for the spinorbitals, where the exchange potential term is not local. However, unlike the HF case, because the ensemble energy is not invariant to rotations of the spinorbitals, the Hamiltonian operator in the single-particle equations is in general not common to all spinorbitals. As a result, the orbitals which solve the ensemble single-particle equations are not automatically orthogonal and it becomes necessary, in general, to enforce their orthogonality, e.g., by employing off-diagonal Lagrange multipliers. The optimized effective potential method provides an elegant solution to the problem of a common Hamiltonian operator in the ensemble single-particle equations, with the additional benefit of a local exchange and correlation potential, as required by ensemble KS DFT.

Over the years, a lot of effort has been put into formulating the theoretical framework of ensemble DFT.[31–35] It has become clear that there is no easy way to obtain a good approximation for the ensemble XC functional and using the ground-state approximations might be the best starting point. Two approaches do not use existing functionals; first is the aforementioned QLDA due to Walter Kohn which was derived for a thermal ensemble and as such is not very successful in describing ensembles of states that do not interact with one another. The other is the recent of work of Yang et al.[36] where the exact KS and XC potentials for given ensemble densities are extracted and applied to the helium atom. The authors speculate that knowledge of the exact potential can be useful in constructing ensemble density functionals. While this is certainly true, this sort of task has been attempted previously for ground-state functionals without much success (see e.g., Ref. 37) and it is not expected to be easier in case of ensemble density.

The use of ground state expressions for the Hartree, exchange and correlation energy introduces a self-interaction error in the ensemble energy. This error has been called "ghost interaction" by Gidopoulos et al.[26] The error lies in the conventional definition of the ensemble Hartree energy that employs the ground state expression $E_H = \frac{1}{2}\int\int \frac{\rho_{\text{ens}}(\mathbf{r})\rho_{\text{ens}}(\mathbf{r}')}{|\mathbf{r}-\mathbf{r}'|} d\mathbf{r}d\mathbf{r}'$. Consider the excitation of an electron occupied in an orbital of the state Ψ_1 to an orbital unoccupied in Ψ_1 but occupied in the excited state Ψ_2. The ensemble density is the weighted sum of the densities of the states Ψ_1, Ψ_2. Hence, the Hartree energy will include the spurious Coulomb interaction of the density from the orbital occupied by the electron before excitation with the density of the orbital hosting it after the excitation. This kind of self-interaction is not encountered in ground state DFT, nor ground state HF theory and the ground state expressions for the exchange energy cannot correct it. The best way to correct for the ghost self-interaction is to write the ensemble Hartree energy as the sum of the Hartree energies of the member configurations of the ensemble.[24,26,38,39]

The spin-symmetry of the ensemble members introduces a further challenge to the ensemble theory for excited states (DFT and HF). Consider the He atom. The ground state is a singlet and the lowest excitation is a triplet. The next lowest excitation is a singlet. So, for the physical He atom, the ensemble with configurations of lowest energy may have one member (the singlet ground state), four members (singlet ground state and triplet excited state) or five members (the above plus the singlet excited state). If we now consider the noninteracting system, the ground state is nondegenerate but if we want to consider the lowest excitation, this is quadruply degenerate. Therefore one may consider just the ground state, or an ensemble of five members. This places restrictions on the choice of ensemble weights for the noninteracting system. For example, the freedom to choose unequal weights is restricted, because if we imagine an adiabatic connection path, where we switch on the electron repulsion, a noninteracting ensemble with randomly unequal weights for the four degenerate excited states cannot in general connect smoothly to an ensemble of the physical system, where the fourfold degeneracy is broken and where the ensemble weights will satisfy the inequalities (2). In Ref. 26, the problem was addressed by choosing the members of the noninteracting ensemble to be eigenstates of S^2 and not single Slater determinants. Such a choice has also led to improved accuracy of the results (see table in Ref. 26). This point is discussed in Ref. 39, where eigenstates of S^2 are also used. Another way of bypassing the problem is to use Boltzmann weights for the members of the ensemble, which guarantee at least the existence of a smooth adiabatic connection between the noninteracting and interacting systems.

Another challenge for ensemble DFT, seemingly of lesser importance but with major impact on the quality of results is the lack of a clear path to obtain the energies of specific states in the ensemble. In the wavefunction approaches, it is fairly reasonable to identify the resulting set of wavefunctions with the true eigenvectors of the Hamiltonian. In the ensemble DFT, however, extracting the energies is not straightforward. Various methods have been proposed[1,20,35,40,41,42] but those which are theoretically justified are not feasible in practical calculations and vice versa (with the exception of Levy's method which involves calculation of the first excitation energy from orbital energies resulting from an ensemble calculation—this, however, limits the method to calculating only the first excitation energy). We will not present all these methods in detail, this problem will be briefly discussed in Section 4.3 in the context of multiconfiguration methods.

A final question which the quantum chemistry community has tried to address concerns the actual capabilities of ensemble methods. It has been shown[25,28,38] that ensemble DFT methods employing the ground-state approximations as the XC functionals can (at least for small molecules) match the accuracy of some less sophisticated TD-DFT methods. However, so far only the multiconfiguration ensemble methods have been able to surpass the limits of standard TD-DFT. In the next sections, we will present basic ideas associated with this group of methods and some of their results.

2. MULTICONFIGURATION WAVEFUNCTION-BASED METHODS FOR ENSEMBLES

Before moving on to multiconfiguration ensemble density methods, we will briefly discuss how multireference effects in excited states can be taken into account within ensemble wavefunction-based methods. As mentioned in Section 1, in the simplest ensemble wavefunction method, the ensemble may be constructed from m single-determinant states $\{\Phi_I\}$. Then, application of the ensemble variational principle leads to the ensemble HF (eHF) method.[42] In this method, the ensemble energy given as a weighted sum of the state energies

$$E^{\text{eHF}} = \sum_{I=1}^{m} \omega_I \langle \Phi_I | \hat{H} | \Phi_I \rangle, \tag{13}$$

is optimized with respect to spinorbitals $\{\varphi_p\}$ building single determinants subject to their orthonormality. Note that since the ensemble energy is in general not invariant to orbital rotations, it is possible to determine optimal spinorbitals uniquely. An exception is the hyper Hartree–Fock (HHF) method, an equiensemble method,[14,43] where the ensemble is composed of all the N-electron Slater determinants that can be constructed from a given basis set of R orthonormal spin orbitals, with $R \geq N$. The inclusion of all Slater determinants in the HHF ensemble restores the invariance of the HHF ensemble energy with respect to orbital rotations and simplifies the HHF equations.[14,44]

Since excited states are described in the eHF by single-determinant functions, the excited state energies predicted by this method are expected to be in error especially for states of multiconfigurational nature. A step beyond ensemble HF method would involve first doing the eHF calculation to obtained a set of eHF spinorbitals and then constructing an ensemble of states formed as linear combination of determinants

$$\forall_{I=1}^{m} \quad \psi_I = \sum_J C_{IJ} \Phi_J^{\text{eHF}}, \tag{14}$$

where the superscript eHF indicates that spinorbitals used to construct a given determinant have been obtained from the eHF method. The corresponding ensemble energy would read

$$E = \sum_{I=1}^{m} \omega_I \sum_{JK} C_{IJ} C_{IK} \left\langle \Phi_J^{\text{eHF}} | \hat{H} | \Phi_K^{\text{eHF}} \right\rangle. \tag{15}$$

Its minimization with respect to the configuration interaction (CI) expansion coefficients $\{C_{IJ}\}$ with the constraint that wavefunctions $\{\psi_I\}$ stay orthonormal leads to a system of equations

$$\omega_I \sum_K \left\langle \Phi_J^{\text{eHF}} | \hat{H} | \Phi_K^{\text{eHF}} \right\rangle C_{IK} = \sum_K \mathbf{\Lambda}_{IK} C_{KJ}, \tag{16}$$

where $\mathbf{\Lambda}$ is a Lagrangian matrix. If all the weights are different, then the minimization problem (16) is equivalent to diagonalisation of Hamiltonian matrix in the space spanned by the states $\{\Phi_I^{\text{eHF}}\}$ and approximating the ground and excited state energies of the system as eigenvalues of the Hamiltonian. On the other hand, when the weights are equal, the ensemble energy (15) is the trace of the Hamiltonian matrix and as such remains invariant to orthogonal rotations among the determinants $\{\Phi_I^{\text{eHF}}\}$. Hence, diagonalisation of the Hamiltonian in this case conserves the energy of the ensemble.

The method based on the ensemble energy given in Eq. (15), introduced in Ref. 42 and called Ens-WF, is then a simple ensemble counterpart of the CI method. The accuracy of the state energies predicted by the Ens-WF method hinges on the choice of the configurations that enter the calculation and the choice of the ensemble weights. In Ref. 42, the configurations are selected based on chemical intuition. Using the fact that the ensemble variational principle is a special case of the Helmholtz free energy principle[42], the weights have been established as Boltzmann factors:

$$\omega_I = \frac{\exp[-\beta E_I]}{\sum_K^m \exp[-\beta E_K]} \tag{17}$$

with approximate state energy values E_I and a tunable β parameter. Obviously, Boltzmann factors are only one possible choice but they have certain attractive properties, such as

- Weights corresponding to degenerate states are equal. This ensures the invariance of the ensemble energy to rotations among degenerate states;
- By choosing β parameter different from zero one can remove the high-lying states from the ensemble (see Fig. 1);

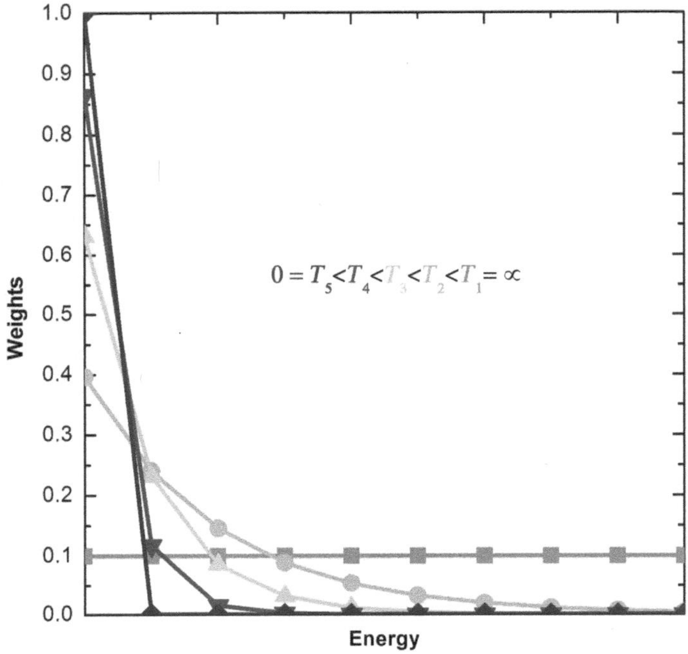

Figure 1 Distribution of Boltzmann weights in a 10-state ensemble for different temperatures. The horizontal axis represents the energies of states in the ensembles.

- It prevents the numerical phenomenon of "root flipping", through keeping the right order of the expectation values of the trial functions throughout the optimization.

When $\beta \rightarrow \infty$ (or, equivalently, the temperature T is 0, see star-marked curve on Fig. 1), only the ground state has a nonzero weight. In such a case, the eHF and HF approaches are equivalent and consequently the Ens-WF method reduces to the CI method in the given space.

Since excited state energies result from a CI calculation anyway, does introducing an ensemble method of the same cost makes sense at all? These methods differ only in spinorbitals employed to built states (spinorbitals come from HF and eHF methods for CI and Ens-WF methods, respectively). In fact, it has been found that the Ens-WF method (with finite values of the β parameter) can produce more accurate energies than CI for systems with degenerate ground states (e.g., beryllium atom, see Fig. 2) and it can improve the shape of the dissociation curves of diatomic molecules, see Ref. 42.

One could go a step further and construct a fully self-consistent method, the ensemble analog of the multiconfiguration self-consistent field (MCSCF) method. In fact, such an approach would be equivalent to the

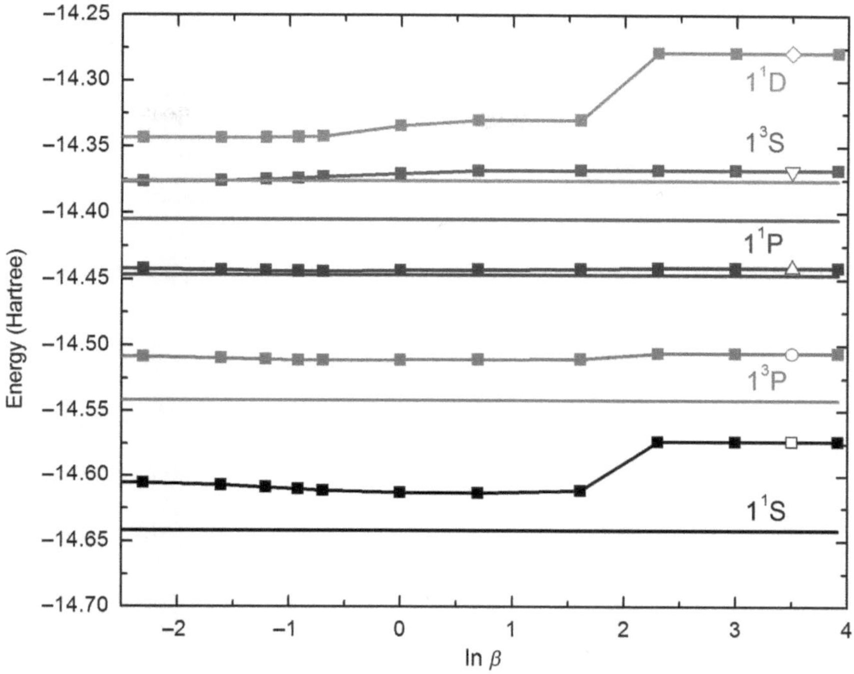

Figure 2 Energies of states of beryllium atom calculated for different values of β parameter. Unmarked solid lines, LR-CCSD reference; marked lines, Ens-WF; open markers, $\beta \to \infty$ (CI) limit.

state-average MCSCF (SA-MCSCF) method proposed by Werner and Meyer[45] without even referring to the ensemble theories. In SA-MCSCF, the orbitals and the expansion coefficients of the wavefunction are optimized in a self-consistent manner to minimize the mean average of energy for all states. The states in this average can be weighted equally but usually their weights reflect the significance of each state. SA-MCSCF was primarily introduced to avoid the problem of root flipping—a phenomenon in which the solution oscillates between two nearly degenerate states. It was also shown that the algorithms employing SA-MCSCF have better convergence properties than those for the standard MCSCF methods. SA-MCSCF is not entirely justified theoretically but it could be made rigorous if the weights were not chosen arbitrarily but rather according to the energy criterion. Interestingly, the SA-MCSCF method has been used to model chemical reaction mechanisms and it was shown that the choice of Boltzmann factors as weights improves in many cases the accuracy[46] of the method.

The accuracy of the Ens-WF method (with or without orbital minimization) is limited by the length of expansion of states in single-determinant function basis. Taking too large a space of determinants would lead to prohibitively expensive calculations. Shorter expansions, on the other hand, will still take into account multireference effects but the corresponding approximations to the state energies will be missing an important amount of the dynamic correlation energy. The remedy of this deficiency is offered by multiconfiguration ensemble density functional theory discussed in the next sections.

3. STATE-AVERAGED SPIN-RESTRICTED KS METHOD

One of the first density functional methods based on the ensemble representation of electron density is the spin-restricted ensemble-referenced Kohn-Sham method (REKS) developed by Filatov et al.[47,48] Although originally formulated for predicting ground states of molecules the electronic structure of which is dominated by static correlation,[47] it has been later expanded to treat both a ground and a first singlet excited state.[49,50]

The REKS method is particularly suited for molecules whose ground state wavefunction is well described using two configurations. In such a case, the ground state density of a physical (interacting) system of interest is represented by a two-state ensemble density of a noninteracting, KS, reference. The multiconfiguration character of the interacting wavefunction is manifested by non-integer occupation of some of the frontier KS orbitals. The underlying assumption in the REKS approach is that if a singlet ground state interacting wavefunction is well represented by a combination of two configurations $|\ldots\varphi_a\bar{\varphi}_a\rangle$ and $|\ldots\varphi_b\bar{\varphi}_b\rangle$, then a corresponding electron density can be obtained by taking an ensemble of two KS determinants. One of the determinants includes HOMO, $\{\varphi_a, \bar{\varphi}_a\}$, orbitals whereas the other one LUMO, $\{\varphi_b, \bar{\varphi}_b\}$. HOMO and LUMO orbitals form an active space. Thus, the REKS ensemble density takes the form

$$\rho^{\mathrm{REKS}}(\mathbf{r}) = 2\sum_i |\varphi_i(\mathbf{r})|^2 + n_a|\varphi_a(\mathbf{r})|^2 + n_b|\varphi_b(\mathbf{r})|^2, \tag{18}$$

where n_a and n_b are the occupation numbers constrained to a range $[0, 2]$ and summing up to 2, and the summation with respect to i runs through all orbitals, which are fully occupied in both determinants. Although the ensemble REKS density only approximately corresponds to the true multiconfiguration wavefunction of the physical system, it is constrained to have the same spin and spatial symmetry as the interacting ground-state wavefunction. Such an approach, despite not being entirely rigorous, is very effective for systems where static correlation plays an important role. The REKS functional for the ensemble density given in Eq. (18) has been proposed to comprise a weighted sum of the KS density functional energies of the two configurations and a coupling term involving singly excited configurations generated out of the active space. It reads[47]

$$E^0 = \frac{n_a}{2}E^{\mathrm{KS}}[\Psi_1] + \frac{n_b}{2}E^{\mathrm{KS}}[\Psi_2] \\ -f(n_a, n_b)\left(E^{\mathrm{KS}}\left[\Psi^T\right] - \frac{1}{2}E^{\mathrm{KS}}\left[\Psi^{S1}\right] - \frac{1}{2}E^{\mathrm{KS}}\left[\Psi^{S2}\right]\right), \tag{19}$$

where Ψ_1, Ψ_2 are the two possibly nearly degenerate configurations, $|\ldots\varphi_a\bar{\varphi}_a\rangle$ and $|\ldots\varphi_b\bar{\varphi}_b\rangle$, and $\Psi^T = |\ldots\varphi_a\varphi_b\rangle$, $\Psi^{S1} = |\ldots\varphi_a\bar{\varphi}_b\rangle$ and $\Psi^{S2} = |\ldots\bar{\varphi}_a\varphi_b\rangle$ are the remaining (respectively, triplet and two singlets) configurations constructed from the active orbitals. The function $f(n_a, n_b)$ has been found empirically and it interpolates between a single-configuration ($n_a \approx 2, n_b \approx 0$) case and a case when static correlation dominates ($n_a \approx n_b \approx 1$). A ground state energy follows by optimization of the functional (19) with respect to the orbitals and the occupation numbers $\{n_a, n_b\}$ under the orthonormality constraint for the former and the non-negativity and the normalization conditions for the latter.

Using orbitals and the occupation numbers obtained from the minimization of the REKS functional it is possible to find an approximation to the energy of the first open-shell singlet excited state and it is given by the following expression[51]

$$E^1 = E^{\mathrm{KS}}\left[\Psi^{S1}\right] + E^{\mathrm{KS}}\left[\Psi^{S2}\right] - E^{\mathrm{KS}}\left[\Psi^T\right]. \tag{20}$$

Kazaryan et al.[49] proposed to obtain a better approximation to the excitation energy from the ground to the first open shell singlet state by considering a weighted sum of E^0 and E^1, namely

$$E^{\mathrm{SA-REKS}} = \omega_0 E^0 + \omega_1 E^1, \tag{21}$$

where $\omega_0 + \omega_1 = 1$; minimizing it with respect to the KS orbitals and the fractional occupation numbers and finally computing the energies of the ground and excited states from Eqs. (19) and (20), respectively. The state-averaged REKS (SA-REKS) functional given in Eq. (21) can be seen as an orbital-dependent approximate realization of the Gross–Oliveira–Kohn (GOK) functional (Eq. (9)). Notice that functionals (19) and (20) are free from the ghost interaction and include spin polarization. Like the SA-MCSCF the choice of weights in the ensemble energy expression (21) is arbitrary. The application of the SA-REKS method is more justified for homosymmetric systems (e.g., molecules dissociating to equal fragments).[48] For heterosymmetric systems, it has been argued that more accurate values of the excitation energies will follow from diagonalization of the Hamiltonian in the two-dimensional space consisting of configuration state functions corresponding to a ground and an excited state[50] resulting in the state-interacting-state-averaged-REKS (SI-SA-REKS) method.

Because only a very small active space is used in the SI-SA-REKS methods, they are fairly computationally inexpensive and can be used for rather large systems.[48,50] They offer a well-balanced treatment of the dynamic and static correlation and their multiconfiguration character allows conical intersections and some potential energy surfaces to be described. On the other hand, they are limited to only ground and open-shell first singlet excited states and the number of active orbitals is limited to 2. Moreover, their use demands some preliminary knowledge of the chemical system under study. Also, the theoretical foundations of SI-SA-REKS contain considerable empiricism—in the choice of the density functionals as well as in the construction of the function $f(n_a,n_b)$.

4. RANGE-SEPARATED APPROACHES

4.1 Generalized Adiabatic Connection for Ensembles

As hinted in Section 1, the XC component of the GOK universal functional (Eq. (12)) can be formulated in the adiabatic connection (AC) framework. Adiabatic connection[52] allows one to find an XC ground state density functional by linking a physical system of interest with a KS (noninteracting) one. It is achieved by gradually turning on electron–electron interaction by increasing a coupling strength parameter and modifying the external potential so that a density of the partially interacting system equals that of the fully interacting one. The adiabatic connection approach has been brought to the realm of ensemble DFT by Nagy[34] who has shown that a Hartree-XC (HXC) ensemble density functional can be obtained by integrating ensemble electron interaction energy of the partially interacting system with respect to the coupling constant λ, namely

$$E^{\mathrm{HXC}}_{m,\omega}[\rho_{\mathrm{ens}}] = \sum_{I=1}^{m} \omega_I \int_0^1 \mathrm{d}\lambda \langle \Psi_I^\lambda | \hat{V}_{\mathrm{ee}} | \Psi_I^\lambda \rangle \tag{22}$$

where the wavefunctions are eigenfunctions of a partially interacting Hamiltonian

$$\left(\hat{T} + \lambda \ \hat{V}_{\mathrm{ee}} + \hat{V}^\lambda \right) | \Psi_I^\lambda \rangle = \mathcal{E}_I^\lambda | \Psi_I^\lambda \rangle \tag{23}$$

with a local potential $\hat{V}^\lambda$ assuring that the density constraint

$$\rho_{\mathrm{ens}}(\mathbf{r}) = \sum_{I=1}^{m} \omega_I \langle \Psi_I^\lambda | \hat{\rho}(\mathbf{r}) | \Psi_I^\lambda \rangle \tag{24}$$

is fulfilled for any value of the coupling strength $\lambda \in [0, 1]$. Therefore, in the Nagy formulation, the density is fixed along the adiabatic connection and the ensemble weights are not varied.

Franck and Fromager[35] proposed a generalized adiabatic connection for ensembles (GACE) by demanding the density to be weight independent along the connection and varying coupling strength together with ensemble weights. This approach naturally leads to a weight-dependent ensemble density functional, in line with the early work of Gross, Oliveira, and Kohn.[17] The greatest advantage of the generalized AC is that for a two-state ensemble the ω-dependent XC functional can be Taylor-expanded with respect to ω. This has led to deriving rigorous conditions satisfied by the exact ensemble functional.

In the GACE approach, an ensemble density $\rho_{\mathrm{ens}}(\mathbf{r})$ is weight independent and one considers two-state ensembles formed from a ground and a first excited state of a partially interacting system (the electron–electron interaction is scaled with λ)

$$\left(\hat{T} + \lambda \hat{V}_{\mathrm{ee}} + \hat{V}^{\lambda,\xi} \right) \left| \Psi_I^{\lambda,\xi} \right\rangle = \mathcal{E}_I^{\lambda,\xi} \left| \Psi_I^{\lambda,\xi} \right\rangle, \tag{25}$$

where a local potential $\hat{V}^{\lambda,\xi}$ fixes the ensemble density

$$\rho_{\mathrm{ens}} = (1 - \xi) \rho_{\Psi_1^{\lambda,\xi}} + \xi \rho_{\Psi_2^{\lambda,\xi}}, \tag{26}$$

The density is constant both for varying coupling parameter $\lambda \in [0, 1]$ and weight parameter $\xi \in [0,\omega]$. For a two-state ensemble with $(1 - \omega)$ and ω being weights of the ground and the excited state, respectively, the generalized adiabatic connection expression for the XC ensemble density functional defined in Eq. (12) can be written as[35]

$$E_{2,\omega}^{\mathrm{XC}}[\rho_{\mathrm{ens}}] = E^{\mathrm{XC}}[\rho_{\mathrm{ens}}] + \int_0^{\omega} \mathrm{d}\xi \left[\left(\mathcal{E}_2^{\lambda=1,\xi} - \mathcal{E}_1^{\lambda=1,\xi} \right) - \left(\mathcal{E}_2^{\lambda=0,\xi} - \mathcal{E}_1^{\lambda=0,\xi} \right) \right] \tag{27}$$

where E^{XC} is the ground state functional. Interestingly, it is evident from (27) that a deviation of the ensemble XC energy from its ground state counterpart results from a difference of excitation energies of the fully interacting ($\lambda = 1$) and noninteracting ($\lambda = 0$) systems integrated with respect to an ensemble weight ξ. Franck and Fromager have shown that a GACE approach can be employed to rigorously define an ω-dependent short-range ensemble density functional, which is a component of the range-separated multideterminant ensemble DFT formalism presented in the next section.

Using the generalized adiabatic connection allows one to attempt to construct exact functionals for simple cases and use them together with the derived exact conditions to obtain approximations for ensemble density functionals in the future.

4.2 Range-Separated Ensemble DFT

While the pragmatic construction of a multiconfiguration method based on ensemble DFT proposed by Filatov et al. leads to encouraging results,[48] other approaches are feasible on firm theoretical ground. One of the possibilities to introduce multiconfiguration effects in ensemble DFT is based on using a range-separated ensemble density functional. Just like in ground-state theories,[53] in order to avoid double counting of the electron–electron interaction, one separates the electron–electron interaction operator $1/r$ into a short-range $\left(v_{\mathrm{ee}}^{\mathrm{SR}}\right)$ and a long-range $\left(v_{\mathrm{ee}}^{\mathrm{LR}}\right)$ part, namely

$$\frac{1}{r} = v_{\mathrm{ee}}^{\mathrm{SR}}(r) + v_{\mathrm{ee}}^{\mathrm{LR}}(r), \tag{28}$$

where

$$\lim_{r \to \infty} r v_{\mathrm{ee}}^{\mathrm{LR}}(r) = 1 \tag{29}$$

and

$$\lim_{r \to 0} r v_{\mathrm{ee}}^{\mathrm{SR}}(r) = 1. \tag{30}$$

The most commonly used range partitioning[54,55] uses the error function and assumes the following expressions for the range components of the electron repulsion operator:

$$v_{\mathrm{ee}}^{\mathrm{SR},\mu}(r) = \frac{1-\mathrm{erf}(\mu r)}{r}, \tag{31}$$

$$v_{\mathrm{ee}}^{\mathrm{LR},\mu}(r) = \frac{\mathrm{erf}(\mu r)}{r}. \tag{32}$$

The parameter μ governs the range separation and for $\mu=0$ the long-range function $v_{\mathrm{ee}}^{\mathrm{LR},\mu}(r)$ vanishes, while in the limit $\mu \to \infty$ it accounts for the full-range electron interaction, i.e., $\lim_{\mu\to\infty} v_{\mathrm{ee}}^{\mathrm{LR}}(r) = 1/r$ and $\lim_{\mu\to\infty} v_{\mathrm{ee}}^{\mathrm{SR}}(r) = 0$. The range-separated ensemble energy functional can be formally defined if the GOK universal functional (Eq. (10)) is decomposed into long- and short-range parts

$$F_{m,\omega}[\rho_{\mathrm{ens}}] = F_{m,\omega}^{\mathrm{LR},\mu}[\rho_{\mathrm{ens}}] + E_{m,\omega}^{\mathrm{SR},\mu}[\rho_{\mathrm{ens}}] \tag{33}$$

and the LR functional is defined by employing a Levy–Lieb constrained-search formulation, i.e.,

$$F_{m,\omega}^{\mathrm{LR},\mu}[\rho_{\mathrm{ens}}] = \min_{\{\Psi_I\}\to\rho_{\mathrm{ens}}} \sum_{I=1}^{m} \omega_I \left\langle \Psi_I \middle| \hat{T} + \hat{V}_{\mathrm{ee}}^{\mathrm{LR},\mu} \middle| \Psi_I \right\rangle. \tag{34}$$

The functional $E_{m,\omega}^{\mathrm{SR},\mu}[\rho_{\mathrm{ens}}]$ is therefore a short-range ensemble density complement, which can be further divided into a classical Hartree part being explicitly given in terms of an ensemble density and the remainder—the XC short-range ensemble density functional

$$E_{m,\omega}^{\mathrm{SR},\mu}[\rho_{\mathrm{ens}}] = E_{m,\omega}^{\mathrm{SR-H},\mu}[\rho_{\mathrm{ens}}] + E_{m,\omega}^{\mathrm{SR-XC},\mu}[\rho_{\mathrm{ens}}]. \tag{35}$$

If the ensemble energy functional is defined as follows

$$E_{m,\boldsymbol{\omega}}[\{\psi_I\}] = \sum_{I=1}^{m} \omega_I \left\langle \psi_I \middle| \hat{T} + \hat{V}_{\mathrm{ee}}^{\mathrm{LR},\mu} + \hat{V}_{ext} \middle| \psi_I \right\rangle + E_{m,\boldsymbol{\omega}}^{\mathrm{SR},\mu}[\rho_{\mathrm{ens}}], \tag{36}$$

with ρ_{ens} yielded by states $\{\psi_I\}_{I=1}^{m}$ by the relation (11) and weights $\{\omega_I\}_{I=1}^{m}$, the GOK variational principle implies in a straightforward manner that a full minimization of the functional (36) under the orthonormality constraint for states leads to obtaining the exact energy of the ensemble

$$\min_{\substack{\{\psi_I\} \\ \forall_{IJ}\ \langle\psi_I|\psi_J\rangle=\delta_{IJ}}} E_{m,\boldsymbol{\omega}}[\{\psi_I\}] = E_{m,\boldsymbol{\omega}}[\{\Psi_I^{\mathrm{LR}}\}] = \sum_{I=1}^{m} \omega_I E_I^{(0)} \tag{37}$$

and the exact ensemble density of a fully interacting system

$$\rho_{\text{ens}}^{(0)}(\mathbf{r}) = \sum_{I=1}^{m} \omega_I \left\langle \Psi_I^{(0)} | \hat{\rho}(\mathbf{r}) | \Psi_I^{(0)} \right\rangle = \sum_{I=1}^{m} \omega_I \left\langle \Psi_I^{\text{LR}} | \hat{\rho}(\mathbf{r}) | \Psi_I^{\text{LR}} \right\rangle. \tag{38}$$

$\{\Psi_I^{\text{LR}}\}_{I=1}^{m}$ are the first m eigenfunctions of the equation with a partially interacting Hamiltonian, i.e.,

$$\left(\hat{T} + \hat{V}_{\text{ee}}^{\text{LR},\mu} + \hat{V}_{m,\omega}^{\text{SR},\mu}[\rho_{\text{ens}}] + \hat{V}\right)\Psi_I^{\text{LR}} = E_I^{\text{LR}}\Psi_I^{\text{LR}} \tag{39}$$

sharing an external potential $\hat{V}$ with the fully interacting Hamiltonian $\hat{H} = \hat{T} + \hat{V}_{\text{ee}} + \hat{V}$ the eigenfunctions and eigenvalues of which are $\{\Psi_I^{(0)}\}$ and $\{E_I^{(0)}\}$, respectively. The short-range local potential $V_{m,\omega}^{\text{SR},\mu}[\rho_{\text{ens}}]$ present in Eq. (39) follows from taking a functional derivative of $E_{m,\boldsymbol{\omega}}^{\text{SR},\mu}$ with respect to the density. Clearly, while a density of a partially interacting system coincides with the fully interacting density, cf. Eq. (38), the ensemble energies of the two systems are different. Thus Eq. (37) implies $\sum_{I=1}^{m} \omega_I E_I^{\text{LR}} \neq E_{m,\omega}\left[\{\Psi_I^{\text{LR}}\}\right]$.

Note that a formal definition of the short-range ensemble functional provided by Eq. (33) does not provide much insight on how to construct approximations. Interestingly, Franck and Fromager[35] derived an expression for $E_{m,\omega}^{\text{SR},\mu}[\rho_{\text{ens}}]$ within a generalized adiabatic connection formalism for ensembles (GACE) and it may serve as a better starting point for finding rigorous conditions satisfied by the short-range functional and developing approximations to it. As it has been mentioned in Section 4.1, in GACE both a coupling constant and weights are varied along the adiabatic connection. In a range-dependent GACE, a range-separation parameter ν takes the coupling constant role c.f. λ in Eqs. (25)–(27). By considering a partially interacting system described by the equation

$$\left(\hat{T} + \hat{V}_{\text{ee}}^{\text{LR},\nu} + \hat{V}^{\nu,\xi}\right)\left|\Psi_I^{\nu,\xi}\right\rangle = \mathcal{E}_I^{\nu,\xi}\left|\Psi_I^{\nu,\xi}\right\rangle, \tag{40}$$

with the external potential $\hat{V}^{\nu,\xi}$ being such that Eq. (26) is satisfied (if λ is replaced with ν) Franck and Fromager obtained the following expression for the short-range XC ensemble functional for a two-state ($m=2$) ensemble

$$\begin{aligned} E_{2,\omega}^{\text{SR}-\text{XC},\mu}[\rho_{\text{ens}}] = E^{\text{SR}-\text{XC},\mu}[\rho_{\text{ens}}] + \int_0^{\omega} \mathrm{d}\xi \Big[&\left(\mathcal{E}_2^{\nu=\infty,\xi} - \mathcal{E}_1^{\nu=\infty,\xi}\right) \\ &- \left(\mathcal{E}_2^{\nu=\mu,\xi} - \mathcal{E}_1^{\nu=\mu,\xi}\right)\Big], \end{aligned} \tag{41}$$

where $E^{\mathrm{SR-XC},\mu}[\rho_{\mathrm{ens}}]$ is simply a ground state short-range XC functional. It is therefore evident from Eq. (41) that the error one introduces by neglecting the ω dependence of the ensemble functional and using a ground state functional instead amounts to neglecting differences between excitation energies of the fully interacting system ($\nu=\infty$) and the partially interacting one ($\nu=\mu$).

4.3 Multiconfiguration Methods Based on Range-Separated Ensemble DFT

The range-separated ensemble DFT has only been introduced as a means to construct a multiconfiguration method and is not generally expected—using the currently available XC approximations—to be able to capture bond-breaking and double excitations. If, however, one allows for the multi-configuration wavefunctions in the ensemble, one can correctly reproduce whole dissociation curves of molecules.[42] Explicitly, if the wavefunctions are expressed as linear combinations of the Slater determinants $\{\Phi_I^{\mathrm{eHF}}\}$ built of optimal orbitals obtained from ensemble HF, then the ensemble energy functional takes the form

$$E_{m,\omega}[\mathbb{C}]=\sum_{I=1}^{m}\omega_I\sum_{JK}C_{IJ}C_{IK}\left\langle\Phi_J^{\mathrm{eHF}}|\hat{T}+\hat{V}_{\mathrm{ee}}^{\mathrm{LR},\mu}+\hat{V}|\Phi_K^{\mathrm{eHF}}\right\rangle+E_{m,\omega}^{\mathrm{SR},\mu}[\rho_{\mathrm{ens}}]. \tag{42}$$

Minimization of the functional (42) with respect to C_{IJ} coefficients on condition that $\mathbb{C}$ is unitary (in fact, $\mathbb{C}$ is orthogonal, as we chose C_{IJ} to be real numbers) is equivalent to a diagonalization of the effective Hamiltonian, defined as

$$H'_{JK}=\left\langle\Phi_J^{\mathrm{eHF}}|\hat{T}+\hat{V}_{\mathrm{ee}}^{\mathrm{LR},\mu}+\hat{V}_{m,\omega}^{\mathrm{SR},\mu}[\rho_{\mathrm{ens}}]+\hat{V}|\Phi_K^{\mathrm{eHF}}\right\rangle, \tag{43}$$

where

$$\hat{V}_{m,\omega}^{\mathrm{SR},\mu}[\rho_{\mathrm{ens}}]=\sum_{i=1}^{N}\frac{\delta E_{m,\omega}^{\mathrm{SR},\mu}[\rho_{\mathrm{ens}}]}{\delta\rho_{\mathrm{ens}}(\mathbf{r}_i)} \tag{44}$$

is the short-range potential. Hence, this gives a method very similar to the above Ens-WF method (cf. Section 2) and therefore we will denote it as Ens-lrWF+srDF.

Of course, one still needs to choose the short-range density functional describing the ensemble. Here, the chosen general form is

$$E_{m,\omega}^{\mathrm{SR},\mu}[\rho_{\mathrm{ens}}] = \frac{1}{2}\iint \rho_{\mathrm{ens}}(\mathbf{r}_1)\rho_{\mathrm{ens}}(\mathbf{r}_2)\frac{\mathrm{erfc}(\mu r_{12})}{r_{12}}d\mathbf{r}_1 d\mathbf{r}_2 + E^{\mathrm{SR-XC},\mu}[\rho_{\mathrm{ens}}], \tag{45}$$

where the functional is split into the short-range Hartree interaction and the XC part of the functional chosen to be a ground state functional. Since there are no XC functionals available designed specifically for ensembles, the short-range PBE functional[56,57] was used in Ref. 42 and the following numerical examples. Although most states described have an open-shell character, a non-spin-polarized version of the functional was used. Hence, spin polarization is only introduced through the wavefunction part of the method.

Another problem in this method is that it is not clear how to calculate the energies of specific states. Identifying the states with the resulting wavefunctions is only justified in the $\mu \to \infty$ limit (which corresponds to the Ens-WF method). Nevertheless, calculating the energies as the expectation values of the (effective) Hamiltonian w.r.t. its eigenvectors seems to produce the best results. A result of such an approach is shown in Fig. 3 presenting the dissociation curves of the three lowest Σ states of BH molecule. When compared to Ens-WF and CI curves, definite improvement is visible—as expected, the method produces lower absolute energies then the purely wavefunction approaches but it does not significantly change the shapes of the curves. It still seems that a significant part of the correlation energy is missing.

Calculating the expectation value of the effective Hamiltonian w.r.t. to its eigenvectors is not the only possible way of calculating the energies. For example, Rebolini et al.[9] suggested taking the differences of the eigenvalues of the effective Hamiltonian, Eq. (43),

$$\Delta E_I = E_I^{\mathrm{LR},\mu} - E_0^{\mathrm{LR},\mu} \tag{46}$$

as a rather crude approximations for excitation energies. In Ref. 9, this possibility has been explored for various systems along the adiabatic connection, with the exact (obtained form a full configuration interaction calculation) ground-state density used in $V^{\mathrm{SR},\mu}$. We, on the other hand, did not perform a full CI calculation in the limit but used CI in the space built of the states included in the ensemble and compared ground-state (see Fig. 4) calculation with an equiensemble ($\beta = 0$) one (Fig. 5) and an ensemble one, where the ground state is dominant—Fig. 6. The errors from the ground-state calculation are very large, up to 4.5 eV. If a full CI calculation was performed, the

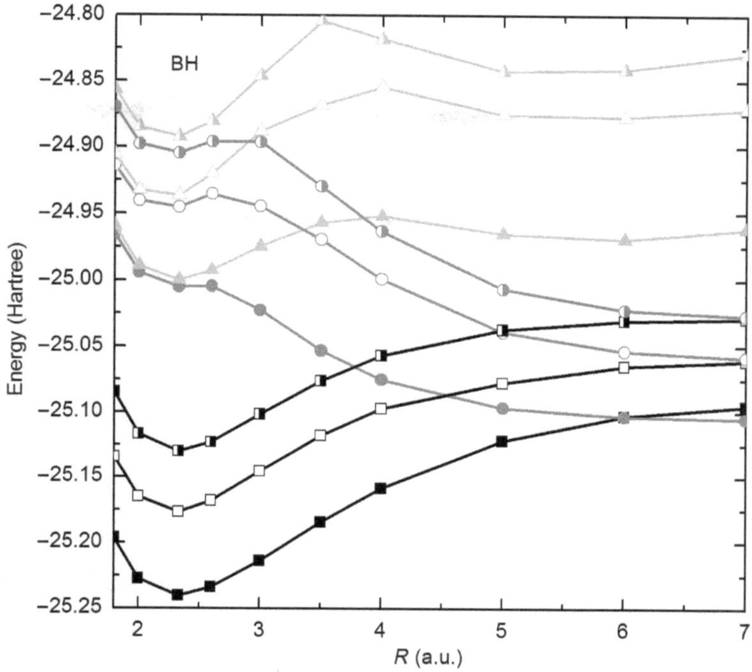

Figure 3 Dissociation curves of the first 3 Σ states of BH Ens-lrWF+srDF ($\beta=1$, $\mu=2$) versus Ens-WF ($\beta=1.0$) compared to CCSD results. Full markers, LR-CCSD; half-filled, Ens-WF; empty, Ens-lrWF+srDF. Squares, $1^1\Sigma^+$ state; circles, $1^3\Sigma^+$; triangles, $2^1\Sigma^+$.

excitation energies would be close to exact for large (10–15 a.u.) values of μ. Because of the small ensemble (i.e., also the configuration space), for large μ, e.g., the excitation energies of Π states are too high in the dissociation limit. Still, for the ground-state calculation for most excitations the CI limit of the method is optimal due to the fact that CI correctly separates the triplet and singlet states. On the other hand, both for the equiensemble case (see Fig. 5) and the $\beta=0.5$ (see Fig. 6) one, the errors for small and medium μ are much smaller than for the ground-state calculation. The optimal value of the range parameter μ both for $\beta=0.5$ and the equiensemble is $\mu=1$, while the optimum for the dissociation limit is shifted toward DFT—$\mu=0.2$ and $\mu=0.3$, respectively. While Rebolini's approach (i.e., Eq. (46)) of obtaining the excitation energies does not seem suitable for Ens-lrWF+srDF method, it is encouraging that the quality of the excitation energies obtained from ensemble calculations is better than those arising from a ground-state one.

More examples of Ens-lrWF+srDF calculations can be found in Ref. 42 but it is clear that while the method includes the description of both static

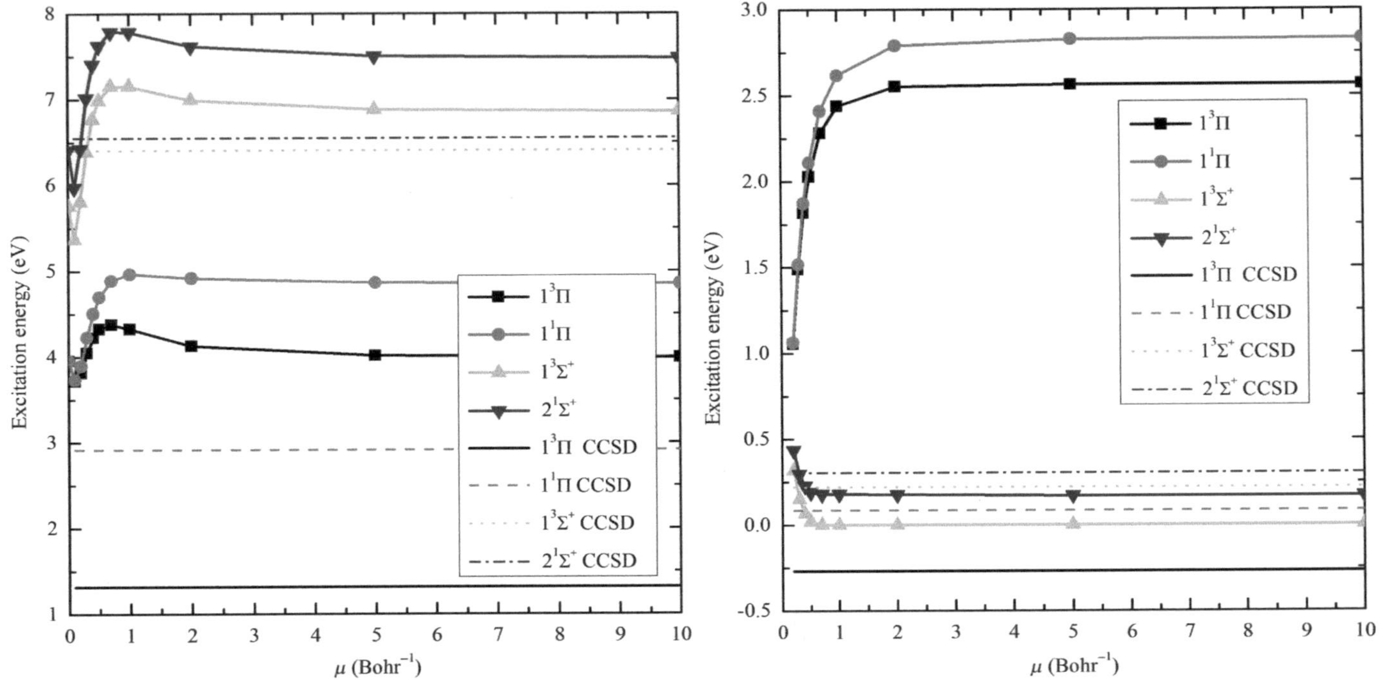

Figure 4 BH ($R_{BH}=2.329$ a.u.—left fig. $R_{BH}=7.0$ a.u.—right fig.) excitation energies along the μ parameter. Ground state (Ens-lrWF+srDF with $\beta \rightarrow \infty$) calculation versus LR-CCSD benchmark.

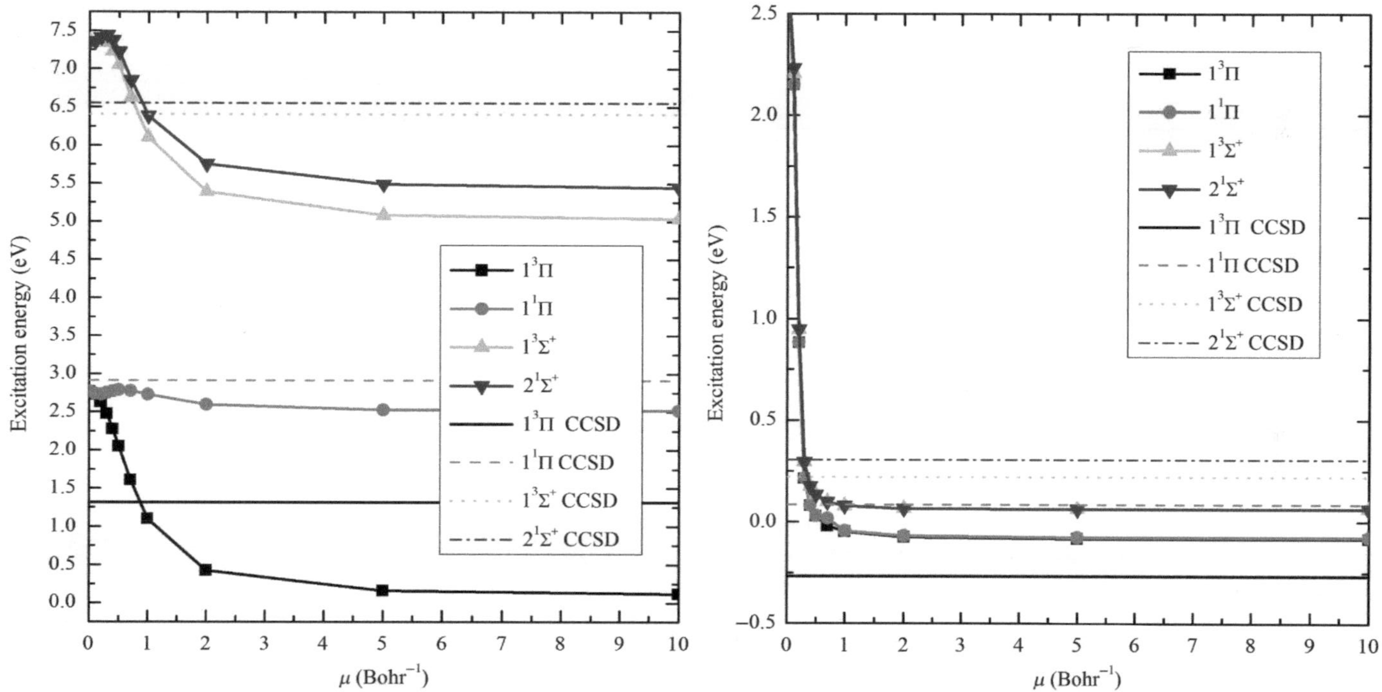

Figure 5 BH ($R_{BH} = 2.329$ a.u.—left fig. $R_{BH} = 7.0$ a.u.—right fig.) excitation energies along the μ parameter. Equiensemble ($\beta = 0$) Ens-lrWF+srDF calculation versus LR-CCSD benchmark.

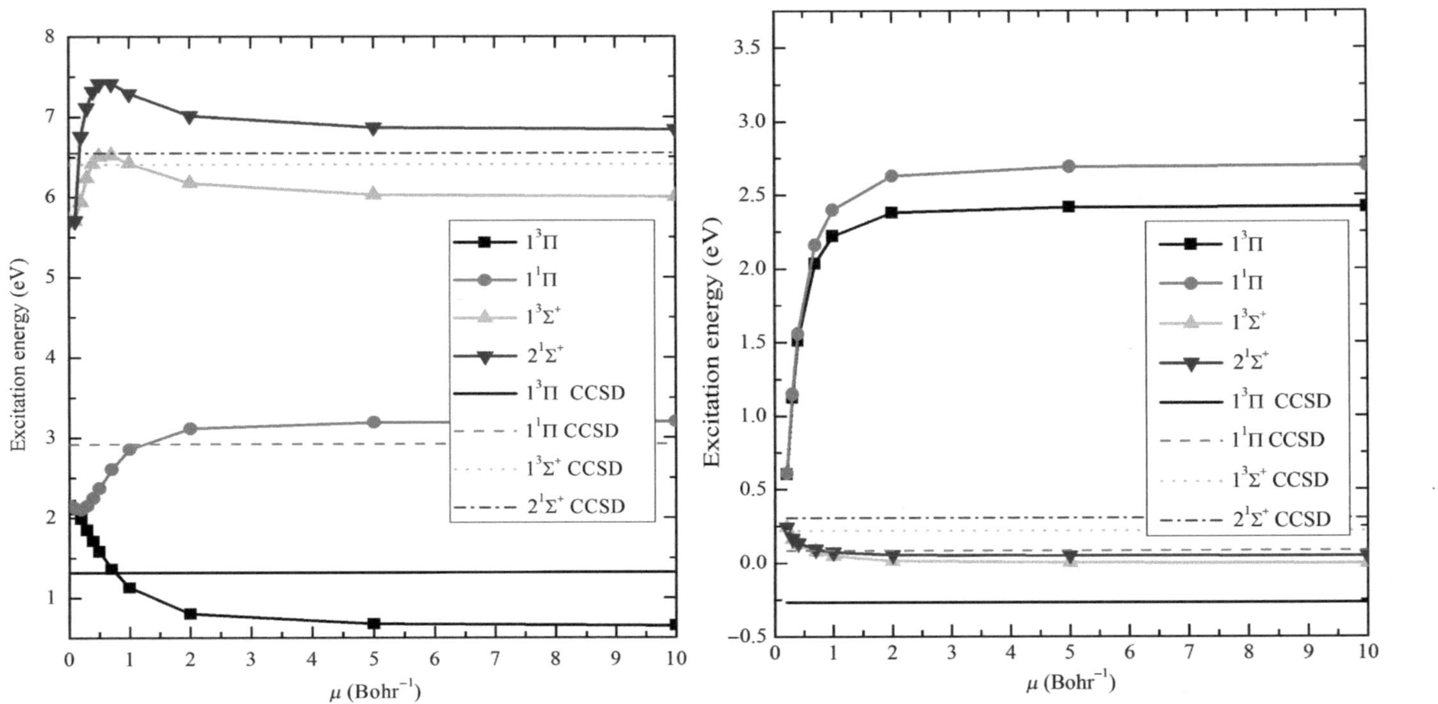

Figure 6 BH ($R_{BH}=2.329$ a.u.—left fig. $R_{BH}=7.0$ a.u.—right fig.) excitation energies along the μ parameter. Ens-lrWF+srDF ($\beta=0.5$) calculation versus LR-CCSD benchmark.

and dynamic correlation at modest cost, it suffers from some serious problems. One of the sources of errors is the previously mentioned issue of describing open-shell states with a density functional without spin polarization. Another is the problem of "ghost interaction", specific to ensemble DFT. To deal with those problems, a different density functional must be employed.

The solution of the first problem is fairly simple—one can just use a spin-polarized short-range functional. This would of course destroy the initial simplicity of the method—with the spin densities instead of the total ensemble density one cannot calculate a common short-range potential and it is impossible to construct an effective Hamiltonian which could then be diagonalized. To minimize the ensemble energy with respect to the expansion coefficients, one needs to use some direct minimization algorithm which is both more costly and more cumbersome.

The problem of the ghost interaction can be solved in a similar manner. As previously mentioned (see Section 1), best results seem to be obtained when the density functional is made ghost-interaction-free by construction.[38,39,58] This leads to the following expression for the ensemble energy

$$E_{m,\omega}[\mathbb{C}] = \sum_{I=1}^{m} \omega_I \left\{ \sum_{JK} C_{IJ} C_{IK} \left\langle \Phi_J \middle| \hat{H}^{\mathrm{LR}} \middle| \Phi_K \right\rangle + E_{m,\omega}^{\mathrm{SR-HXC}} \left[\rho_I^{\alpha}, \rho_I^{\beta} \right] \right\}. \tag{47}$$

where $\hat{H}^{\mathrm{LR}}$, the long-range Hamiltonian, consists of the one-electron operator and the long-range part of the two-electron interaction and $\rho_I^{\alpha}, \rho_I^{\beta}$ are the state spin densities. The possibility of building such a method will be explored in more detail in future work but the result of a test calculation for the hydrogen molecule, involving minimization of the expression (47) with respect to the expansion coefficients and the orbitals is presented in Table 1 along with results calculated by other ensemble methods and a standard TD-DFT calculation. The new method, labeled here MCeDFT, produces more accurate excitation energies (ΔE) than TD-DFT, SA-REKS, and Ens-WF methods but slightly worse than Ens-lrWF+srDF, which can be explained by the introduction of some spin contamination during the optimization of the expansion coefficients. Results for the equilibrium geometry of the hydrogen molecule are not presented because of the lack of data for SA-REKS but also because all the methods are expected to perform well for that case. It has to be emphasized, however, that in MCeDFT a spin adaptation scheme is needed to improve quality of the results.

Table 1 Energies of the First Two Singlet States of H_2 Molecule and the Difference Between them (ΔE_{exc}) in Hartree, Obtained with Different Methods for Two Separations of Hydrogen Nuclei

	R (a.u.)	Ens-WF[a]	Ens-lrWF+srDF[a]	SA-REKS[b]	MCeDFT[a]	TD-DFT[b]	Exact[c]
$1^1\Sigma_g^+$	3	−1.0327	−1.0343	−1.0522	−1.0584	−1.0532	−1.0573
	7	−0.9798	−0.7212	−0.9994	−1.0188	−0.9197	−1.0002
$1^1\Sigma_u^+$	3	−0.5631	−0.9686	−0.8329	−0.6841	−0.7891	−0.7525
	7	−0.5681	−0.6229	−0.8210	−0.5969	−0.8461	−0.6772
ΔE	3	0.4696	0.3131	0.2193	0.3743	0.2640	0.3048
	7	0.4018	0.3457	0.1784	0.4219	0.0736	0.3230

[a]Aug-cc-pVDZ basis set,
[b]Data from Ref. 49,
[c]Data from Refs. 59,60

5. SUMMARY AND OUTLOOK

Despite being somewhat older than the—currently most popular method for the excited states calculations—time-dependent density functional theory, the ensemble methods for the excited states are still in their infancy.

This group of methods is founded upon the ensemble variational (minimum) principle, which states that a weighted sum of *m* expectation values of the Hamiltonian of the system is greater or equal than the sum of *m* lowest eigenvalues of this Hamiltonian, provided that the expectation values are taken with respect to mutually orthogonal trial functions and weights in the sum form a nongrowing sequence with respect to the eigenvalues. This simple inequality is essential in the proof of a HK-type theorem for ensembles that maps the ensemble electron density to an external potential. The ensemble density functional theory established on the basis of the HK theorem is the most promising ensemble method, as it shares DFT's favorable scaling with the size of the system and is potentially able to describe electron correlation.

While ensemble DFT does not share some of TD-DFT's shortcomings caused by, e.g., the use of adiabatic approximations, it has a number of its own challenges and problems. Among them, lack of suitable HXC ensemble functional approximations is the gravest obstacle. Over the years, a number of exact conditions and some clues pointing toward such approximations have been formulated but still most of the available expressions are only slightly modified versions of the ground-state functionals. In addition, imposing the orthogonality condition on the wavefunctions is a challenge in practical calculations and the task of extracting the state energies from the result of an ensemble calculation has no apparent solution.

All these problems inhibited the development and exploration of ensemble methods to theoretical considerations and calculations on small systems which made it difficult to evaluate their true potential. On the other hand, the experience of the ground-state DFT development suggests that even when good XC functional approximations are available, modeling multiconfiguration effects—which is the key motivation to developing time-independent DFT for excited states—is not necessarily an easy task when using a single-reference wavefunction. One can expect similar problems when dealing with the excited states.

Through adapting the ground state DFT's solution to this problem, two types of multiconfiguration approaches to ensemble DFT have been developed. One, based on spin-restricted ensemble-referenced KS methods designed originally for strongly correlated systems, is indeed able to reproduce conical intersections of fairly large systems but is restricted to calculation of the first excited singlet state energy and contains a substantial portion of empiricism. The other path is a range-separated method combining an ensemble wavefunction configuration-interaction-like method and an ensemble density functional. It is able to reproduce dissociation curves of diatomic molecules but it also slightly overestimates the absolute energies (especially of the ground state) and requires some preliminary knowledge about the system. Currently, work is in progress to use ghost-interaction-free, spin-polarized density functionals in the multiconfiguration methods.

Summing up, ensemble DFT, in particular its multiconfiguration realizations have the potential to become an alternative to TD-DFT, especially for systems where multiconfiguration effects come into play. None of the existing approaches, however, has a "black-box" character, nor it is able to treat a wide class of systems with uniform accuracy, or is computationally efficient in its current implementation. These are obstacles that can be overcome in the future by formulating conditions regulating the size of the ensemble, the weights, using more suitable density functional approximations and formulating more efficient and fully self-consistent implementations. Until then, TD-DFT will most likely continue to dominate the density functional approach to the world of excited state calculations.

ACKNOWLEDGMENTS

This work was supported by the National Science Centre of Poland under Grant No. DEC-2012/05/B/ST4/01200.

REFERENCES

1. Theophilou, A. K. Energy Density Functional Formalism for Excited States. *J. Phys. C* **1979**, *12* (24), 5419–5430. http://dx.doi.org/10.1088/0022-3719/12/24/013.
2. Casida, M. E.; Jamorski, C.; Casida, K. C.; Salahub, D. R. Molecular Excitation Energies to High-Lying Bound States from Time-Dependent Density-Functional Response Theory: Characterization and Correction of the Time-Dependent Local Density Approximation Ionization Threshold. *J. Chem. Phys.* **1998**, *108* (11), 4439–4449.
3. Dreuw, A.; Head-Gordon, M. Failure of Time-Dependent Density Functional Theory for Long-Range Charge-Transfer Excited States: The Zincbacteriochlorin-Bacteriochlorin and Bacteriochlorophyll-Spheroidene Complexes. *J. Am. Chem. Soc.* **2004**, *126* (12), 4007–4016.
4. Eriksen, J. J.; Sauer, S. P.; Mikkelsen, K. V.; Christiansen, O.; Jensen, H. J. A.; Kongsted, J. Failures of TDDFT in Describing the Lowest Intramolecular Charge-Transfer Excitation in Para-Nitroaniline. *Mol. Phys.* **2013**, *111* (9-11), 1235–1248.
5. Neugebauer, J.; Baerends, E. J.; Nooijen, M. Vibronic Coupling and Double Excitations in Linear Response Time-Dependent Density Functional Calculations: Dipole-Allowed States of N_2. *J. Chem. Phys.* **2004**, *121* (13), 6155–6166.
6. Görling, A. Density-Functional Theory for Excited States. *Phys. Rev. A* **1996**, *54* (5), 3912.
7. Ziegler, T.; Seth, M.; Krykunov, M.; Autschbach, J.; Wang, F. On the Relation Between Time-Dependent and Variational Density Functional Theory Approaches for the Determination of Excitation Energies and Transition Moments. *J. Chem. Phys.* **2009**, *130* (15), 154102.
8. Seidu, I.; Krykunov, M.; Ziegler, T. The Formulation of a Constricted Variational Density Functional Theory for Double Excitations. *Mol. Phys.* **2014**, *112*, 661–668.
9. Rebolini, E.; Toulouse, J.; Teale, A. M.; Helgaker, T.; Savin, A. Excitation Energies Along a Range-Separated Adiabatic Connection. *J. Chem. Phys.* **2014**, *141* (4), 044123.
10. Evangelista, F. A.; Shushkov, P.; Tully, J. C. Orthogonality Constrained Density Functional Theory for Electronic Excited States. *J. Phys. Chem. A* **2013**, *117* (32), 7378–7392.
11. Rebolini, E.; Toulouse, J.; Teale, A. M.; Helgaker, T.; Savin, A. Calculating Excitation Energies by Extrapolation Along Adiabatic Connections. *Phys. Rev. A* **2015**, *91* (3), 032519.
12. Grimme, S.; Waletzke, M. A Combination of Kohn-Sham Density Functional Theory and Multi-Reference Configuration Interaction Methods. *J. Chem. Phys.* **1999**, *111* (13), 5645–5655.
13. Gross, E. K. U.; Oliveira, L. N.; Kohn, W. Rayleigh-Ritz Variational Principle for Ensembles of Fractionally Occupied States. *Phys. Rev. A* **1988**, *37* (8), 2805–2808. http://dx.doi.org/10.1103/PhysRevA.37.2805.
14. Gidopoulos, N. I.; Theophilou, A. K. Hartree-Fock Equations Determining the Optimum Set of Spin Orbitals for the Expansion of Excited-States. *Philos. Mag. B* **1994**, *69* (5), 1067–1074. http://dx.doi.org/10.1080/01418639408240176.
15. Stoddart, J. C.; Davis, K. Generalization of the Kohn Sham Rho-4/3 Result for Excited States. *Solid State Commun.* **1982**, *42* (2), 147–148.
16. Gidopoulos, N. I.; Papaconstantinou, P. G.; Gross, E. K. U. Ensemble-Hartree-Fock Scheme for Excited States. The Optimized Effective Potential Method. *Physica B* **2002**, *318* (4), 328–332. http://dx.doi.org/10.1016/S0921-4526(02)00799-8.

17. Gross, E. K. U.; Oliveira, L. N.; Kohn, W. Density-Functional Theory for Ensembles of Fractionally Occupied States. I. Basic Formalism. *Phys. Rev. A* **1988**, *37* (8), 2809–2820. http://dx.doi.org/10.1103/PhysRevA.37.2809.
18. Kryachko, E. S. On the Original Proof by Reductio ad Absurdum of the Hohenberg-Kohn Theorem for Many-Electron Coulomb Systems. *Int. J. Quant. Chem.* **2005**, *103* (6), 818–823.
19. Katriel, J.; Zahariev, F. Hamiltonian Formulation of Weighted-Ensemble Density-Functional Theory. *Phys. Rev. A* **2002**, *65* (2), 024501.
20. Gross, E. K. U.; Oliveira, L. N.; Kohn, W. Density-Functional Theory for Ensembles of Fractionally Occupied States. II. Application to the He Atom. *Phys. Rev. A* **1988**, *37* (8), 2821–2833. http://dx.doi.org/10.1103/PhysRevA.37.2821.
21. Vosko, S. H.; Wilk, L.; Nusair, M. Accurate Spin-Dependent Electron Liquid Correlation Energies for Local Spin Density Calculations: A Critical Analysis. *Can. J. Phys.* **1980**, *58* (8), 1200. http://dx.doi.org/10.1139/p80-159.
22. Kohn, W. Density-Functional Theory for Excited States in a Quasi-Local-Density Approximation. *Phys. Rev. A* **1986**, *34* (2), 737–741. http://dx.doi.org/10.1103/PhysRevA.34.737.
23. Glushkov, V. N.; Levy, M. Optimized Effective Potential Method for Individual Low-Lying Excited States. *J. Chem. Phys.* **2007**, *126* (17), 174106.
24. Nagy, A. An Alternative Optimized Potential Method for Ensembles of Excited States. *J. Phys. B* **2001**, *34* (12), 2363. URL: http://stacks.iop.org/0953-4075/34/i=12/a=305.
25. Andrejkovics, I.; Nagy, A. Excitation Energies in Density Functional Theory: Comparison of Several Methods for the H2O, N2, CO and C2H4 Molecules. *Chem. Phys. Lett.* **1998**, *296* (5-6), 489–493. http://dx.doi.org/10.1016/S0009-2614(98)0175-6.
26. Gidopoulos, N. I.; Papaconstantinou, P. G.; Gross, E. K. U. Spurious Interactions, and Their Correction, in the Ensemble-Kohn-Sham Scheme for Excited States. *Phys. Rev. Lett.* **2002**, *88* (3), 033003.
27. Hadjisavvas, N.; Theophilou, A. Rigorous Formulation of Slater's Transition-State Theory for Excited States. *Phys. Rev. A* **1985**, *32* (2), 720–724. http://dx.doi.org/10.1103/PhysRevA.32.720.
28. Glushkov, V. N.; Theophilou, A. K. Application of the Subspace Density Functional Theory to the Excitation Energies of Molecules. *J. Phys. B* **2002**, *35* (10), 2313–2324. http://dx.doi.org/10.1088/0953-4075/35/10/310.
29. Paragi, G.; Gyemant, I. K.; Van Doren, V. E. Investigation of Exchange Potentials for Excited States by Parameter Fitting. *Chem. Phys. Lett.* **2000**, *324* (5-6), 440–446. http://dx.doi.org/10.1016/S0009-2614(00)00613-8.
30. Paragi, G.; Gyemant, I. K.; Van Doren, V. E. Investigation of Exchange-Correlation Potentials in Ensemble Density Functional Theory: Parameter Fitting and Excitation Energy. *J. Mol. Struct.(Theochem)* **2001**, *571* (5-6), 153–161. http://dx.doi.org/10.1016/S0166-1280(01)00561-9.
31. Nagy, Á. Excited States in Density Functional Theory. *Int. J. Quant. Chem.* **1998**, *70* (4-5), 681–691.
32. Joubert, D. P. Exact Expressions for Ensemble Functionals from Particle Number Dependence. *J. Chem. Phys.* **2012**, *136* (17), 174113.
33. Joubert, D. P. Some Formal Properties of Ensemble Density Functionals. *Int. J. Quant. Chem.* **2013**, *113* (8), 1076–1085.
34. Nagy, A. Coordinate Scaling and Adiabatic Connection Formula for Ensembles of Fractionally Occupied Excited States. *Int. J. Quant. Chem.* **1995**, *56* (4), 225.
35. Franck, O.; Fromager, E. Generalised Adiabatic Connection in Ensemble Density-Functional Theory for Excited States: Example of the H_2 Molecule. *Mol. Phys.* **2014**, *112* (12), 1684–1701.

36. Yang, Z.-h.; Trail, J. R.; Pribram-Jones, A.; Burke, K.; Needs, R. J.; Ullrich, C. A. Exact and Approximate Kohn-Sham Potentials in Ensemble Density-Functional Theory. *Phys. Rev. A* **2014**, *90* (4), 042501.
37. Umrigar, C. J.; Gonze, X. Accurate Exchange-Correlation Potentials and Total-Energy Components for the Helium Isoelectronic Series. *Phys. Rev. A* **1994**, *50* (5), 3827.
38. Pastorczak, E.; Pernal, K. Ensemble Density Variational Methods with Self- and Ghost-Interaction-Corrected Functionals. *J. Chem. Phys.* **2014**, *140* (18), 18A514.
39. Pribram-Jones, A.; Yang, Z.-h.; Trail, J. R.; Burke, K.; Needs, R. J.; Ullrich, C. A. Excitations and Benchmark Ensemble Density Functional Theory for Two Electrons. *J. Chem. Phys.* **2014**, *140* (18), 18A541.
40. Oliveira, L. N.; Gross, E. K. U.; Kohn, W. Ensemble-Density Functional Theory for Excited States. *Int. J. Quant. Chem.* **1990**, *24*, 707.
41. Levy, M. Excitation Energies from Density-Functional Orbital Energies. *Phys. Rev. A* **1995**, *52* (6), R4313–R4315. http://dx.doi.org/10.1103/PhysRevA.52.R4313.
42. Pastorczak, E.; Gidopoulos, N. I.; Pernal, K. Calculation of Electronic Excited States of Molecules Using the Helmholtz Free-Energy Minimum Principle. *Phys. Rev. A* **2013**, *87* (6), 062501.
43. Slater, J. C. *The Self-Consistent Field for Molecules and Solids. Quantum Theory of Molecules and Solids*, Vol. 4; McGraw-Hill Book Inc., 1974.
44. Blair, A. I.; Kroukis, A.; Gidopoulos, N. I. A Correction for the Hartree-Fock Density of States for Jellium Without Screening. *J. Chem. Phys.* **2015**, *142* (8), 084116.
45. Werner, H. J.; Meyer, W. A Quadratically Convergent MCSCF Method for the Simultaneous Optimization of Several States. *J. Chem. Phys.* **1981**, *74* (10), 5794.
46. Deskevich, M. P.; Nesbitt, D. J.; Werner, H. J. Dynamically Weighted Multiconfiguration Self-Consistent Field: Multistate Calculations for $F + H_2O \rightarrow HF + OH$ Reaction Paths. *J. Chem. Phys.* **2004**, *120* (16), 7281–7289.
47. Filatov, M.; Shaik, S. A Spin-Restricted Ensemble-Referenced Kohn-Sham Method and Its Application to Diradicaloid Situations. *Chem. Phys. Lett.* **1999**, *304* (429-437).
48. Filatov, M. Spin-Restricted Ensemble-Referenced Kohn-Sham Method: Basic Principles and Application to Strongly Correlated Ground and Excited States of Molecules. *WIREs Comput. Mol. Sci.* **2015**, *5* (1), 146–167.
49. Kazaryan, A.; Heuver, J.; Filatov, M. Excitation Energies from Spin-Restricted Ensemble-Referenced Kohn-Sham Method: A State-Average Approach. *J. Phys. Chem. A* **2008**, *112* (50), 12980–12988.
50. Huix-Rotllant, M.; Filatov, M.; Gozem, S.; Schapiro, I.; Olivucci, M.; Ferré, N. Assessment of Density Functional Theory for Describing the Correlation Effects on the Ground and Excited State Potential Energy Surfaces of a Retinal Chromophore Model. *J. Chem. Theory Comput.* **2013**, *9* (9), 3917–3932.
51. Ziegler, T.; Rauk, A.; Baerends, E. J. On the Calculation of Multiplet Energies by the Hartree-Fock-Slater Method. *Theor. Chim. Acta* **1977**, *43* (3), 261–271. http://dx.doi.org/10.1007/BF00551551. URL: http://dx.doi.org/10.1007/BF00551551.
52. Harris, J. Adiabatic-Connection Approach to Kohn-Sham Theory. *Phys. Rev. A* **1984**, *29* (4), 1648.
53. Toulouse, J.; Colonna, F.; Savin, A. Long-Range-Short-Range Separation of the Electron-Electron Interaction in Density-Functional Theory. *Phys. Rev. A* **2004**, *70* (6), 062505.
54. Savin, A. On Degeneracy, Near-Degeneracy and Density Functional. In: Seminario, J. M., Ed.; Recent Developments and Applications of Modern Density Functional Theory;; Elsevier: Amsterdam, 1996; pp 327–357.
55. Stoll, H.; Savin, A. Density Functionals for Correlation Energies of Atoms and Molecules. In: Dreizler, R. M., da Providencia, J., Eds.; Density Functional Methods in Physics; Plenum: New York, 1985; pp 177–207.

56. Perdew, J. P.; Burke, K.; Ernzerhof, M. Generalized Gradient Approximation Made Simple. *Phys. Rev. Lett.* **1996**, 77 (18), 3865.
57. Goll, E.; Werner, H. J.; Stoll, H. A Short-Range Gradient-Corrected Density Functional in Long-Range Coupled-Cluster Calculations for Rare Gas Dimers. *Phys. Chem. Chem. Phys.* **2005**, 7, 3917–3923.
58. Nagy, A. Optimized Potential Method for Ensembles of Excited States. *Int. J. Quant. Chem.* **1998**, *69* (3), 247–254.
59. Kołos, W.; Wolniewicz, L. Potential-Energy Curves for the X $^1\Sigma_g^+$, B $^3\Sigma_u^+$, and C $^1\pi_u$ States of the Hydrogen Molecule. *J. Chem. Phys.* **1965**, *43* (7), 2429.
60. Kołos, W.; Wolniewicz, L. Potential-Energy Curve for the B $^1\Sigma_u^+$ State of the Hydrogen Molecule. *J. Chem. Phys.* **1966**, *45* (2), 509.

CHAPTER TWELVE

Application of the Space-Pseudo-Time Method to Density Functional Theory

Daniel H. Gebremedhin[1], **Charles A. Weatherford**[1]
Department of Physics, Florida A&M University, Florida, USA
[1]Corresponding authors: e-mail address: daniel1.gebremedhin@famu.edu; charles.weatherford@famu.edu

Contents

Abstract

A numerical solution of the Kohn–Sham (KS) differential equation within the local density approximation is presented. The present method involves solving for the Hartree potential from its differential form which is the Poisson equation. Radial differential equations are derived for closed-shell atoms and are solved as initial value problems. A self-consistent procedure for solving the resulting radial KS equations based on our new algorithm for solving differential equations is also discussed. Numerical tests are done on the Helium atom and comparison with results obtained from spectral exponential and Gaussian type basis functions are shown.

1. INTRODUCTION

Density functional theory (DFT) is a very powerful technique which is applicable to atomic, molecular, and condensed matter systems. DFT employs a one-particle equation, albeit nonlinear since the effective potential depends on the electron density. This effective potential is generally not known exactly and obtaining a good approximation for it is a major area of research interest. In contrast, Hartree–Fock (HF) theory also leads to a

Advances in Quantum Chemistry, Volume 73
ISSN 0065-3276
http://dx.doi.org/10.1016/bs.aiq.2015.06.004

similar one-particle equation, but all of the potentials are well defined. This has allowed for the exploration of its full capacity and the extent of its limitations are well known.

Various approximations for the exchange-correlation (XC) functional have been defined. In this chapter, we will employ the most common XC functional, namely, the local density approximation (LDA) and present an efficient numerical solution to the Kohn–Sham DFT (KSDFT) equation for atoms based on the resulting effective potential. This work has been motivated by the availability of a new numerical solver for ordinary differential equations (DEs).[1] We have generally called this algorithm the space-time method where the time can be real, imaginary, or pseudo, as in the present case for the radial variable, which we treat as a pseudo-time. In addition, a recent article by Bagayoko[2] makes the point that in order to have a truly *ab initio* KSDFT, one must attain orbital energies that are a variationally minimum, and only then can an evaluation of the quality of the XC be made. In practice, Bagayoko et al. methodically increased their basis set size by adding more and more functions into their basis until the occupied orbital energies reach their minimum. In addition, on the physical systems they considered, once the basis set error has been eliminated, an LDA is sufficient to produce numerical results that are in good agreement with experiment. The present work is in a sense a partial test of these conclusions for the most simple multi-electron atom since there is essentially no basis set error in the present method. That is, instead of a diagonalization, very accurate KSDFT eigenvalues are obtained by a shooting method using a spectral element method with first derivative continuity across the element boundaries.

In the next section, we will present a summary of the KSDFT. Specifically, the KS equation, the Poisson equation and the LDA are discussed. Then a radial differential equation, appropriate for the XC functional, is derived and a self-consistent solution is explained. Finally, numerical results, comparisons with optimized exponential and Gaussian type basis function results, and conclusions are presented.

2. KOHN–SHAM EQUATIONS

In this section, will setup the necessary notation and formalism that will lead to derivation of ordinary DEs which will be numerically solved later. Thus the main equations of KSDFT will be briefly summarized.

We will use spherical coordinates and the notation used in Ref. 3 will be followed.

The Kohn–Sham (KS) equation is[4] (atomic units will be used though out)

$$\left(-\frac{1}{2}\nabla_{\mathbf{r}}^2 + v_{\text{eff}}(\mathbf{r})\right)\Psi_i(\mathbf{r}) = \epsilon_i\Psi_i(\mathbf{r}) \tag{1}$$

where, the KS effective potential is given by:

$$v_{\text{eff}}(\mathbf{r}) = v_{\text{ext}}(\mathbf{r}) + \mathbf{\Phi}(\mathbf{r}) + v_{\text{xc}}(\mathbf{r}) \tag{2}$$

The first term is the external potential due to the nuclei

$$v_{\text{ext}}(\mathbf{r}) = -\sum_{A=1}^{M}\frac{Z_A}{|\mathbf{r}-\mathbf{r}_A|} \tag{3}$$

where, M is the number label of the nucleus in the system (atom, molecule, extended system). Z_A and $\mathbf{r}_A$ are the atomic number and fixed positions respectively of each of the nuclei. $\mathbf{\Phi}$ is commonly referred to as the Hartree potential (HP) defined by

$$\mathbf{\Phi}(\mathbf{r}) = \int \frac{\rho(\mathbf{x})}{|\mathbf{r}-\mathbf{x}|}\,\mathrm{d}^3x \tag{4}$$

The last term in Eq. (2) is the XC potential defined as

$$v_{\text{xc}}[\rho](\mathbf{r}) = \frac{\delta E_{\text{xc}}[\rho]}{\delta\rho(\mathbf{r})} \tag{5}$$

with E_{xc} being exchange-correlation energy. δ represents the functional derivative with respect to the electron density ρ. The density function is expressed as a sum over the spatial orbital densities as

$$\rho(\mathbf{r}) = \sum_{i=1}^{N}|\Psi_i(\mathbf{r})|^2 \tag{6}$$

with a number fixed N of electrons.

$$N = \int \rho(\mathbf{r})\,\mathrm{d}^3r \tag{7}$$

The parameters M and N completely specify the system in question. The ground state energy E_0 is then given by:

$$E_0 = \sum_{i=1}^{N} \epsilon_i - \frac{1}{2} \int \Phi(\mathbf{r}) \rho(\mathbf{r})\, \mathrm{d}^3 r + E_{\mathrm{xc}}[\rho](\mathbf{r}) - \int v_{\mathrm{xc}}[\rho](\mathbf{r}) \rho(\mathbf{r})\, \mathrm{d}^3 r \tag{8}$$

The integral equation given in Eq. (4) is not necessarily the most convenient way of calculating the HP.

2.1 Poisson's Equation

The HP can also be evaluated from its equivalent differential form which is the Poisson's equation given below.

$$\nabla_{\mathbf{r}}^2 \Phi(\mathbf{r}) = -4\pi\rho(\mathbf{r}) \tag{9}$$

In this chapter, we are going to propagate solutions to both Eq. (1) and Eq. (9), and hence, knowledge about the boundary conditions of the respective solution functions at the origin is required. The form of the HF potential at the origin can be derived directly from Eq. (4). To this end, we employ the Laplace expansion shown below with the first two terms written explicitly.

$$\begin{aligned} \frac{1}{|\mathbf{r} - \mathbf{x}|} &= \frac{4\pi}{r_>} \left[\frac{1}{4\pi} + \frac{r_<}{3 r_>} \sum_{m=-1}^{1} Y_1^{m*}(\hat{x}) Y_1^m(\hat{r}) \right. \\ &\left. + \sum_{l=2}^{\infty} \frac{1}{2l+1} \left(\frac{r_<}{r_>} \right)^l \sum_{m=-l}^{l} Y_l^{m*}(\hat{x}) Y_l^m(\hat{r}) \right] \end{aligned} \tag{10}$$

where, Y is a spherical harmonic. Hence, after taking the required derivatives of the above equation with respect to $r_<$, we can let $r_< \to 0$ and substitute the resulting expression into Eq. (4) to get the following boundary conditions of the HP at the origin.

$$\Phi(0) = \int \mathrm{d}\hat{x} \int_0^\infty x \rho(\mathbf{x})\, \mathrm{d}x \tag{11}$$

$$\left. \frac{\mathrm{d}}{\mathrm{d}r} \Phi(\mathbf{r}) \right|_{r=0} = \frac{4\pi}{3} \sum_{m=-1}^{1} Y_1^m(\hat{r}) \int \mathrm{d}\hat{x}\, Y_1^{m*}(\hat{x}) \int_0^\infty \rho(\mathbf{x})\, \mathrm{d}x \tag{12}$$

Φ can now be formally expanded in spherical harmonics as shown below, which is best suited for atoms due to symmetry.

$$\Phi(\mathbf{r}) = \sum_{l=0}^{\infty} \sum_{m=-l}^{l} \phi_l^m(r) Y_l^m(\hat{r}) \tag{13}$$

Substituting this into Eq. (9) and making the necessary projections we get,

$$\left(\frac{\mathrm{d}^2}{\mathrm{d}r^2}+\frac{2}{r}\frac{\mathrm{d}}{\mathrm{d}r}-\frac{l(l+1)}{r^2}\right)\phi_l^m(r)=-4\pi\int\rho(\mathbf{r})Y_l^{m*}(\hat{r})\,\mathrm{d}\hat{r}. \tag{14}$$

For the important case when the density is spherically symmetric, i.e., $\rho(\mathbf{r})\equiv\rho(r)=R(r)\left|Y_0^0(\hat{r})\right|^2=R(r)/4\pi$, we have

$$\left(\frac{\mathrm{d}^2}{\mathrm{d}r^2}+\frac{2}{r}\frac{\mathrm{d}}{\mathrm{d}r}-\frac{l(l+1)}{r^2}\right)\phi_l^m(r)=-R(r)\int Y_l^{m*}(\hat{r})\,\mathrm{d}\hat{r}. \tag{15}$$

R will be further specified in Section 3. Using the identity

$$\int Y_l^{m*}(\hat{r})\,\mathrm{d}\hat{r}=\sqrt{4\pi}\,\delta_{l0}\delta_{m0} \tag{16}$$

only $l=m=0$ term survives and we have

$$\left(\frac{\mathrm{d}^2}{\mathrm{d}r^2}+\frac{2}{r}\frac{\mathrm{d}}{\mathrm{d}r}\right)\phi(r)=-\sqrt{4\pi}R(r) \tag{17}$$

where we have dropped the zero quantum numbers on ϕ. From the expansion shown in Eq. (13), we can further identify $\Phi(\mathbf{r})\equiv\Phi(r)=\phi(r)/\sqrt{4\pi}$ and rewrite the above equation as

$$\left(\frac{\mathrm{d}^2}{\mathrm{d}r^2}+\frac{2}{r}\frac{\mathrm{d}}{\mathrm{d}r}\right)\Phi(r)=-R(r) \tag{18}$$

In the present case, the boundary conditions given in Eqs. (11) and (12) will simplify to

$$\Phi(0)=\int_0^\infty xR(x)\,\mathrm{d}x \tag{19}$$

$$\left.\frac{\mathrm{d}}{\mathrm{d}r}\Phi(r)\right|_{r=0}=0. \tag{20}$$

The above three equations will be used below to calculate the HP. For the sake of completeness, we multiply Eq. (18) from left by r^2 and turn the DE into the following self-adjoint form.

$$\frac{\mathrm{d}}{\mathrm{d}r}\left(r^2\frac{\mathrm{d}}{\mathrm{d}r}\right)\Phi(r)=-r^2R(r) \tag{21}$$

Integrating the above equation we get the following familiar electrostatics equation regarding electric field intensity,[5] that is also consistent with Eq. (20).

$$-\frac{\mathrm{d}}{\mathrm{d}r}\Phi(r) = \frac{1}{r^2}\int_0^r x^2 R(x)\,\mathrm{d}x \tag{22}$$

Next, we shall outline a similar one-variable representation of the last term of the effective potential in Eq. (2).

2.2 Local Density Approximation

The LDA, which is based on a homogeneous electron gas, is the approximation of the XC functional that will be implemented in this chapter. This uniformity leads to an XC potential that is a functional of only the density, unlike, say, inhomogeneous models that would generally have dependence on higher order local variations. More specifically, the LDA approximation of an XC energy functional is represented as

$$E_{\mathrm{xc}}[\rho](\mathbf{r}) = \int \epsilon_{\mathrm{xc}}[\rho](\mathbf{r})\rho(\mathbf{r})\,\mathrm{d}^3 r \tag{23}$$

where, ϵ_{xc} is the XC energy per electron of a homogeneous electron gas with density ρ. Hence, from the definition given in Eq. (5) and variational calculus, we get the following equivalent relation for the corresponding XC potential.

$$v_{\mathrm{xc}}[\rho](\mathbf{r}) = \epsilon_{\mathrm{xc}}(\rho) + \rho\frac{\mathrm{d}}{\mathrm{d}\rho}\epsilon_{\mathrm{xc}}(\rho) \tag{24}$$

A homogeneous electron gas must also imply spherical symmetry so the above two equations can be written as

$$E_{\mathrm{xc}}[\rho](r) = \int_0^\infty r^2 R(r)\epsilon_{\mathrm{xc}}[\rho](r)\,\mathrm{d}r \tag{25}$$

and

$$v_{\mathrm{xc}}[\rho](r) = \epsilon_{\mathrm{xc}}(\rho) + \rho\frac{\delta}{\delta\rho}\epsilon_{\mathrm{xc}}(\rho) \tag{26}$$

Now, the XC energy density is usually separated into exchange (ϵ_{x}) and correlation (ϵ_{c}) parts as,

$$\epsilon_{\mathrm{xc}}(\rho) = \epsilon_{\mathrm{x}}(\rho) + \epsilon_{\mathrm{c}}(\rho) \tag{27}$$

with analogous quantities for the potential v_{xc} as well. For the homogeneous electron gas, the exchange part is known exactly. It is

$$\epsilon_{\mathrm{x}}(\rho) = -\frac{3}{4}\left(\frac{3\rho}{\pi}\right)^{1/3} \tag{28}$$

Substituting into Eq. (26), we have

$$v_{\mathrm{x}}(\rho) = -\left(\frac{3\rho}{\pi}\right)^{1/3} \tag{29}$$

For the correlation part, we shall adopt the well-known analytic expressions which were derived in Ref. 6. This particular formula is the result of an interpolation based on data generated using a numerical quantum Monte Carlo method published in Ref. 7. The formulas are given in terms of a spherical radial parameter r_s, known as the Wigner-sphere radius.

$$r_s(\rho) = \left(\frac{3}{4\pi\rho}\right)^{1/3} \tag{30}$$

The above function maps a given density into a coordinate position as shown. Once we have obtained an r_s from our density function ρ, then (both inside and outside of a unit Wigner-sphere) the energy density can be evaluated using

$$\epsilon_{\mathrm{c}}(r_s) = \begin{cases} A \ln r_s + B + C r_s \ln r_s + D r_s, & \text{if } r_s < 1 \\ \dfrac{\gamma}{1 + \beta_1\sqrt{r_s} + \beta_2 r_s}, & \text{if } r_s \geq 1. \end{cases} \tag{31}$$

Note that the inside region has a higher value of the density. The correlation part of Eq. (26) is now equivalent to the following (using a chain rule and Eq. (30)):

$$v_{\mathrm{c}}(r_s) = \left(1 - \frac{r_s}{3}\frac{\delta}{\delta r_s}\right)\epsilon_{\mathrm{c}}(r_s) \tag{32}$$

Then upon application on Eq. (31), we get

$$v_{\mathrm{c}}(r_s) = \begin{cases} A \ln r_s + (B - \frac{1}{3}A) + \frac{2}{3} C r_s \ln r_s + \frac{1}{3}(2D - C) r_s, & \text{if } r_s < 1 \\ \dfrac{\gamma\left(1 + \frac{7}{6}\beta_1\sqrt{r_s} + \frac{4}{3}\beta_2 r_s\right)}{\left(1 + \beta_1\sqrt{r_s} + \beta_2 r_s\right)^2}, & \text{if } r_s \geq 1 \end{cases} \tag{33}$$

For closed-shell atoms, the correlation potential for an unpolarized uniform electron gas is used and the corresponding numerical values of the above coefficients are

$$\begin{aligned} &A=0.0311, \quad B=-0.048, \quad C=0.0020, \quad D=-0.0116, \\ &\gamma=-0.1423, \quad \beta_1=1.0529, \quad \beta_2=0.3334. \end{aligned} \tag{34}$$

Complete detailed results are available in Ref. 6.

3. RADIAL EQUATIONS

We are now in a position to write the KS radial equation for atoms. Following the same derivation as that of Section 2.1, and supplying the LDA approximation discussed above, we arrive at the following one-variable form of Eq. (1).

$$\left[-\frac{1}{2}\frac{d^2}{dr^2}-\frac{1}{r}\frac{d}{dr}-\frac{Z}{r}+\Phi(r)+v_x(r)+\lambda v_c(r)\right]\psi_i(r)=\epsilon_i\psi_i(r) \tag{35}$$

Note that an expansion similar to Eq. (13) is assumed to have been made on the KS orbital in Eq. (1), naturally leading to the notation $\Psi(\mathbf{r})\equiv\Psi(r)=\psi(r)/\sqrt{4\pi}$. Z is the atomic number of the atom. The last three terms in the above Hamiltonian are functionals of a density function R which, for closed-shell atoms (with even number of electrons), is defined as

$$R(r)=2\sum_{i=1}^{N/2}|\psi_i(r)|^2 \tag{36}$$

where, ψ_i is an eigenfunction of Eq. (35). For open-shell atoms (odd number of electrons), on the other hand, the spin of the electrons is important and the density must be grouped into two sets accordingly, say, $R=R\uparrow+R\downarrow$. Only closed-shell atoms are discussed in this chapter. N is the total number of electrons such that

$$N=\int_0^\infty r^2R(r)\,dr. \tag{37}$$

The HF potential Φ emerges as a solution to Eq. (18), while v_x and v_c are to be calculated from Eqs. (29) and (33) respectively. This dependence of the Hamiltonian on its own eigenfunction makes Eq. (35) a nonlinear differential equation (DE) that can only be solved by iteration using a self-consistent method.

In this chapter, the coefficient λ will be assigned the values of 0 and 1 in order to examine the contribution of the correlation potential to the ground state energy which is explicitly written below

$$E_0 = 2\sum_{i=1}^{N/2}\epsilon_i - \frac{1}{2}\int_0^{\infty} r^2 R(r)\Phi(r)\,\mathrm{d}r + \int_0^{\infty} r^2 R(r)\{\epsilon_{\mathrm{x}}[\rho](r) - v_{\mathrm{x}}[\rho](r)\}\,\mathrm{d}r + \lambda\int_0^{\infty} r^2 R(r)\{\epsilon_{\mathrm{c}}[\rho](r) - v_{\mathrm{c}}[\rho](r)\}\,\mathrm{d}r \tag{38}$$

ϵ_{x} and ϵ_{c} are given in Eqs. (28) and (31), respectively, while ϵ_i is obtained from Eq. (35). This completes the discussion of the set of equations necessary to perform numerical evaluation, which mainly involves solving Eqs. (35) and (18) using our space-pseudo-time algorithm (SPTa).

The SPTa is a finite element-based propagation method that introduces a new set of local basis functions that enforce both function and first derivative continuity across two consecutive elements. It leads to a system of simultaneous equations which can be constructed by a collocation technique that only involves evaluation of the potentials at grid points. This avoids evaluation of projection integrals completely, which makes it ideally suited for any forms of the XC potentials that might arise in DFT. Here, we will only mention aspects of the SPTa required for the current content referring further details to Ref. 1.

For a particular finite element in r, let τ be a corresponding local variable with domain $-1 \le \tau \le 1$ defined by the linear transformation

$$r = q\tau + p \tag{39}$$

where, p is the mid-point and q half the size of the element.

Then the SPTa involves decomposing the required function $\bar{F}(\tau)$ into a local basis functions as shown below:

$$\bar{F}(\tau) = \sum_{\mu=0}^{M-1} u_\mu(\tau)B_\mu + s_0(\tau)\dot{\bar{F}}(-1) + \bar{F}(-1) \tag{40}$$

In the present case, $\bar{F}$ can represent either of Φ or ψ as a function of τ. The overhead dot means derivative with respect to the argument τ. The basis functions s and u are defined in terms of Legendre polynomials of the first kind.[1] The values of the function and its first derivative are required at the beginning of each element. In particular, at the origin, those values with regard to the HF potential, are given in Eqs. (19) and (20). For the KS orbital ψ, however, we can arbitrarily fix $\psi_i(r=0)=1$ and renormalize later according to Eq. (37). Since, the value of the derivative of the KS orbital at the origin is unknown, only for the very first element, we have to expand in terms of the s functions as shown below.

$$\bar{F}(\tau) = \sum_{\mu=0}^{M-1} s_\mu(\tau) B_\mu + \psi(r=0) \tag{41}$$

The rest of the propagation can then resume normally using Eq. (40).

As is mentioned above, Eq. (35) is a nonlinear DE that can only be solved iteratively. We seeded the first iteration by ignoring the terms in the potential that depend on the density function and solving the resulting hydrogen-like atom using the exact known eigenvalues. An $r_{\max}$, which limits the domain of propagation must also be set. The rest of the iterative procedure is summarized below.

1. For a given eigenvalues ϵ_i, solve Eq. (35) for $\psi_i(r), \forall r \in [0, r_{\max}]$ using the SPTa. For each orbital ψ_i, store all the necessary parameters $\left(p, q, B, \dot{\bar{\psi}}_i(-1), \bar{\psi}_i(-1)\right)$ of all the elements so that ψ can be readily computed at any desired point in $[0, r_{\max}]$.
2. Calculate and save the coefficients that normalize each ψ_i to unity so that Eq. (37) is obeyed.
3. Utilizing the above two steps, setup a subroutine that can compute R using Eq. (37) at any required point inside $[0, r_{\max}]$.
4. Initialize the HF potential according to Eqs. (19) and (20) and solve Eq. (18) up to $\Phi(r_{\max})$ using the SPTa. Similarly, store the parameters $\left(p, q, B, \dot{\bar{\Phi}}(-1), \bar{\Phi}(-1)\right)$ for all elements.
5. Now that we can evaluate all the terms in the effective potential, locate all of the eigenvalues (ϵ_i) of the resulting system in Eq. (35) by employing the SPTa.
6. Check for convergence and go back to step (1) if the eigenvalues are not within the required tolerance. Otherwise, exit the loop and calculate the total ground state energy given in Eq. (38).

4. NUMERICAL RESULTS AND CONCLUSIONS

We have tested the present method of solving the KS equations for closed-shell atoms which is based on the LDA to calculate the ground state of the Helium atom ($N = Z = 2$). We used $M = 2$ to 9 local basis functions in the expansion shown in Eq. (40) and (41) and the extent of propagation has been limited to $r_{max} = 30$. For the He atom, we only need to calculate the ground state ϵ_1 of the radial KS equation. We solved for a hydrogen-like atom with $Z = 2$ and $\epsilon_1 = -Z^2/2$, by employing the SPTa, as the 0th iteration in order to generate the first iterate of the density R.

At first, we considered only the exchange potential, which is obtained by setting $\lambda = 0$. Table 1 shows the numerical values of ϵ_1 and E_0 for the basis set sizes shown. The iteration has been allowed to converge to near working (machine) precision. The calculated value for the ground state energy, $E_0 = -2.72363979$, is variationally higher than that of the HF theory limit -2.861.[8] This discrepancy between the two energies is precisely due to the effect of the exchange potential of the LDA versus the exchange potential in HF[9] since, the one particle Hamiltonians of the two theories are identical up to this particular term.

Inclusion of the correlation potential v_c of the LDA produces $E_0 = -2.83429$ which is still a significantly higher value than that of the HF approximation. Table 2 shows ϵ_1 and E_0 for $\lambda = 1$ with increasing basis set size M. Notice that the KS eigenvalue ϵ_1 has converged to about 6 figures confirming the validity the total energy. Apparently, the complete effect of dynamic correlation between the two electrons is not effectively captured by the XC potential of the LDA, which is supposed to include self-interaction corrections.[6] We are confident that the eigenvalues $E_0 = 2.72363979$ and $E_0 = -2.83429$ we have calculated for the values of $\lambda = 0$ and $\lambda = 1$

Table 1 Orbital Energy ϵ_1 and Total Ground State Energy E_0 for $\lambda = 0$

M	ϵ_1	E_0
2	−0.511 612 794 935 716	−2.656 040 136 128 63
3	−0.517 189 483 598 956	−2.725 065 470 961 80
4	−0.516 983 385 518 174	−2.723 623 747 467 99
5	−0.516 967 547 076 162	−2.723 632 478 418 80
6	−0.516 968 117 350 820	−2.723 639 408 499 91
7	−0.516 968 186 463 829	−2.723 639 780 508 34
8	−0.516 968 194 146 871	−2.723 639 793 048 20
9	−0.516 968 192 325 768	−2.723 639 792 907 71

M is the basis set size.

Table 2 Orbital Energy ϵ_1 and Total Ground State Energy E_0 for $\lambda = 1$

M	ϵ_1	E_0
2	−0.564 833 394 205 826	−2.764 292 854 372 73
3	−0.570 462 360 128 714	−2.835 788 416 526 66
4	−0.570 226 127 791 036	−2.834 278 425 069 06
5	−0.570 208 694 947 266	−2.834 281 846 721 41
6	−0.570 207 352 780 398	−2.834 283 505 070 53
7	−0.570 209 655 883 562	−2.834 290 756 201 68
8	−0.570 209 889 841 234	−2.834 291 314 320 30
9	−0.570 209 221 532 859	−2.834 289 260 480 18

M is the basis set size.

respectively in the DFT LDA approximation discussed in Section 2.2, are accurate. This assertion is due to our experience with the efficiency of SPTa. We thus assert that the SPTa has essentially zero basis set error.

For comparison and as a check, we have repeated the above calculation by expanding the orbital $\psi_i(r)$ in exponential type basis functions as shown below:

$$\psi_i(r) = \sum_{n=1}^{n_{\max}} S_n(r)\, C_{ni} \tag{42}$$

where, $S_n(r)$ is the radial part of a Coulomb Sturmian with the quantum number $l = 0$.[10,11] Namely,

$$S_n(\alpha, r) = \frac{2}{n}\alpha^{3/2} e^{-\alpha r} L^1_{n-1}(2\alpha r). \tag{43}$$

The screening parameter α is a real number which can be optimized to produce the lowest ϵ_1 for a given basis set size $n_{\max}$. L is an associated Laguerre polynomial.[12] The Sturmians satisfy the following DE.

$$\left(-\frac{1}{2}\frac{\mathrm{d}^2}{\mathrm{d}r^2} - \frac{1}{r}\frac{\mathrm{d}}{\mathrm{d}r} - \frac{\alpha n}{r} + \frac{\alpha^2}{2}\right) S_n(\alpha, r) = 0 \tag{44}$$

The coefficients C will then be eigenvectors of a generalized eigenvalue problem that follows after the projection integrals have been carrying out. Here, we will only present the numerical results since this procedure, which

Table 3 For $\lambda = 1$, Values of Orbital Energy ϵ_1 and Total Ground State Energy E_0 for Exponential Type Functions

n_{max}	α	ϵ_1	E_0
1	1.155 632 396 609 60	−0.667 743 118 046 828	−2.559 998 755 378 62
2	1.060 664 695 184 11	−0.562 504 795 905 156	−2.729 769 113 031 13
3	1.058 178 984 335 56	−0.559 871 381 444 722	−2.801 739 900 062 80
4	1.064 030 819 011 96	−0.566 080 791 903 628	−2.824 298 775 808 12
5	1.066 755 577 297 80	−0.568 983 730 847 985	−2.831 250 644 759 26
6	1.067 605 793 358 97	−0.569 891 065 006 820	−2.833 394 742 696 95
7	1.067 834 654 149 27	−0.570 135 424 301 047	−2.834 042 432 807 70
8	1.067 889 282 026 37	−0.570 193 759 333 401	−2.834 227 471 751 42
9	1.067 900 835 634 65	−0.570 206 097 374 588	−2.834 275 162 788 13

n_{max} is basis set size and α is the corresponding optimal screening parameter.

is also self-consistent, can be inferred from books such as Ref. 9. Table 3 shows the values of ϵ_1 and E_0 for $\lambda = 1$ for the shown number of terms n_{max} in Eq. (42). The optimized value of screening parameter α that results in the lowest possible ϵ_1 are also given. All iterations are terminated after the eigenvalues were all well converged. Apparently, the resulting E_0 is very close to but still less accurate than the SPTa shown in Table 2.

Further insight can be gained by repeating the above calculation using Gaussian type basis functions. We used radial, isotropic, harmonic oscillator wavefunctions, G, which correspond to the $l = 0$ solutions (in spherical coordinate system) of the Schrödinger equation with a radial harmonic oscillator potential.[13]

$$\psi_i(r) = \sum_{\mu=0}^{\mu_{max}-1} G_\mu(r)\, D_{\mu i} \tag{45}$$

The isotropic functions shown above are explicitly given by:

$$G_\mu(\beta, r) = \left[\sqrt{\pi}(\mu+1)\binom{2\mu+1}{\mu}\right]^{-\frac{1}{2}} 2^{\mu+1}(2\beta)^{3/4} e^{-\beta r^2} L_\mu^{1/2}(2\beta r^2) \tag{46}$$

and they satisfy

$$\left[-\frac{1}{2}\frac{d^2}{dr^2} - \frac{1}{r}\frac{d}{dr} + 2\beta^2 r^2 - \beta(4\mu+3)\right] G_\mu(\beta, r) = 0 \tag{47}$$

Table 4 For $\lambda = 1$, Values of Orbital Energy ϵ_1 and Total Ground State Energy E_0 for Exponential Type Functions

μ_{max}	β	ϵ_1	E_0
1	0.290 447 768 645 546	−0.467 889 875 416 040	−1.965 364 524 158 78
2	0.221 265 648 674 719	−0.425 830 864 263 759	−1.955 325 533 651 76
3	0.500 978 293 189 388	−0.513 236 447 032 326	−2.451 831 952 677 98
4	0.471 754 814 487 701	−0.492 451 569 218 084	−2.474 395 396 809 76
5	0.682 244 037 402 693	−0.535 023 873 031 428	−2.635 511 450 729 97
6	0.662 556 746 819 686	−0.524 831 647 983 934	−2.647 101 654 925 80
7	0.824 647 395 928 554	−0.547 091 062 005 167	−2.715 231 493 677 95
8	0.805 705 340 174 161	−0.541 428 084 424 167	−2.721 025 728 409 96
9	0.943 749 038 429 923	−0.554 169 864 939 769	−2.756 108 179 892 80
10	0.923 820 407 376 820	−0.550 722 835 180 090	−2.759 216 904 515 40
11	1.048 288 713 183 22	−0.558 580 400 233 415	−2.779 572 818 642 79
12	1.027 054 034 209 52	−0.556 337 918 666 443	−2.781 351 299 608 03
13	1.143 106 014 456 49	−0.561 475 241 730 046	−2.794 165 798 209 13
14	1.120 603 588 891 21	−0.559 940 885 687 982	−2.795 237 202 842 40
15	1.230 904 063 982 90	−0.563 459 632 642 412	−2.803 797 356 234 13

μ_{max} is basis set size and β is the corresponding optimal screening parameter.

Table 4 shows the output of the KSDFT calculation using these Gaussian functions. Comparison with Table 3 confirms the fact that exponential type orbitals are much better basis functions compared to Gaussians.

In Figs. 1 and 2, we have plotted the last row ($M = 9$) of the correlation and the XC potentials, respectively, given in Eqs. (29) and (33) for $\lambda = 1$. The exchange potential has a relatively more prominent presence near the origin than the correlation potential. Otherwise, both have a smooth and monotonically decreasing magnitude. The ground state $\psi_1(r)$ of Eq. (35) that correspond to the results given in the last rows of Tables 2 and 4 are also shown in Fig. 3. An analogous plot of the Sturmian basis set (not shown) completely overlaps with that of the SPTa with no visible difference. It is interesting to see the Gaussian plot intercepting the vertical axis with zero derivative, and hence, failing to produce the cusp condition at the origin.

In conclusion, we have outlined a very robust set of one-variable equations that are derived from the KS equations. We have also implemented our

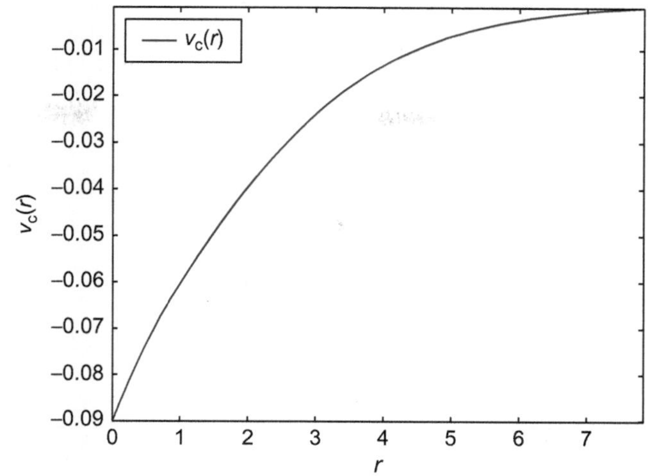

Figure 1 The correlation potential $v_c(r)$ for $\lambda = 1$.

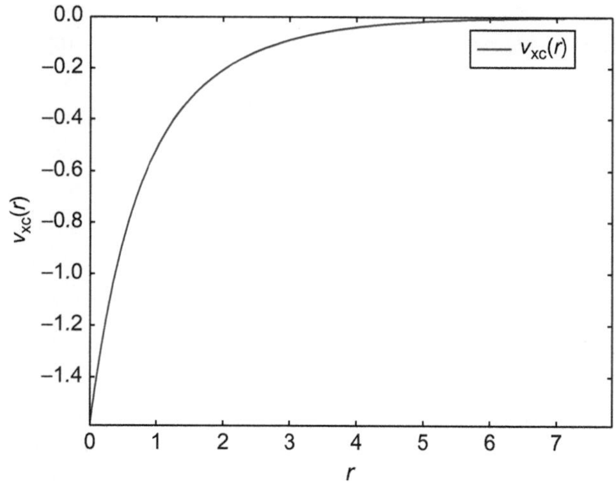

Figure 2 The exchange-correlation potential $v_{xc}(r)$ for $\lambda = 1$.

SPTa to solve the resulting equations efficiently in a self-consistent fashion. Direct comparison with commonly available methods[14] shows a clear superiority of the SPTa. We also have repeated the same calculation by employing exponential and Gaussian type basis functions by manually optimizing the screening parameters to produce the lowest possible orbital energy. The KS eigenvalues of the Sturmian and isotropic basis functions exhibit slower convergence and are variationally higher than that of the

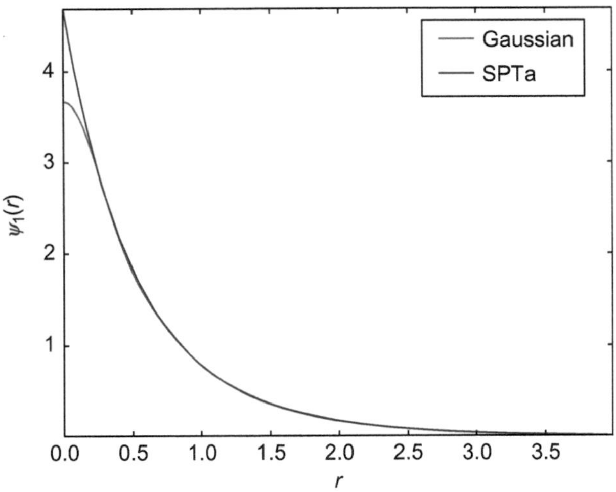

Figure 3 The ground state KS wave function $\psi_1(r)$ for $\lambda = 1$.

SPTa, which is $\epsilon_1 = -0.570209$. Although ϵ_1 is converged sufficiently, the total ground state energy $E_0 = -2.834289$ is still less accurate even compared to the HF theory. Our results are hence inconclusive with regard to the findings of Bagayoko[2] which were essentially that with accurate eigenvalues, the ground state energy is relatively insensitive to the correlation potential and that an LDA is typically sufficient to give highly accurate results. Further testes must be conducted on more complicated systems than the He atom. The SPTa has the potential to handle larger systems and to extract benchmark results employing other approximations for the XC functional.

ACKNOWLEDGMENTS

D.H.G. and C.A.W. were partially supported by the Department of Energy/National Nuclear Security Agency under award number DE-NA0002630. C.A.W. was also supported in part by the Defense Threat Reduction Agency under award number HDTRA1-12-1-0036.

REFERENCES

1. Gebremedhin, D. H.; Weatherford, C. A. *Calculations for the One-Dimensional Soft Coulomb Problem and the Hard Coulomb Limit. Phys. Rev. E* **2014**, *89*, 053319. http://dx.doi.org/10.1103/PhysRevE.89.053319. http://link.aps.org/doi/10.1103/PhysRevE.89.053319.
2. Bagayoko, D. *Understanding Density Functional Theory (DFT) and Completing it in Practice. AIP Adv.* **2014**, *4*(12). http://dx.doi.org/10.1063/1.4903408 http://scitation.aip.org/content/aip/journal/adva/4/12/10.1063/1.4903408. 127104.

3. Gebremedhin, D.; Weatherford, C. *Canonical Two-Range Addition Theorem for Slater-Type Orbitals. Int. J. Quant. Chem.* **2013**, *113*(1), 71–75. http://dx.doi.org/10.1002/qua.24319. http://dx.doi.org/10.1002/qua.24319.
4. Parr, R. G.; Yang, W. *Density-Functional Theory of Atoms and Molecules (International Series of Monographs on Chemistry)*; Oxford University Press: Oxford, NY, 1994.
5. Bromley, D.; Greiner, W. *Classical Electrodynamics*; Springer: New York, NY, 1998 *https://books.google.com/books?id=pov6H2QGNWAC.*
6. Perdew, J. P.; Zunger, A. *Self-Interaction Correction to Density-Functional Approximations for Many-Electron Systems. Phys. Rev. B* **1981**, *23*, 5048–5079. http://link.aps.org/doi/10.1103/PhysRevB.23.5048http://dx.doi.org/10.1103/PhysRevB.23.5048.
7. Ceperley, D. M.; Alder, B. J. *Ground State of the Electron Gas by a Stochastic Method. Phys. Rev. Lett.* **1980**, *45*, 566–569. http://dx.doi.org/10.1103/PhysRevLett.45.566. http://link.aps.org/doi/10.1103/PhysRevLett.45.566.
8. Davis, C. L.; Jensen, H. A.; Monkhorst, H. J. *Accurate Hartree? Fock Wave Functions without Exponent Optimization. J. Chem. Phys.* **1984**, *80*(2), 840–855. http://dx.doi.org/10.1063/1.446738. http://scitation.aip.org/content/aip/journal/jcp/80/2/10.1063/1.446738.
9. Szabo, A.; Ostlund, N. *Modern Quantum Chemistry: Introduction to Advanced Electronic Structure Theory*; Dover Publications: Mineola, NY, 1996.
10. Avery, J.; Avery, J. *Generalized Sturmians and Atomic Spectra.* World Scientific Printers: Singapore, 2006.
11. Weniger, E. J. *Weakly Convergent Expansions of a Plane Wave and Their Use in Fourier Integrals. J. Math. Phys.* **1985**, *26*(2), 276–291. http://dx.doi.org/10.1063/1.526970. http://scitation.aip.org/content/aip/journal/jmp/26/2/10.1063/1.526970.
12. Arfken, G. B.; Weber, H. J.; Harris, F. E. *Mathematical Methods for Physicists, Sixth Edition: A Comprehensive Guide*; Academic Press: New York, 2005.
13. Liboff, R. *Introductory Quantum Mechanics*; Addison-Wesley: San Francisco, CA, 2003.
14. Thijssen, J. *Computational Physics*, 2nd; Cambridge University Press: Cambridge, UK, 2007.

CHAPTER THIRTEEN

Potential Energy Curves of NaK Molecule from All-Electron Multireference-Coupled Cluster Calculations

Monika Musiał[1], Patrycja Skupin, Anna Motyl
Institute of Chemistry, University of Silesia, Katowice, Poland
[1]Corresponding author: e-mail address: musial@ich.us.edu.pl

Contents

Abstract

The multireference-coupled cluster theory formulated in the (2,0) sector of the Fock space provides a description of the states obtained by attachment of two electrons to the reference system. If the reference is assumed to be a doubly ionized cation, then the results relate to the neutral molecule. In the current work, the above scheme is applied to the calculations of the potential energy curves of the NaK molecule adopting as a reference system the doubly ionized NaK^{2+} structure. Such computational strategy relies on the fact that the closed-shell NaK^{2+} reference dissociates into the closed-shell fragments ($NaK^{2+} \Rightarrow Na^{+} + K^{+}$). This is advantageous since the restricted Hartree–Fock function can be used as the reference in the whole range of interatomic distances. This scheme offers *first-principle* method without any model or effective potential parameters for the description of the bond-breaking processes.

Advances in Quantum Chemistry, Volume 73
ISSN 0065-3276
http://dx.doi.org/10.1016/bs.aiq.2015.04.003

1. INTRODUCTION

The growing interest in using cold and ultra-cold techniques to study molecular collision dynamics and to produce molecules in a specific quantum state creates a demand for accurate experimental and theoretical studies of the intramolecular potential of alkali diatomic molecules.[1,2]

A description of the dissociation of the standard single bond formed by a pair of electrons is still not a trivial task. The ideal situation occurs when the closed-shell molecule dissociates into closed-shell fragments. In this case, the RHF (restricted Hartree–Fock) function provides a suitable reference for the whole range of interatomic distances, but such situations do not occur, however, for real chemical bonds. They can be encountered in the Van der Waals molecules, e.g., between noble gas atoms. However, the conditions mentioned above may be met in the cases when we dissociate a double positive ion (in our case NaK^{2+}) instead of the neutral molecule (i.e., NaK) since in the former case we obtain closed-shell products (Na^+ and K^+), isoelectronic with noble gas atoms, Ne and Ar, respectively. It can be done on condition that we have at our disposal a computational scheme able to treat with sufficient accuracy the states with two attached electrons, in other words, able to describe the neutral molecule on the basis of the doubly ionized reference. Thus, the DEA (double-electron attachment) strategy avoids the use of the unrestricted Hartree–Fock reference. The latter has well-known disadvantages connected with the broken space (spin) symmetry problems as well as with difficulties in reaching converged solutions around the critical geometries.

In the framework of the coupled cluster (CC)[3–12] theory, we can consider DEA schemes[13–21]: one formulated within the EOM (equation-of-motion) theory[13,21] and the other based on the multireference-coupled cluster (MRCC) method formulated in the (2, 0) sector of the Fock space (FS).[14–18] The FS-CC technique seems particularly suitable for the bond-breaking problem owing to its rigorous size extensivity which is crucial for the correct reproduction of the atomic excitation energies in the dissociation limit. In our previous work, this method was applied successfully for the alkali metal diatomics: Li_2,[15,17] Na_2,[17] K_2,[19] Rb_2,[16] NaLi,[18] and KLi.[20] Contrary to the previous NaK studies,[22–24] based on the effective potential methods, in this work we apply the size-extensive FS-CCSD (2, 0) method with all electrons correlated. This is a *first-principles* computational scheme able to produce potential energy curves (PECs) for selected states of NaK without introduction of any auxiliary parameters. By using the intermediate Hamiltonian (IH) formulation of the FS-CC approach, we eliminate also the convergence problems connected with so-called intruder states which were a pain in the standard formulations of the Fock space multireference CC scheme. Our calculations converge smoothly for all interatomic distances from equilibrium to the dissociation limit.

2. SYNOPSIS OF THE THEORY

The DEA calculations are usually associated with the evaluation of the double-electron affinity. However, in the cases where we assume the unit deprived of two electrons as a reference, the DEA provides the description of the neutral state. Thus, in general we may write $AB^{+n} \rightarrow AB^{+n-2}$; $E_{DEA} = E_{AB^{+n}} - E_{AB^{+n-2}}$; $n = 0, 2$, where $n = 0$ refers to the regular double-electron affinity and $n = 2$ represents the case under study.

As mentioned in Section 1, there are two principal computational schemes used so far as the correlated methods fulfilling the DEA conditions: EOM-CC[13,21] and FS-CC.[14] In this work, we used the latter which is rigorously size-extensive.

We seek the solution of the Schrödinger equation:

$$H\Psi_k = E_k\Psi_k \tag{1}$$

where the E_k and Ψ_k are the energy and wave function for the k-state.

The CC ground state function is expressed via the exponential Ansatz[3,4]

$$\Psi_0 = e^T\Phi_0 \tag{2}$$

where T is a cluster operator in the current work approximated by the sum of the operators responsible for the single (T_1) and double (T_2) excitations[25]:

$$T = T_1 + T_2 \tag{3}$$

The essence of the multireference approach[26–35] is represented by the equation:

$$H_{eff}\widetilde{\Psi}_k = E_k\widetilde{\Psi}_k \tag{4}$$

i.e., the exact eigenvalues E_k can be obtained by diagonalization of the effective Hamiltonian operator within the model space with $H_{\text{eff}} = PH\Omega P$, where P is a projection operator and Ω is the valence universal wave operator, $\Omega = \{e^S\}P$. In the current case, the model space is formed by $\Phi^{\alpha\beta}$

configurations obtained by distribution of the two added electrons among α,β active particles which are a subset of all virtual one-particle levels.

The hierarchical structure of the FS solutions requires, to solve the FS equations for the (k,l) sector, all the solutions for lower rank sectors (i,j) with $i \le k$ and $j \le l$ be known. In the current work this requirement enforces solving the FS equations for the (0, 0) and (1, 0) sectors. The (0, 0) sector corresponds to the single reference solution for the reference system, i.e., NaK^{2+} ion. The (1, 0) sector is—in general—responsible for calculations of the electron affinity; however in the current case, it describes the process $NaK^{2+} \Rightarrow NaK^{+}$.

Thus, the energy values of the double-electron attached states are obtained by the diagonalization of the effective Hamiltonian within the $\Phi^{\alpha\beta}$ configurational space:

$$H_{\text{eff}}^{(2,0)} = P^{(2,0)} H e^{S^{(0,0)} + S^{(1,0)} + S^{(2,0)}} P^{(2,0)} \tag{5}$$

where the operator $P^{(2,0)}$ is defined as

$$P^{(2,0)} = \sum_{\alpha\beta} |\Phi^{\alpha\beta}\rangle\langle\Phi^{\alpha\beta}| \tag{6}$$

The $S^{(0,0)}(\equiv T)$, $S^{(1,0)}$, and $S^{(2,0)}$ are cluster operators for sectors indicated by the superscripts.

Thus, we may summarize steps in the Fock space CC calculations:

1. solve the (0,0) sector (do CC calculations for the reference system)
2. construct $\bar{H}$ elements using converged T amplitudes
3. select active space $\rightarrow$ model space
4. solve iteratively the (1,0) sector constructing in each iteration $H_{eff}^{(1,0)}$
5. if (1,0) sector done solve iteratively equations for the (2,0) sector constructing in each iteration $H_{eff}^{(2,0)}$
6. once the amplitude got converged, diagonalize the $\mathbf{H}_{\text{eff}}^{(2,0)}$ matrix to obtain the sought eigenvalues

It should be mentioned that in the practical realization, the IH[36–43] variant of the FS-CCSD approach[14] is used (see our previous work[14] for details). The latter avoids so-called intruder state problems plaguing the standard FS-CC approach and limiting its use to small model spaces.

The first three steps in this formulation are the same as in the standard H_{eff} formalism. Due to step 4, we have all amplitudes from the one-valence sector and we can construct the $Y^{(2,0)}$ operator which is needed to obtain the $H_{\mathrm{I}}^{(2,0)}$ operator (i.e., IH operator):

$$H_{\mathrm{I}}^{(2,0)} = P_{\mathrm{o}}^{(2,0)} \bar{H} P^{(2,0)} + P_{\mathrm{o}}^{(2,0)} \bar{H} Y^{(2,0)} P^{(2,0)} \tag{7}$$

where

$$Y^{(2,0)} = Q_{\mathrm{o}}^{(2,0)} X^{(2,0)} P^{(2,0)} \tag{8}$$

$$X^{(2,0)} = \{e^{\tilde{S}(2,0)} - 1\} P^{(2,0)} \tag{9}$$

with

$$\tilde{S}^{(2,0)} = S^{(1,0)} + S^{(2,0)} \tag{10}$$

Thus, the eigenvalues of the FS-CC effective Hamiltonian can be obtained as eigenvalues of the simpler IH operator, H_{I}, which is a matrix representation of $\bar{H}$ in the reference and intermediate space (i.e., M_{o}) with the $P_{\mathrm{o}} - P$ part modified by a "dressing" which is constructed from $\bar{H}$ and the cluster operators S known from the lower sector. The intermediate space (part of the orthogonal one) $M_{\mathrm{I}}^{(2,0)}$ is spanned by all determinants which are obtained by the direct action of the $S^{(2,0)}$ operator on $M^{(2,0)}$ (model space). The remaining part of the orthogonal space is associated with $Q_{\mathrm{o}}^{(2,0)}$ projection. Moreover, since equations in the one-valence sector are linear in S, we may use the EOM-CC technique to obtain (1, 0) solutions[34,37,44,45] (the same applies to the (0, 1) case.[34,37,45,46]).

3. RESULTS AND DISCUSSION

All calculations are done using the ACES II[47] program system supplemented with the FS-CCSD (2,0) module.[14] The results are collected in Figs. 1–3 and in Tables 1 and 2 along with the available experimental data. In all DEA calculations, the orbitals used are obtained by the RHF solution for the NaK^{2+} system. We used the POL1[48] basis set with the spherical harmonic polarization functions and all electrons correlated. The size of the active space for the FS-CCSD (2, 0) part has been set to 34 (i.e., 34 lowest virtual orbitals have been selected as active). The further increase in the size of the active space does not affect meaningfully the results.

In the first row of Table 1, we compare the total ground state energy of the NaK molecule (last column) with the sum of the ground state energies for Na and K atoms. The results are identical. In the next rows, we compare the

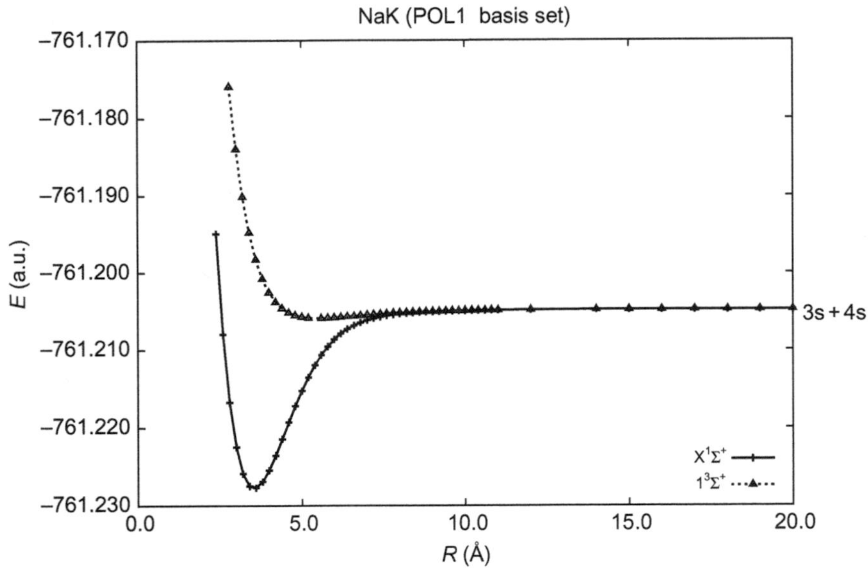

Figure 1 Potential energy curves of the NaK molecule with the FS-CCSD (2, 0) method for Na(3s) + K(4s) dissociation limit in the POL1 basis set.

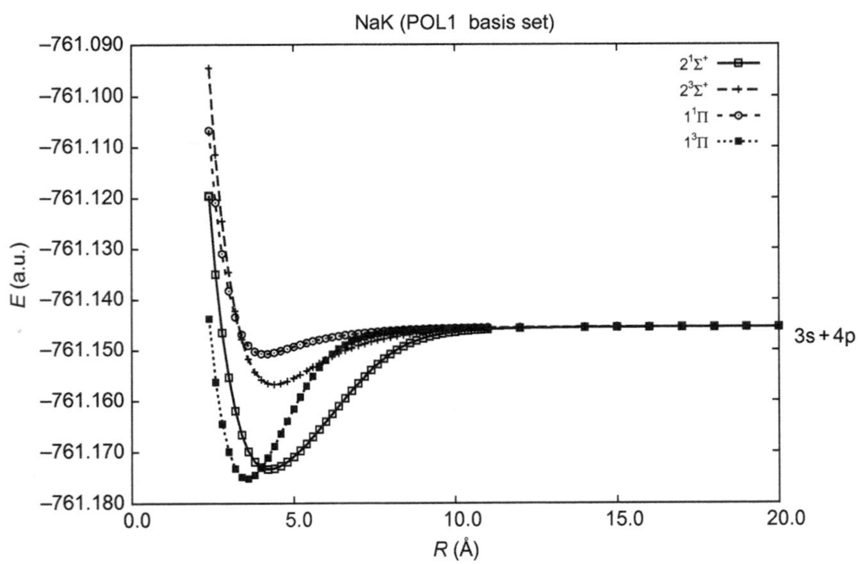

Figure 2 Potential energy curves of the NaK molecule with the FS-CCSD (2, 0) method for Na(3s) + K(4p) dissociation limit in the POL1 basis set.

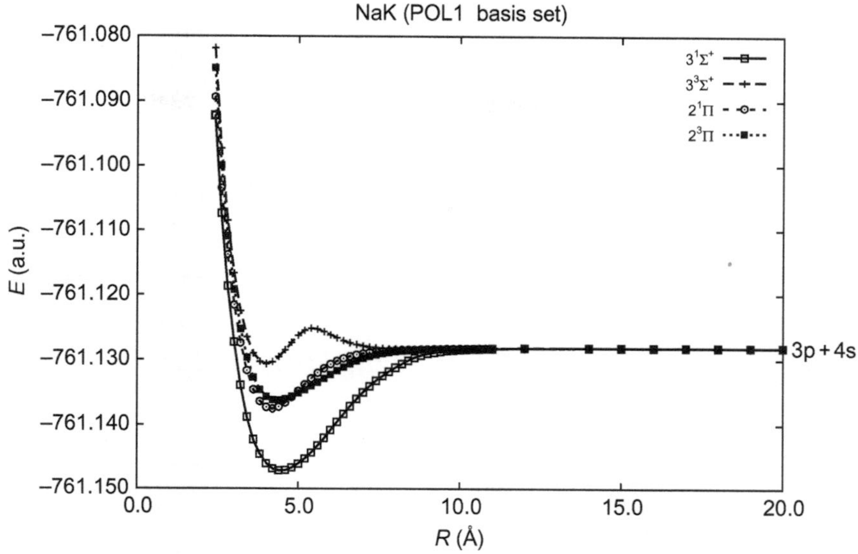

Figure 3 Potential energy curves of the NaK molecule with the FS-CCSD (2, 0) method for Na(3p) + K(4s) dissociation limit in the POL1 basis set.

Table 1 Energies of the Electronic States at the Dissociation Limit of the NaK Molecule Compared to the Atomic Energies (POL1 Basis Set)

Na[a]		**K[a]**		**Na + K**	**NaK($R=\infty$)**
Config.	E (a.u.)	Config.	E (a.u.)	E (a.u.)	E (a.u.)
[Ne]3s	− 161.929180	[Ar]4s	− 599.275607	− 761.204787	− 761.204787
	ΔE (eV)		ΔE (eV)	ΔE (eV)	ΔE (eV)
[Ne]3s	0	[Ar]4p	1.614215	1.614215	1.614215
[Ne]3p	2.084961	[Ar]4s	0	2.084961	2.084961
[Ne]3s	0	[Ar]5s	2.555957	2.555957	2.555957

[a]Energy calculated using EA-EOM-CCSD(≡ FS-CCSD (1,0)) method.

excitation energies to the three lowest states of the NaK molecule. We see that in the supermolecular calculations, the excitation energy of the NaK molecule is exactly equal to the atomic excitation energies at infinite separation. Note that in all cases, the process of DEA involves both centers since the two added electrons are placed each on one atom. This observation is important since it means that the size-intensive methods (like DEA-EOM) would fail to give the size-extensive results contrary to the FS-MRCC approach.

Table 2 Spectroscopic Constants of the NaK Molecule (FS-CCSD (2, 0) Method, POL1 Basis Set)

Sym.	R_e (Å)	D_e(cm^{-1})	T_e(cm^{-1})	w_e(cm^{-1})	w_ex_e(cm^{-1})	References
			Na (3s)+K(4s)			
$1^1\Sigma^+$	3.537	5034	0	119.60	0.51	This work
	3.499	5274	0	124.03	0.50	Exp.[49,50]
$1^3\Sigma^+$	5.498	253	4781	21.52	0.47	This work
	5.448	208	–	22.82	0.59	Exp.[49,51]
			Na(3s)+K(4p)			
$2^1\Sigma^+$	4.236	6121	11,933	78.50	0.21	This work
	4.196	6220	12,137	81.25	–	Exp.[52]
$2^3\Sigma^+$	4.393	2478	15,576	70.18	0.48	This work
	4.308	2443	15,751	73.40	0.48	Exp.[53,54]
$1^1\Pi$	4.105	1169	16,885	64.80	1.16	This work
	4.013	1324	16,993	71.46	1.15	Exp.[55]
$1^3\Pi$	3.555	6528	11,526	117.36	0.36	This work
	3.502	6698	11,562	120.37	0.33	Exp.[53,56]
			Na(3p)+K(4s)			
$3^1\Sigma^+$	4.457	4174	17,677	67.51	0.06	This work
	4.438	4443	17,787	69.61	0.06	Exp.[57]
$3^3\Sigma^+$	3.993	548	21,303	93.58	0.65	This work
$2^1\Pi$	4.180	2057	19,794	82.05	0.54	This work
	4.155	2247	20,090	82.76	0.45	Exp.[58,59]
$2^3\Pi$	4.333	1769	20,082	62.30	0.52	This work
	4.225	2001	20,248	67.38	0.79	Exp.[60]
MAE[a]	0.057	159	156	3.26	0.07	

[a]Mean absolute error.

In this work, we have studied PECs for the lowest-lying electronic states of the NaK molecule. These are the states correlating to the three dissociation limits: Na(3s) + K(4s)—two states, Na(3s) + K(4p)—four states, and Na(3p) + K(4s)—four states. In the cases where the p atomic levels are engaged, some of the molecular states (Π) are degenerated; hence, the total number of state eigenfunctions amounts to 14.

The computed PECs are shown in Figs. 1–3, and the respective spectroscopic constants (obtained with help of the LEVEL-8 program of Le Roy[61]) are collected in Table 2. The computed constants are compared with the experimental values wherever the latter are available.

3.1 Na(3s) + K(4s) Dissociation Limit

In Fig. 1, we present two curves: one representing the ground state $X^1\Sigma^+$ and the other—the triplet excited state $1^3\Sigma^+$. The respective values of the molecular constants are listed in Table 2. The computed dissociation energy, D_e, compares quite well with the experimental value (differing by 240 cm^{-1}) The energy curve representing the triplet state is almost repulsive. Upon closer inspection, we observe a shallow potential well, 253 cm^{-1} deep, agreeing very well with the observed value of 208 cm^{-1}. For both states, the equilibrium bond lengths obtained here are off the experiment by 0.04–0.05 Å.

3.2 Na(3s) + K(4p) Dissociation Limit

The four states correlate to the Na(3s) + K(4p) limit: $2^1\Sigma^+$, $2^3\Sigma^+$, $1^1\Pi$, and $1^3\Pi$ and the respective PECs are shown in Fig. 2. All of them represent bound states with the largest dissociation energy observed for the triplet Π and singlet $2^1\Sigma^+$ states. The depths of the potential wells are comparable for both states and equal to 6528 and 6121 cm^{-1}, which differ from the experiment by 170 and 99 wave-numbers, respectively (see Table 2). The computed D_e values for other two states, $2^3\Sigma^+$ and $1^1\Pi$, are much lower but again they correlate well with the experiment, giving deviations of 35 and 155 wave-numbers, respectively. Another important quantity characterizing excited states is an excitation energy, T_e, which in our case is reproduced quite well with the errors of 204, 175, 108, and 36 wave-numbers, respectively, for the four states considered. Relatively large deviations from the experiment are observed for the equilibrium bond length with the average of ca. 0.07 Å.

For the singlet $2^1\Sigma^+$ state, this work gives deviations of 0.04 Å for the bond length, 99 cm^{-1} for the dissociation energy, and 2.75 cm^{-1} for the harmonic frequency. Results of the same quality are obtained for the $^1\Pi$ state.

3.3 Na(3p) + K(4s) Dissociation Limit

The next group dissociating to the Na(3p) + K(4s) limit includes four states: $3^1\Sigma^+$, $3^3\Sigma^+$, $2^1\Pi$, and $2^3\Pi$ (see Fig. 3). The largest bonding energy is observed for the $3^1\Sigma^+$ state with the depth of the potential well equal to 4174 cm^{-1} (experimental value: 4443 cm^{-1}). Both Π states, $2^1\Pi$ and $2^3\Pi$, are represented by regular Morse-type curve with relatively shallow minima of ca. 2000 cm^{-1} which are off the experimental values by about 200 wave-numbers. The most interesting shape is observed in the case of $3^3\Sigma^+$ curve with the potential barrier of 1201 cm^{-1} occurring around 5.4 Å (reported also in the other theoretical papers, e.g., Ref. 24). In addition, our calculations indicate that this state has a very shallow Van der Waals well at large separation around 8.47 Å with D_e equal to 43 cm^{-1}. Unfortunately, the comparison with experimental data is impossible since the latter are unavailable. For both singlet states, the equilibrium bond length is reproduced with accuracy of ca. 0.02 Å, contrary to the $2^3\Pi$ states for which the R_e value if off by 0.108 Å.

It should also be mentioned that the harmonic frequency values, also listed in Table 2, remain in satisfactory agreement with the experiment. The average deviation for all considered states is equal to 3.26 cm^{-1} (see the bottom of Table 2).

4. CONCLUSIONS

First-principles quantum chemical calculations are applied in the theoretical study of PECs for the lowest-lying electronic states of NaK molecule. The results obtained are of comparable quality to other theoretical values available in the literature, obtained—in a majority of cases—with the adjustable model or effective potentials (parameterized for a given molecule). In the whole range of interatomic distances, the calculations were done for the RHF reference function with all 30 electrons correlated. The crucial strategy in obtaining a smooth and correct curve is to adopt a DEA strategy relying (i) on the removal of a pair of electrons from the NaK molecule to obtain a convenient RHF reference and (ii) on using the DEA scheme to recover the original neutral structure by attaching two electrons to the NaK^{2+} ion. The method best suited to the second step is a rigorously size-extensive MRCC scheme formulated within the (2, 0) sector of the Fock space.

In spite of the fact that present calculations are based on the rather modest basis set and that relativistic effects (large, particularly for the K atom) are neglected the results are quite satisfactory. This demonstrates the power and practical utility of the FS-MRCC (2,0) method in the study of the dissociation of the alkali metal diatomic molecules. The most important perspective of future work is to include relativistic effects in order to increase the accuracy and also to broaden the range of molecules which can be treated with this FS method.

ACKNOWLEDGMENT

This work has been supported by the National Science Centre, Poland under Grant No. 2013/11/B/ST4/02191.

REFERENCES

1. Jones, K. M.; Tiesinga, E.; Lett, P. D.; Julienne, P. S. Ultracold Photoassociation Spectroscopy: Long-Range Molecules and Atomic Scattering. *Rev. Mod. Phys. A* **2006**, *78*, 483.
2. Tomza, M.; Goerz, M. H.; siał, M.; Moszynski, R.; Koch, C. P. Optimized Production of Ultracold Ground-State Molecules: Stabilization Employing Potentials with Ion-Pair Character and Strong Spin-Orbit Coupling. *Phys. Rev. A* **2012**, *86*, 043424.
3. Čizek, J. On the Correlation Problem in Atomic and Molecular Systems. Calculation of Wavefunction Components in Ursell-Type Expansion Using Quantum-Field Theoretical Methods. *J. Chem. Phys.* **1966**, *45*, 4256; Čizek, J. On the Use of the Cluster Expansion and the Technique of Diagrams in Calculations of Correlation Effects in Atoms and Molecules. *Adv. Chem. Phys.* **1969**, *14*, 15; Paldus, J.; Čizek, J.; Shavitt, I. Correlation Problems in Atomic and Molecular Systems. IV. Extended Coupled-Pair Many-Electron Theory and Its Application to the BH_3 Molecule. *Phys. Rev. A* **1974**, *5*, 50.
4. Bartlett, R. J. Coupled-Cluster Approach to Molecular Structure and Spectra: A Step Toward Predictive Quantum Chemistry. *J. Phys. Chem.* **1989**, *93*, 1697; Bartlett, R. J. Coupled-Cluster Theory: An Overview of Recent Developments. In: *Modern Electronic Structure Theory, Part 2*; Yarkony, D. R. Ed.; World Scientific Publishing: New York, 1995; pp 1047–1131; Bartlett, R. J.; Stanton, J. F. Applications of Post-Hartree-Fock Methods: A Tutorial. In: Lipkowitz, K. B., Boyd, D. B., Eds.; Reviews in Computational Chemistry, Vol. 5; VCH Publishers: New York, 1994; pp 65–169.
5. Paldus, J.; Li, X. A Critical Assessment of Coupled Cluster Method in Quantum Chemistry. *Adv. Chem. Phys.* **1999**, *110*, 1.
6. Harris, F. E. Coupled-Cluster Method for Excitation Energies. *Int. J. Quantum Chem. Symp.* **1977**, *11*, 403.
7. Noga, J.; Bartlett, R. J.; Urban, M. Towards a Full CCSDT Model for Electron Correlation. CCSDT-n Models. *Chem. Phys. Lett.* **1987**, *134*, 126.
8. Noga, J.; Bartlett, R. J. The Full CCSDT Model for Molecular Electronic Structure. *J. Chem. Phys.* **1987**, *86*, 7041.
9. Kucharski, S. A.; Bartlett, R. J. The Coupled-Cluster Single, Double, Triple, and Quadruple Excitation Method. *J. Chem. Phys.* **1992**, *97*, 4282.
10. Musiał, M.; Kucharski, S. A.; Bartlett, J. B. T_5 Operator in Coupled Cluster Calculations. *Chem. Phys. Lett.* **2000**, *320*, 542; MusiaŁ, M.; Kucharski, S. A.; Bartlett, R. J. Formulation and Implementation of the Full Coupled-Cluster Method Through Pentuple Excitations. *J. Chem.* Phys. **2002**, *116*, 4382.
11. Bartlett, R. J.; Musiał, M. Coupled-Cluster theory in Quantum Chemistry. *Rev. Mod. Phys.* **2007**, *79*, 291; Bartlett, R. J.; Musiał, M. Addition by Subtraction in Coupled-Cluster Theory: A Reconsideration of the CC and CI Interface and the nCC Hierarchy. *J. Chem. Phys.* **2006**, *125*, 204105.

12. Piecuch, P.; Kowalski, K.; Pimienta, I. S. O.; Fan, P. D.; Lodriguito, M.; McGuire, M. J.; Kucharski, S. A.; Kuś, T.; Musiał, M. Method of Moments of Coupled-Cluster Equations: A New Formalism for Designing Accurate Electronic Structure Methods for Ground and Excited States. *Theor. Chem. Acc.* **2004**, *112*, 349.
13. Musiał, M.; Kucharski, S. A.; Bartlett, R. J. Multireference Double Electron Attached Coupled Cluster Method with Full Inclusion of the Connected Triple Excitations: MR-DA-CCSDT. *J. Chem. Theory Comput.* **2011**, 7, 3088.
14. Musiał, M. Multireference Fock Space Coupled Cluster Method in the Effective and Intermediate Hamiltonian Formulation for the (2,0) Sector. *J. Chem. Phys.* **2012**, *136*, 134111.
15. Musiał, M.; Kucharski, S. A. First Principle Calculations of the Potential Energy Curves for Electronic States of the Lithium Dimer. *J. Chem. Theory Comput.* **2014**, *10*, 1200.
16. Tomza, M.; Skomorowski, W.; Musiał, M.; Gonzales-Ferez, R.; Koch, C. P.; Moszynski, R. Interatomic Potentials, Electric Properties, and Spectroscopy of the Ground and Excited States of the Rb_2 Molecule: Ab Initio Calculations and Effects of a Non-Resonant Field. *Mol. Phys.* **2013**, *111*, 1781.
17. Musiał, M.; Kowalska-Szojda, K.; Lyakh, D. I.; Bartlett, R. J. Potential Energy Curves via Double Electron Affinity Calculations: Dissociation of Alkali Metal Dimers. *J. Chem. Phys.* **2013**, *138*, 194553.
18. Mieszczanin, P.; Musiał, M.; Kucharski, S. A. Potential Energy Curves via Double Electron Affinity Calculations: Example of NaLi Molecule. *Mol. Phys.* **2013**, *112*, 726.
19. Musiał, M.; Kucharski, S. A. Double Electron Affinity for Potential Energy Curves of Closed Shell Molecules. In: *Lecture Notes in Engineering and Computer Science 2014: Proceedings of the World Congress on Engineering and Computer Science 2014, WCECS, 22–24 October 2014, San Francisco, USA*, 2014.
20. Musial, M.; Motyl, A.; Skupin, P.; Kucharski, S. A. Potential Energy Curves for the Low Lying Electronic States of Kli with Fock Space Coupled Cluster Method. *Adv. Quantum Chem.* (in press)
21. Shen, J.; Piecuch, P. Doubly Electron-Attached and Doubly Ionized Equation-Of-Motion Coupled-Cluster Methods with 4-Particle-2-Hole and 4-Hole-2-Particle Excitations and Their Active Space Extensions. *J. Chem. Phys.* **2013**, *138*, 194102.
22. Magnier, S.; Aubert-Frécon, M.; Millié, P. Potential Energies, Permanent and Transition Dipole Moments for Numerous Electronic Excited States of NaK. *J. Mol. Spectrosc.* **2000**, *200*, 96.
23. Magnier, S.; Millié, P. Potential Curves for the Ground and Numerous Highly Excited Electronic States of K_2 and NaK. *Phys. Rev. A Gen. Phys.* **1996**, *54*, 204.
24. Stevens, W. J.; Konowalow, D. D.; Ratcliff, L. B. Electronic Structure and Spectra of the Lowest Five $^1\Sigma^+$ End $^3\Sigma^+$ States, and Lowest Three $^1\pi$, $^3\pi$, $^1\Delta$ and $^3\Delta$ States of NaK. *J. Chem. Phys.* **1984**, *80*, 1215.
25. Purvis, G. D.; Bartlett, R. J. A Full Coupled-Cluster Singles and Doubles Model: The Inclusion of Disconnected Triples. *J. Chem. Phys.* **1982**, *76*, 1910.
26. Mukherjee, D.; Pal, S. Use of Cluster-Expansion Methods in the Open-Shell Correlation-Problem. *Adv. Quantum Chem.* **1989**, *20*, 292, and references therein; Lindgren, I.; Mukherjee, D. On the Connectivity Criteria in the Open-Shell Coupled-Cluster Theory for General Model Spaces. *Phys. Rep.* **1987**, *151*, 93; Pal, S. Fock Space Multi-Reference Coupled-Cluster Method for Energies and Energy Derivatives. *Mol. Phys.* **2010**, *108*, 3033.
27. Sinha, D.; Mukhopadhyay, S.; Mukherjee, D. A Note on the Direct Calculation of Excitation Energies by Quasi-Degenerate MBPT and Coupled-Cluster Theory. *Chem. Phys. Lett.* **1986**, *129*, 369.
28. Chaudhuri, R.; Mukhopadhyay, D.; Mukherjee, D. Applications of Open-Shell Coupled Cluster Theory Using an Eigenvalue-Independent Partitioning Technique: Approximate Inclusion of Triples in IP Calculations. *Chem. Phys. Lett.* **1989**, *162*, 393.

29. Kaldor, U. Open-Shell Coupled-Cluster Method: Electron Affinities of Li and Na. *J. Comput. Chem.* **1987**, *8*, 448.
30. Pal, S.; Rittby, M.; Bartlett, R. J. Multi-Reference Coupled-Cluster Methods for Ionization Potentials with Partial Inclusion of Triple Excitations. *Chem. Phys. Lett.* **1989**, *160*, 212.
31. Bernholdt, D. E.; Bartlett, R. J. A Critical Assessment of Multireference Fock-Space CCSD and Perturbative Third-Order Triples Approximations for Photoelectron Spectra and Quasidegenerate Potential Energy Surfaces. *Adv. Quantum Chem.* **1999**, *34*, 271.
32. Vaval, N.; Pal, S.; Mukherjee, D. Fock Space Multireference Coupled Cluster Theory: Noniterative Inclusion of Triples for Excitation Energies. *Theor. Chem. Acc.* **1998**, *99*, 100.
33. Meissner, L.; Bartlett, R. J. Transformation of the Hamiltonian in Excitation Energy Calculations: Comparison between Fockspace Multireference Coupled-Cluster and Equation-of-Motion Coupled-Cluster Methods. *J. Chem. Phys.* **1991**, *94*, 6670; Meissner, L.; Bartlett, R. J. A Dressing for the Matrix Elements of the Singles and Doubles Equation-of-Motion Coupled-Cluster Method That Recovers Additive Separability of Excitation Energies. *J. Chem. Phys.* **1995**, *102*, 7490.
34. Musiał, M.; Bartlett, R. J. Fock Space Multireference Coupled Cluster Method with Full Inclusion of Connected Triples for Excitation Energies. *J. Chem. Phys.* **2004**, *121*, 1670.
35. Lyakh, D. I.; Musiał, M.; Lotrich, V.; Bartlett, R. J. Multireference Nature of Chemistry: the Coupled-Cluster View. *Chem. Rev.* **2012**, *112*, 182.
36. Malrieu, J. P.; Durand, P.; Daudey, J. P. Intermediate Hamiltonians as a New Class of Effective Hamiltonians. *J. Phys. A Math. Gen.* **1985**, *18*, 809.
37. Meissner, L. Fock-Space Coupled-Cluster Method in the Intermediate Hamiltonian Formulation: Model with Singles and Doubles. *J. Chem. Phys.* **1998**, *108*, 9227.
38. Musiał, M.; Meissner, L.; Kucharski, S. A.; Bartlett, J. B. Molecular Applications of Intermediate Hamiltonian Fock-Space Coupled-Cluster Method for Calculation of Excitation Energies. *J. Chem. Phys.* **2005**, *122*, 224110.
39. Musiał, M.; Meissner, L. The Fock-Space Coupled-Cluster Method in the Calculation of Excited State Properties. *Collect. Czech. Chem. Commun.* **2005**, *70*, 811.
40. Meissner, L.; Malinowski, P. Intermediate Hamiltonian Formulation of the Valence-Universal Coupled-Cluster Method for Atoms. *Phys. Rev. A* **2000**, *61*, 062510.
41. Musial, M.; Bartlett, R. J. Multireference Fock-Space Coupled-Cluster and Equation-of-Motion Coupled-Cluster Theories: The Detailed Interconnections. *J. Chem. Phys.* **2008**, *129*, 134105; Musial, M.; Bartlett, R. J. Charge-Transfer Separability and Size-Extensivity in the Equation-of-Motion Coupled Cluster Method: EOM-CCx. *J. Chem. Phys.* **2011**, *134*, 034106.
42. Musiał, M.; Bartlett, R. J. Benchmark Calculations of the Fock-Space Coupled Cluster Single, Double, Triple Excitation Method in the Intermediate Hamiltonian Formulation for Electronic Excitation Energies. *Chem. Phys. Lett.* **2008**, *457*, 267.
43. Musial, M.; Bartlett, R. J. Intermediate Hamiltonian Fock-Space Multireference Coupled-Cluster Method with Full Triples for Calculation of Excitation Energies. *J. Chem. Phys.* **2008**, *129*, 044101; Musiał, M. Efficient Realization of the Fock-Space Coupled-Cluster Method with Connected Triple Excitations. *Chem. Phys. Lett.* **2009**, *470*, 358; Musial, M.; Bartlett, R. J. Spin-Free Intermediate Hamiltonian Fock-Space Coupled-Cluster Theory with Full Inclusion of Triple Excitations for Restricted Hartree-Fock Based Triplet States. *J. Chem. Phys.* **2008**, *129*, 244111; Musial, M.; Bratlett, R. J. Multi-Reference Fock Space Coupled-Cluster Method in the Intermediate Hamiltonian Formulation for Potential Energy Surfaces. *J. Chem. Phys.* **2011**, *135*, 044121.
44. Musiał, M.; Bartlett, R. J. Equation-of-Motion Coupled Cluster Method with Full Inclusion of the Connected Triple Excitations for Electron Attached States: EA-EOM-CCSDT. *J. Chem. Phys.* **2003**, *119*, 1901.

45. Musiał, M.; Kucharski, S. A.; Bartlett, R. J. Approximate Inclusion of the T_3 and R_3 Operators in the Equation-of-Motion Coupled-Cluster Method. *Adv. Quantum. Chem.* **2004**, *47*, 209.
46. Musiał, M.; Kucharski, S. A.; Bartlett, R. J. Equation-of-Motion Coupled Cluster Method with Full Inclusion of the Connected Triple Excitations for Ionized States: IP-EOM-CCSDT. *J. Chem. Phys.* **2003**, *118*, 1128.
47. ACES II program is a product of the Quantum Theory Project University of Florida Authors: Stanton, J. F.; Gauss, J.; Watts, J. D.; Nooijen, M.; Oliphant, N.; Perera, S. A.; Szalay, P. G.; Lauderdale, W. J.; Kucharski, S. A.; Gwaltney, S. R.; Beck, S.; Balková, A.; Musial, M.; Bernholdt, D. E.; Baeck, K. K.; Sekino, H.; Rozyczko, P.; Huber, C.; Bartlett, R. J.. Bartlett and Integral packages included are VMOL (Almlöf, J.; Taylor, P.); VPROPS (Taylor, P. R.); A modified version of ABACUS integral derivative package (Helgaker, T. U.; Jensen, H. J. Aa.; Olsen, J.; Jørgensen, P.; Taylor, P. R.).
48. Sadlej, A. J. Medium-Size Polarized Basis Sets for High-Level Correlated Calculations of Molecular Electric Properties. *Collect. Czech. Chem. Commun.* **1988**, *53*, 1995.
49. Gerdes, A.; Hobein, M.; Knöckel, H.; Tiemann, E. Ground State Potentials of the NaK Molecule. *Eur. Phys. J. D.* **2008**, *49*, 67.
50. Wormsbecher, R. F.; Hessel, M. M.; Lovas, F. J. Microwave Optical Double Resonance of $X^1\Sigma^+$ of NaK. *J. Chem. Phys.* **1981**, *74*, 6983.
51. Breford, E. J.; Engelke, F. The $a^3\Sigma^+$ State of NaK. High Resolution Spectroscopy Using Laser-Induced Fluorescence (LIF). *J. Chem. Phys.* **1979**, *71*, 1994.
52. Ross, A. J.; Clements, R. M.; Barrow, R. F. The $A(2)^1\Sigma^+$ State of NaK. *J. Mol. Spectrosc.* **1988**, *127*, 546.
53. Ferber, R.; Pazyuk, E. A.; Stolyarov, A. V.; Zaitevskii, A.; Kowalczyk, P.; Chen, H.; He, W.; Stwalley, W. C. The $c^3\Sigma^+$, $b^3\pi$, and $a^3\Sigma^+$ States of NaK Revisited. *J. Chem. Phys.* **2000**, *112*, 5740.
54. Kowalczyk, P. High Resolution Laser Spectroscopy of the $c^3\Sigma^+ \Leftarrow X^1\Sigma^+$ and $b^3\pi \Leftarrow X^1\Sigma^+$ Forbidden Transitions in NaK. *J. Chem. Phys.* **1989**, *91*, 2779.
55. Kasahara, S.; Baba, M.; Katô, H. High Resolution Laser Spectroscopy up to the Dissociation Limit of the NaK $B^1\pi$ State, and Predissociation near the Dissociation Limit. *J. Chem. Phys.* **1991**, *94*, 7713.
56. Ross, A. J.; Effantin, C.; D'Incan, J.; Barrow, R. F. Laser-Induced Fluorescence of NaK: The $b(1)^3\pi$ State. *J. Phys. B At. Mol. Opt. Phys.* **1986**, *19*, 1449.
57. Ross, A. J.; Crozet, P.; Russier-Antoine, I.; Grochola, A.; Kowalczyk, P.; Jastrze¸bski, W.; Kortyka, P. On the $C^1\Sigma^+$ State of NaK. *Mol. Spectrosc.* **2004**, *226*, 95.
58. Adohi-Krou, A.; Jastrzebski, W.; Kowalczyk, P.; Stolyarov, A. V.; Ross, A. J. Investigation of the $D^1\pi$ State of NaK by Polarisation Labelling Spectroscopy. *J. Mol. Spectrosc.* **2008**, *250*, 27.
59. Pazyuk, A. E.; Stolyarov, A. V.; Zaitevskii, A.; Ferber, R.; Kowalczyk, P.; Teichteil, C. Spin-Orbit Coupling in the $d^1\pi \sim d^3\pi$ Complex of $^{23}Na^{39}K$. *Mol. Phys.* **1999**, *96*, 955.
60. Kowalczyk, P. Perturbation Facilitated Observation of the $d^3\pi$ State in NaK. *J. Mol. Spectrosc.* **1989**, *136*, 1.
61. Le Roy, R.J. *LEVEL 8.0: A Computer Program for Solving the Radial Schrödinger Equation for Bound and Quasibound Levels.* University of Waterloo Chemical Physics Research Report CP-663, 2007. http://leroy.uwaterloo.ca/programs/.

CHAPTER FOURTEEN

The Correlation Effects in Density Functional Theory Along the Dissociation Path

Adam Buksztel*, Szymon Śmiga*, Ireneusz Grabowski*,[1]
*Institute of Physics, Faculty of Physics, Astronomy and Informatics, Nicolaus Copernicus University, Torun, Poland
[1]Corresponding author: e-mail address: ig@fizyka.umk.pl

Contents

1. INTRODUCTION

The density functional theory (DFT) within the Kohn–Sham (KS) scheme[1] is the most widely used approach in electronic structure calculations. The use of KS-DFT methods for theoretical description of physical and chemical properties of quantum systems consisting of up to thousands of electrons has become routine, since it provides considerable accuracy at a reasonable computational cost. However, the accuracy of these methods strongly depends on the approximations used in the exchange-correlation functionals and corresponding exchange-correlation potentials.

Advances in Quantum Chemistry, Volume 73
ISSN 0065-3276
http://dx.doi.org/10.1016/bs.aiq.2015.07.002

Unfortunately, despite its great success, standard (density-dependent) KS-DFT still lacks a theoretically correct, nonempirical, and versatile method. Typical problems of most of the conventional KS-DFT functionals are the self-interaction error, qualitatively incorrect (exchange-)correlation potentials (in particular, the wrong asymptotic behavior), poor description of long-range interactions, troublesome description of charge-transfer excited states, or incorrect ionization energy.

In recent years, the *ab initio* DFT approach[2] has raised considerable interest, since it overcomes most of the common issues of standard DFT. More importantly, it provides a systematic means of improving accuracy, ensuring convergence to the right answer. One possible realization of the *ab initio* DFT is the use of explicitly orbital-dependent functionals, which leads directly to the optimized effective potential (OEP) method[3–5] within the KS formalism. The exchange-only OEP method (OEPx or EXX), derived by Sharp and Horton[6] and Talman and Shadwick,[3] is now well established in calculations for molecular and solid-state systems. Many correlation OEP functionals have been defined so far,[2,7–12] although the simplest ones (based on the second-order many-body perturbation theory) seem to be most interesting, because of their relatively low computational cost of including correlation effects. Among them the most promising, in this context, are the recently developed OEP2-SOS functionals,[13,14] based on the spin-resolved concept originally proposed by Grimme.[15] As already shown,[16–18] the OEP methods provide qualitatively good exchange-correlation potentials and electronic densities compared to exact methods, while the exchange-correlation potentials obtained with standard DFT are usually substantially different.

In this chapter, we extend our analysis to weakly interacting systems focused on nonequilibrium geometries, up to the dissociation limit. Several standard DFT (semi-)local functionals are considered, i.e., SVWN5, BLYP, and PBE, as well as two orbital-dependent second-order OEP functionals (OEP2-sc and OEP2-SOSb). The quality of each approach is assessed in terms of the (exchange-)correlation potentials and correlated densities. We conclude the analysis with inspection of interaction energy surfaces for the Ne dimer, which provides valuable information about physical implications of using each approximation.

Throughout the chapter, $i,j,\ldots$ indices indicate occupied orbitals, $a,b,\ldots$ indicate unoccupied ones, and $p,q,\ldots$ are used for any orbitals. For the sake of simplicity of the considerations, all the formulas are given in the spin-restricted form, and the atomic units are assumed.

2. THEORY

Within the KS implementation of DFT,[1] the electronic density

$$\rho(\mathbf{r}) = \sum_{i}^{N} |\varphi_i(\mathbf{r})|^2 \tag{1}$$

is computed from the N occupied KS orbitals $\varphi_i(\mathbf{r})$, which are the solutions of the one-particle KS equations

$$\left[-\frac{1}{2}\nabla^2 + v(\mathbf{r}) + \int \frac{\rho(\mathbf{r}')}{|\mathbf{r}-\mathbf{r}'|}\mathrm{d}\mathbf{r}' + v_{\mathrm{xc}}[\rho](\mathbf{r})\right]\varphi_{\mathrm{p}}(\mathbf{r}) = \varepsilon_{\mathrm{p}}\varphi_{\mathrm{p}}(\mathbf{r}), \tag{2}$$

where ε_{p} are the KS eigenvalues, $v(\mathbf{r})$ is the external potential due to the nuclei, and the third term is the Coulomb potential. The last term ($v_{\mathrm{xc}}[\rho](\mathbf{r})$) is the local exchange-correlation potential, formally defined as the functional derivative of the exchange-correlation functional $E_{\mathrm{xc}}[\rho]$ with respect to the electronic density $\rho(\mathbf{r})$

$$v_{\mathrm{xc}}[\rho](\mathbf{r}) = \frac{\delta E_{\mathrm{xc}}[\rho]}{\delta \rho(\mathbf{r})}. \tag{3}$$

In practice, the exchange-correlation functional is usually expressed as a sum ($E_{xc}[\rho] = E_x[\rho] + E_c[\rho]$) of exchange-only ($E_x[\rho]$) and correlation ($E_c[\rho]$) functionals, which are defined separately. This formulation is convenient and crucial for constructing the approximations of both functionals.

In the case where exchange-correlation functionals are explicitly density dependent, the forms of the exchange and correlation potentials are easily obtained by direct differentiation. In this work, we will reduce our interest and discussion to the local (SVWN5[1,19]) and gradient-dependent (BLYP,[20,21] PBE[22]) functionals only, since they still represent the most popular choice among DFT functionals for both applications, and development of new approximations.[23–27] Moreover, the exchange-correlation potentials corresponding to these functionals can be relatively easily plotted and further compared to the reference ones, which is an important point of current analysis.

Alternatively, the explicitly orbital-dependent (i.e., based on the KS orbitals and eigenvalues) exchange-correlation functionals can be used. However in this case, in order to define a local multiplicative KS exchange-correlation potential $v_{\mathrm{xc}}[\rho](\mathbf{r})$,[7,9,28,29] one must additionally employ the OEP method.[3,6] Formally, there exist several different approaches to determine the functional derivative in Eq. (3) in the orbital-dependent case, and all of them lead to the same working OEP equations. The simplest one is to apply the chain rule for functional differentiation.[7,9,29,30] Another option is to follow the original suggestion of Sharp and Horton and determine $v_{xc}(\mathbf{r})$ by direct minimization of the total energy of the system with respect to variations in this potential,[31] or use the KS density condition,[32] which constitutes the foundation of the *ab initio* DFT approach[2] and allows to define a range of orbital-dependent exchange-correlation energy functionals and potentials.[2,33,34]

In the orbital-dependent case, the exchange energy functional has the form of the standard Hartree–Fock exchange energy

$$E_x[\{\phi_q\}] = -\sum_{ij} (ij|ji), \tag{4}$$

with $(pq|rs)$ being a two-electron integral in the Mulliken notation computed from KS orbitals. For the implementation of the full self-consistent-field (SCF)-correlated OEP, the orbital-dependent correlation functionals have been defined in the context of perturbation theory, mainly restricted to the use of second-order correlation. Among others the most common are those based on the second-order Görling–Levy Perturbation Theory (OEP-GL2).[2,8,9,35–37] However, OEP-GL2 was shown to be unsatisfactory,[2,9,35–37] yielding highly overestimated correlation effects, i.e., correlation energies, potentials, and correlated densities. Moreover, the OEP-GL2 functional also causes convergence problems, so that in many cases an SCF solution cannot be obtained.[8,10]

To avoid the above-mentioned problems and improve stability and accuracy, new approaches in that context have been proposed. One of these is based on a modified MBPT(2) functional with a semi-canonical transformation of the Hamiltonian (OEP2-sc).[2,10,37] This method has been found to provide a stable alternative to the OEP-GL2-type perturbation theories. Recently, new methods based on the spin-component-scaled (SCS) second-order functional (OEP2-SCS)[13,14] have been proposed which also overcome the shortcomings of OEP-GL2 functional. In these methods, the OEP2-SOS (scaled-opposite-spin) variant, based on the simplified case[38] of the more general approach, namely the spin-resolved second-order correlation energy expression,[15] seems to be most effective. In the OEP2-SOS case, the correlation functional takes a very simple form:

$$E_{\mathrm{OEP2-SOS}} = c_{\mathrm{OS}} \sum_{ijab} \frac{(ia|jb)^2}{\epsilon_i + \epsilon_j - \epsilon_a - \epsilon_b}, \tag{5}$$

where c_{OS} is the opposite-spin (OS) system-dependent coefficient.

We have proposed few practical ansatzes to calculate the OS coefficient without using any empirical parameters. As shown in Ref. 14, a reasonable description of the correlation KS potentials: the HOMO-LUMO gap (the difference between highest occupied molecular orbital (HOMO) and lowest unoccupied molecular orbital (LUMO) energies) and correlated densities is obtained in most cases by using a system-dependent value of c_{OS},

$$c_{\mathrm{OS}}^{\mathrm{b}} = \frac{E_{\mathrm{OS}}@\mathrm{HF}}{E_{\mathrm{OS}}@\mathrm{OEPx}}, \tag{6}$$

where E_{OS}@HF and E_{OS}@OEPx denote the opposite spin part (see Eq. 5) of the second-order correlation energy expression computed with Hartree–Fock and OEP exchange-only orbitals, respectively. This value of the scaling parameter can thus be used to define the OEP2-SOSb method, which is, by construction, optimized for the description of correlation-potential-related properties—the main focus of this chapter. We emphasize, that in practice, there is no additional numerical cost necessary for calculating the value of the c_{OS} parameter. These OEP2-SOS methods provide an accurate OEP correlation functional that largely outperforms the OEP-GL2 approach, providing much better (with almost CCSD(T) quality) total and correlation energies (in the case of OEP2-SOSh method[14]) correlation KS potentials, correlated densities, HOMO-LUMO gaps, and other quantities.[13,14] Moreover, the OEP2-SOS method demonstrated a clear advantage over the OEP-GL2 method, converging also in difficult cases (e.g., the beryllium atom). Finally, because only the opposite-spin correlation is used in its formulation, the computational cost of the method can be reduced[38] with respect to other second-order approaches. ($\mathcal{O}(N^4)$ vs. $\mathcal{O}(N^5)$).

For the detailed orbital-dependent equations corresponding to the OEP-GL2, OEP2-sc, and OEP2-SOS functionals used in this study, we refer the reader to Refs. 2,10,13,14,39,40.

3. COMPUTATIONAL DETAILS

To assess the behavior of the KS-DFT method with different approximations for the exchange-correlation functionals at the dissociation limit, we performed the KS calculations using either density-dependent or orbital-dependent functionals and potentials. For the standard, density-dependent functionals, we used local and gradient ones, i.e., the SVWN5,[1,19] BLYP,[20,21] and PBE,[22] respectively. In the category of orbital-dependent functionals, we studied the OEP2-sc[2] method and recently developed OEP2-SOSb.[14]

In order to solve the OEP equations and determine the exchange-correlation potential, which is then used in the iterative SCF solution of the KS equations (see Eq. 2), we use the finite basis set implementation of the OEP method from Refs. 30,41. Since we use in this procedure finite Gaussian-type basis sets, which at large r decays much faster than $1/r$, an incorrect asymptotic behavior of our exchange-correlation potentials would be obtained. Therefore to preserve the correct $(-1/r)$ asymptotic behavior, we use the Colle–Nesbet[42] idea and seed the exchange-correlation OEP potential[14,18] by the Slater potential.[43] Additionally, when a final solution is obtained, all eigenvalues as well as exchange-correlation potential are shifted to satisfy the so-called HOMO condition on exchange[5] and correlation[9] level, respectively. To minimize the numerical instabilities[10,40,44–52] and properly represent the exchange-correlation potential, in all calculations, fully uncontracted basis sets were used. In case of the He_2, an even tempered 20s10p2d basis set was employed and for Ne_2 and ArNe molecules, the ROOS-ATZP[53] one.

In all calculations (HF, OEP, and KS-DFT), the change of at most 10^{-8} a.u. in density matrix elements was taken as the convergence criterion. All calculations have been performed in our development version of ACES II[54] quantum chemistry package.

4. NUMERICAL RESULTS AND DISCUSSION

In this section, we present the results of our best second-order OEP methods, i.e., OEP2-sc and OEP2-SOSb,[14] and compare them with results of a few (semi-)local standard (density-dependent) KS-DFT functionals, i.e., SVWN5, BLYP, and PBE. Performance of all these methods is investigated in case of stretching interatomic distances for a few weakly interacting systems (He_2, Ne_2, and ArNe). For clarity of this chapter, we limit the discussion in Sections 4.1 and 4.2 to the case of equilibrium and dissociated geometries; however, we have analyzed all intermediate geometries with a step of 1 Å (1 a.u. for He_2). We can state that neither correlation potentials nor correlated densities show any significant deviations from what is presented here.

Our analysis is split into three parts. We start with a direct comparison of the correlation and exchange-correlation potentials in Section 4.1. Next, in Section 4.2, electron density is a subject of a detailed investigation, using the idea of correlated density defined as

$$\rho_{c}^{\text{method}}(\mathbf{r}) = \rho_{xc}^{\text{method}}(\mathbf{r}) - \rho_{x}^{\text{method}}(\mathbf{r}), \tag{7}$$

where $\rho_{xc}^{\text{method}}(\mathbf{r})$ is the total electron density calculated on exchange-correlation level of *method* and $\rho_{x}^{\text{method}}(\mathbf{r})$ is the total electron density calculated on exchange-only level of the same *method* (e.g., for BLYP, it is $\rho_{c}^{\text{BLYP}}(\mathbf{r}) = \rho_{xc}^{\text{BLYP}}(\mathbf{r}) - \rho_{x}^{\text{B88}}(\mathbf{r})$). Finally, we take a closer look at the interaction energy curve for Ne dimer in Section 4.3 to complete the comprehensive analysis of the problem.

4.1 Correlation and Exchange-Correlation Potentials

The first step in our investigation is making a direct comparison of KS correlation and exchange-correlation potentials, which (by definition) are key quantities in KS-DFT. It is well known[14,18] that the quality of the KS (exchange-)correlation potentials has a direct impact on other quantities calculated in the KS-DFT procedure, e.g., total and correlation energies, orbital energies, HOMO energies, HOMO-LUMO gaps, or excitation energies. In this context, the analysis of the KS potentials is an important test for the new and existing exchange-correlation KS-DFT functionals, especially in description of different systems and environments, e.g., equilibrium and stretched geometries, dissociation limits, quasidegenerate systems.

The correlation potentials for the He_2 system are presented in Fig. 1 for both, equilibrium ($R = 5.6$ a.u., left panel) and near dissociation ($R = 9.0$ a.u., right panel) interatomic distances. In the upper plots, we report curves calculated using standard DFT functionals (VWN5, LYP, and PBEc), while in the bottom ones there are the correlation potentials calculated from second-order OEP methods (OEP2-sc, OEP2-SOSb). All correlation potentials (calculated from either density-dependent and orbital-dependent functionals) are stable (i.e., they are almost unchanged) with respect to the increase of interatomic distance up to the dissociation limit, showing the same qualitative and quantitative behavior. Moreover, we can see that the shapes of the potentials originating from individual atoms are similar to those obtained from single-atom calculations.

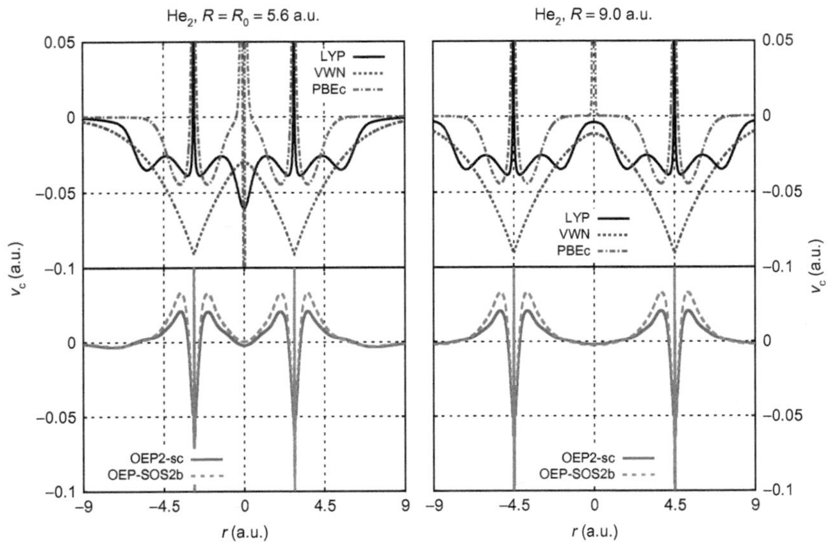

Figure 1 Correlation potential for He_2 system at the equilibrium geometry (left panel) and stretched geometry (right panel), obtained from standard KS-DFT (LYP, VWN, PBEc—upper plots) and second-order OEP calculations (bottom panels).

In the more detailed behavior analysis of different KS-DFT potentials, we compare them to those obtained from OEP2-sc calculations. As was shown in many test calculations, the correlation potentials and also correlated densities obtained from the OEP2-sc method are very close to the CCSD(T) results[14,16–18] and the exact ones,[55] thus we will use them as reference quantities in our analysis.

Comparing to results obtained from the OEP2-SOSb calculations, we can see in Fig. 1 that the latter ones show excellent agreement with the OEP2-sc correlation potentials, even though they have slightly higher amplitudes. The oscillations at nuclei are a manifestation of a well-known numerical artifact resulting from the use of a finite basis set expansion of the potentials in the procedure of solving OEP equations.

In contrast to the orbital-dependent methods, all investigated standard DFT correlation potentials present the wrong qualitative and quantitative behavior, either in the equilibrium or near dissociation limit distance. In particular, VWN5 and LYP correlation potentials do not vanish fast enough either in asymptotic or interatomic regions. Furthermore, both of them are negative in the whole range, as opposed to the reference results. In the case of GGA (PBE and BLYP) calculations, we also encountered strong oscillations of correlation potentials in the region between atoms. A detailed discussion of this unphysical behavior of the GGA potential can be found in the Appendix.

Similar observations can be made for ArNe system. The correlation potentials for this system are reported on Fig. 2 for both, equilibrium ($R = 3.489$ Å, left panel) and near dissociation interatomic distances ($R = 8$ Å, right panel). Again, the orbital-dependent potentials are very close to the reference ones. However, in the case of OEP2-SOSb potentials note only subtle changes of the correlation potentials but they are visible not only in the interatomic region but also in the outer regions (see, e.g., near $r \approx -9$ a.u. and $r \approx 7$ a.u. in the equilibrium geometry panel of Fig. 2). The standard density-dependent functionals present a completely erroneous behavior also for the ArNe system. Also unphysical, large oscillations of the GGA correlation potentials are observed in the region between atoms. The analysis and explanation of such behavior can also be found in the Appendix.

We also analyzed the total exchange-correlation potentials. The results are presented in Figs. 3 (He_2) and 4 (ArNe). Again, note that all of the density-dependent and orbital-dependent KS potentials are very stable when the interatomic distance is increased. As can be observed for all tested methods, the total KS potentials are dominated by their exchange parts.

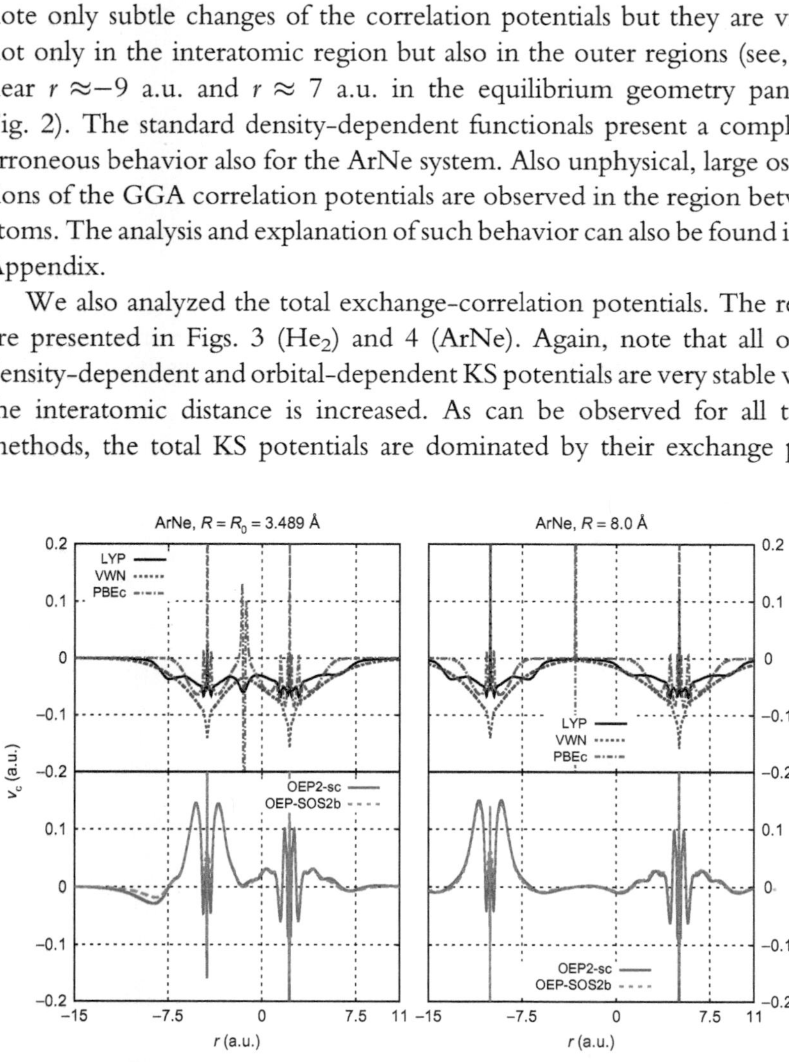

Figure 2 Correlation potential for ArNe system at the equilibrium geometry (left panel) and stretched geometry (right panel), obtained from standard KS-DFT (LYP, VWN, PBEc—upper plots) and second-order OEP calculations (bottom panels).

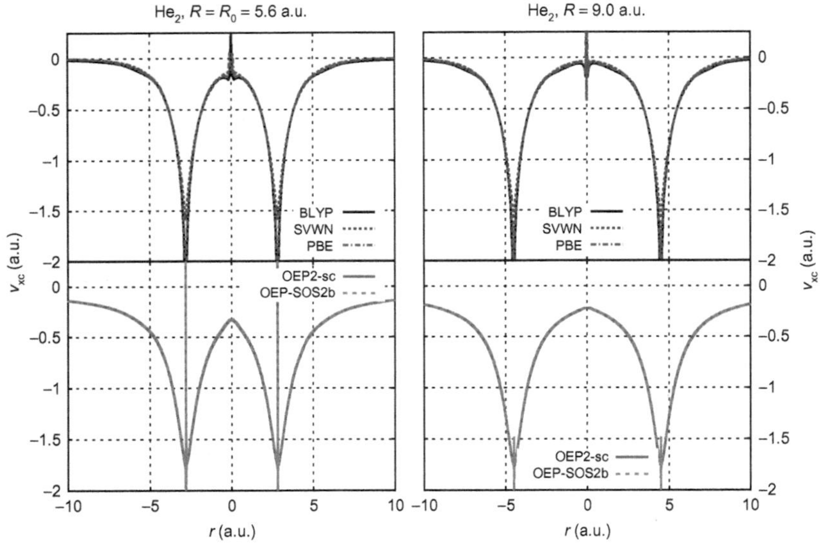

Figure 3 Exchange-correlation potential for He_2 system at the equilibrium geometry (left panel) and stretched geometry (right panel), obtained from standard KS-DFT (BLYP, SVWN, PBE—upper plots) and second-order OEP calculations (bottom panels).

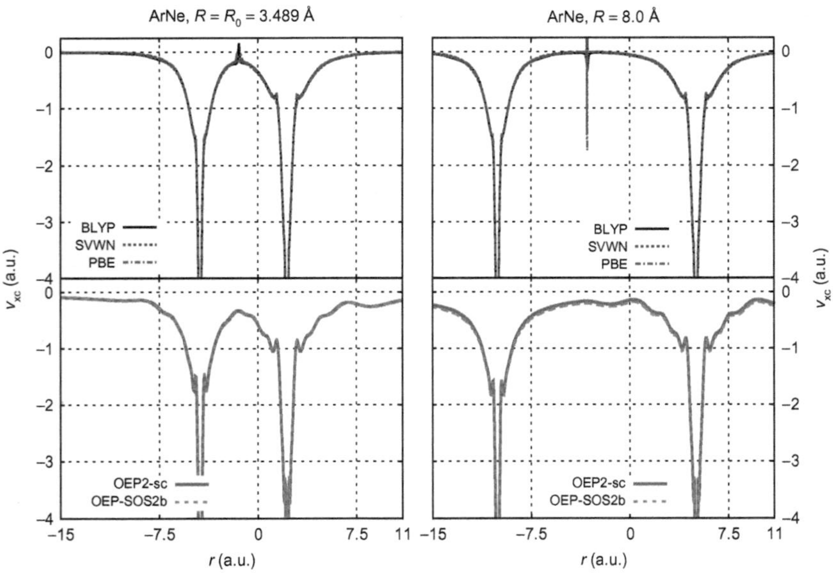

Figure 4 Exchange-correlation potential for ArNe system at the equilibrium geometry (left panel) and stretched geometry (right panel), obtained from standard KS-DFT (BLYP, SVWN, PBE—upper plots) and second-order OEP calculations (bottom panels).

The OEP exchange-correlation potentials calculated from OEP2-SOSb and OEP2-sc are almost indistinguishable on the scale presented. In the case of density-dependent potentials, there is no significant deviation between the BLYP and SVWN5 potentials, whereas the PBE shows deeper minima at nuclei and oscillations in the region between atoms, which are similar, as in the correlation-only case. The reason for these oscillations is the same as in case of correlation potentials, i.e., the way of constructing the GGA functional and potential (see Appendix).

In the case of OEP exchange-correlation potentials obtained for the ArNe dimer, characteristic bumps related to the shell structure can be observed. In a contrast, no such behavior occurs for standard DFT methods, except from PBE potential near the Ar atom, although the effect is much less pronounced. Moreover, the OEP exchange-correlation potentials have correct asymptotic behavior, which results in nonzero values of the potential even in the middle between separated atoms (right panels on Figs. 3 and 4). This is not the case for the standard density-dependent DFT exchange-correlation potentials, which vanish too fast, particularly in the interatomic region.

4.2 Density-Based Analysis

Another important quantity which should be included in the analysis of correlation effects is the correlated density defined by Eq. (7). This comparison provides another effective tool for assessing the quality of the functionals used in KS-DFT calculations. In our analysis, the density is calculated along the main symmetry axis and reported on Figs. 5 (He_2) and 6 (ArNe) for different KS-DFT approaches. For both systems and all interatomic distances, the correlated densities calculated using OEP-KS methods show similar qualitative behavior, although that of OEP2-SOSb is slightly underestimated with respect to the OEP2-sc one, the curves are still almost indistinguishable.

The correlated densities obtained with standard DFT methods are either underestimated (BLYP) or overestimated (SVWN5, PBE) compared to the reference OEP2-sc results. A closer examination shows that the PBE curves have reversed sign. This might result from a strong repulsive character of correlation potential at the nuclei (Figs. 1 and 2), which slightly pushes the electron density away. However, it turns out that the changes in electron density due to correlation effects are dominated by internal atomic correlation, while the interatomic correlation influence on the density is not visible

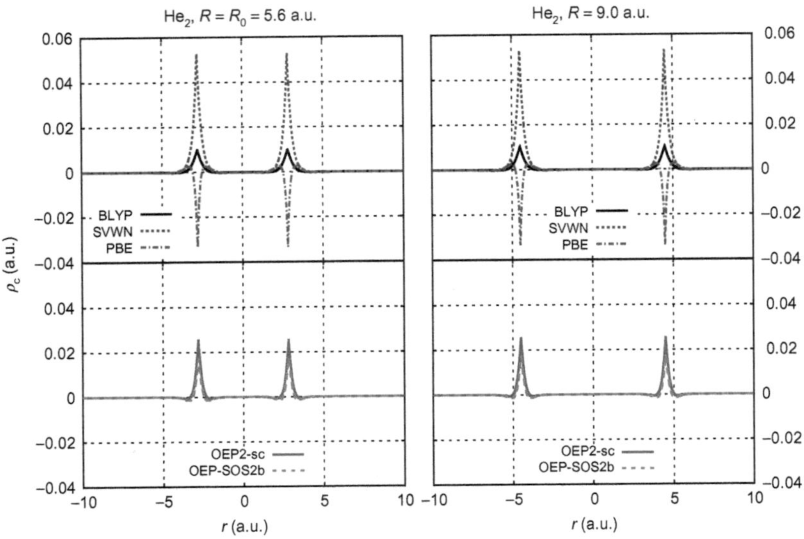

Figure 5 Correlated density (ρ_c) for He_2 system at the equilibrium geometry (left panel) and stretched geometry (right panel), obtained from standard KS-DFT (BLYP, SVWN, PBE—upper plots) and second-order OEP calculations (bottom panels).

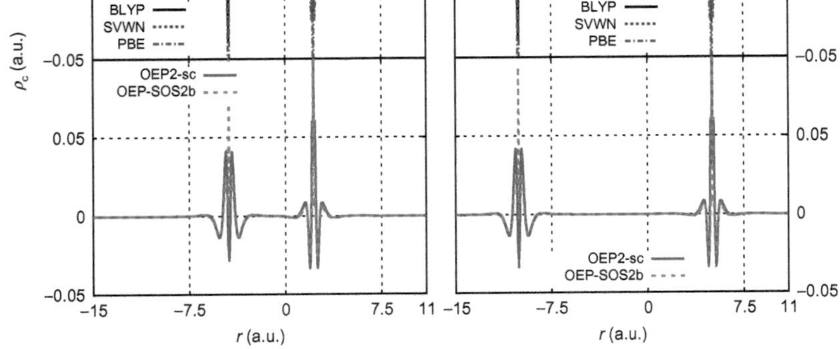

Figure 6 Correlated density (ρ_c) for ArNe system at the equilibrium geometry (left panel) and stretched geometry (right panel), obtained from standard KS-DFT (BLYP, SVWN, PBE—upper plots) and second-order OEP calculations (bottom panels).

at the given scale. These wrong results for the correlated densities calculated from the standard KS-DFT functionals, together with the correlation potentials, evidently show that LYP, PBEc, and VWN5 correlation potentials do not show any correlation effects on those quantities, which should be taken into account especially in the context of the use of these functionals to define new methods in KS-DFT.

Again as in the KS potential analysis, we can see that there are almost no changes in shape, size, and qualitative behavior of the correlated densities along the dissociation path. Here we present only plots for the two distances (at equilibrium geometry and at the dissociation limit), but as we mentioned before, the correlated density curves remain almost unchanged for all geometries defined by stretched interatomic distances.

4.3 Intermolecular Interaction Energy Curves

As a final analysis, we have also calculated the intermolecular interaction energy curves for the Ne dimer, which is the usual test for checking the quality of the methods in the dissociation-related problems. The results are presented in Fig. 7. The CCSD(T) curve is considered as a reference one here. As we can see, the OEP2-sc intermolecular interaction energy curve is close

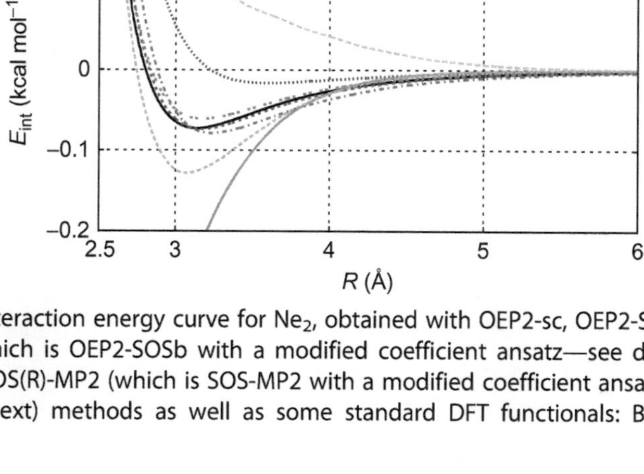

Figure 7 Interaction energy curve for Ne_2, obtained with OEP2-sc, OEP2-SOSb, OEP2-SOS(R)b (which is OEP2-SOSb with a modified coefficient ansatz—see discussion in text), and SOS(R)-MP2 (which is SOS-MP2 with a modified coefficient ansatz—see discussion in text) methods as well as some standard DFT functionals: BLYP, SVWN, and PBE.

to the reference CCSD(T) one, as was already observed by Lotrich et al.[56] Most of the standard KS-DFT functionals completely fail in describing the binding energies of the Ne dimer. The curves computed using SVWN5 and BLYP functionals are completely wrong, being totally attractive (SVWN5) or totally repulsive (BLYP). Thus, they do not predict formation of a bound state. The PBE method provides better interaction energy curve, but unfortunately much too deep.

An interesting case is the interaction energy curve calculated using the OEP2-SOSb method, which is excessively shallow and flat, with its minimum displaced by about 0.5 Å (we assume the equilibrium interatomic distance $R_0 = 3.1$ Å). This seems to be in contradiction with the observed very good quality of the exchange-correlation potentials and correlated densities along the dissociation path. However, this is the characteristic behavior of the SCS method applied directly to the noncovalently interacting systems,[57–59] where the "standard" parametrization of the SCS method does not work. In this case, the different ways of calculating scaling coefficients (e.g., for SOS-MP2 and in our case of OEP2-SOSb) must be employed.[58] The scaling coefficient is obtained then by using the observed proportionality of interaction correlation energy curves. After applying this procedure, the two plotted curves, i.e., SOS(R)-MP2 and OEP2-SOS(R)b, are both qualitatively and quantitatively correct and close to the reference CCSD(T) results, although the minimum of OEP2-SOS(R)b interaction curve is slightly shifted to longer interatomic distances. It shall be emphasized that OEP2-SOS(R)b is essentially the same method as OEP2-SOSb, yet with a different ansatz for the scaling coefficient, which is calculated after the OEP2-SOSb method is converged ($c_{OS} = 2.009$) and applied to scale the interaction correlation energy. Thus, the improvement is significant.

5. FINAL REMARKS

To enable us to compare the performance of conventional and *ab initio* DFT methods, we have used them to calculate a number of quantities, i.e., the correlation potentials, correlation densities, and interaction potential energy curves for a few weakly interacting systems in stretched geometries. In our research, we have found that all the investigated methods are highly stable (i.e., providing correlation potentials and correlated densities), when the interatomic distances are increased. In addition, they are very similar to those obtained from single-atom calculations. The standard DFT results, however stable (in the same context), are in general of poor quality in all examined areas. Above all, they do not properly predict the formation of a bound state in the weakly interacting systems. Moreover, they provide either underestimated or overestimated correlated densities. The correlation potentials are qualitatively incorrect as compared to the reference ones, even having the wrong sign. We also observed some disturbing, unphysical oscillations of the GGA exchange and correlation potentials in the half-bond area. We have found and explained the reason of this behavior, pointing out that it is directly connected to the form of the exchange-correlation GGA potential (see Appendix).

On the other hand, note that the orbital-dependent OEP2-SOSb method yields results very close to the reference OEP2-sc ones. This is a key observation because this approximate method allows comparable accuracy to be obtained at the much lower computational cost.

The above KS potential and density analysis of the KS-DFT methods are complemented by the interaction energy curve inspection. The standard density-dependent KS-DFT functionals completely failed in this area and do not predict formation of a proper bound state. OEP2-SOSb does not provide the correct interaction energy curve, even though it has correct exchange-correlation potential and correlated density. However, it can be fixed by redefining the scaling coefficient at the correlated interaction energy level.[58]

ACKNOWLEDGMENTS

This work was partially supported by the National Science Center under Grant No. DEC-2013/11/B/ST4/00771 and by research fellowship within project Enhancing Educational Potential of Nicolaus Copernicus University in the Disciplines of Mathematical and Natural Sciences (project no. POKL.04.01.01-00-081/10).

CONFLICT OF INTEREST

The authors declare no competing financial interest.

APPENDIX. ANALYSIS OF THE BEHAVIOR OF THE INVESTIGATED GGA POTENTIALS NEAR THE MIDDLE OF THE BOND

In order to explain the oscillating behavior of the PBE and BLYP potentials between the atoms (see Section 4.1), we performed an analysis of their components, focusing on the half-bond area, where the oscillations were most noticeable (see Figs. 1–4). To simplify the analysis, we limit our discussion to the exchange parts of PBE (PBEx) and BLYP (B88) functionals. Nevertheless, the extension of the following argument to correlation functionals is straightforward, yet technically much more complicated.

Each of the exchange GGA functionals can be written in a general form as

$$E_{\mathrm{X}}^{\mathrm{GGA}} = C_{\mathrm{X}} \int d^3\mathbf{r}\ \rho^{4/3}(\mathbf{r}) F_{\mathrm{x}}(s), \tag{A.1}$$

where $C_{\mathrm{x}} = -\frac{3}{4}\left(\frac{3}{\pi}\right)^{1/3}$ and $F_{\mathrm{x}}(s)$ is the exchange enhancement factor expressed as the function of reduced density gradient $s = |\nabla\rho| / \left(2(3\pi^2)^{1/3}\rho^{4/3}\right)$. In fact, $F_{\mathrm{x}}(s)$ defines a specific approximate GGA functional. The corresponding GGA exchange potential is obtained by taking the functional derivative of the GGA functional (Eq. A.1) and can be written as follows[60]

$$v_{\mathrm{X}}^{\mathrm{GGA}}(\mathbf{r}) = v_1(\mathbf{r}) + v_2(\mathbf{r}) + v_3(\mathbf{r}). \tag{A.2}$$

The terms on the right-hand side are defined as

$$v_1(\mathbf{r}) = \frac{4}{3} C_{\mathrm{x}} \rho^{1/3}(\mathbf{r}) F_{\mathrm{x}}, \tag{A.3}$$

$$v_2(\mathbf{r}) = -C_{\mathrm{x}} \rho^{1/3}(\mathbf{r}) \frac{q}{s} \frac{\mathrm{d}F_x}{\mathrm{d}s}, \tag{A.4}$$

$$v_3(\mathbf{r}) = -C_{\mathrm{x}} \rho^{1/3}(\mathbf{r}) \left(u - \frac{4}{3}s^3\right) \frac{\mathrm{d}}{\mathrm{d}s} \left(\frac{1}{s}\frac{\mathrm{d}F_{\mathrm{x}}}{\mathrm{d}s}\right), \tag{A.5}$$

where $u = \nabla\rho\nabla(|\nabla\rho|)/[24\pi^2\rho^3]$ and $q = \nabla^2\rho/[4(3\pi^2)^{2/3}\rho^{5/3}]$ is the reduced Laplacian.

As one can see, the $v_1(\mathbf{r})$ term (Eq. A.3) has a form of the exchange-only LDA potential multiplied by GGA enhancement factor $F_{\mathrm{x}}(s)$, and thus, it depends only on the density and its gradient. The second term (Eq. A.4) contains q, which depends also on the Laplacian of the density, and the third one (Eq. A.5) depends again only on the density and the gradient of the density.

In the case of PBEx functional, the enhancement factor has the form

$$F_{\mathrm{x}}^{\mathrm{PBEx}}(s) = 1 + \kappa - \frac{\kappa^2}{\kappa + \mu s^2}, \tag{A.6}$$

where κ and μ are constants which are equal to 0.804 and 0.21951,[22] respectively. Around the middle of the bond where $s \longrightarrow 0$ (stationary point of ρ), the components of PBEx potential behave as follows

$$\begin{aligned} v_1^{\mathrm{PBEx}}(\mathbf{r}) &\xrightarrow[s\to 0]{} C_{\mathrm{x}}\frac{4}{3}\rho^{1/3}(\mathbf{r}) \\ v_2^{\mathrm{PBEx}}(\mathbf{r}) &\xrightarrow[s\to 0]{} -C_{\mathrm{x}}\rho^{1/3}(\mathbf{r})2\mu q \\ v_3^{\mathrm{PBEx}}(\mathbf{r}) &\xrightarrow[s\to 0]{} 0. \end{aligned}$$

A similar observation can be made for the B88 potential. In this case, the enhancement factor is defined as[20]

$$F_{\mathrm{x}}^{\mathrm{B88}}(s) = 1 - \frac{\beta(c_1 s)^2}{1 + 6\beta c_1 s \ln\left(c_1 s + \sqrt{1 + (c_1 s)^2}\right)}, \tag{A.7}$$

where $c_1 = 2(3\pi^2)^{1/3}$ and $\beta = 0.0042$. When $s \longrightarrow 0$, the v_1^{B88} and v_3^{B88} parts have the same limits as the corresponding parts of PBEx potential. The difference occurs only for v_2^{B88}

$$v_2^{\mathrm{B88}}(\mathbf{r}) \xrightarrow[s\to 0]{} C_{\mathrm{x}}\rho^{1/3}(\mathbf{r})2c_1^2\beta q.$$

Nevertheless, the behavior of both GGA potentials is dominated in this region by v_2 component.

In the case of noncovalently bound systems, such as Ne_2 being investigated here, the value of q at the intermolecular distances is large,[61] even at the equilibrium geometry. It occurs particularly in the neighborhood of the stationary point of ρ, where q reaches its local maximum. It is caused by the fact that for weakly interacting systems, the density in this area is very low, much lower than its Laplacian, and in consequence $q \gg 0$ (see the definition of q). This can be clearly observed in the bottom panels of Fig. A.1, where the value of reduced Laplacian q is plotted for the Ne_2 system along the symmetry axis for equilibrium and stretched geometries.

To analyze the individual parts of the GGA exchange potentials, we plotted them for PBEx at top panels of Fig. A.1. As one can see, the v_2 component, which is q-dependent, gives a significant contribution to the total PBEx potential in the area around the middle of the bond and thus causes huge unphysical oscillations. Similar behavior of GGA potentials can be observed for other types of bonding; however, it is more pronounced for the nonequilibrium geometries (low values of density at the stationary point are required).

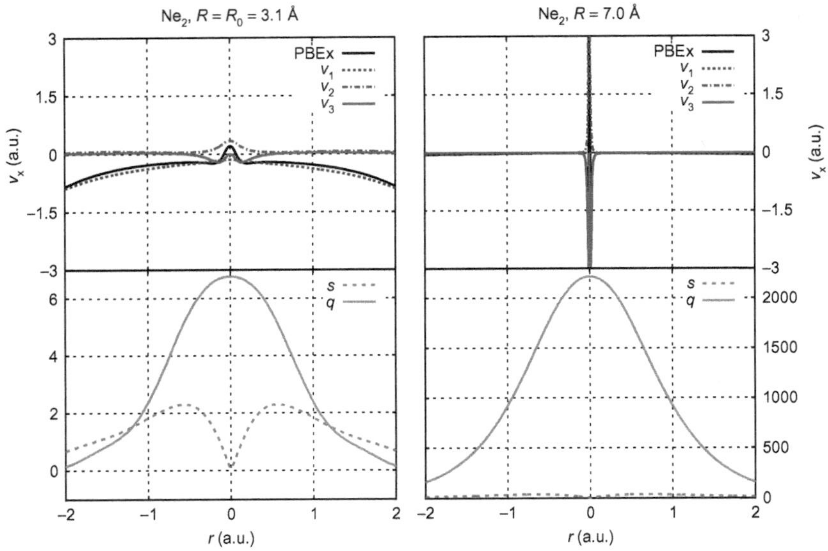

Figure A.1 Components of the PBE exchange potential at the equilibrium geometry (upper left panel) and stretched geometry (upper right panel) of the Ne_2. The bottom panels present reduced gradient *s* and reduced Laplacian *q* (both dimensionless, see Appendix) in function of interatomic distance.

REFERENCES

1. Kohn, W.; Sham, L. J. Self-Consistent Equations Including Exchange and Correlation Effects. *Phys. Rev.* **1965**, *140*, A1133–A1138.
2. Bartlett, R. J.; Grabowski, I.; Hirata, S.; Ivanov, S. The Exchange-Correlation Potential in *Ab Initio* Density Functional Theory. *J. Chem. Phys.* **2005**, *122*, 034104.
3. Talman, J. D.; Shadwick, W. F. Optimized Effective Atomic Central Potential. *Phys. Rev. A* **1976**, *14*, 36–40.
4. Grabo, T.; Kreibich, T.; Kuth, S.; Gross, E. K. U. Orbital Functionals in Density Functional Theory: The Optimized Effective Potential Method. In: Anisimov, V. I., Ed.; Strong Coulomb Correlations in Electronic Structure Calculations; Gordon and Breach: London, 1999; pp 203–311.
5. Krieger, J. B.; Li, Y.; Iafrate, G. J. Construction and Application of an Accurate Local Spin-Polarized Kohn-Sham Potential with Integer Discontinuity: Exchange-Only Theory. *Phys. Rev. A* **1992**, *45*, 101–126.
6. Sharp, R. T.; Horton, G. K. A Variational Approach to the Unipotential Many-Electron Problem. *Phys. Rev.* **1953**, *90*, 317.
7. Görling, A.; Levy, M. Exact Kohn-Sham Scheme Based on Perturbation Theory. *Phys. Rev. A* **1994**, *50*, 196–204.
8. Jiang, H.; Engel, E. Second-Order Kohn-Sham Perturbation Theory: Correlation Potential for Atoms in a Cavity. *J. Chem. Phys.* **2005**, *123*, 224102.
9. Grabowski, I.; Hirata, S.; Ivanov, S.; Bartlett, R. J. *Ab initio* Density Functional Theory: OEP-MBPT(2). A New Orbital-Dependent Correlation Functional. *J. Chem. Phys.* **2002**, *116*, 4415.

10. Grabowski, I.; Lotrich, V.; Bartlett, R. J. *Ab Initio* Density Functional Theory Applied to Quasidegenerate Problems. *J. Chem. Phys.* **2007**, *127*, 154111.
11. Bartlett, R. J. *Ab Initio* DFT and Its Role in Electronic Structure Theory. *Mol. Phys.* **2010**, *108*, 3299–3311.
12. Hirata, S.; Ivanov, S.; Bartlett, R. J.; Grabowski, I. Exact-Exchange Time-Dependent Density-Functional Theory for Static and Dynamic Polarizabilities. *Phys. Rev. A* **2005**, *71*, 032507. http://dx.doi.org/10.1103/PhysRevA.71.032507. URL: http://link.aps.org/doi/10.1103/PhysRevA.71.032507.
13. Grabowski, I.; Fabiano, E.; Della Sala, F. Optimized Effective Potential Method Based on Spin-Resolved Components of the Second-Order Correlation Energy in Density Functional Theory. *Phys. Rev. B* **2013**, *87*, 075103.
14. Grabowski, I.; Fabiano, E.; Teale, A. M.; Śmiga, S.; Buksztel, A.; Della Sala, F. Orbital-Dependent Second-Order Scaled-Opposite-Spin Correlation Functionals in the Optimized Effective Potential Method. *J. Chem. Phys* **2014**, *141*, 024113.
15. Grimme, S. Improved Second-Order Møller-Plesset Perturbation Theory by Separate Scaling of Parallel- and Antiparallel-Spin Pair Correlation Energies. *J. Chem. Phys.* **2003**, *118*, 9095–9102.
16. Grabowski, I.; Teale, A. M.; Śmiga, S.; Bartlett, R. J. Comparing *Ab Initio* Density-Functional and Wave Function Theories: The Impact of Correlation on the Electronic Density and the Role of the Correlation Potential. *J. Chem. Phys.* **2011**, *135*, 114111.
17. Grabowski, I.; Lotrich, V.; Hirata, S. *Ab Initio* DFT—The Seamless Connection Between WFT and DFT. *Mol. Phys.* **2010**, *108*, 3313.
18. Grabowski, I.; Teale, A. M.; Fabiano, E.; Śmiga, S.; Buksztel, A.; Sala, F. D. A Density Difference Based Analysis of Orbital-Dependent Exchange-Correlation Functionals. *Mol. Phys.* **2014**, *112* (5–6), 700–710. http://dx.doi.org/10.1080/00268976.2013.854424. http://www.tandfonline.com/doi/pdf/10.1080/00268976.2013.854424, URL: http://www.tandfonline.com/doi/abs/10.1080/00268976.2013.854424.
19. Vosko, S. H.; Wilk, L.; Nusair, M. Accurate Spin-Dependent Electron Liquid Correlation Energies for Local Spin Density Calculations: A Critical Analysis. *Can. J. Phys.* **1980**, *55*, 1200–1211.
20. Becke, A. Density-Functional Exchange-Energy Approximation with Correct Asymptotic Behavior. *Phys. Rev. A* **1988**, *38*, 3098.
21. Lee, C.; Yang, W.; Parr, R. G. Development of the Colle-Salvetti Correlation-Energy Formula into a Functional of the Electron Density. *Phys. Rev. B* **1988**, *37*, 785.
22. Perdew, J. P.; Burke, K.; Ernzerhof, M. Generalized Gradient Approximation Made Simple. *Phys. Rev. Lett.* **1996**, 77, 3865–3868.
23. Tao, J.; Perdew, J. P.; Staroverov, V. N.; Scuseria, G. E. Climbing the Density Functional Ladder: Nonempirical Meta Generalized Gradient Approximation Designed for Molecules and Solids. *Phys. Rev. Lett.* **2003**, *91*, 146401.
24. Perdew, J. P.; Ruzsinszky, A.; Csonka, G. I.; Constantin, L. A.; Sun, J. Workhorse Semilocal Density Functional for Condensed Matter Physics and Quantum Chemistry. *Phys. Rev. Lett.* **2009**, *103*, 026403.
25. Constantin, L. A.; Fabiano, E.; Sala, F. D. Semilocal Dynamical Correlation with Increased Localization. *Phys. Rev. B* **2012**, *86*, 035130.
26. Constantin, L. A.; Fabiano, E.; Della Sala, F. Meta-GGA Exchange-Correlation Functional with a Balanced Treatment of Nonlocality. *J. Chem. Theory Comput.* **2013**, *9*(5), 2256–2263.
27. Ruzsinszky, A.; Sun, J.; Xiao, B.; Csonka, G. I. A Meta-GGA Made Free of the Order of Limits Anomaly. *J. Chem. Theory Comput.* **2012**, *8*(6), 2078–2087.
28. Krieger, J. B.; Li, Y.; Iafrate, G. J. Derivation and Application of an Accurate Kohn-Sham Potential with Integer Discontinuity. *Phys. Lett. A* **1990**, *146*, 256–260.

29. Görling, A.; Levy, M. DFT Ionization Formulas and a DFT Perturbation Theory for Exchange and Correlation. *Int. J. Quantum Chem. Symp.* **1995**, *29*, 93.
30. Ivanov, S.; Hirata, S.; Bartlett, R. J. Exact Exchange Treatment for Molecules in Finite-Basis-Set Kohn-Sham Theory. *Phys. Rev. Lett.* **1999**, *83*, 5455.
31. Yang, W.; Wu, Q. Direct Method for Optimized Effective Potentials in Density-Functional Theory. *Phys. Rev. Lett.* **2002**, *89*, 143002.
32. Levy, M. Excitation Energies from Density-Functional Orbital Energies. *Phys. Rev. A* **1995**, *52*, R4313.
33. Verma, P.; Bartlett, R. J. Increasing the Applicability of Density Functional Theory. II. Correlation Potentials from the Random Phase Approximation and Beyond. *J. Chem. Phys.* **2012**, *136*, 044105. http://dx.doi.org/10.1063/1.3678180.
34. Grabowski, I. Comparison of Second-Order Orbital-Dependent DFT Correlation Functionals. *Int. J. Quantum Chem.* **2008**, *108*, 2076.
35. Facco Bonetti, A.; Engel, E.; Schmid, R. N.; Dreizler, R. M. Investigation of the Correlation Potential from Kohn-Sham Perturbation Theory. *Phys. Rev. Lett.* **2001**, *86*, 2241–2244. http://dx.doi.org/10.1103/PhysRevLett.86.2241.
36. Mori-Sánchez, P.; Wu, Q.; Yang, W. Orbital-Dependent Correlation Energy in Density-Functional Theory Based on a Second-Order Perturbation Approach: Success and Failure. *J. Chem. Phys.* **2005**, *123*, 062204.
37. Schweigert, I. V.; Lotrich, V. F.; Bartlett, R. J. Ab Initio Correlation Functionals from Second-Order Perturbation Theory. *J. Chem. Phys.* **2006**, *125*, 104108.
38. Jung, Y.; Lochan, R. C.; Dutoi, A. D.; Head-Gordon, M. Scaled Opposite-Spin Second Order Møller-Plesset Correlation Energy: An Economical Electronic Structure Method. *J. Chem. Phys.* **2004**, *121*(20), 9793–9802. http://dx.doi.org/10.1063/1.1809602.
39. Bartlett, R. J.; Lotrich, V.; Schweigert, I. V. *Ab Initio* Density Functional Theory: The Best of Both Worlds? *J. Chem. Phys.* **2005**, *123*, 062205.
40. Grabowski, I.; Lotrich, V. Accurate Orbital-Dependent Correlation and Exchange-Correlation Potentials from Non-iterative *Ab Initio* DFT Calculations. *Mol. Phys.* **2005**, *103*, 2087.
41. Görling, A. New KS Method for Molecules Based on an Exchange Charge Density Generating the Exact Local KS Exchange Potential. *Phys. Rev. Lett.* **1999**, *83*, 5459.
42. Colle, R.; Nesbet, R. K. Optimized Effective Potential in Finite-Basis-Set Treatment. *J. Phys. B* **2001**, *34*, 2475.
43. Slater, J. C. A Simplification of the Hartree-Fock Method. *Phys. Rev.* **1951**, *81*, 385.
44. Hirata, S.; Ivanov, S.; Grabowski, I.; Bartlett, R. J.; Burke, K.; Talman, J. D. Can Optimized Effective Potentials Be Determined Uniquely? *J. Chem. Phys.* **2001**, *115*, 1635.
45. Staroverov, V. N.; Scuseria, G. E.; Davidson, E. R. Effective Local Potentials for Orbital-Dependent Density Functionals. *J. Chem. Phys.* **2006**, *125*, 081104.
46. Heßelmann, A.; Götz, A. W.; Della Sala, F.; Görling, A. Numerically Stable Optimized Effective Potential Method with Balanced Gaussian Basis Sets. *J. Chem. Phys.* **2007**, *127*(5), 054102. http://dx.doi.org/10.1063/1.2751159.
47. Joubert, D. P. Uniqueness of the Iterative Solution of the Optimized Effective Potential Equation. *J. Chem. Phys.* **2007**, *127*, 244104.
48. Kollmar, C.; Filatov, M. Optimized Effective Potential Method: Is It Possible to Obtain an Accurate Representation of the Response Function for Finite Orbital Basis Sets? *J. Chem. Phys.* **2007**, *127*, 114104.
49. Peach, M. J. G.; Kattirtzi, J. A.; Teale, A. M.; Tozer, D. J. Shielding Constants and Chemical Shifts in DFT: Influence of Optimized Effective Potential and Coulomb-Attenuation. *J. Chem. Phys.* **2010**, *114*, 7179.
50. Glushkov, V. N.; Fesenko, S. I.; Polatoglou, H. M. On Finite Basis Set Implementation of the Exchange-Only Optimized Effective Potential Method. *Theor. Chem. Acc.* **2009**, *124*, 365.

51. Glushkov, V. N.; Theophilou, A. K. Density-Functional Theory with Effective Potential Expressed as a Mapping of the External Potential: Applications to Open-Shell Molecules. *J. Chem. Phys.* **2006**, *124*, 034105.
52. Bulat, F. A.; Heaton-Burgess, T.; Cohen, A. J.; Yang, W. Optimized Effective Potentials from Electron Densities in Finite Basis Sets. *J. Chem. Phys.* **2007**, *127*, 174101.
53. Widmark, P. O.; Malmqvist, P. A.; Roos, B. Density Matrix Averaged Atomic Natural Orbital (ANO) Basis Sets for Correlated Molecular Wave Functions. *Theor. Chim. Acta* **1990**, 77, 291.
54. ACES II (Quantum Theory Project, University of Florida); Written by J. S. Stanton, J. Gauss, J. D. Watts, M. Nooijen, N. Oliphant, S. A. Perera, P. G. Szalay, W. J. Lauderdale, S. A. Kucharski, S. R. Gwaltney, S. Beck, A. Balkov, D. E. Bernholdt, K. K. Baeck, P. Rozyczko, H. Sekino, C. Hober and R. J. Bartlett. Containing Contributions from VMOL (J. Almlöf and P. R. Taylor); VPROPS (P. Taylor); ABACUS; (T. Helgaker, H. J. Aa. Jensen, P. Jørgensen, J. Olsen, and P. R. Taylor).
55. Umrigar, C. J.; Gonze, X. Accurate Exchange-Correlation Potentials and Total-Energy Components for the Helium Isoelectronic Series. *Phys. Rev. A* **1994**, *50*, 3827–3837.
56. Lotrich, V.; Bartlet, R. J.; Grabowski, I. Intermolecular Potential Energy Surfaces of Weakly Bound Dimers Computed from Ab Initio dft: The Right Answer for the Right Reason. *Chem. Phys. Lett.* **2005**, *405*, 43–48.
57. Distasio, R. A., Jr.; Head-Gordon, M. Optimized Spin-Component Scaled Second-Order Møller-Plesset Perturbation Theory for Intermolecular Interaction Energies. *Mol. Phys.* **2007**, *105*, 1073–1083.
58. Grabowski, I.; Fabiano, E.; Sala, F. D. Simple Non-empirical Procedure for Spin-Component-Scaled MP2 Methods Applied to the Calculation of Dissociation Energy Curve of Noncovalently-Interacting Systems. *Phys. Chem. Chem. Phys.* **2013**, *15*(37), 15485–15493.
59. Grimme, S.; Goerigk, L.; Fink, R. F. Spin-Component-Scaled Electron Correlation Methods. *Comput. Mol. Sci.* **2012**, *2*, 886–906.
60. Perdew, J. P.; Yue, W. Accurate and Simple Density Functional for the Electronic Exchange Energy: Generalized Gradient Approximation. *Phys. Rev. B* **1986**, *33*, 8800–8802. http://dx.doi.org/10.1103/PhysRevB.33.8800. URL: http://link.aps.org/doi/10.1103/PhysRevB.33.8800.
61. Laricchia, S.; Constantin, L. A.; Fabiano, E.; Della Sala, F. Laplacian-Level Kinetic Energy Approximations Based on the Fourth-Order Gradient Expansion: Global Assessment and Application to the Subsystem Formulation of Density Functional Theory. *J. Chem. Theory Comput.* **2014**, *10*(1), 164–179. http://dx.doi.org/10.1021/ct400836s.

CHAPTER FIFTEEN

Introduction to the Variational and Diffusion Monte Carlo Methods

Julien Toulouse*,†,1, Roland Assaraf*,†, Cyrus J. Umrigar‡

*Sorbonne Universités, UPMC Univ Paris 06, UMR 7616, Laboratoire de Chimie Théorique, Paris, France
†CNRS, UMR 7616, Laboratoire de Chimie Théorique, Paris, France
‡Laboratory of Atomic and Solid State Physics, Cornell University, Ithaca, New York, USA
[1]Corresponding author: e-mail address: julien.toulouse@upmc.fr

Contents

Abstract

We provide a pedagogical introduction to the two main variants of real-space quantum Monte Carlo methods for electronic structure calculations: variational Monte Carlo (VMC) and diffusion Monte Carlo (DMC). Assuming no prior knowledge on the subject, we review in depth the Metropolis–Hastings algorithm used in VMC for sampling the square of an approximate wave function, discussing details important for applications to electronic systems. We also review in detail the more sophisticated DMC algorithm within the fixed-node approximation, introduced to avoid the infamous Fermionic sign problem, which allows one to sample a more accurate approximation to the ground-state wave function. Throughout this review, we discuss the statistical methods used for evaluating expectation values and statistical uncertainties. In particular, we show how to estimate nonlinear functions of expectation values and their statistical uncertainties.

Advances in Quantum Chemistry, Volume 73
ISSN 0065-3276
http://dx.doi.org/10.1016/bs.aiq.2015.07.003

This chapter provides a pedagogical introduction to the two main variants of real-space quantum Monte Carlo (QMC) methods for electronic structure calculations: variational Monte Carlo (VMC) and diffusion Monte Carlo (DMC). For more details of these methods, see, e.g., Refs. 1–6. For reviews on applications of QMC methods in chemistry and condensed-matter physics, see, e.g., Refs. 7–8.

1. VARIATIONAL MONTE CARLO

1.1 Basic Idea

The idea of the VMC method[9,10] is simply to calculate the multidimensional integrals appearing in quantum mechanics using a Monte Carlo numerical integration technique.[a] The quantity of greatest interest is the variational energy associated with a Hamiltonian $\hat{H}$ and a wave function Ψ, which can be written as

$$E_{\mathrm{v}} = \frac{\langle \Psi | \hat{H} | \Psi \rangle}{\langle \Psi | \Psi \rangle} = \frac{\int \mathrm{d}\mathbf{R}\, \Psi(\mathbf{R})^2 E_{\mathrm{L}}(\mathbf{R})}{\int \mathrm{d}\mathbf{R}\, \Psi(\mathbf{R})^2} = \int \mathrm{d}\mathbf{R}\, \rho(\mathbf{R}) E_{\mathrm{L}}(\mathbf{R}), \tag{1}$$

where $E_{\mathrm{L}}(\mathbf{R}) = (H\Psi(\mathbf{R}))/\Psi(\mathbf{R})$ is the *local energy* depending on the $3N$ coordinates $\mathbf{R}$ of the N electrons, and $\rho(\mathbf{R}) = \Psi(\mathbf{R})^2 / \int \mathrm{d}\mathbf{R}\Psi(\mathbf{R})^2$ is the normalized probability density. For simplicity of notation, we have assumed that $\Psi(\mathbf{R})$ is real valued; the extension to complex $\Psi(\mathbf{R})$ is straightforward. The variational energy can be estimated as the average value of $E_{\mathrm{L}}(\mathbf{R})$ on a sample of M points $\mathbf{R}_k$ sampled from the probability density $\rho(\mathbf{R})$,

$$E_{\mathrm{v}} \approx \bar{E}_{\mathrm{L}} = \frac{1}{M} \sum_{k=1}^{M} E_{\mathrm{L}}(\mathbf{R}_k). \tag{2}$$

In practice, the points $\mathbf{R}_k$ are sampled using the Metropolis–Hastings algorithm.[12,13]

The advantage of this approach is that it does not use an analytical integration involving the wave function, and thus does not impose severe constraints on the form of the wave function. The wave functions usually used in QMC are of the Jastrow–Slater form,

[a] To the best of our knowledge, the first calculation of multidimensional integrals appearing in quantum mechanics by using Monte Carlo methods was done by Conroy.[11]

$$\Psi(\mathbf{R}) = J(\mathbf{R})\Phi(\mathbf{R}), \tag{3}$$

where $J(\mathbf{R})$ is a Jastrow factor and $\Phi(\mathbf{R})$ is a Slater determinant or a linear combination of Slater determinants.[b] The Jastrow factor is generally of the form $J(\mathbf{R}) = e^{f(\mathbf{R})}$. It depends explicitly on the interparticle distances r_{ij}, allowing for an efficient description of the so-called electron "dynamic" correlation.

In practice, the VMC method has two types of errors:

- a *systematic error*, due to the use of an approximate wave function (as in other wave-function methods) and
- a *statistical uncertainty*, due to the sampling of finite size M (which is specific to Monte Carlo methods).

Of course, the variational energy is an upper bound of the exact ground-state energy, but the systematic error is generally unknown, since its determination requires knowing the exact solution. By contrast, the statistical uncertainty can be easily estimated by the usual statistical techniques. For this, let us examine more closely the meaning of Eq. (2). The average of the local energy $\bar{E}_{\rm L}$ on a finite sample is itself a random variable, taking different values on different samples. The central limit theorem establishes that, if $E_{\rm L}(\mathbf{R}_k)$ are random variables that are *independent* (i.e., not correlated) and *identically distributed*, with finite expected value $\mathrm{E}[E_{\rm L}]$ and finite variance, $\mathrm{V}[E_{\rm L}] = \mathrm{E}[(E_{\rm L} - E_{\rm v})^2]$, then in the large M limit the probability distribution of the random variable $\bar{E}_{\rm L}$ converges (in the mathematical sense of convergence in distribution) to a Gaussian (or normal) distribution of expected value $\mathrm{E}[E_{\rm L}]$ and variance $\mathrm{V}[E_{\rm L}]/M$,

$$\mathrm{E}[\bar{E}_{\rm L}] = \mathrm{E}[E_{\rm L}] = E_{\rm v}, \tag{4a}$$

$$\mathrm{V}[\bar{E}_{\rm L}] = \frac{\mathrm{V}[E_{\rm L}]}{M}. \tag{4b}$$

This means that $\bar{E}_{\rm L}$ is an *estimator* of $E_{\rm v}$ with a statistical uncertainty which can be defined by the standard deviation of its Gaussian distribution

$$\sigma[\bar{E}_{\rm L}] = \sqrt{\mathrm{V}[\bar{E}_{\rm L}]} = \sqrt{\frac{\mathrm{V}[E_{\rm L}]}{M}}. \tag{5}$$

The meaning of this standard deviation is that the desired expected value $E_{\rm v}$ has a probability of 68.3% of being in the interval $[\bar{E}_{\rm L} - \sigma, \bar{E}_{\rm L} + \sigma]$,

[b] In QMC, it is convenient to use wave functions in which the values of the spin coordinates have been fixed, so Ψ is a function of the spatial coordinates $\mathbf{R}$ only.

a probability of 95.5% of being in the interval $[\bar{E}_L - 2\sigma, \bar{E}_L + 2\sigma]$, and a probability of 99.7% of being in the interval $[\bar{E}_L - 3\sigma, \bar{E}_L + 3\sigma]$. Note that, if the variance $V[E_L]$ is infinite but the expected value $E[E_L]$ is finite, then the law of large numbers guarantees the convergence of $\bar{E}_L$ to $E[E_L]$ when $M \to \infty$ but with a statistical uncertainty which is more difficult to estimate and which decreases more slowly than $1/\sqrt{M}$.

It is important to note that the statistical uncertainty decreases as $1/\sqrt{M}$ *independently of the dimension of the problem*. This is in contrast to deterministic numerical integration methods for which the convergence of the integration error deteriorates with the spatial dimension d. For example, Simpson's integration rule converges as $1/M^{(4/d)}$ (provided the integrand has up to fourth-order derivatives) so that for $d > 8$, Monte Carlo methods are more efficient for large M.

The statistical uncertainty is reduced if the variance of the local energy $V[E_L]$ is small. In the limit that Ψ is an exact eigenfunction of $\hat{H}$, the local energy E_L becomes exact, independent of $\mathbf{R}$, and thus, its variance $V[E_L]$ and the statistical uncertainty of $\bar{E}_L$ vanish. This is known as the *zero-variance* property. Since the systematic error (or bias) of the variational energy $\Delta E = E_v - E_0$ (where E_0 is the exact energy) also vanishes in this limit, there is a zero-bias property as well. For these reasons, a great deal of effort has been expended on developing robust and efficient wave-function optimization methods.

1.2 Estimation of the Statistical Uncertainty

In practice, the probability density $\rho(\mathbf{R})$ is sampled with the Metropolis–Hastings algorithm which provides a sequence of points $\mathbf{R}_k$ correctly distributed according to $\rho(\mathbf{R})$ but *sequentially* (*or serially*) *correlated* (i.e., non-independent). This is a consequence of each point being sampled from a probability distribution conditional on the previous point. One can define an *autocorrelation time* (defined more precisely later) that is, roughly speaking, the average time for points to decorrelate. This sequential correlation must be taken into account when using the central limit theorem for evaluating the statistical uncertainty. This is done using the *blocking* technique, which is described next.

Let us consider a sequence of M realizations X_k (sequentially correlated) of a random variable X of expected value $E[X]$ and of variance $V[X]$. For example, X could be the local energy E_L. We divide this sequence into M_b successive blocks of M_s steps each. The *block average* $\bar{X}_b$ is

$$\bar{X}_{\mathrm{b}} = \frac{1}{M_{\mathrm{s}}} \sum_{k=1}^{M_{\mathrm{s}}} X_k. \tag{6}$$

The expected value of $\bar{X}_{\mathrm{b}}$ is also the expected value of X, i.e., $E[\bar{X}_{\mathrm{b}}] = E[X]$, but its variance is not simply $V[X]/M_{\mathrm{s}}$ since the variables X_k are not independent. We can now define the *global average* $\bar{X}$ of the whole sample as the average over all the blocks of the block averages

$$\bar{X} = \frac{1}{M_{\mathrm{b}}} \sum_{b=1}^{M_{\mathrm{b}}} \bar{X}_b, \tag{7}$$

where $\bar{X}_b$ with a math subscript "b" indicates the block average for the bth block (whereas $\bar{X}_{\mathrm{b}}$ with a Roman subscript "b" indicates the generic random variable). The global average $\bar{X}$ is another random variable with the same expected value as X, i.e., $\mathrm{E}[\bar{X}] = \mathrm{E}[\bar{X}_{\mathrm{b}}] = \mathrm{E}[X]$. If the length of the blocks is large compared to the autocorrelation time, then the block averages $\bar{X}_b$ can be considered as being independent, and the variance of the global average is simply

$$\mathrm{V}[\bar{X}] = \frac{\mathrm{V}[\bar{X}_{\mathrm{b}}]}{M_{\mathrm{b}}}, \tag{8}$$

which leads to the statistical uncertainty of $\bar{X}$

$$\sigma[\bar{X}] = \sqrt{\mathrm{V}[\bar{X}]} = \sqrt{\frac{\mathrm{V}[\bar{X}_{\mathrm{b}}]}{M_{\mathrm{b}}}}. \tag{9}$$

In practice, the statistical uncertainty on a finite sample is calculated as

$$\sigma[\bar{X}] \approx \sqrt{\frac{1}{M_{\mathrm{b}} - 1} \left(\frac{1}{M_{\mathrm{b}}} \sum_{b=1}^{M_{\mathrm{b}}} \bar{X}_b{}^2 - \left(\frac{1}{M_{\mathrm{b}}} \sum_{b=1}^{M_{\mathrm{b}}} \bar{X}_b \right)^2 \right)}, \tag{10}$$

where the $M_{\mathrm{b}} - 1$ term appearing instead of M_{b} is necessary to have an unbiased estimator of the standard deviation on the sample (see the Appendix). It takes into account the fact that the computed variance is relative to the sample average rather than the true expected value.

Finally, let us examine the variance $\mathrm{V}[\bar{X}_{\mathrm{b}}]$. Since the variables X_k are not independent, the expansion of $\mathrm{V}[\bar{X}_{\mathrm{b}}]$ involves the covariances between the variables

$$\mathrm{V}[\bar{X}_\mathrm{b}] = \frac{1}{M_\mathrm{s}^2} \sum_{k,l} \mathrm{Cov}[X_k, X_l] = \frac{\mathrm{V}[X]}{M_\mathrm{s}} + \frac{2}{M_\mathrm{s}^2} \sum_{k<l} \mathrm{Cov}[X_k, X_l] = T_\mathrm{c} \frac{\mathrm{V}[X]}{M_\mathrm{s}}, \tag{11}$$

defining the autocorrelation time of X

$$T_\mathrm{c} = 1 + \frac{2}{\mathrm{V}[X] M_\mathrm{s}} \sum_{k<l} \mathrm{Cov}[X_k, X_l]. \tag{12}$$

The autocorrelation time is equal to 1 in the absence of correlation between the variables, i.e., $\mathrm{Cov}[X_k,X_l] = 0$ for $k \neq l$, but can be large in the presence of sequential correlation. It is instructive to express the statistical uncertainty as a function of T_c

$$\sigma[\bar{X}] = \sqrt{T_\mathrm{c} \frac{\mathrm{V}[X]}{M_\mathrm{s} M_\mathrm{b}}} = \sqrt{T_\mathrm{c} \frac{\mathrm{V}[X]}{M}}, \tag{13}$$

where $M = M_\mathrm{s} M_\mathrm{b}$ is the total size of the sample. The expression (13) allows one to interpret T_c as a factor giving the number of effectively independent points in the sample, $M_\mathrm{eff} = M/T_\mathrm{c}$. In practice, it is useful to calculate the autocorrelation time as $T_\mathrm{c} = M_\mathrm{s} \mathrm{V}[\bar{X}_\mathrm{b}]/\mathrm{V}[X]$ and check whether the length of the blocks is large enough for a correct estimation of the statistical uncertainty, e.g., $M_\mathrm{s} > 100\ T_\mathrm{c}$. If M_s is not much greater than T_c, then the statistical uncertainty $\sigma[\bar{X}]$ and the autocorrelation time T_c will be underestimated.

In the Appendix, we further explain how to estimate the statistical uncertainty of nonlinear functions of expectation values, which often occur in practice.

1.3 Calculation Cost

The calculation cost required to reach a given statistical uncertainty $\sigma[\bar{X}]$ is

$$t = t_\mathrm{s} M = t_\mathrm{s} \frac{T_\mathrm{c} \mathrm{V}[X]}{\sigma[\bar{X}]^2} \tag{14}$$

where t_s is the calculation time per iteration. The $1/\sigma[\bar{X}]^2$ dependence implies that decreasing the statistical uncertainty by a factor of 10 requires to increase the computational time by a factor of 100. This quadratic dependence directly stems from the central limit theorem and seems unavoidable.[c] However, one can play with the three other parameters:

- T_c depends on the sampling algorithm and on the random variable X. For efficient algorithms such as Umrigar's one,[3,15] the autocorrelation time of the local energy is close to 1 and little further improvement seems possible;
- t_s is usually dominated by the cost of evaluating X. For the local energy, the evaluation cost depends on the form of the wave function;
- $V[X]$ depends on the choice of the random variable X with its associated probability distribution, the only constraint being that the expected value $E[X]$ must equal the expectation value of the observable (otherwise, this is a biased estimator). The choice of a good probability distribution is usually called *importance sampling*. Even for a fixed probability distribution, it is possible to use various estimators for X, some of which have smaller variance than others, since one has the freedom to add any quantity with zero expectation value. This has been exploited to construct improved estimators for diverse observables.[16–20] There is often a compromise to be found between a low computation time per iteration t_s and a low variance $V[X]$.

1.4 Sampling Technique

The probability density, $\rho(\mathbf{R}) = \Psi(\mathbf{R})^2 / \int d\mathbf{R}\Psi(\mathbf{R})^2$, is generally complicated and cannot be sampled by direct methods such as the transformation method or the rejection method. Instead, the Metropolis–Hastings (or generalized Metropolis) algorithm, which can be used to sample any known probability density, is used. It employs a stochastic process, more specifically, a Markov chain.

1.4.1 Stochastic Process

A *stochastic process* represents the evolution—say in "time"—of a random variable. It is described by a trajectory of successive points $\mathbf{R}_1, \mathbf{R}_2, \ldots, \mathbf{R}_M$ with an associated probability distribution $P(\mathbf{R}_M, \ldots, \mathbf{R}_2, \mathbf{R}_1)$. The idea of evolution in time can be made more explicit by decomposing the probability of the whole trajectory into products of the conditional probability of having

[c] Quasi Monte Carlo methods [14] can in some cases achieve a convergence rate of $\mathcal{O}(\ln(M)/M)$ rather than $\mathcal{O}(1/\sqrt{M})$. However, they have not been used for QMC applications, in part because in QMC the sampled distributions, for systems with more than a few electrons, are very highly peaked.

a particular point knowing that all the previous points have already been realized. For example, for $M = 3$, the probability of the trajectory is

$$P(\mathbf{R}_3,\mathbf{R}_2,\mathbf{R}_1) = P(\mathbf{R}_3|\mathbf{R}_2,\mathbf{R}_1)P(\mathbf{R}_2|\mathbf{R}_1)P(\mathbf{R}_1). \tag{15}$$

1.4.2 Markov Chain

A *Markov chain* is a stochastic process for which the conditional probability for the transition to a new point $\mathbf{R}_k$ depends only on the previous point $\mathbf{R}_{k-1}$

$$P(\mathbf{R}_k|\mathbf{R}_{k-1},\ldots,\mathbf{R}_1) = P(\mathbf{R}_k|\mathbf{R}_{k-1}), \tag{16}$$

i.e., the process "forgets" the way it arrived at point $\mathbf{R}_{k-1}$. The probability of a trajectory can thus be simply written as, e.g., for $M = 3$,

$$P(\mathbf{R}_3,\mathbf{R}_2,\mathbf{R}_1) = P(\mathbf{R}_3|\mathbf{R}_2)P(\mathbf{R}_2|\mathbf{R}_1)P(\mathbf{R}_1), \tag{17}$$

and $P(\mathbf{R}_\mathrm{f}|\mathbf{R}_\mathrm{i})$ is called the transition probability from point $\mathbf{R}_\mathrm{i}$ to point $\mathbf{R}_\mathrm{f}$. Note that, in general, the transition probability can depend on time (measured by the index k). We will consider here only the case of a stationary Markov chain for which the transition probability is time independent.

In the following, we will use notation corresponding to the case of states $\mathbf{R}_k$ in a continuous space ("integrals" instead of "sums"), but we will ignore the possibly subtle mathematical differences between the continuous and discrete cases, and we will often use the vocabulary of the discrete case (e.g., "matrix"). The transition probability matrix, P, is a *stochastic matrix*, i.e., it has the following two properties:

$$P(\mathbf{R}_\mathrm{f}|\mathbf{R}_\mathrm{i}) \geq 0 \quad \text{(nonnegativity)}, \tag{18a}$$

$$\int \mathrm{d}\mathbf{R}_\mathrm{f}\, P(\mathbf{R}_\mathrm{f}|\mathbf{R}_\mathrm{i}) = 1 \quad \text{(column normalization)}. \tag{18b}$$

The second property expresses the fact that the probability that a point $\mathbf{R}_\mathrm{i}$ is somewhere at the next step must be 1. The eigenvalues of a stochastic matrix are between 0 and 1, and there is at least one eigenvalue equal to 1. The latter property is a consequence of the fact that, for a column-normalized matrix, the vector with all components equal to one is a left eigenvector with eigenvalue 1. The target probability distribution $\rho(\mathbf{R})$ is sampled by constructing a Markov chain converging to $\rho(\mathbf{R})$. A necessary condition is that the distribution $\rho(\mathbf{R})$ is a (right) eigenvector of $P(\mathbf{R}_\mathrm{f}|\mathbf{R}_\mathrm{i})$ with the eigenvalue 1

$$\int d\mathbf{R}_i\, P(\mathbf{R}_f|\mathbf{R}_i)\rho(\mathbf{R}_i) = \rho(\mathbf{R}_f) = \int d\mathbf{R}_i\, P(\mathbf{R}_i|\mathbf{R}_f)\rho(\mathbf{R}_f) \;\; \forall \mathbf{R}_f, \quad (19)$$

where the second equality simply comes from the normalization condition (18b). Equation (19) is a *stationarity condition* for $\rho(\mathbf{R})$. It means that if we start from the target distribution $\rho(\mathbf{R})$, then we will continue to sample the same distribution by applying the Markov chain. However, we need more than that. We want that any initial distribution $\rho_{\mathrm{ini}}(\mathbf{R})$, e.g., a delta function at some initial point, evolves to the target stationary distribution $\rho(\mathbf{R})$ by repeated applications of the transition matrix

$$\begin{aligned}
&\lim_{M\to\infty} \int d\mathbf{R}_1\, P^M(\mathbf{R}|\mathbf{R}_1)\rho_{\mathrm{ini}}(\mathbf{R}_1) = \\
&\lim_{M\to\infty} \int d\mathbf{R}_1 d\mathbf{R}_2 \ldots d\mathbf{R}_M\, P(\mathbf{R}|\mathbf{R}_M)P(\mathbf{R}_M|\mathbf{R}_{M-1})\ldots P(\mathbf{R}_2|\mathbf{R}_1)\rho_{\mathrm{ini}}(\mathbf{R}_1) \\
&\quad = \rho(\mathbf{R}),
\end{aligned} \quad (20)$$

i.e., $\rho(\mathbf{R})$ must be the dominant eigenvector of P (the unique eigenvector of largest eigenvalue). If the stationarity condition (19) is satisfied, then this will always be the case except if P has several eigenvectors with eigenvalue 1. One can show that the matrix P has only one eigenvector of eigenvalue 1 if P is a primitive matrix, i.e., if there is an integer $n \geq 1$ such that all the elements of the matrix P^n are strictly positive, $P^n(\mathbf{R}_k|\mathbf{R}_l) > 0, \forall\, \mathbf{R}_k, \mathbf{R}_l$. This means that it must be possible to move between any pair of states $\mathbf{R}_k$ and $\mathbf{R}_l$ in n steps. This ensures that all states can be visited, and that the Markov chain converges to the unique stationary distribution $\rho(\mathbf{R})$. The Markov chain is then said to be *ergodic*.

In practice, instead of imposing the stationarity condition (19), the Markov matrix is constructed by imposing the more stringent *detailed balance* condition,

$$P(\mathbf{R}_f|\mathbf{R}_i)\rho(\mathbf{R}_i) = P(\mathbf{R}_i|\mathbf{R}_f)\rho(\mathbf{R}_f), \quad (21)$$

which forces the probability flux between the two states $\mathbf{R}_i$ and $\mathbf{R}_f$ to be the same in both directions. This is a sufficient (but not necessary) condition for $\rho(\mathbf{R})$ to be the stationary distribution. A Markov chain satisfying condition (21) is said to be reversible.

In practice, a Markov chain is realized by a *random walk*. Starting from an initial point $\mathbf{R}_1$ (or walker)—i.e., a delta-function distribution $\delta(\mathbf{R} - \mathbf{R}_1)$—sample the second point $\mathbf{R}_2$ by drawing from the probability distribution

$P(\mathbf{R}_2|\mathbf{R}_1)$, then a third point $\mathbf{R}_3$ by drawing from $P(\mathbf{R}_3|\mathbf{R}_2)$, and so on. After disregarding a certain number of iterations M_{eq} corresponding to a transient phase called *equilibration*, the random walk samples the stationary distribution $\rho(\mathbf{R})$ in the sense that $\rho(\mathbf{R}) = \text{E}[\delta(\mathbf{R}-\mathbf{R}_k)] \approx (1/M)\sum_{k=1}^{M}\delta(\mathbf{R}-\mathbf{R}_k)$ and the averages of the estimators of the observables of interest are calculated. The rate of convergence to the stationary distribution $\rho(\mathbf{R})$ and the autocorrelation times of the observables are determined by the second largest eigenvalue of the matrix P (see, e.g., Ref. 21). The random walk must be sufficiently long so as to obtain a representative sample of the states, making a nonnegligible contribution to the expected values. If the transitions between states belonging to two contributing regions of the space of states are too improbable, as may happen for example for dissociated atoms, then there is a risk that the random walk remains stuck in a region of space and seems converged, even though the true stationary distribution is not yet reached. To avoid this problem, smart choices for the transition matrix can be crucial in various applications of Monte Carlo methods.[22,23]

1.4.3 Metropolis–Hastings Algorithm

In the Metropolis–Hastings algorithm,[12,13] one realizes a Markov chain with the following random walk. Starting from a point $\mathbf{R}_\text{i}$, a new point $\mathbf{R}_\text{f}$ is determined in two steps:

1. a temporary point $\mathbf{R}'_\text{f}$ is proposed with the probability $P_{\text{prop}}(\mathbf{R}'_\text{f}|\mathbf{R}_\text{i})$,
2. the point $\mathbf{R}'_\text{f}$ is accepted (i.e., $\mathbf{R}_\text{f} = \mathbf{R}'_\text{f}$) with probability $P_{\text{acc}}(\mathbf{R}'_\text{f}|\mathbf{R}_\text{i})$ or rejected (i.e., $\mathbf{R}_\text{f} = \mathbf{R}_\text{i}$) with probability $P_{\text{rej}}(\mathbf{R}'_\text{f}|\mathbf{R}_\text{i}) = 1 - P_{\text{acc}}(\mathbf{R}'_\text{f}|\mathbf{R}_\text{i})$.

The corresponding transition probability can be written as

$$P(\mathbf{R}_\text{f}|\mathbf{R}_\text{i}) = \begin{cases} P_{\text{acc}}(\mathbf{R}_\text{f}|\mathbf{R}_\text{i})P_{\text{prop}}(\mathbf{R}_\text{f}|\mathbf{R}_\text{i}) & \text{if } \mathbf{R}_\text{f} \neq \mathbf{R}_\text{i} \\ 1 - \int \text{d}\mathbf{R}'_\text{f}\, P_{\text{acc}}(\mathbf{R}'_\text{f}|\mathbf{R}_\text{i})P_{\text{prop}}(\mathbf{R}'_\text{f}|\mathbf{R}_\text{i}) & \text{if } \mathbf{R}_\text{f} = \mathbf{R}_\text{i} \end{cases} \tag{22}$$

or, in a single expression,

$$\begin{aligned} P(\mathbf{R}_\text{f}|\mathbf{R}_\text{i}) = {} & P_{\text{acc}}(\mathbf{R}_\text{f}|\mathbf{R}_\text{i})P_{\text{prop}}(\mathbf{R}_\text{f}|\mathbf{R}_\text{i}) \\ & + \left[1 - \int \text{d}\mathbf{R}'_\text{f}\, P_{\text{acc}}(\mathbf{R}'_\text{f}|\mathbf{R}_\text{i})P_{\text{prop}}(\mathbf{R}'_\text{f}|\mathbf{R}_\text{i})\right] \\ & \times \delta(\mathbf{R}_\text{i} - \mathbf{R}_\text{f}). \end{aligned} \tag{23}$$

The proposal probability $P_{\text{prop}}(\mathbf{R}_\text{f}|\mathbf{R}_\text{i})$ is a stochastic matrix, i.e., $P_{\text{prop}}(\mathbf{R}_\text{f}|\mathbf{R}_\text{i}) \geq 0$ and $\int \text{d}\mathbf{R}_\text{f} P_{\text{prop}}(\mathbf{R}_\text{f}|\mathbf{R}_\text{i}) = 1$, ensuring that $P(\mathbf{R}_\text{f}|\mathbf{R}_\text{i})$ fulfills the nonnegativity condition (18a). The second term in Eq. (23) with the delta function ensures that $P(\mathbf{R}_f|\mathbf{R}_i)$ fulfills the normalization condition (18b). The acceptance probability is chosen so as to fulfill the detailed balance condition (21), for $\mathbf{R}_f \neq \mathbf{R}_i$,

$$\frac{P_{\text{acc}}(\mathbf{R}_f|\mathbf{R}_i)}{P_{\text{acc}}(\mathbf{R}_i|\mathbf{R}_f)} = \frac{P_{\text{prop}}(\mathbf{R}_i|\mathbf{R}_f)\rho(\mathbf{R}_f)}{P_{\text{prop}}(\mathbf{R}_f|\mathbf{R}_i)\rho(\mathbf{R}_i)}. \tag{24}$$

Several choices are possible. The choice of Metropolis et al.[12] maximizes the acceptance probability

$$P_{\text{acc}}(\mathbf{R}_f|\mathbf{R}_i) = \min\left\{1, \frac{P_{\text{prop}}(\mathbf{R}_i|\mathbf{R}_f)\rho(\mathbf{R}_f)}{P_{\text{prop}}(\mathbf{R}_f|\mathbf{R}_i)\rho(\mathbf{R}_i)}\right\}. \tag{25}$$

The acceptance probability is not a stochastic matrix, even though both the proposal and the total Markov matrices are stochastic. Since only the ratio $\rho(\mathbf{R}_f)/\rho(\mathbf{R}_i)$ is involved in Eq. (25), it is not necessary to calculate the normalization constant of the probability density $\rho(\mathbf{R})$. It is clear that the acceptance probability of Eq. (25) is optimal, but there is considerable scope for ingenuity in choosing a proposal probability $P_{\text{prop}}(\mathbf{R}_f|\mathbf{R}_i)$ that leads to a small autocorrelation time.

1.4.4 Choice of the Proposal Probability

The original paper of Metropolis et al.[12] employed a symmetric proposal matrix, in which case the proposal matrix drops out of the formula for the acceptance. The advantage of having a nonsymmetric proposal matrix was pointed out by Hastings.[13] One has a lot of freedom in the choice of the proposal probability $P_{\text{prop}}(\mathbf{R}_f|\mathbf{R}_i)$. The only constraints are that $P_{\text{prop}}(\mathbf{R}_f|\mathbf{R}_i)$ must be a stochastic matrix leading to an ergodic Markov chain and that it must be possible to efficiently sample $P_{\text{prop}}(\mathbf{R}_f|\mathbf{R}_i)$ with a direct sampling method. The proposal probability determines the average size of the proposed moves $\mathbf{R}_i \rightarrow \mathbf{R}_f$ and the average acceptance rate of these moves. In order to reduce sequential correlation, one has to make moves as large as possible but with a high acceptance rate. In practice, for a given form of the proposal matrix, there is a compromise to be found between the average size of the proposed moves and the average acceptance rate.

The simplest choice for $P_{\text{prop}}(\mathbf{R}_f|\mathbf{R}_i)$ is a distribution that is uniform inside a small cube $\Omega(\mathbf{R}_i)$ centered in $\mathbf{R}_i$ and of side length Δ and zero outside

$$P_{\text{prop}}(\mathbf{R}_{\text{f}}|\mathbf{R}_{\text{i}}) = \begin{cases} \dfrac{1}{\Delta^{3N}} & \text{if } \mathbf{R}_{\text{f}} \in \Omega(\mathbf{R}_{\text{i}}) \\ 0 & \text{elsewhere.} \end{cases} \tag{26}$$

In practice, a move according to Eq. (26) is proposed,

$$\mathbf{R}_{\text{f}} = \mathbf{R}_{\text{i}} + \frac{\Delta}{2}\boldsymbol{\chi}, \tag{27}$$

where $\boldsymbol{\chi}$ is a vector of $3N$ random numbers drawn from the uniform distribution between -1 and 1. The size of the cube Δ can be adjusted so as to minimize the autocorrelation time of the local energy, but the latter remains large and the sampling is inefficient.

Clever choices use information from the distribution $\rho(\mathbf{R})$, in particular its local gradient, to guide the sampling. A choice for $P_{\text{prop}}(\mathbf{R}_{\text{f}}|\mathbf{R}_{\text{i}})$ which would lead to large moves with an acceptance probability equal to 1 would be $P_{\text{prop}}(\mathbf{R}_{\text{f}}|\mathbf{R}_{\text{i}}) = \rho(\mathbf{R}_{\text{f}})$, independently from $\mathbf{R}_{\text{i}}$, but we would then be back to the initial problem of sampling a complicated distribution $\rho(\mathbf{R})$. A good choice for $P_{\text{prop}}(\mathbf{R}_{\text{f}}|\mathbf{R}_{\text{i}})$ is the Green function of the Fokker–Planck equation in the short-time approximation

$$P_{\text{prop}}(\mathbf{R}_{\text{f}}|\mathbf{R}_{\text{i}}) = \frac{1}{(2\pi\tau)^{3N/2}} e^{-\frac{(\mathbf{R}_{\text{f}} - \mathbf{R}_{\text{i}} - \mathbf{v}(\mathbf{R}_{\text{i}})\tau)^2}{2\tau}}, \tag{28}$$

where $\mathbf{v}(\mathbf{R}) = \nabla\Psi(\mathbf{R})/\Psi(\mathbf{R})$ is called the *drift velocity* of the wave function and τ is the time step which can be adjusted so as to minimize the autocorrelation time of the local energy. In practice, a move according to Eq. (28) is proposed

$$\mathbf{R}_{\text{f}} = \mathbf{R}_{\text{i}} + \mathbf{v}(\mathbf{R}_{\text{i}})\tau + \boldsymbol{\eta}, \tag{29}$$

where $\boldsymbol{\eta}$ is a vector of $3N$ random numbers drawn from the Gaussian distribution of average 0 and standard deviation $\sqrt{\tau}$. The term $\boldsymbol{\eta}$ describes an isotropic Gaussian diffusion process (or Wiener process). The term $\mathbf{v}(\mathbf{R}_{\text{i}})\tau$ is a drift term which moves the random walk in the direction of increasing $|\Psi(\mathbf{R})|$.

The optimal size of the move is smaller in regions where $\mathbf{v}(\mathbf{R})$ is changing rapidly. For example, $\mathbf{v}(\mathbf{R})$ has a discontinuity at the nuclear positions. Hence, it is more efficient to make smaller moves for electrons in the core than for electrons in the valence regions. In doing this, care must be taken to ensure the detailed balance condition. An elegant solution is provided in the

VMC algorithm of Refs. 3,15 where the electron moves are made in spherical coordinates centered on the nearest nucleus and the size of radial moves is proportional to the distance to the nearest nucleus. In addition, the size of the angular moves gets larger as one approaches a nucleus. This algorithm allows one to achieve, in many cases, an autocorrelation time of the local energy close to 1.

1.4.5 Expectation Values

The expectation value of an operator $\hat{O}$ can be computed by averaging the corresponding local value $O(\mathbf{R}_f) = \langle \mathbf{R}_f | \hat{O} | \Psi \rangle / \Psi(\mathbf{R}_f)$ at the Monte Carlo points $\mathbf{R}_f$ after the accept/reject step. A somewhat smaller statistical error can be achieved by instead averaging

$$P_{\mathrm{acc}}(\mathbf{R}_f|\mathbf{R}_i)\ O(\mathbf{R}_f) + (1 - P_{\mathrm{acc}}(\mathbf{R}_f|\mathbf{R}_i))\ O(\mathbf{R}_i), \tag{30}$$

regardless of whether the proposed move is accepted or rejected.

1.4.6 Moving the Electrons All At Once or One by One?

So far we have assumed that, for a many-electron system, all the electrons are moved and then this move is accepted or rejected in a single step. In fact, it is also possible to move the electrons one by one, i.e., move the first electron, accept or reject this move, then move the second electron, accept or reject this move, and so on. In this case, the transition probability for N electrons can be formally decomposed as

$$\begin{aligned} P(\mathbf{R}_f|\mathbf{R}_i) = \ & P(\mathbf{r}_{1,f}\mathbf{r}_{2,f}\ldots\mathbf{r}_{N,f}|\mathbf{r}_{1,f}\mathbf{r}_{2,f}\ldots\mathbf{r}_{N,i}) \\ & \times\cdots\times P(\mathbf{r}_{1,f}\mathbf{r}_{2,f}\ldots\mathbf{r}_{N,i}|\mathbf{r}_{1,f}\mathbf{r}_{2,i}\ldots\mathbf{r}_{N,i}) \\ & \times P(\mathbf{r}_{1,f}\mathbf{r}_{2,i}\ldots\mathbf{r}_{N,i}|\mathbf{r}_{1,i}\mathbf{r}_{2,i}\ldots\mathbf{r}_{N,i}), \end{aligned} \tag{31}$$

where each one-electron transition probability (knowing that the other electrons are fixed) is made of a proposal probability and an acceptance probability just as before. If each one-electron transition probability satisfies the stationary condition (19), then the global transition probability satisfies it as well.

Moving the N electrons one by one requires more calculation time than moving the electrons all at once, since the wave function must be recalculated after each move to calculate the acceptance probability. The calculation time does not increase by a factor of N as one may naively think but typically by a factor of 2 if the value of the wave function is recalculated in a clever way after an one-electron move. For example, for Slater determinants, one can use the matrix determinant lemma in conjunction with the Sherman–Morrison formula (see, e.g., Ref. 24) to efficiently recalculate the values of the determinants when only one row or column has been changed. In spite of the increase in the calculation time, it has been repeatedly shown in the literature (see, e.g., Refs. 10,15,25,26) that, for systems with many electrons, moving the electrons one by one leads to a more efficient algorithm: larger moves can be made for the same average acceptance, so the points $\mathbf{R}_k$ are less sequentially correlated and the autocorrelation time of the local energy is smaller (by a factor larger than the one necessary for compensating the increase of the calculation time per iteration).

2. DIFFUSION MONTE CARLO

2.1 Basic Idea

While the VMC method is limited by the use of an approximate wave function Ψ, the idea of the DMC method [5,27–30] is to sample from the exact wave function Ψ_0 of the ground state of the system. If we have this exact wave function Ψ_0, then the associated exact energy E_0 can be obtained from the mixed expectation value using the trial wave function Ψ,

$$E_0 = \frac{\langle\Psi_0|\hat{H}|\Psi\rangle}{\langle\Psi_0|\Psi\rangle} = \frac{\displaystyle\int \mathrm{d}\mathbf{R}\,\Psi_0(\mathbf{R})\Psi(\mathbf{R})E_\mathrm{L}(\mathbf{R})}{\displaystyle\int \mathrm{d}\mathbf{R}\,\Psi_0(\mathbf{R})\Psi(\mathbf{R})}, \tag{32}$$

since Ψ_0 is an eigenfunction of the Hamiltonian $\hat{H}$. The advantage of the mixed expectation value (32) is that it does not require calculating the action of $\hat{H}$ on Ψ_0. The integral in Eq. (32) involves the local energy of the trial wave function, $E_\mathrm{L}(\mathbf{R}) = (H\Psi(\mathbf{R}))/\Psi(\mathbf{R})$, and can be estimated in a similar way as in VMC by calculating the average of $E_\mathrm{L}(\mathbf{R})$ on a sample of points $\mathbf{R}_k$ representing the mixed distribution $\Psi_0(\mathbf{R})\Psi(\mathbf{R})/\int \mathrm{d}\mathbf{R}\,\Psi_0(\mathbf{R})\Psi(\mathbf{R})$.

But how to access to the exact wave function Ψ_0? Let us consider the action of the *imaginary-time evolution operator* ($t \rightarrow -\mathrm{i}t$) on an arbitrary wave function such as the trial wave function Ψ

$$|\Psi(t)\rangle = \mathrm{e}^{-(\hat{H}-E_\mathrm{T})t}|\Psi\rangle, \tag{33}$$

where E_T is for now an undetermined trial energy. Using the spectral decomposition of the evolution operator (written with the eigenstates Ψ_i

and the eigenenergies E_i of $\hat{H}$), we see that the limit of an infinite propagation time is dominated by the state Ψ_0 with the lowest energy having a nonzero overlap with Ψ

$$\lim_{t\to\infty} |\Psi(t)\rangle = \lim_{t\to\infty} \sum_i \mathrm{e}^{-(E_i-E_\mathrm{T})t} |\Psi_i\rangle\langle\Psi_i|\Psi\rangle = \lim_{t\to\infty} \mathrm{e}^{-(E_0-E_\mathrm{T})t} |\Psi_0\rangle\langle\Psi_0|\Psi\rangle, \tag{34}$$

since all the other states of energies $E_i > E_0$ decay exponentially faster. The exponential $\mathrm{e}^{-(E_0-E_\mathrm{T})t}$ can be eliminated by adjusting E_T to E_0, and we then obtain that $\Psi(t)$ becomes proportional to Ψ_0

$$\lim_{t\to\infty} |\Psi(t)\rangle \propto |\Psi_0\rangle. \tag{35}$$

In position representation, Eq. (33) is written as

$$\Psi(\mathbf{R}_\mathrm{f}, t) = \int \mathrm{d}\mathbf{R}_\mathrm{i}\; G(\mathbf{R}_\mathrm{f}|\mathbf{R}_\mathrm{i};t)\Psi(\mathbf{R}_\mathrm{i}), \tag{36}$$

where $G(\mathbf{R}_\mathrm{f}|\mathbf{R}_\mathrm{i};t) = \langle\mathbf{R}_\mathrm{f}|\mathrm{e}^{-(\hat{H}-E_\mathrm{T})t}|\mathbf{R}_\mathrm{i}\rangle$ is called the *Green function* (or the imaginary-time propagator from $\mathbf{R}_\mathrm{i}$ to $\mathbf{R}_\mathrm{f}$). Multiplying and dividing by $\Psi(\mathbf{R}_\mathrm{f})$ and $\Psi(\mathbf{R}_\mathrm{i})$, we obtain the evolution equation of the mixed distribution $f(\mathbf{R},t) = \Psi(\mathbf{R},t)\Psi(\mathbf{R})$

$$f(\mathbf{R}_\mathrm{f}, t) = \int \mathrm{d}\mathbf{R}_\mathrm{i}\; \tilde{G}(\mathbf{R}_\mathrm{f}|\mathbf{R}_\mathrm{i};t)\, \Psi(\mathbf{R}_\mathrm{i})^2, \tag{37}$$

where $\tilde{G}(\mathbf{R}_\mathrm{f}|\mathbf{R}_\mathrm{i};t)$ is the *importance-sampling* Green function,

$$\tilde{G}(\mathbf{R}_\mathrm{f}|\mathbf{R}_\mathrm{i};t) = \Psi(\mathbf{R}_\mathrm{f})\; G(\mathbf{R}_\mathrm{f}|\mathbf{R}_\mathrm{i};t)\, \frac{1}{\Psi(\mathbf{R}_\mathrm{i})}, \tag{38}$$

i.e., $\tilde{G}(\mathbf{R}_\mathrm{f}|\mathbf{R}_\mathrm{i};t)$ is $G(\mathbf{R}_\mathrm{f}|\mathbf{R}_\mathrm{i};t)$ similarity-transformed by the diagonal matrix that has the values of Ψ along the diagonal. In the limit of infinite time, the mixed distribution becomes proportional to the target stationary distribution: $f(\mathbf{R}) = \lim_{t\to\infty} f(\mathbf{R}, t) \propto \Psi_0(\mathbf{R})\Psi(\mathbf{R})$.

In practice, an analytical expression of the Green function is known only in the limit of a short propagation time, $\tilde{G}(\mathbf{R}_\mathrm{f}|\mathbf{R}_\mathrm{i};\tau)$, where τ is a small time step, and one must thus iterate to obtain the stationary distribution

$$f(\mathbf{R}) = \lim_{M\to\infty} \int d\mathbf{R}_1 d\mathbf{R}_2 \ldots d\mathbf{R}_M \; \tilde{G}(\mathbf{R}|\mathbf{R}_M;\tau) \; \tilde{G}(\mathbf{R}_M|\mathbf{R}_{M-1};\tau) \\ \ldots \; \tilde{G}(\mathbf{R}_2|\mathbf{R}_1;\tau) \; \Psi(\mathbf{R}_1)^2. \tag{39}$$

A short-time approximation to the Green function is obtained by applying the Trotter–Suzuki formula, $\mathrm{e}^{-(\hat{T}+\hat{V})\tau} = \mathrm{e}^{-\hat{V}\tau/2}\mathrm{e}^{-\hat{T}\tau}\mathrm{e}^{-\hat{V}\tau/2} + O(\tau^3)$, where $\hat{T}$ and $\hat{V}$ are the kinetic and potential energy operators. In position representation, this approximation leads to the following expression

$$G(\mathbf{R}_\mathrm{f}|\mathbf{R}_\mathrm{i};\tau) \approx \frac{1}{(2\pi\tau)^{3N/2}} \mathrm{e}^{-\frac{(\mathbf{R}_\mathrm{f}-\mathbf{R}_\mathrm{i})^2}{2\tau}} \mathrm{e}^{-\left(\frac{V(\mathbf{R}_\mathrm{f})+V(\mathbf{R}_\mathrm{i})}{2} - E_\mathrm{T}\right)\tau}, \tag{40}$$

where $V(\mathbf{R})$ is the potential energy. Similarly, assuming for now that the trial wave function is of the same sign in $\mathbf{R}_\mathrm{i}$ and $\mathbf{R}_\mathrm{f}$, i.e., $\Psi(\mathbf{R}_\mathrm{f})/\Psi(\mathbf{R}_\mathrm{i}) > 0$, a short-time approximation to the importance-sampling Green function is [5,31]

$$\tilde{G}(\mathbf{R}_\mathrm{f}|\mathbf{R}_\mathrm{i};\tau) \approx \frac{1}{(2\pi\tau)^{3N/2}} \mathrm{e}^{-\frac{(\mathbf{R}_\mathrm{f}-\mathbf{R}_\mathrm{i}-\mathbf{v}(\mathbf{R}_\mathrm{i})\tau)^2}{2\tau}} \mathrm{e}^{-\left(\frac{E_\mathrm{L}(\mathbf{R}_\mathrm{f})+E_\mathrm{L}(\mathbf{R}_\mathrm{i})}{2} - E_\mathrm{T}\right)\tau}, \tag{41}$$

where the drift velocity $\mathbf{v}(\mathbf{R}) = \nabla\Psi(\mathbf{R})/\Psi(\mathbf{R})$ and the local energy $E_\mathrm{L}(\mathbf{R})$ were assumed constant between $\mathbf{R}_\mathrm{i}$ and $\mathbf{R}_\mathrm{f}$. This short-time approximation implies a *finite time-step error* in the calculation of all observables, which should in principle be eliminated by extrapolating the results to $\tau = 0$ (see Refs. 32–34 for proofs that the time-step error vanishes in the $\tau \to 0$ limit).

2.2 Stochastic Realization

The stochastic realization of Eq. (39) is less trivial than for VMC. The Green function $\tilde{G}(\mathbf{R}_\mathrm{f}|\mathbf{R}_\mathrm{i};\tau)$ is generally not a stochastic matrix, since it does not conserve the normalization of the probability density: $\int d\mathbf{R}_\mathrm{f} \; \tilde{G}(\mathbf{R}_\mathrm{f}|\mathbf{R}_\mathrm{i};\tau) \neq 1$. We can nevertheless write the elements of $\tilde{G}$ as the product of the corresponding elements of a stochastic matrix P and a weight matrix W,

$$\tilde{G}(\mathbf{R}_\mathrm{f}|\mathbf{R}_\mathrm{i};\tau) = P(\mathbf{R}_\mathrm{f}|\mathbf{R}_\mathrm{i})\,W(\mathbf{R}_\mathrm{f}|\mathbf{R}_\mathrm{i}), \tag{42}$$

where, in the short-time approximation, $P(\mathbf{R}_\mathrm{f}|\mathbf{R}_\mathrm{i}) = (2\pi\tau)^{-3N/2} \mathrm{e}^{-(\mathbf{R}_\mathrm{f}-\mathbf{R}_\mathrm{i}-\mathbf{v}(\mathbf{R}_\mathrm{i})\tau)^2/2\tau}$ and $W(\mathbf{R}_\mathrm{f}|\mathbf{R}_\mathrm{i}) = \mathrm{e}^{-((E_\mathrm{L}(\mathbf{R}_\mathrm{f})+E_\mathrm{L}(\mathbf{R}_\mathrm{i}))/2-E_\mathrm{T})\tau}$. Note that $\tilde{G}$ reduces to a stochastic matrix in the limit $\Psi \to \Psi_0$. The stochastic realization is then a weighted random walk. Start from a walker at an initial position $\mathbf{R}_1$ with a weight $w_1 = 1$, i.e., a distribution $w_1\delta(\mathbf{R} - \mathbf{R}_1)$. Sample the position $\mathbf{R}_2$ of the walker at the next iteration from the probability distribution $P(\mathbf{R}_2|\mathbf{R}_1)$ [according to Eq. (29)] and give it weight $w_2 = W(\mathbf{R}_2|\mathbf{R}_1) \times w_1$, sample the third position $\mathbf{R}_3$ from the probability distribution $P(\mathbf{R}_3|\mathbf{R}_2)$ and give it weight $w_3 = W(\mathbf{R}_3|\mathbf{R}_2) \times w_2$, and so on. After an equilibration phase, the random walk should sample the stationary distribution $f(\mathbf{R}) \propto \mathrm{E}[w_k\delta(\mathbf{R} - \mathbf{R}_k)] \approx (1/M)\sum_{k=1}^{M} w_k\delta(\mathbf{R} - \mathbf{R}_k)$. In reality, this procedure is terribly inefficient. Because the weights w_k are products of a large number of random variables, they can become very large at some iterations and very small at other iterations. Consequently, the averages are dominated by a few points with large weights, even though the calculation of any point of the Markov chain takes the same computational time regardless of its weight. This problem can be alleviated by keeping the product of the weights for only a finite number n of consecutive iterations [35]

$$w_k = \prod_{l=k-n+1}^{k} W(\mathbf{R}_l|\mathbf{R}_{l-1}). \tag{43}$$

However, using a finite n introduces a bias in the sampled stationary distribution. In practice, for an n large enough to have a reasonably small bias, this procedure still remains inefficient.

The solution is to use at each iteration k a population of M_k walkers, with positions $\mathbf{R}_{k,\alpha}$ and weights $w_{k,\alpha}$ (where $\alpha = 1,2,\ldots,M_k$), performing random walks with a *branching or birth–death process* designed to make the weights $w_{k,\alpha}$ vary in only a small range from walker to walker in a given iteration, and from iteration to iteration, while still sampling the correct distribution $f(\mathbf{R}) \propto \mathrm{E}[\sum_{\alpha=1}^{M_k} w_{k,\alpha}\delta(\mathbf{R} - \mathbf{R}_{k,\alpha})] \approx (1/M)\sum_{k=1}^{M}\sum_{\alpha=1}^{M_k} w_{k,\alpha}\delta(\mathbf{R} - \mathbf{R}_{k,\alpha})$. Various unbiased variants are possible, characterized by a population size M_k that either varies or is constant from iteration to iteration, and by weights $w_{k,\alpha}$ that can either be equal or different for each walker.

The simplest variant uses a varying population size M_k and weights all equal to one, $w_{k,\alpha} = 1$. At each iteration k, each walker α is replaced by $m_{k,\alpha}$ unit-weight copies of itself, where $m_{k,\alpha}$ is an integer equal on average to what should be the current weight $W_{k,\alpha} = W(\mathbf{R}_{k,\alpha}|\mathbf{R}_{k-1,\alpha})$. For example, if the walker α should have the weight $W_{k,\alpha} = 2.7$ at iteration k, this walker is replaced by $m_{k,\alpha} = 3$ copies of itself with a probability 0.7 or replaced by $m_{k,\alpha} = 2$ copies of itself with a probability 0.3. More generally, $m_{k,\alpha} = \lfloor W_{k,\alpha} \rfloor + 1$ with probability $W_{k,\alpha} - \lfloor W_{k,\alpha} \rfloor$ and $m_{k,\alpha} = \lfloor W_{k,\alpha} \rfloor$ otherwise, where $\lfloor W_{k,\alpha} \rfloor$ is the nearest integer smaller than $W_{k,\alpha}$. If $m_{k,\alpha} = 0$, the walker is terminated.

This procedure does not change the sampled stationary distribution.[d] This variant has the disadvantage that the integerization of the weights results in unnecessary duplications of walkers, leading to more correlated walkers and thus to a smaller number of statistically independent points in the sample. Another disadvantage is that it leads to unnecessary fluctuations in the sum of the weights, a quantity that is relevant for computing the growth estimator of the energy.

A better solution is the split–join algorithm[6] which limits the duplication of walkers by keeping residual noninteger weights $w_{k,\alpha}$. At each iteration k, after updating the weights according to $w_{k,\alpha} = W(\mathbf{R}_{k,\alpha}|\mathbf{R}_{k-1,\alpha}) \times w_{k-1,\alpha}$, each walker α with a weight $w_{k,\alpha} > 2$ is split into $\lfloor w_{k,\alpha} \rfloor$ walkers, each being attributed the weight $w_{k,\alpha}/\lfloor w_{k,\alpha} \rfloor$. If walkers α and β each have weight $< 1/2$, keep walker α with probability $w_{k,\alpha}/(w_{k,\alpha} + w_{k,\beta})$ and walker β otherwise. In either case, the surviving walker gets weight, $w_{k,\alpha} + w_{k,\beta}$. This algorithm has the advantage that it conserves the total weight of the population of walkers $W_k = \sum_{\alpha=1}^{M_k} w_{k,\alpha}$ for a given iteration. Yet another possibility is the stochastic reconfiguration algorithm,[36,37] which uses a fixed population size M_k, and walkers of equal noninteger weights within each iteration, though the weights of the walkers fluctuate from one iteration to the next.

To avoid the explosion or extinction of the population of walkers (or their weights if M_k is kept fixed), the value of E_T can be adjusted during the iterations. For example, a choice for E_T at iteration $k+1$ is $E_T(k+1) = E_0^{\text{est}}(k) - C\log(W_k/W_0)$ where $E_0^{\text{est}}(k)$ is an estimate of E_0 at iteration k, C is a constant, W_k is the total weight of the population of walkers, and W_0 is the target total weight. Because of fluctuations, E_T thus slightly varies with respect to E_0 during the iterations, which introduces a systematic bias on the weights and thus on the stationary distribution $f(\mathbf{R})$. The adjustment of E_T makes $f(\mathbf{R})$ too small in regions where $E_L(\mathbf{R}) < E_0$ and too large in regions where $E_L(\mathbf{R}) > E_0$. Both of these have the effect of raising the energy. This is called *population-control error.* This error is generally small and decreases with increasing number of walkers as $1/M_k$.[6] Besides, it is possible to eliminate almost completely this error by undoing the modification of weights introduced by the variation of E_T for the last several iterations.[6,38]

[d] One can write: $\mathrm{E}\left[\sum_{\alpha=1}^{M_k} W_{k,\alpha}\delta(\mathbf{R}-\mathbf{R}_{k,\alpha})\right] = \mathrm{E}\left[\sum_{\alpha=1}^{M_k} m_{k,\alpha}\delta(\mathbf{R}-\mathbf{R}_{k,\alpha})\right] = \mathrm{E}\left[\sum_{\alpha=1}^{M_{k+1}} \delta(\mathbf{R}-\mathbf{R}_{k+1,\alpha})\right]$, where $\mathbf{R}_{k+1,\alpha}$ are the positions of the $M_{k+1} = \sum_{\alpha=1}^{M_k} m_{k,\alpha}$ walkers used for the next iteration $k+1$ obtained after making $m_{k,\alpha}$ copies of the αth walker.

In the limit of an infinitesimal time step, the transition matrix $P(\mathbf{R}_f|\mathbf{R}_i)$ has a stationary distribution $\Psi(\mathbf{R})^2$, and the weight term $W(\mathbf{R}_f|\mathbf{R}_i)$ converts this distribution into the mixed distribution $\Psi_0(\mathbf{R})\Psi(\mathbf{R})$. One can get rid of the finite time-step error in the transition matrix $P(\mathbf{R}_f|\mathbf{R}_i)$ by introducing an accept/reject step as in the Metropolis–Hastings algorithm.[5] For this, the transition matrix is redefined as $P(\mathbf{R}_f|\mathbf{R}_i) = P_{\text{acc}}(\mathbf{R}_f|\mathbf{R}_i)P_{\text{prop}}(\mathbf{R}_f|\mathbf{R}_i)$, for $\mathbf{R}_i \neq \mathbf{R}_f$, with the proposal probability

$$P_{\text{prop}}(\mathbf{R}_f|\mathbf{R}_i) = \frac{1}{(2\pi\tau)^{3N/2}} e^{-\frac{(\mathbf{R}_f - \mathbf{R}_i - \mathbf{v}(\mathbf{R}_i)\tau)^2}{2\tau}}, \tag{44}$$

and the acceptance probability

$$P_{\text{acc}}(\mathbf{R}_f|\mathbf{R}_i) = \min\left\{1, \frac{P_{\text{prop}}(\mathbf{R}_i|\mathbf{R}_f)\Psi(\mathbf{R}_f)^2}{P_{\text{prop}}(\mathbf{R}_f|\mathbf{R}_i)\Psi(\mathbf{R}_i)^2}\right\}. \tag{45}$$

With this modification, $P(\mathbf{R}_f|\mathbf{R}_i)$ has the stationary distribution $\Psi(\mathbf{R})^2$ even for a finite time step. Of course, the finite time-step error persists in the term $W(\mathbf{R}_f|\mathbf{R}_i)$. Since certain moves are rejected, $P(\mathbf{R}_f|\mathbf{R}_i)$ corresponds now to a process of diffusion with drift with an effective time step $\tau_{\text{eff}} < \tau$. This effective time step can be estimated during the calculation from the average acceptance rate, and it is consistent to use it in the term $W(\mathbf{R}_f|\mathbf{R}_i)$ in place of τ. In practice, just as in VMC, it is also more efficient in DMC to move the electrons one by one, i.e., to decompose $P(\mathbf{R}_f|\mathbf{R}_i)$ according to Eq. (31). We then arrive at a DMC algorithm very similar to the VMC algorithm, with weights in addition. Note, however, that since a relatively small time step must be used in DMC, the average moves are smaller than in VMC and the autocorrelation time of the local energy is larger than in VMC.

The energy is calculated as the average of the local energy over the distribution $f(\mathbf{R})/\int d\mathbf{R} f(\mathbf{R})$. For M iterations (after the equilibration phase) and M_k walkers, we have

$$E_0 \approx \bar{E}_L = \frac{\sum_{k=1}^{M}\sum_{\alpha=1}^{M_k} w_{k,\alpha} E_L(\mathbf{R}_{k,\alpha})}{\sum_{k=1}^{M}\sum_{\alpha=1}^{M_k} w_{k,\alpha}}. \tag{46}$$

Just as in VMC, there is a zero-variance principle on the energy in DMC. In the limit that the trial wave function Ψ is an exact eigenfunction of the Hamiltonian, E_L is independent of $\mathbf{R}$, the weights reduce to 1, and the variance on $\bar{E}_L$ vanishes.

Note that for an observable $\hat{O}$ that does not commute with the Hamiltonian, the average $\bar{O}_L$ over the mixed DMC distribution is an estimator of $\langle\Psi_0|\hat{O}|\Psi\rangle/\langle\Psi_0|\Psi\rangle$ which is only an approximation to the exact expectation value $\langle\Psi_0|\hat{O}|\Psi_0\rangle/\langle\Psi_0|\Psi_0\rangle$ with an $\mathcal{O}(||\Psi-\Psi_0||)$ error. Since the average $\bar{O}_L$ over the VMC distribution also has an error that is linear in $||\Psi-\Psi_0||$ but with a prefactor that is twice as large, an $\mathcal{O}(||\Psi-\Psi_0||^2)$ approximation is provided by twice the average of O_L over the mixed DMC distribution minus the average of O_L over the VMC distribution.[39] For a recent survey of exact methods for sampling the pure distribution Ψ_0^2, see Ref. 40, and for a discussion of the techniques used for calculating pure expectation values of various classes of operators, see Ref. 2.

2.3 Fermionic Sign Problem

In Eq. (41), we have assumed that the trial wave function $\Psi(\mathbf{R})$ is always of the same sign, i.e., it does not have any nodes (points $\mathbf{R}$ such that $\Psi(\mathbf{R}) = 0$). This is the case for the ground-state wave function of a Bosonic system, and for a few simple electronic systems (two electrons in a spin-singlet state, such as the ground state of the He atom or of the H_2 molecule). In this case, the algorithm presented above allows one to obtain the exact energy of the system, after elimination of the finite time-step error and the population-control error. If the wave function of the Fermionic ground state has nodes, then there is always at least one Bosonic state of lower energy, and the true ground state of the Hamiltonian is a Bosonic state for which the wave function $\Psi_B(\mathbf{R})$ can be chosen strictly positive. If one applied the Green function exactly, starting from the distribution $\Psi(\mathbf{R})^2$ the distribution would correctly converge to $\Psi_0(\mathbf{R})\Psi(\mathbf{R})$ since the trial wave function is antisymmetric (with respect to the exchange of two electrons) and has a zero overlap with all the Bosonic states which are symmetric. However, in reality one applies the Green function using a finite sampling in position space which does not allow one to impose the antisymmetry. Starting from an antisymmetric wave function Ψ, a small component of Ψ_B can thus appear, and it grows and eventually dominates. The distribution tends to $\Psi_B(\mathbf{R})\Psi(\mathbf{R})$ and the energy formula in Eq. (46) only gives 0/0 (the positive and negative weights cancel out) with statistical noise. Even if one imposed antisymmetry and eliminated the Bosonic states, e.g., by considering all electron permutations in each walker, the problem persists because different paths between the same points in this antisymmetrized space can contribute with opposite sign. Since Ψ_0 and $-\Psi_0$ are equally good solutions of the Schrödinger equation, the algorithm would sample each with approximately equal probability, leading again to the cancellation of positive and negative weight contributions. These are different manifestations of the infamous *Fermionic sign problem.*

2.4 Fixed-Node Approximation

To avoid the sign problem in DMC, the *fixed-node approximation* (FN) [28,29,41] is introduced. The idea is to force the convergence to a wave function approximating the Fermionic ground state by fixing its nodes to be the same as those of the trial wave function $\Psi(\mathbf{R})$. Formally, one can define the FN Hamiltonian, $\hat{H}_{\mathrm{FN}}$, by adding to the true Hamiltonian $\hat{H}$ infinite potential barriers at the location of the nodes of $\Psi(\mathbf{R})$.[42] The ground-state wave function of this Hamiltonian is called the FN wave function Ψ_{FN} and its energy is the FN energy E_{FN},

$$\hat{H}_{\mathrm{FN}}|\Psi_{\mathrm{FN}}\rangle = E_{\mathrm{FN}}|\Psi_{\mathrm{FN}}\rangle. \tag{47}$$

In the $3N$-dimensional space of positions $\mathbf{R}$, the nodes of $\Psi(\mathbf{R})$ define hypersurfaces of dimension $3N-1$. The position space is then partitioned in nodal pockets of $\Psi(\mathbf{R})$, delimited by nodal surfaces, in which the wave function has a fixed sign. In each nodal pocket, the FN wave function is the solution to the Schrödinger equation satisfying vanishing boundary conditions on the nodal surface. The FN Green function corresponding to the Hamiltonian $\hat{H}_{\mathrm{FN}}$ is

$$G_{\mathrm{FN}}(\mathbf{R}_{\mathrm{f}}|\mathbf{R}_{\mathrm{i}};t) = \langle \mathbf{R}_{\mathrm{f}}|\mathrm{e}^{-(\hat{H}_{\mathrm{FN}}-E_{\mathrm{T}})t}|\mathbf{R}_{\mathrm{i}}\rangle, \tag{48}$$

and only permits moves $\mathbf{R}_{\mathrm{i}} \to \mathbf{R}_{\mathrm{f}}$ inside a nodal pocket. The importance-sampling FN Green function,

$$\tilde{G}_{\mathrm{FN}}(\mathbf{R}_{\mathrm{f}}|\mathbf{R}_{\mathrm{i}};t) = \Psi(\mathbf{R}_{\mathrm{f}})\; G_{\mathrm{FN}}(\mathbf{R}_{\mathrm{f}}|\mathbf{R}_{\mathrm{i}};t)\, \frac{1}{\Psi(\mathbf{R}_{\mathrm{i}})}, \tag{49}$$

also confines the moves inside a nodal pocket, and it is thus always greater or equal to zero. A short-time approximation to $\tilde{G}_{\mathrm{FN}}(\mathbf{R}_{\mathrm{f}}|\mathbf{R}_{\mathrm{i}};t)$ is then again given by Eq. (41). The stochastic algorithm previously described can thus be applied directly. Thanks to the FN approximation, the weights always remain positive, and the stationary mixed distribution $f(\mathbf{R})$ is proportional to $\Psi_{\mathrm{FN}}(\mathbf{R})\Psi(\mathbf{R})$.

The largest contributions to the finite time-step error come from singularities of the drift velocity $\mathbf{v}(\mathbf{R}) = \nabla\Psi(\mathbf{R})/\Psi(\mathbf{R})$ and of the local energy $E_{\mathrm{L}}(\mathbf{R})$ in the Green function of Eq. (41). Since the gradient of the trial wave function $\nabla\Psi(\mathbf{R})$ (and of the exact wave function) does not generally vanish at the location of the nodes, the drift velocity $\mathbf{v}(\mathbf{R})$ diverges at the nodes, which leads to too large moves near the nodes for finite time steps. The drift velocity has discontinuities also at particle coalescences (both electron–nucleus and electron–electron). Similarly, for an approximate trial wave function $\Psi(\mathbf{R})$, the local energy $E_L(\mathbf{R})$ also diverges at the nodes and at particle coalescences (unless the Kato cusp conditions [43,44] are imposed). The finite time-step error can be greatly reduced by replacing $\mathbf{v}(\mathbf{R})$ and $E_L(\mathbf{R})$ in the Green function by approximate integrals of these quantities over the time step τ.[6]

If importance sampling is not used, it is necessary to kill walkers that cross the nodes of Ψ to impose the FN boundary condition. In practice, importance sampling is almost always used. In that case, it is better to reject the moves of walkers crossing the nodes, which is consistent with the FN approximation, but even this is not necessary since the number of walkers that cross the node per unit time goes to zero as $\tau \to 0$ (Ref. 6).[e] For a finite time step, there are node crossing events, but these are just part of the finite time-step error and in practice essentially the same time-step error is obtained whether the walkers are allowed to cross nodes or not.

We may wonder whether the walkers have to sample all the nodal pockets. The tiling theorem [45] establishes that all the nodal pockets of the ground-state wave function of a many-electron Hamiltonian with a reasonable local potential are equivalent, i.e., the permutations of any nodal pocket are sufficient to cover the entire space. This means that, for ground-state calculations, the distribution of the walkers over the nodal pockets is irrelevant.

By applying the variational principle, it is easy to show that the FN energy is an upper bound to the exact energy

$$E_{\rm FN} = \frac{\langle \Psi_{\rm FN} | \hat{H}_{\rm FN} | \Psi_{\rm FN} \rangle}{\langle \Psi_{\rm FN} | \Psi_{\rm FN} \rangle} = \frac{\langle \Psi_{\rm FN} | \hat{H} | \Psi_{\rm FN} \rangle}{\langle \Psi_{\rm FN} | \Psi_{\rm FN} \rangle} \geq E_0, \tag{50}$$

the second equality coming from the fact that the infinite potential barriers in $\hat{H}_{\rm FN}$ do not contribute to the expectation value since $\Psi_{\rm FN}$ is zero on the

---

[e] The drift velocity moves electrons away from the nodal surface, but for small τ the diffusion term dominates and can cause walkers to cross nodes. The density of walkers goes quadratically to zero near nodes and walkers that are roughly within a distance $\sqrt{\tau}$ can cross. Hence, the number that cross per Monte Carlo step goes as $\int_0^{\sqrt{\tau}} x^2 dx \sim \tau^{3/2}$, and so the number that cross per unit time goes to zero as $\sqrt{\tau}$.

nodal surface. Since the wave function Ψ_{FN} is an eigenfunction of $\hat{H}_{\text{FN}}$, the FN energy can also be expressed using the mixed expectation value

$$E_{\text{FN}} = \frac{\langle \Psi_{\text{FN}} | \hat{H}_{\text{FN}} | \Psi \rangle}{\langle \Psi_{\text{FN}} | \Psi \rangle} = \frac{\langle \Psi_{\text{FN}} | \hat{H} | \Psi \rangle}{\langle \Psi_{\text{FN}} | \Psi \rangle}, \tag{51}$$

where the Hamiltonian $\hat{H}_{\text{FN}}$ has been replaced by $\hat{H}$ for essentially the same reason as before, viz., both Ψ and Ψ_{FN} are zero where $\hat{H}_{\text{FN}}$ is infinite. In practice, the FN energy is thus obtained by the same energy formula (46).

The accuracy of the DMC results with the FN approximation thus depends on the quality of the nodal surface of the trial wave function. For a trial wave function with a single Slater determinant, the error due to the FN approximation can often be large, even for energy differences. For example, for the C_2 molecule, the FN error for a single-determinant trial wave function is 1.6 eV for the total energy and 0.8 eV for the dissociation energy.[46] In order to reduce this error, one can use several Slater determinants and optimize the parameters of the wave function (Jastrow parameters, coefficients of determinants, coefficients that express the orbitals in terms of the basis functions, and exponents of the basis functions) in VMC (see Refs. 46–53). This allows one to reach near chemical accuracy ($\sim$ 1 kcal/mol) in DMC for calculations of energy differences such as molecular atomization energies.[54]

ACKNOWLEDGMENTS

CJU was supported in part by grants NSF CHE-1112097 and DOE-CMCSN DE-SC0006650.

APPENDIX. STATISTICAL ESTIMATOR OF NONLINEAR FUNCTIONS OF EXPECTATION VALUES

We often need to estimate nonlinear functions of expectation values. The simplest example is the variance,

$$\mathrm{V}[X] = \mathrm{E}[X^2] - \mathrm{E}[X]^2, \tag{A.1}$$

which is a quadratic function of the expectation values of two random variables X^2 and X. Another example is the calculation of the energy in DMC using weights [see Eq. (46)], with simplified notation,

$$E_0 = \frac{\mathrm{E}[wE_\mathrm{L}]}{\mathrm{E}[w]}, \tag{A.2}$$

involving a ratio of two expectation values.

Consider a nonlinear function, $f(\mathrm{E}[X],\mathrm{E}[Y])$, of two expectation values, $\mathrm{E}[X]$ and $\mathrm{E}[Y]$. The usual simple estimator of $f(\mathrm{E}[X],\mathrm{E}[Y])$ is $f(\bar{X},\bar{Y})$ where

$$\bar{X} = \frac{1}{M_\mathrm{b}} \sum_{b=1}^{M_\mathrm{b}} \bar{X}_b, \tag{A.3}$$

and

$$\bar{Y} = \frac{1}{M_\mathrm{b}} \sum_{b=1}^{M_\mathrm{b}} \bar{Y}_b, \tag{A.4}$$

are averages over a finite number of blocks M_b, and $\bar{X}_b$ and $\bar{Y}_b$ are the block averages of X and Y, respectively [see Eq. (6)]. As discussed before, each block average is itself an average over a sufficiently large number of steps, M_s, so that the block averages can be assumed to be independent of each other. The simple estimator can be justified as follows. (i) When the law of large numbers holds, $\bar{X}$ and $\bar{Y}$ converge, with increasing M_b, almost surely to $\mathrm{E}[X]$ and $\mathrm{E}[Y]$, respectively. (ii) Hence, $f(\bar{X},\bar{Y})$ converges to $f(\mathrm{E}[X],\mathrm{E}[Y])$ provided that f is continuous at the point $(\mathrm{E}[X],\mathrm{E}[Y])$. However, because f is nonlinear, $f(\bar{X},\bar{Y})$ has a systematic error, i.e., $E[f(\bar{X},\bar{Y})] \neq f(\mathrm{E}[X],\mathrm{E}[Y])$, that vanishes only in the limit of infinite sample size, $M_\mathrm{b} \to \infty$. Though not necessary, in the following, for the sake of simplicity, we assume that $f(\bar{X},\bar{Y})$ has a finite expectation value and a finite variance.[f]

A.1 Systematic Error

Let us first consider the case of a nonlinear function $f(x)$ of a single variable. By definition, the systematic error of the estimator $f(\bar{X})$ is $\mathrm{E}[f(\bar{X})] - f(\mathrm{E}[X])$. The systematic error can be evaluated using a second-order Taylor expansion of the function $f(\bar{X})$ around $\mathrm{E}[X]$ (assuming that f is at least a C^2 function in the neighborhood of $\mathrm{E}[X]$)

[f] $E[f(\bar{X},\bar{Y})]$ can be undefined when f has a point at which it diverges, e.g., $f(x,y) = x/y$. In this case, this definition of the systematic error does not have a strict meaning. In practice, this is not a problem for this f provided that the absolute value of the expectation value of Y over a block is larger than a few times the square root of its variance, say, $|E[\bar{Y}_\mathrm{b}]| > 5\sqrt{V[\bar{Y}_\mathrm{b}]}$.

$$f(\bar{X}) = f(\mathrm{E}[X]) + \left(\frac{\mathrm{d}f}{\mathrm{d}x}\right)(\bar{X} - \mathrm{E}[X]) + \frac{1}{2}\left(\frac{\mathrm{d}^2 f}{\mathrm{d}x^2}\right)(\bar{X} - \mathrm{E}[X])^2 + \cdots, \tag{A.5}$$

where the derivatives of f are evaluated at $\mathrm{E}[X]$. If we take the expectation value of this expression, the linear term vanishes

$$\mathrm{E}[f(\bar{X})] = f(\mathrm{E}[X]) + \frac{1}{2}\left(\frac{\mathrm{d}^2 f}{\mathrm{d}x^2}\right)\mathrm{E}\left[(\bar{X} - \mathrm{E}[X])^2\right] + \cdots. \tag{A.6}$$

Assuming the random variable X has a finite variance and that the higher-order terms can be neglected, the systematic error is thus

$$\mathrm{E}[f(\bar{X})] - f(\mathrm{E}[X]) = \frac{1}{2}\left(\frac{\mathrm{d}^2 f}{\mathrm{d}x^2}\right)\mathrm{V}[\bar{X}] + \cdots = \frac{1}{2}\left(\frac{\mathrm{d}^2 f}{\mathrm{d}x^2}\right)\frac{\mathrm{V}[\bar{X}_\mathrm{b}]}{M_\mathrm{b}} + \cdots. \tag{A.7}$$

Hence, the estimator $f(\bar{X})$ has a systematic error with a leading term proportional to $1/M_\mathrm{b}$. Note that if the hypotheses (especially the finite variance) are not satisfied, the systematic error can decrease more slowly than $1/M_\mathrm{b}$. Equation (A.7) can easily be generalized to a function of several variables. For example, for two variables, the systematic error is

$$\begin{aligned}\mathrm{E}[f(\bar{X},\bar{Y})] - f(\mathrm{E}[X],\mathrm{E}[Y]) &= \frac{1}{2}\left(\frac{\partial^2 f}{\partial x^2}\right)\frac{\mathrm{V}[\bar{X}_\mathrm{b}]}{M_\mathrm{b}} + \frac{1}{2}\left(\frac{\partial^2 f}{\partial y^2}\right)\frac{\mathrm{V}[\bar{Y}_\mathrm{b}]}{M_\mathrm{b}} \\ &\quad + \left(\frac{\partial^2 f}{\partial x \partial y}\right)\frac{\mathrm{Cov}[\bar{X}_\mathrm{b},\bar{Y}_\mathrm{b}]}{M_\mathrm{b}} + \cdots,\end{aligned} \tag{A.8}$$

where the second-order derivatives are evaluated at $(\mathrm{E}[X],\mathrm{E}[Y])$. The leading neglected term is $O(1/M_\mathrm{b}^2)$ if the third moments of X and Y are finite. The second-order derivatives in Eq. (A.8) can in practice be evaluated at $(\bar{X},\bar{Y})$ without changing the order of the approximation. Hence, an estimator for $f(\mathrm{E}[X],\mathrm{E}[Y])$ with an $O(1/M_\mathrm{b}^2)$ error is

$$\begin{aligned}f(\mathrm{E}[X],\mathrm{E}[Y]) &\approx f(\bar{X},\bar{Y}) - \frac{1}{2}\left(\frac{\partial^2 f}{\partial x^2}\right)\frac{\mathrm{V}[\bar{X}_\mathrm{b}]}{M_\mathrm{b}} - \frac{1}{2}\left(\frac{\partial^2 f}{\partial y^2}\right)\frac{\mathrm{V}[\bar{Y}_\mathrm{b}]}{M_\mathrm{b}} \\ &\quad - \left(\frac{\partial^2 f}{\partial x \partial y}\right)\frac{\mathrm{Cov}[\bar{X}_\mathrm{b},\bar{Y}_\mathrm{b}]}{M_\mathrm{b}} + \cdots,\end{aligned} \tag{A.9}$$

where the second-order derivatives are evaluated at $(\bar{X},\bar{Y})$.

This approach is general and can be used to recover some well-known unbiased estimators. For example, let us consider the covariance of two random variables

$$\mathrm{Cov}[X,Y] = \mathrm{E}[XY]-\mathrm{E}[X]\mathrm{E}[Y] = f(\mathrm{E}[XY],\mathrm{E}[X],\mathrm{E}[Y]), \tag{A.10}$$

for which $f(x,y,z) = x - yz$. In this case, the generalization of Eq. (A.8) to three variables with $\bar{X}=(1/M)\sum_{i=1}^{M} X_i$ and $\bar{Y}=(1/M)\sum_{i=1}^{M} Y_i$ where X_i and Y_i are M uncorrelated realizations of X and Y, respectively, gives

$$\mathrm{E}[\bar{XY} - \bar{X}\,\bar{Y}] - \mathrm{Cov}[X,Y] = -\frac{\mathrm{Cov}[X,Y]}{M}, \tag{A.11}$$

which leads to the usual unbiased estimator for the covariance

$$\mathrm{Cov}[X,Y] \approx \frac{M}{M-1}(\bar{XY} - \bar{X}\,\bar{Y}) = \frac{1}{M-1}\sum_{i=1}^{M}(X_i-\bar{X})(Y_i-\bar{Y}). \tag{A.12}$$

A.2 Statistical Uncertainty

First consider a function of a single variable. The statistical uncertainty of $f(\bar{X})$ is given by $\sigma[f(\bar{X})]=\sqrt{\mathrm{V}[f(\bar{X})]}$ where the variance of $f(\bar{X})$ is $\mathrm{V}[f(\bar{X})]=\mathrm{E}\left[(f(\bar{X})-\mathrm{E}[f(\bar{X})])^2\right]$. Subtracting Eq. (A.6) from Eq. (A.5) gives

$$f(\bar{X})-\mathrm{E}[f(\bar{X})]=\left(\frac{\mathrm{d}f}{\mathrm{d}x}\right)(\bar{X}-\mathrm{E}[X])+\cdots. \tag{A.13}$$

Taking the square of this equation and the expectation value leads to the leading term in the variance of $f(\bar{X})$

$$\mathrm{V}[f(\bar{X})]=\left(\frac{\mathrm{d}f}{\mathrm{d}x}\right)^2 \mathrm{V}[\bar{X}]+\cdots. \tag{A.14}$$

This equation can be generalized to a function of several variables. For example, for two variables, the variance of $f(\bar{X},\bar{Y})$ is

$$\mathrm{V}[f(\bar{X},\bar{Y})]=\left(\frac{\partial f}{\partial x}\right)^2 \mathrm{V}[\bar{X}]+\left(\frac{\partial f}{\partial y}\right)^2 \mathrm{V}[\bar{Y}]+2\left(\frac{\partial f}{\partial x}\right)\left(\frac{\partial f}{\partial y}\right)\mathrm{Cov}[\bar{X},\bar{Y}]+\cdots. \tag{A.15}$$

Equation (A.15) can be used for estimating the variance of $f(\bar{X}, \bar{Y})$ at the cost of evaluating the variances $\mathrm{V}[\bar{X}]$ and $\mathrm{V}[\bar{Y}]$ and the covariance $\mathrm{Cov}[\bar{X}, \bar{Y}]$. Note, however, that it can give a severe underestimate of the error if $\partial f/\partial x$ and $\partial f/\partial y$ are small and M_{b} is not sufficiently large.

There is a simple alternative for estimating the variance of f that does not suffer from this limitation and does not require calculating covariances. Consider again the case of a single variable. Instead of defining the block average of f in the obvious way, i.e., $\bar{f}_b = f(\bar{X}_b)$, we define the block average of f as

$$\begin{aligned} &\bar{f}_1 = f(\bar{X}_1) \text{ for the first block } b = 1 \\ &\bar{f}_b = b f(\bar{X}(b)) - (b-1) f(\bar{X}(b-1)) \text{ for any block } b \geq 2, \end{aligned} \tag{A.16}$$

where $\bar{X}(b)$ is the running global average up to block b

$$\bar{X}(b) = \frac{1}{b} \sum_{b'=1}^{b} \bar{X}_{b'}. \tag{A.17}$$

With this definition of the block average, it is easy to check that

$$f(\bar{X}) = \frac{1}{M_{\mathrm{b}}} \sum_{b=1}^{M_{\mathrm{b}}} \bar{f}_b, \tag{A.18}$$

i.e., we have written $f(\bar{X})$ as an average of random variables $\bar{f}_b$. Provided that the variance of X is finite, the block average $\bar{f}_b$ introduced in Eq. (A.16) can be expanded as

$$\bar{f}_b = f(\mathrm{E}[X]) + \left(\frac{\mathrm{d}f}{\mathrm{d}x}\right) (\bar{X}_b - \mathrm{E}[X]) + \cdots. \tag{A.19}$$

Assuming that f has a second-order Taylor expansion, the neglected term converges to zero in probability for large b, at least as $1/(bM_{\mathrm{s}})$. Therefore, according to Eq. (A.19), for large b, the random variables $\bar{f}_b$ converge to independent and equidistributed random variables (since the random variables $\bar{X}_b$ are).[g] Consequently, the variance of $f(\bar{X})$ can be estimated with the usual formula

[g] The naive definition of the block average as $\bar{f}_b = f(\bar{X}_b)$ would also lead to Eq. (A.19) but the neglected term would not converge to zero for large b.

$$\mathrm{V}[f(\bar{X})] \approx \frac{\mathrm{V}[\bar{f}_b]}{M_\mathrm{b}} \approx \frac{1}{M_\mathrm{b}-1}\left(\frac{1}{M_\mathrm{b}}\sum_{b=1}^{M_\mathrm{b}} \bar{f}_b^2 - f(\bar{X})^2\right). \tag{A.20}$$

This formula applies similarly for functions of several variables. The advantage of Eq. (A.20) over Eq. (A.15) for estimating the variance is that it is much simpler to implement and compute, especially for functions of many variables. The estimation of the variance can be simply updated at each block, just as for the expectation value.

REFERENCES

1. Hammond, B. L., Jr.; Lester, W. A.; Reynolds, P. J. *Monte Carlo Methods in Ab Initio Quantum Chemistry*; World Scientific: Singapore, 1994.
2. Nightingale, M. P.; Umrigar, C. J. Monte Carlo Eigenvalue Methods in Quantum Mechanics and Statistical Mechanics. In: *Monte Carlo Methods in Chemical Physics*; Ferguson, D. M., Siepmann, J. I., Truhlar, D. G., Eds.; Advances in Chemical Physics, 105; Wiley: New York, 1998. Chapter 4.
3. Umrigar, C. J. Variational Monte Carlo Basics and Applications to Atoms and Molecules. In: *Quantum Monte Carlo Methods in Physics and Chemistry*; Nightingale, M. P., Umrigar, C. J., Eds.; NATO ASI Ser. C, 525; Kluwer: Dordrecht, 1999; p 129.
4. Foulkes, W. M. C.; Mitas, L.; Needs, R. J.; Rajagopal, G. Quantum Monte Carlo Simulations of Solids. *Rev. Mod. Phys.* **2001**, *73*, 33.
5. Reynolds, P. J.; Ceperley, D. M.; Alder, B. J.; Lester, W. A. Fixed-Node Quantum Monte Carlo for Molecules. *J. Chem. Phys.* **1982**, 77, 5593.
6. Umrigar, C. J.; Nightingale, M. P.; Runge, K. J. A Diffusion Monte Carlo Algorithm with Very Small Time-Step Errors. *J. Chem. Phys.* **1993**, *99*, 2865.
7. Austin, B. M.; Zubarev, D. Y.; Lester, W. A. Quantum Monte Carlo and Related Approaches. *Chem. Rev.* **2012**, *112*, 263.
8. Kolorenč, J.; Mitas, L. Applications of Quantum Monte Carlo Methods in Condensed Systems. *Rep. Prog. Phys.* **2011**, *74*, 026502.
9. McMillan, W. L. Ground State of Liquid He^4. *Phys. Rev.* **1965**, *138*, A442.
10. Ceperley, D.; Chester, G. V.; Kalos, M. H. Monte Carlo Simulation of a Many-Fermion Study. *Phys. Rev. B* **1977**, *16*, 3081.
11. Conroy, H. Molecular Schrödinger Equation. II. Monte Carlo Evaluation of Integrals. *J. Chem. Phys.* **1964**, *41*, 1331.
12. Metropolis, N.; Rosenbluth, A. W.; Rosenbluth, M. N.; Teller, A. H.; Teller, E. Equations of State Calculations by Fast Computing Machines. *J. Chem. Phys.* **1953**, *21*, 1087.
13. Hastings, W. K. Monte Carlo Sampling Methods Using Markov Chains and Their Applications. *Biometrika* **1970**, *57*, 97.
14. Caflisch, R. E. Monte Carlo and Quasi-Monte Carlo Methods. *Acta Numer.* **1998**, *1998*, 1.
15. Umrigar, C. J. Accelerated Metropolis Method. *Phys. Rev. Lett.* **1993**, *71*, 408.
16. Assaraf, R.; Caffarel, M. Zero-Variance Principle for Monte Carlo Algorithms. *Phys. Rev. Lett.* **1999**, *83*, 4682.
17. Assaraf, R.; Caffarel, M. Computing Forces with Quantum Monte Carlo. *J. Chem. Phys.* **2000**, *113*, 4028.
18. Assaraf, R.; Caffarel, M. Zero-Variance Zero-Bias Principle for Observables in Quantum Monte Carlo: Application to Forces. *J. Chem. Phys.* **2003**, *119*, 10536.
19. Assaraf, R.; Caffarel, M.; Scemama, A. Improved Monte Carlo Estimators for the One-Body Density. *Phys. Rev. E* **2007**, *75*, 035701(R).

20. Toulouse, J.; Assaraf, R.; Umrigar, C. J. Zero-Variance Zero-Bias Quantum Monte Carlo Estimators of the Spherically and System-Averaged Pair Density. *J. Chem. Phys.* **2007**, *126*, 244112.
21. Gilks, W. R.; Richardson, S.; Spiegelhalter, D. J. *Markov Chain Monte Carlo in Practice*; Chapman & Hall: London, 1996.
22. Wolff, U. *Collective Monte Carlo Updating for Spin Systems. Phys. Rev. Lett.* **1989**, *62*, 361–364. http://link.aps.org/doi/10.1103/PhysRevLett.62.361.
23. Melko, R. G.; Sandvik, A. W. *Stochastic Series Expansion Algorithm for the* S = *12*XY *Model with Four-Site Ring Exchange. Phys. Rev. E* **2005**, *72*, 026702. http://link.aps.org/doi/10.1103/PhysRevE.72.026702.
24. Press, W.; Teukolsky, S.; Vetterling, W.; Flannery, B. *Numerical Recipes*; Cambridge University Press: Cambridge, 1992.
25. López-Rios, P.; Ma, A.; Drummond, N. D.; Towler, M. D.; Needs, R. J. Inhomogeneous Backflow Transformations in Quantum Monte Carlo Calculations. *Phys. Rev. E* **2006**, *74*, 066701.
26. Lee, R. M.; Conduit, G. J.; Nemec, N.; López-Rios, P.; Drummond, N. D. Strategies for Improving the Efficiency of Quantum Monte Carlo Calculations. *Phys. Rev. E* **2011**, *83*, 066706.
27. Grimm, R.; Storer, R. G. Monte-Carlo Solution of Schrödinger's Equation. *J. Comput. Phys.* **1971**, 7, 134.
28. Anderson, J. B. A Random-Walk Simulation of the Schrödinger Equation: H_3^+. *J. Chem. Phys.* **1975**, *63*, 1499.
29. Anderson, J. B. Quantum Chemistry by Random Walk. H ^2P, H_3^+ $D_{3h}{}^1A_1'$, $H_2{}^3\Sigma_u^+$, $H_4{}^1\Sigma_g^+$, Be^1S. *J. Chem. Phys.* **1976**, *65*, 4121.
30. Moskowitz, J. W.; Schmidt, K. E.; Lee, M. A.; Kalos, M. H. A New Look at Correlation Energy in Atomic and Molecular Systems. II. The Application of the Green's Function Monte Carlo Method to LiH. *J. Chem. Phys.* **1982**, 77, 349.
31. Schmidt, K. E. Variational and Green's Function Monte Carlo Calculations of Few-Body Systems. In: *Models and Methods in Few-Body Physics*; Ferreira, L. S., Fonseca, A. C., Streit, L., Eds.; Lecture Notes in Physics, 273; Springer: Berlin Heidelberg, 1987; pp 363–407.
32. Rothstein, S. M.; Vrbik, J. A Green's Function Used in Diffusion Monte Carlo. *J. Chem. Phys.* **1987**, *87*, 1902.
33. Anderson, J. B.; Garmer, D. R. Validity of Random Walk Methods in the Limit of Small Time Steps. *J. Chem. Phys.* **1987**, *87*, 1903.
34. Reynolds, P. J.; Owen, R. K.; Lester, W. A. Is There a Zeroth Order Timestep Error in Diffusion Quantum Monte Carlo?*J. Chem. Phys.* **1987**, *87*, 1905.
35. Caffarel, M.; Claverie, P. Development of a Pure Diffusion Quantum Monte Carlo Method Using a Full Generalized Feynman-Kac Formula. II. Applications to Simple Systems. *J. Chem. Phys.* **1988**, *88* (2), 1100–1109.
36. Buonaura, M. C.; Sorella, S. Numerical Study of the Two-Dimensional Heisenberg Model Using a Green Function Monte Carlo Technique with a Fixed Number of Walkers. *Phys. Rev. B* **1998**, *57*, 11446.
37. Assaraf, R.; Caffarel, M.; Khelif, A. Diffusion Monte Carlo Methods with a Fixed Number of Walkers. *Phys. Rev. E* **2000**, *61*, 4566.
38. Nightingale, M. P.; Blöte, H. W. J. Gap of the Linear Spin-1 Heisenberg Antiferromagnet: A Monte Carlo Calculation. *Phys. Rev. B* **1986**, *33*, 659.
39. Ceperley, D. M.; Kalos, M. H. Quantum Many-Body Problem. In: Binder, K. Ed.; Monte Carlo Methods in Statistical Physics; Springer: Berlin, 1979; pp 145–194.
40. Rothstein, S. M. A Survey on Pure Sampling in Quantum Monte Carlo Methods. *Can. J. Chem.* **2013**, *91*, 505.
41. Klein, D. J.; Pickett, H. M. Nodal Hypersurfaces and Anderson's Random-Walk Simulation of the Schrödinger Equation. *J. Chem. Phys.* **1976**, *64*, 4811.

42. Badinski, A.; Haynes, P. D.; Needs, R. J. Nodal Pulay Terms for Accurate Diffusion Quantum Monte Carlo Forces. *Phys. Rev. B* **2008**, 77, 085111.
43. Kato, T. On the Eigenfunctions of Many-Particle Systems in Quantum Mechanics. *Commun. Pure Appl. Math.* **1957**, *10*, 151.
44. Pack, R. T.; Byers-Brown, W. Cusp Conditions for Molecular Wavefunctions. *J. Chem. Phys.* **1966**, *45*, 556.
45. Ceperley, D. M. Fermion Nodes. *J. Stat. Phys.* **1991**, *63*, 1237.
46. Toulouse, J.; Umrigar, C. J. Full Optimization of Jastrow-Slater Wave Functions with Application to the First-Row Atoms and Homonuclear Diatomic Molecules. *J. Chem. Phys.* **2008**, *128*, 174101.
47. Umrigar, C. J.; Wilson, K. G.; Wilkins, J. W. Optimized Trial Wave Functions for Quantum Monte Carlo Calculations. *Phys. Rev. Lett.* **1988**, *60*, 1719.
48. Nightingale, M. P.; Melik-Alaverdian, V. Optimization of Ground- and Excited-State Wave Functions and van der Waals Clusters. *Phys. Rev. Lett.* **2001**, *87*, 043401.
49. Schautz, F.; Filippi, C. Optimized Jastrow-Slater Wave Functions for Ground and Excited States: Application to the Lowest States of Ethene. *J. Chem. Phys.* **2004**, *120*, 10931.
50. Umrigar, C. J.; Filippi, C. Energy and Variance Optimization of Many-Body Wave Functions. *Phys. Rev. Lett.* **2005**, *94*, 150201.
51. Sorella, S. Wave Function Optimization in the Variational Monte Carlo Method. *Phys. Rev. B* **2005**, *71*, 241103.
52. Umrigar, C. J.; Toulouse, J.; Filippi, C.; Sorella, S.; Hennig, R. G. Alleviation of the Fermion-Sign Problem by Optimization of Many-Body Wave Functions. *Phys. Rev. Lett.* **2007**, *98*, 110201.
53. Toulouse, J.; Umrigar, C. J. Optimization of Quantum Monte Carlo Wave Functions by Energy Minimization. *J. Chem. Phys.* **2007**, *126*, 084102.
54. Petruzielo, F. R.; Toulouse, J.; Umrigar, C. J. Approaching Chemical Accuracy with Quantum Monte Carlo. *J. Chem. Phys.* **2012**, *136*, 124116.

CHAPTER SIXTEEN

Configuration Interaction Monte Carlo with Coupled Clusters Wave Functions

Alessandro Roggero*, Paolo Mori†, Abishek Mukherjee‡,§, Francesco Pederiva†,¶,1

*Institute for Nuclear Theory, University of Washington, Seattle, Washington, USA
†Dipartimento di Fisica, Università di Trento, Trento, Italy
‡ECT*, Strada delle Tabarelle 286, Trento, Italy
§ClusterVision B.V., Nieuw-Zeelandweg 15B, Amsterdam, Netherlands
¶INFN-TIFPA, Trento Institute for Fundamental Physics and Applications, Trento, Italy
[1]Corresponding author: e-mail address: pederiva@science.unitn.it

Contents

Abstract

Quantum Monte Carlo algorithms in Fock space have gained popularity in the last few years. Here we review the Configuration Interaction Monte Carlo (CIMC) algorithm. CIMC provides a way to implement the imaginary time propagation projecting the ground state of a given Hamiltonian in a model Hilbert space that (1) makes use of an importance function, and in particular of the wave function computed in a Coupled Cluster calculation, and (2) exploits a continuous time algorithm to eliminate the approximations due to the use of a finite imaginary time step. Some results and discussions from the implementation in the three-dimensional electron gas and first row atoms are also presented.

Advances in Quantum Chemistry, Volume 73
ISSN 0065-3276
http://dx.doi.org/10.1016/bs.aiq.2015.09.003

1. INTRODUCTION

Quantum Monte Carlo (QMC) simulations provide great flexibility in terms of the Hilbert space in which the system under investigation is represented.

The long standing algorithms have all been developed in the coordinate space representation. The simulation would in this case consist of the generation of a random walk in the positions of the particle sampled either from an *ansatz* for the many-body wave function (variational Monte Carlo, VMC) or from a propagator that implements an imaginary time evolution of an initial state, projecting out its lowest energy component (e.g., diffusion Monte Carlo, DMC). Within this framework a very important key ingredient for a successful computation of energies and other properties is the availability of an accurate wave function to be used as an "importance function" for generating the random walks. In coordinate space, accurate wave functions are constructed by including of explicit correlations either as a generalized Jastrow factor or building the antisymmetric wave function as a Slater determinant of correlated functions.

In chemistry, the necessity of dealing with fermionic particles forces one to face another important problem, that is the instability of the calculation due to the presence of the so-called Fermion Sign Problem, which refers to the exponential decay of the signal to noise ratio in the evaluation of estimators when one tries to project out (e.g., by DMC) a state which is not the absolute ground state of the Hamiltonian, such as the lowest energy antisymmetric state. Also in this case the availability of an accurate wave function provides a way to partially circumvent the problem by means of the so-called fixed-node approximation. In this approximation, the projected wave function is forced to vanish at the nodal surface of some approximate wave function. The more accurate the latter is, the lower the impact of this approximation.

More recently, a number of imaginary time projection algorithms have been developed that consider a random walk in Fock space (configuration space)[1–3] rather than in coordinate space. In this approach, the main ingredient is the basis set on which the wave function has to be expanded and the random walk eventually provides stochastic overlaps that give the closest representation of the ground state in the Hilbert space generated by that specific set of basis states. While this approach is reminiscent of more standard quantum chemistry algorithms, it has the advantage that sampling techniques allow one to operate on very large basis sets.

However, in this kind of algorithms, the concept of importance sampling is not as straightforward as it is in coordinate space. Some progress has recently been achieved by Kolodrubetz and Clark.[4] Alternative methods to accelerate convergence are based on the semistochastic approach of Umrigar et al.[5] In this chapter, we will review our progress[6,7] in introducing the overlaps generated in a Coupled Cluster calculation as the importance function in a QMC calculation in Fock space based on a continuous (imaginary) time propagator *á la* Green's Function Monte Carlo, namely the Configuration Interaction Monte Carlo (CIMC).[8] In particular, we will review some of our results for bulk systems and comment the procedure we are implementing in order to extend the algorithm to the study of finite systems.

2. METHOD

We start by considering a generic Hamiltonian including up to two-body interactions (a generalization to a three-body force is possible):

$$H \equiv \sum_{\alpha \in \mathcal{S}} \epsilon_\alpha a^\dagger_\alpha a_\alpha + \sum_{\alpha\beta\gamma\delta \in \mathcal{S}} V_{\alpha\beta\gamma\delta} a^\dagger_\alpha a^\dagger_\beta a_\gamma a_\delta \tag{1}$$

Here, $\mathcal{S}$ is the set of single particle states labeled by the Greek letter indices. Such a set is assumed to be finite and of size $\mathcal{N}_\mathcal{S}$. It should be noticed that at this point no assumption is made on the locality of the interaction. This implies that possible nonlocal interactions (arising, e.g., form the spin–orbit terms or more general relativistic corrections) are automatically included in the formalism.

As in most of the QMC methods, the idea is to start from an arbitrary state containing a nonzero component along the ground state of H and to project out all the components orthogonal to the ground state. This can be achieved by means of the power method.

If one considers the operator

$$\mathcal{P} = 1 - \Delta\tau H, \tag{2}$$

it is easy to see that given some arbitrary initial state $|\Phi\rangle$ a repeated application of $\mathcal{P}$ to $|\Phi\rangle$ leads to:

$$\mathcal{P}^M|\Phi\rangle = \left[1 - \Delta\tau H + (\Delta\tau H)^2 + \cdots\right]|\Phi\rangle \tag{3}$$

which, for a sufficiently small $\Delta\tau$ (see later), leads to an exponential suppression of the contribution from the excited states to the quantity on the right-hand side.

We illustrate this for a special case (which will be important for the discussion that follows). This is the limit of $M \to \infty$ while keeping $\Delta\tau M = \tau$ finite. In this case the above equation becomes,

$$\exp(-\tau H)|\Phi\rangle = \sum_n c_n e^{-\tau E_n} |\phi_n\rangle. \tag{4}$$

In the right-hand side of the equation, the initial state has been expanded in eigenstates of the Hamiltonian. The propagation leads to an exponential decay of the ratio of the coefficient of any excited state to that of the ground state.

By adding a "reference energy" E_T, it is possible, in principle, to make the normalization of the ground state component independent of M (or τ). The propagator then becomes:

$$\mathcal{P} = 1 - \Delta\tau(H - E_T). \tag{5}$$

In order for the power method to work, the eigenvalues of the linear propagator in (2) need to be in [−1,1]. This put an additional condition on the time step $\Delta\tau$:

$$\Delta\tau \leq \frac{2}{E_{\max} - E_{\min}} \tag{6}$$

where $E_{\max}$ and $E_{\min}$ are, respectively, the highest and lowest eigenvalues of the many-body Hamiltonian H.

For an N-fermion system, we label the Slater determinants constituting the Hilbert space by $\mathbf{n} \equiv \{n_1 \cdots n_{N_b}\}$ where $n_\alpha = 0, 1$. It is clear from Eq. (6) that the projection of the ground state needs to be reached by iterating the propagation over a short imaginary time. Each propagation cycle is defined as an imaginary "time step." In the configuration representation, the imaginary time step can be explicitly written as:

$$\langle \mathbf{m}|\Psi(\tau + \Delta\tau)\rangle = \sum_{\mathbf{n}} \langle \mathbf{m}|\mathcal{P}|\mathbf{n}\rangle \langle \mathbf{n}|\Psi(\tau)\rangle. \tag{7}$$

As it is commonly done also in standard coordinate space DMC, we rewrite the matrix elements of the propagator as:

$$\langle \mathbf{m}|\mathcal{P}|\mathbf{n}\rangle = g(\mathbf{n})p(\mathbf{m},\mathbf{n}) \tag{8}$$

where

$$g(\mathbf{n}) = \sum_{\mathbf{m}} \langle \mathbf{m}|\mathcal{P}|\mathbf{n}\rangle \quad p(\mathbf{m},\mathbf{n}) = \frac{\langle \mathbf{m}|\mathcal{P}|\mathbf{n}\rangle}{g(\mathbf{n})}. \tag{9}$$

If all the matrix elements of $\mathcal{P}$ were non-negative, then we could interpret $p(\mathbf{m}, \mathbf{n})$ as a transition probability between the states $\mathbf{n}$ and $\mathbf{m}$, while $g(\mathbf{n})$, which is the normalization of the transition probability from any state to the state $|\mathbf{n}\rangle$, can be interpreted as a weight factor for $|\mathbf{n}\rangle$.

In this case, the propagation can be performed stochastically. Transitions would be sampled according to the magnitude of the matrix elements $p(\mathbf{m}, \mathbf{n})$, while the effect of the weighing factors $g(\mathbf{n})$ might be implemented exactly as in standard DMC algorithms as a branching process: after making the transition to a new state, the walkers in the Hilbert space are replicated using the integer part of the weight:

$$n_w(\mathbf{n}) = \mathrm{int}(g(\mathbf{n}) + \xi) \quad \text{with } \xi \text{ sampled uniformly in } (0, 1).$$

This algorithm leads to exact ground state energies (to within statistical fluctuations) and is free from any bias arising from the finite elementary time step ($\Delta\tau$).

However, this primitive implementation suffers from two serious drawbacks. Firstly, for a generic Hamiltonian, the matrix elements $\langle \mathbf{n}|H|\mathbf{m}\rangle$ may have arbitrary sign. This fact prevents the interpretation $p(\mathbf{m}, \mathbf{n})$ (which can now be negative) as probabilities and the algorithm, thus, breaks down.

Another issue with this simplistic approach is that, as the number of states and/or particles increases, the upper bound (6) for $\Delta\tau$ becomes smaller and smaller. The resulting random walk will then be affected by increasingly long autocorrelations, resulting in a dramatic reduction in efficiency.

2.1 Importance Sampling and the Fixed-Node Approximation

The "sign problem" resulting from the undefined sign of the matrix elements of a generic Hamiltonian cannot be circumvented for any interaction including a diverging repulsive term (in particular for Coulomb systems), which leads to arbitrarily large and negative matrix elements of the propagator $\mathcal{P}$.

It is, however, possible to perform stable calculations by implementing an approximate algorithm similar to the fixed-node approximation in standard DMC. As in standard DMC, we need to rely on an importance function $|\Phi_G\rangle$ which in general includes dynamical correlations. Notice that the importance function needs to be defined as its expansion in the Hilbert space we are using. For later convenience, we introduce here a "sign" function:

$$\mathfrak{s}(\mathbf{m},\mathbf{n}) = \mathrm{sgn}[\langle\Phi_G|\mathbf{m}\rangle\langle\mathbf{m}|H|\mathbf{n}\rangle\langle\mathbf{n}|\Phi_G\rangle]. \tag{10}$$

We then define a family of parameterized Hamiltonians $\mathcal{H}_\gamma$. The off diagonal matrix elements ($\mathbf{n}\neq\mathbf{m}$), are given by:

$$\langle\mathbf{m}|\mathcal{H}_\gamma|\mathbf{n}\rangle = \begin{cases} -\gamma\langle\mathbf{m}|H|\mathbf{n}\rangle & \mathfrak{s}(\mathbf{m},\mathbf{n})>0 \\ \langle\mathbf{m}|H|\mathbf{n}\rangle & \text{otherwise} \end{cases}, \tag{11}$$

while the diagonal elements are:

$$\langle\mathbf{n}|\mathcal{H}_\gamma|\mathbf{n}\rangle = \langle\mathbf{n}|H|\mathbf{n}\rangle + (1+\gamma)\sum_{\substack{\mathbf{n}\neq\mathbf{m}\\ \mathfrak{s}(\mathbf{m},\mathbf{n})>0}}\langle\mathbf{m}|H|\mathbf{n}\rangle \tag{12}$$

It can be easily seen that for $\gamma=-1$ the original Hamiltonian is recovered. For any $\gamma\geq 0$, the propagator

$$\langle\mathbf{n}|\mathcal{P}_\gamma|\mathbf{n}\rangle = 1 - \Delta\tau\frac{\langle\Phi_G|\mathbf{m}\rangle}{\langle\mathbf{n}|\Phi_G\rangle}\langle\mathbf{m}|\mathcal{H}_\gamma - E_T|\mathbf{n}\rangle \tag{13}$$

is free from the sign problem, and the ground state of $\mathcal{H}_\gamma$ is an upper bound of the exact ground state energy of H (see, e.g., Refs. 9,10). Moreover, the expectation value of $\mathcal{H}_\gamma$ over its projected ground state is lower than the expectation of the original Hamiltonian over $|\Phi_G\rangle$. This procedure is essentially equivalent to the fixed-node approximation in standard Diffusion Monte Carlo.

Energies can be easily estimated for any positive value of γ via the usual "mixed" estimate, exploiting the hermiticity of the Hamiltonian:

$$\mathcal{E}_\gamma = \frac{\sum_{\mathbf{n}}\mathcal{E}_\gamma^L(\mathbf{n})\Phi_G(\mathbf{n})\Psi_\gamma(\mathbf{n})}{\sum_{\mathbf{n}}\Phi_G(\mathbf{n})\Psi_\gamma(\mathbf{n})} \tag{14}$$

where $\mathcal{E}_\gamma^L$ is the local energy defined by:

$$\mathcal{E}_\gamma^L = \frac{\langle\Phi_G|\mathcal{H}_\gamma|\mathbf{n}\rangle}{\langle\Phi_G|\mathbf{n}\rangle} = \frac{\langle\Phi_G|H|\mathbf{n}\rangle}{\langle\Phi_G|\mathbf{n}\rangle}. \tag{15}$$

Equation (14) is numerically evaluated by adding up the local energy from the samples generated in the random walk:

$$\mathcal{E}_\gamma \sim \frac{1}{N_E}\sum_{i=1}^{N_E}\mathcal{E}_\gamma^L(\mathbf{n}_i). \tag{16}$$

It also possible to extrapolate the energy to $\gamma = -1$ in order to obtain an even tighter upper bound. This can easily be achieved, for instance, by computing the energy for $\gamma = 1$ and $\gamma = 0$ and evaluating:

$$E_{\mathrm{CIMC}} = 2\mathcal{E}_{\gamma=0} - \mathcal{E}_{\gamma=1}. \tag{17}$$

This is the standard energy estimation we define for a CIMC calculation.

2.2 Sampling an Exponential Propagator

We cope with the second problem by using the algorithm firstly introduced by Trivedi and Ceperley,[11] which has some similarities with the Domains Green's Function Monte Carlo by Kalos, Levesque, and Verlet.[12]

The basic idea is to use the limiting case discussed in Eq. (4). The probability of making K consecutive diagonal moves ($\mathbf{n} \rightarrow \mathbf{n}$) is given by,

$$P^K_{\mathrm{diag}}(\mathbf{n}) = p(\mathbf{n},\mathbf{n})^K \tag{18}$$

which in the limit $\Delta\tau \rightarrow 0$ goes to

$$P^K_{\mathrm{diag}}(\mathbf{n}) \rightarrow e^{-K\Delta\tau \sum_{\mathbf{m}\neq\mathbf{n}} \langle \mathbf{m}|H-E_T|\mathbf{n}\rangle} \tag{19}$$

Therefore, the time elapsed between consecutive off-diagonal moves ($K\Delta\tau$) follows a truncated exponential distribution (the distribution is truncated because $K \leq M$).

In practice, one sets a target time step $\delta\tau$ and samples a time $\delta\tau_{od}$ for the off-diagonal move. If $\delta\tau < \delta\tau_{od}$ no off-diagonal move happened for the entire time span of $\delta\tau$ and the state gets reweighted by $e^{-\delta\tau \sum_{\mathbf{m}\neq\mathbf{n}} \langle \mathbf{m}|H-E_T|\mathbf{n}\rangle}$. If instead, we have $\delta\tau > \delta\tau_{od}$ an off-diagonal move is performed selecting a new state $|\mathbf{n}'\rangle$ from the probability

$$\frac{\langle \mathbf{n}'|H-E_T|\mathbf{n}\rangle}{\sum_{\mathbf{m}\neq\mathbf{n}} \langle \mathbf{m}|H-E_T|\mathbf{n}\rangle} \tag{20}$$

This process is then repeated by setting $\delta\tau \leftarrow \delta\tau - \delta\tau_{od}$ and $\mathbf{n} \leftarrow \mathbf{n}'$ until $\delta\tau < \delta\tau_{od}$. This procedure defined above exactly samples:

$$\langle \mathbf{m}|\mathcal{P}|\mathbf{n}\rangle = \langle \mathbf{m}|e^{-\Delta\tau(H-E_T)}|\mathbf{n}\rangle \tag{21}$$

without any time step error.

2.3 Use of Coupled Cluster Wave Functions

As in the standard fixed-node DMC, the choice of importance function is crucial in order to obtain accurate estimates for the energy and other observables. In this framework an excellent choice for the importance function is given by the wave function generated in a Coupled Cluster (CC) calculation. The Coupled Cluster procedure generates a correlated state $|\mathrm{CC}\rangle$ starting from the Hartree–Fock solution of the problem $|\Phi_{\mathrm{HF}}\rangle$:

$$|\mathrm{CC}\rangle = e^{\hat{T}}|\Phi_{\mathrm{HF}}\rangle \tag{22}$$

The operator $\hat{T}$ is hierarchically divided as $\hat{T} = \hat{T}_1 + \hat{T}_2 + \hat{T}_3 + \cdots$, where:

$$\hat{T}_1 = \sum_{ia} t_i^a a_a^\dagger a_i \quad \hat{T}_2 = \frac{1}{4}\sum_{ijab} t_{ij}^{ab} a_a^\dagger a_b^\dagger a_i a_j \quad \cdots \tag{23}$$

In order to use the CC wave function in our calculations, we need an efficient evaluation of its overlap with the an arbitrary element of the Hilbert space we are employing. For a many-fermion calculation, the elements may be Slater determinants describing an M particle-M hole state

$$|\mathbf{m}\rangle = a_{p_1}^\dagger, \ldots, a_{p_M}^\dagger a_{h_1}, \ldots, a_{h_M}|\Phi_{\mathrm{HF}}\rangle \equiv |\Phi_{h_1,\ldots,h_M}^{p_1,\ldots,p_M}\rangle. \tag{24}$$

Let us consider for simplicity the wave function for a homogeneous system (for which the lowest order is coupled cluster doubles or CCD due to translational invariance). The required amplitude can be expressed in turn as a superposition of $M-2$ particle/hole states that can be generated from $\mathbf{m}$. Eventually (the proof is tedious but straightforward) one obtains:

$$\langle \mathbf{m}|\Phi_{\mathrm{CCD}}\rangle = \sum_{\gamma=2}^{M}\sum_{\mu<\nu}^{M}(-1)^{\gamma+\mu+\nu} t_{h_1,h_\gamma}^{p_\mu p_\nu} \Phi_{\mathrm{CCD}}^{M-2}\begin{pmatrix} p_1,p_2,\ldots,p_{\mu-1},p_{\mu+1},\ldots,p_{\nu-1},p_{\nu+1},\ldots,p_M \\ h_2\ldots,h_{\gamma-1},h_{\gamma+1},\ldots,h_M \end{pmatrix} \tag{25}$$

assuming $p_1 < p_2 < \cdots < p_M$ and $h_1 < h_2 < \cdots < h_M$. The normalization is fixed in such a way that $\langle\Phi_{\mathrm{HF}}|\Phi_{\mathrm{CCD}}\rangle = 1$. The extension to singlets (ph states) and triplets (3p-3h states) is straightforward.

In the CC approach, the coefficients t_{ij}^{ab} appearing in the previous equations are to be obtained as the self-consistent solutions of the following nonlinear equation:

$$\langle\Phi_{ij}^{ab}|\hat{H}\left(1+\hat{T}_2+\frac{1}{2}\hat{T}_2^2\right)|\Phi_{\mathrm{HF}}\rangle = \left(\frac{1}{4}\sum_{ijab}\langle ij||ab\rangle t_{ij}^{ab}\right) t_{ij}^{ab} \tag{26}$$

where $\langle ij||ab\rangle \equiv V_{ijab} - V_{ijba}$ are the antisymmetrized two-body matrix elements of the interaction (cf. Eq. (1)).

This procedure is, however, computationally very expensive and in the fixed-node approach all that matters are the signs in Eq. (10). It is then possible to find cheaper approximate ways to determine the doubles coefficients t_{ij}^{ab} while still preserving a good quality in the fixed-node approximation. In the following we explore the use of an importance function, dubbed CCD (1) in what follows, where the coefficients are obtained within second order Moeller–Plesset perturbation theory:

$$t_{ij}^{ab} = \frac{\langle ij||ab\rangle}{\eta_i + \eta_j - \eta_a - \eta_b} \quad \text{with} \quad \eta_i = \epsilon_i + \sum_k \langle ik||ik\rangle. \tag{27}$$

This is equivalent to truncating the self-consistent solution of (26) after the first iteration.

3. RESULTS IN HOMOGENEOUS SYSTEMS

3.1 Electron Gas

The first tests of the algorithm have been performed on the homogeneous electron gas.

In this case, as for any homogeneous many-Fermion system, there is a natural choice of the Hilbert space. For N electrons at density $n = 3/4\pi r_s^3$, the Hartree–Fock solution is given by a Slater determinant of plane waves with wave-vector multiple of $2\pi/L$, where $L = r_s(4\pi N/3)^{1/3}$. The one-body part of the Hamiltonian represented in this Hilbert space will simply be the diagonal kinetic energy $\epsilon_i = \mathbf{k}_i^2/2m$, while the interaction matrix element is given by:

$$V_{ijab} \equiv \langle ij|\hat{V}|ab\rangle = (1 - \delta(\mathbf{k}_a - \mathbf{k}_i))\delta(\mathbf{k}_i + \mathbf{k}_j - \mathbf{k}_a - \mathbf{k}_b)\frac{4\pi}{L^3}\frac{1}{(\mathbf{k}_a - \mathbf{k}_i)^2} \tag{28}$$

where the first factor in parenthesis is needed to exclude the divergence at zero momentum transfer. In Table 1, we compare the results obtained by projecting the ground state from an initial fully converged CCD wave function and from the wave function obtained in MP2 perturbation theory for a system of 14 electrons at various densities.

It is very interesting to see how the projected correlation energy has a very small residual dependence on the initial wave function. As already stated in the previous section, this result suggests that it might not be

Table 1 CCD and CCD(1) Correlation Energies, Calculated Using Conventional CC Theory, Along with the Corresponding Monte Carlo Energies of a 3DEG System with $N = 14$ and $\mathcal{N}_s = 342$, for $r_s = 0.5$, 1.0, and 2.0

r_s	CCD	+ CIMC	CCD(1)	+ CIMC
0.5	−0.572682	−0.5729(3)	−0.659641	−0.5733(2)
1.0	−0.506701	−0.5021(3)	−0.657347	−0.5025(2)
2.0	−0.417946	−0.40317(2)	−0.665071	−0.4029(3)

necessary to reach the (very expensive) full convergence in a CC calculation to obtain a reasonable accuracy in the CIMC result. The sign structure of the CC wave function seems to be sufficiently well established already at the perturbation theory level. Also, notice that the perturbed result, which does not benefit from the upper bound property, is correctly brought closer to the expected correct result.

A consequence of this observation is that CIMC might represent a viable way to substantially reduce the time needed to reach results with an accuracy at least comparable to that of a CC calculation with a limited loss in precision.

In Table 2, we analyzed the dependence of the results on the system size at different densities. In this case, all calculations have been performed starting from the CCD(1) perturbed wave function.

All CIMC calculations in Table 2 have been performed at a fixed basis size with $\mathcal{N}_S = 2378$. As can be seen, the discrepancy between CIMC results and other QMC results increases with the system size. The agreement also deteriorates when decreasing the density. This disagreement is strongly related to the intrinsic quality of the CCD wave function. In fact, it should be pointed out once more that CIMC can project the ground state only within the Hilbert space chosen. In this case, it is not just the size of the basis that limits the Hilbert spac but limiting the CC calculation at the doublet level not only neglects important contributions from the overlaps with 6 particle-6 hole excitations and higher but also excludes from the explored Hilbert space excitations with an odd number of particle holes. A CIMC calculation based on a CCD(T) should compare better with standard QMC results. However, the situation might not be as simple as it appears.

In Table 3, we compare the preliminary results obtained from two sets of CIMC calculations for $N = 14$ electrons at $r_s = 3$. The first set includes contributions from triplets at perturbative level, while the second set is obtained from a fully converged CCD wave function. At a first glance,

Table 2 Correlation Energies (in a.u.) Obtained from CIMC Calculations with the Importance Function Evaluated at Second Order Perturbation Theory, CCD(1), Compared to Previous Results of Sheperd et al.[13] and Lopez Rios et al.[14]

r_s	N	CIMC	Other
0.5	14	−0.5875(6)	−0.5959(7)[13]
	38	−1.809(4)	−1.849(1)[13]
	54	−2.354(2)	−2.435(7)[13] −2.387(2)[14]
1.0	14	−0.5114(5)	−0.5316(4)[13]
	38	−1.521(4)	−1.590(1)[13]
	54	−2.053(4)	−2.124(3)[13] −2.125(2)[14]
2.0	14	−0.4103(6)	−0.444(1)[12]
	38	−1.134(7)	−1.225(2)[12]
3.0	14	−0.333(1)	

Table 3 Preliminary CIMC Results Obtained in the 3DEG Starting from a CCD(T) Wave Function

CCD(T)		
	Energy	**Error**
$K^2=5$	−0.3570	0.0016
$K^2=9$	−0.3865	0.0018
CCD		
	Energy	**Error**
$K^2=5$	−0.3538	0.0009
$K^2=9$	−0.3851	0.0020

The first column indicates the maximum momentum shell included in the single particle basis. Calculations have been performed at $r_s = 3$ for $N = 14$ electrons.

one notices two facts. The first is that the nodal structure defined by the MP2 wave function is substantially worse at such low densities. This is expected due to the increasing role of the correlations. The second fact is that the CCD(T) wave function (perturbation) provides a nodal structure that brings

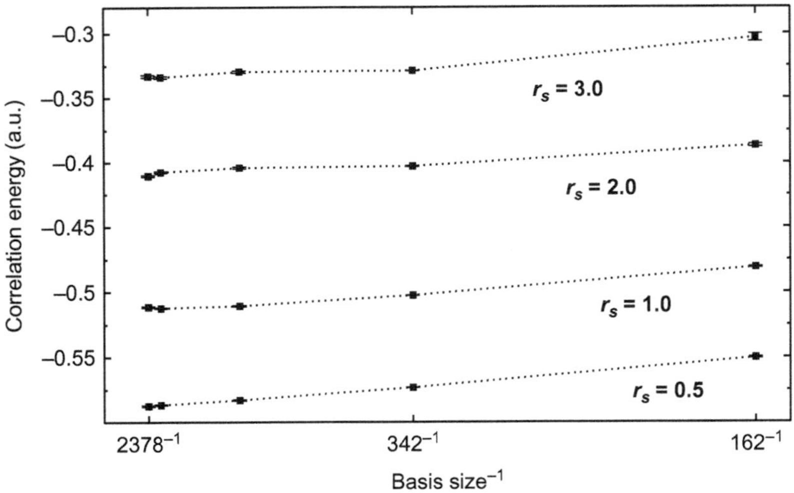

Figure 1 CIMC results at different densities as a function of the inverse basis size $[\mathcal{N}_s]^{-1}$. Extrapolation for $\mathcal{N}_s \to \infty$ is nontrivial for all the cases analyzed. Calculations have been performed for $N = 14$ electrons.

the result essentially at the level of the CCD wave function one, but not much further. Differences in the correlation energy are of order 1%, to be compared with the difference of order 3–8% between CIMC results from CCD(1) and other QMC calculations. This means that in absence of comparisons the analysis of the results has to be very careful.

Another interesting issue concerns the extrapolation in basis size. In Fig. 1, we report the results obtained in CIMC calculations for different values of r_s and for different values of $\mathcal{N}_s$.

The figure displays very clearly that there is no well defined linear extrapolation for large values of $\mathcal{N}_S$. The situation is similar for calculations we performed with $N = 32$ and 54. Thus, at least up to our largest basis size $\mathcal{N}_s = 2378$, we cannot safely do an extrapolation to $\mathcal{N}_s \to \infty$ with a reasonably low χ^2.

As previously said, one of the advantages of introducing the CIMC algorithm is that importance sampling helps to accelerate convergence. In Table 4, we compare the computational cost of our CIMC method to that of similar *i*-FCIQMC calculations by Sheperd et al.[13] for different values of r_s, N, and $\mathcal{N}_s$.

As can be seen, introduction of importance sampling reduces the computational cost by a substantial factor of ≈ 2 orders of magnitude with an efficient parallel implementation.

Table 4 Comparison of the Computational Cost of CIMC and *i*-FCIQMC[13] for Some Selected Cases

			Computational Time (Cpu Hours)	
r_s	N	$\mathcal{N}_s$	CIMC	*i*-FCIQMC
0.5	14	1850	384(0.6)	200
1.0	14	1850	768(1.6)	2500
2.0	14	1850	768(8.8)	2500
2.0	38	922	4608(–)	16000

The number in parenthesis are obtained with a more efficient recent version of the code.

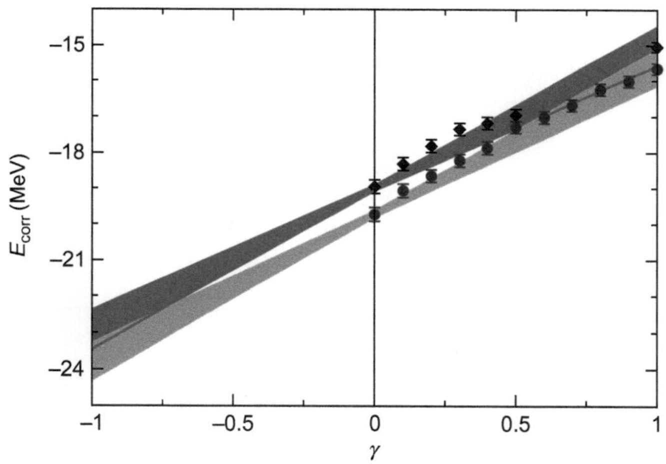

Figure 2 A CIMC calculation for 66 neutrons in a periodic box. Dark grey dots are results obtained from MP2 and black diamonds are obtained varying the MP2 coefficients by random amount (cf. (29)). The energy are given as a function of the parameter γ characterizing the fixed-node Hamiltonian.

3.2 Tests on Other Systems

In this final part, we will show some results obtained for a homogeneous system of nucleons by making use of modern nucleon–nucleon interactions which are intrinsically nonlocal, and therefore very hard to use in a standard QMC calculation. We will show two aspects that were not explored in 3DEG but which are illustrative of the quality of the method. In Fig. 2, we show the correlation energy computed for 66 neutrons in a periodic box, modeled as nonrelativistic particles interacting with the nonlocal potential $\mathrm{NNLO_{opt}}$[15] developed in the framework of Chiral Effective Field Theory (see, e.g., Ref. 16).

In this plot, we show the energies computed starting from an MP2 wave function and a wave function in which the overlap coefficients have been randomly varied

$$\tilde{t}^{ab}_{ij} = t^{ab}_{ij}(0.5 + \xi) \quad \text{with} \quad \xi \ \text{sampled uniformly in} \ (0,1) \tag{29}$$

The bands corresponds to the propagated error of the linear fit that uses the two points $\gamma = 0,1$.

The interesting fact here is that results are quite robust with respect to the variation of the coefficients of the CC calculations. Extrapolation to $\gamma = -1$ is still overlapping within error bars.

Finally, in Fig. 3, we want to show the accuracy that can be reached in computing the momentum distribution of neutrons. In general, observables that do not commute with the Hamiltonian of the system are difficult to compute accurately since local estimators available in the calculation

$$\mathcal{O}^{L}_{\text{mixed}} = \frac{\langle \Phi_G | \mathcal{O}_\gamma | \mathbf{n} \rangle}{\langle \Phi_G | \mathbf{n} \rangle} = \frac{\langle \Phi_G | O | \mathbf{n} \rangle}{\langle \Phi_G | \mathbf{n} \rangle} \tag{30}$$

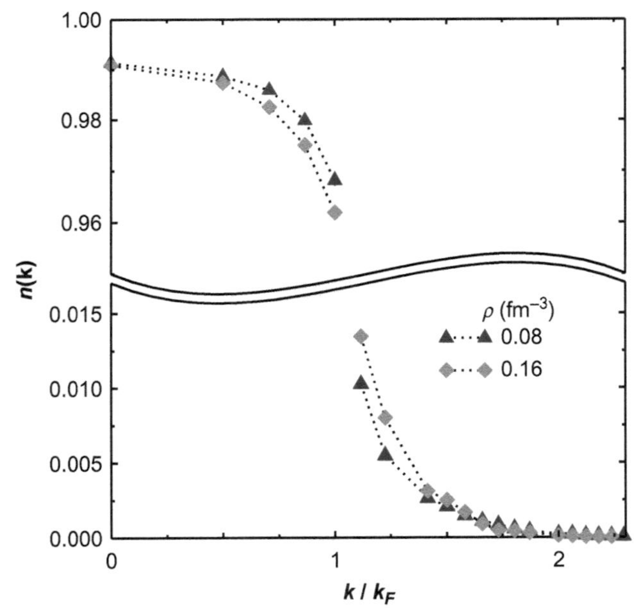

Figure 3 Momentum distribution of 66 neutrons interacting via the $NNLO_{opt}$ potential evaluated by CIMC at two densities: $\rho = 0.08\ \text{fm}^{-3}$ (dark grey triangles) and $\rho = 0.16\ \text{fm}^{-3}$ (light grey squares). Dotted lines are a guide to the eye.

inevitably depend on the chosen guiding function Φ_G. For one-body operators that are diagonal in the basis (momentum space in our case) where the random walk is performed, the unbiased expectation value can be estimated using the technique proposed by Gaudoin and Pitarke.[17] The insets in Fig. 3 show the importance of considering the unbiased estimator for the momentum distribution, in fact the mixed estimator obtained from (30) considerably overestimates the depletion at the Fermi surface. Also in this case the results have been obtained for a system of 66 neutrons interacting with the $\mathrm{NNLO_{opt}}$ interaction.

4. FIRST ROW ATOMS

We have recently extended the method to deal with finite systems. The first test concerned first row atoms with an even number of electrons. There are no major technical issues implied by such extension. Obviously, it is necessary to include one-particle one-hole excitations in $|CC\rangle$ but as already stated in Section 2 this is completely straightforward.

Another important point concerns the basis set to be used. The method is completely general, and it can be used in conjunction with any choice of the representation of the Hilbert space. In principle, we might insist on the choice of a basis diagonalizing the kinetic energy. However, it can be foreseen that convergence would in this case be quite slow. Hence, we decided to turn to a few Gaussian basis sets. In this chapter, we will include only results obtained using the cc-pVDZ, and a set that we will denote as aug-cc-pVDZ*, which is analogous to the aug-cc-pVDZ set, but only including (1s,1p) contractions (with no contractions with the d orbital). In order to perform CC and CIMC calculations all orbitals are orthonormalized.

Calculations were performed for Be, C, O, and Ne atoms. The main point of this preliminary investigation is to check how CIMC can recover CCSD results even starting from the amplitudes of a less than converged CC calculation and even starting from MP2, at the expense of some accuracy but with a substantial speedup of the calculation. As an example we report in Table 5, the amplitudes in the cc-pVDZ basis set obtained in a CCSD and in a MP2 calculation.

In Table 6, we report the correlation energy (in atomic units) for the atoms considered in our calculations. All the reported results were extrapolated to an infinite number of walkers and to $\gamma = -1$.

As can be seen, CIMC calculations starting from MP2 amplitudes and partially converged CCSD amplitudes strongly reduce the differences in

Table 5 Amplitudes for Carbon Atom Wave Function in a MP2 and CCSD Calculation Using the cc-pVDZ Basis

amplitude	MP2 value	CC value
t_1^9	−0.000419584968505	−0.000401806290974
t_1^{13}	0.000726032756716	0.000733616388008
t_2^{10}	−0.000419584968505	−0.000507490573453
t_2^{14}	0.000726032756716	0.000769255545851
$t_{1\,3}^{7\,8}$	−0.000933006312113	−0.000424639373447
$t_{1\,3}^{11\,12}$	−0.000023957069557	0.000049419081067

CC amplitudes have been taken after 10 iterations, when full convergence has not been reached yet.

Table 6 Preliminary Results for the Correlation Energies (in Hartree) for a Selected Set of First Row Atoms Computed by CIMC

basis	atom	MP2	CIMC	CCSD	CIMC
cc-pVDZ	Be	−0.0217	−0.040(1)	−0.04073	−0.04090(1)
	C	−0.03458	−0.0625(5)	−0.06474	−0.0628(3)
	O	−0.00456	−0.06741(2)	−0.06107	−0.0714(1)
	Ne	0.0895	−0.0589(2)	−0.0095	−0.0589(1)
aug-cc-pVDZ*	Be	−0.0228	−0.0437(3)	−0.0395	−0.0436(1)
	C	−0.0455	−0.0860(1)	−0.0712	−0.0791(3)
	O	−0.0393	−0.1300(3)	−0.1001	−0.1287(2)
	Ne	0.0179	−0.1900(4)	−0.0103	−0.1839(6)

Calculations were performed starting from MP2 and partially converged CCSD amplitudes.

correlation energy in all the range of atoms considered. The residual difference has to be attributed to the fixed-node error, and for these preliminary calculations it is always less than 8%. For the Be atom, the substantially exact CC value is reached in both cases. For C, O, and Ne atoms, the CCSD calculation have only partially converged, therefore it is not possible to have a straightforward comparison. All results for the cc-pVDZ are in any case in qualitative agreement with available tabulated results.[18] Particularly interesting is the fact that the correlation energy recovered projecting from the MP2 wave function is essentially 100% in Ne, with no dependence on the wave function for the cc-pVDZ basis set, and a relatively small dependence on the wave function for the augmented basis set (about 3%).

We believe that, though preliminary, these results demonstrate the feasibility of CIMC calculations also in systems of interest in quantum chemistry, and open the way to a more widespread application in conjunction with state-of-the-art Coupled Clusters calculations.

5. CONCLUSIONS

We developed a protocol, the Configuration Interaction Monte Carlo (CIMC), to combine projection Monte Carlo and the Coupled Clusters methods to perform QMC calculations in momentum space. The main ingredients that have been introduced are the Domains GF propagator, importance sampling based on Coupled Clusters wave functions and an efficient recursive algorithm to evaluate the CC overlaps. We reported some results obtained in the 3DEG. While some extrapolation issues remain, a strong reduction in the computational cost was achieved. We also showed some results proving that nonlocal Hamiltonians (in particular for nuclear physics applications) can be treated, contrarily to what happens for standard QMC. A first preliminary investigation of a selected set of first row atoms was presented. This demonstrates the applicability of the method to generalized quantum chemistry problems, in conjunction with Coupled Clusters calculations. Another interesting future direction will be the implementation of spin Hamiltonians (Hubbard model, topological insulators, but also spin propagation for nuclear Hamiltonians).

REFERENCES

1. Booth, G. H.; Thom, A. J. W.; Alavi, A. Fermion Monte Carlo without fixed nodes: A game of life, death and annihilation in Slater determinant space. *J. Chem. Phys.* **2009**, *131*, 054106.
2. Cleland, D.; Booth, G. H.; Alavi, A. Survival of the fittest: Accelerating convergence in full configuration-interaction quantum Monte Carlo. *J. Chem. Phys.* **2010**, *132*, 041103.
3. Booth, G. H.; GrDquoteuneis, A.; Kresse, G.; Alavi, A. Towards an exact description of electronic wavefunctions in real solids. *Nature* **2013**, *493*, 365.
4. Kolodrubetz, M.; Clark, B. K. FCI-QMC approach to the Fermi polaron. *Phys. Rev. B* **2012**, *86*, 075109.
5. Petruzielo, F. R.; Holmes, A. A.; Changlani, H. J.; Nightingale, M. P.; Umrigar, C. J. Semistochastic projector Monte Carlo method. *Phys. Rev. Lett.* **2012**, *109*, 230201.
6. Roggero, A.; Mukherjee, A.; Pederiva, F. Quantum Monte Carlo with coupled-cluster wave functions. *Phys. Rev. B* **2013**, *88*, 115138.
7. Roggero, A.; Mukherjee, A.; Pederiva, F. Quantum Monte Carlo calculations of neutron matter with non-local chiral interactions. *Phys. Rev. Lett.* **2014**, *112*, 221103.
8. Mukherjee, A.; Alhassid, Y. Configuration-interaction Monte Carlo method and its application to the trapped unitary Fermi gas. *Phys. Rev. A* **2013**, *88*, 053622.

9. ten Haaf, D. F. B.; van Bemmel, H. J. M.; van Leeuwen, J. M. J.; van Saarloos, W.; Ceperley, D. M. Proof for an upper bound in fixed-node Monte Carlo for lattice fermions. *Phys. Rev. B* **1995**, *51*, 13039.
10. Sorella, S.; Capriotti, L. Green function Monte Carlo with stochastic reconfiguration: An effective remedy for the sign problem. *Phys. Rev. B* **2000**, *61*, 2599.
11. Trivedi, N.; Ceperley, D. M. Ground-state correlations of quantum antiferromagnets: A Green-function Monte Carlo study. *Phys. Rev. B* **1990**, *41*, 4552.
12. Kalos, M. H.; Levesque, D.; Verlet, L. Helium at zero temperature with hard-sphere and other forces. *Phys. Rev. A* **1974**, *9*, 2178.
13. Shepherd, J. J.; Booth, G. H.; Alavi, A. Investigation of the full configuration interaction quantum Monte Carlo method using homogeneous electron gas models. *J. Chem. Phys.* **2012**, *136*, 244101.
14. Lopez Rios, P.; Ma, A.; Drummond, N. D.; Towler, M. D.; Needs, R. J. Inhomogeneous backflow transformations in quantum Monte Carlo calculations. *Phys. Rev. E* **2006**, *74*, 066701.
15. Ekström, A.; Baardsen, G.; Forssén, C.; Hagen, G.; Hjorth-Jensen, M.; Jansen, G. R.; Machleidt, R.; Nazarewicz, W.; Papenbrock, T.; Sarich, J.; Wild, S. M. Optimized chiral nucleon-nucleon interaction at next-to-next-to-leading order. *Phys. Rev. Lett.* **2013**, *110*, 192502.
16. Machleidt, R.; Entem, D. R. Chiral effective field theory and nuclear forces. *Phys. Rep.* **2011**, *503*, 1.
17. Gaudoin, R.; Pitarke, J. M. Hellman-Feynman operator sampling in diffusion Monte Carlo calculations. *Phys. Rev. Lett.* **2007**, *99*, 126406.
18. *NIST Computational Chemistry Comparison and Benchmark Database. NIST standard reference database number 101*. 2013, (Release 16a, August 2013. Editor: Russell D. Johnson III, and references therein). http://cccbdb.nist.gov/.

CHAPTER SEVENTEEN

X-Ray Constrained Wave Functions: Fundamentals and Effects of the Molecular Orbitals Localization

Alessandro Genoni[*,†,1], **Benjamin Meyer**[*,†]
[*]CNRS, Laboratoire SRSMC, UMR 7565, Vandoeuvre-lès-Nancy, France
[†]Université de Lorraine, Laboratoire SRSMC, UMR 7565, Vandoeuvre-lès-Nancy, France
[1]Corresponding author: e-mail address: alessandro.genoni@univ-lorraine.fr

Contents

Abstract

The determination of wave functions from experimental data is an appealing objective that has stimulated many researchers over the years. Among the various strategies developed to accomplish this task, the X-ray constrained wave function fitting proposed by Jayatilaka is probably the most promising and it is currently used to determine experimental electron distributions in crystals. To directly introduce an easy chemical interpretation in terms of the traditional Lewis molecular picture without resorting to *a posteriori* techniques, this approach has recently been extended in order to obtain molecular orbitals strictly localized on small molecular fragments (e.g., atoms, bonds, or functional groups). In this chapter, after reviewing in detail both the historical development of the "experimental" wave function methods and the theoretical foundations of the X-ray constrained wave function techniques, we analyze for the first time the effect of introducing an *a priori* localization of molecular orbitals in the framework of

Advances in Quantum Chemistry, Volume 73
ISSN 0065-3276
http://dx.doi.org/10.1016/bs.aiq.2015.05.008

Jayatilaka's approach. Our results mainly show that, when large and flexible basis-sets are used in the calculations, the strong initial assumption represented by the predefined localization has limited influence due to information provided by the experimental data.

1. INTRODUCTION

Two major families of quantum mechanical techniques are generally used to compute the physical properties of molecular and crystalline systems: the *ab initio* wave function-based methods and the Density Functional Theory (DFT) strategies. While the former group comprises a series of techniques able to provide wave functions corresponding to increasingly sophisticated and accurate approximations of the Schrödinger equation solutions, the latter relies on the Hohenberg and Kohn theorem[1] according to which the electron density is a crucial physical quantity allowing determination of all ground-state properties of physical systems.

However, although not very well known in the quantum chemistry community, a plausible alternative to the *ab initio* and DFT techniques exists: the approach of the experimentally constrained wave functions or density matrices. These strategies can be considered as semi-empirical methods in which an assumed form for the wave function (or the density matrix) of the systems is fitted to experimental measurements according to some well-defined procedures. There are no limitations in the choice of the experimental data to accomplish this task, even if the X-ray diffraction measurements are the most commonly used both for their general abundance and, especially, for their direct connection to the electron density. For this reason, these techniques are better known as X-ray constrained wave function strategies.

There are mainly three reasons why this theme has been an active research field for about 50 years. The most important one is the necessity to recap all the experimental observations into an object of unquestionable physical meaning. In this sense, the wave function is the best available theoretical option since in quantum mechanics it is the fundamental entity that intrinsically contains all the information about a physical system. As a consequence, unlike other arbitrary chosen model objects (e.g., the multipole expansions[2–4]), an experimentally constrained wave function also allows computation of many properties that are not directly related to those used to determine the wave function itself.

Furthermore, it is worth pointing out that extracting an "experimental" wave function can be extremely useful in all those cases in which it is not possible to perform highly accurate quantum chemistry calculations. In those situations, the available approximate wave functions may provide values of physical properties that deviate from the observed data, and therefore, directly obtaining a wave function from experimental measurements is a powerful technique at our disposal.

Finally, restricting our discussion to the case of constraints consisting in X-ray diffraction data, another crucial reason for determining "experimental" wave functions is represented by the possibility of investigating the Hohenberg and Kohn mapping between electron densities and ground-state wave functions. Of course, solving this mapping problem would be of paramount importance since it would finally allow to perform quantum mechanical calculations exploiting only the simple three-dimensional electron density function.

The first attempt to extract a wave function from experimental measurements dates back to about 50 years ago when Mukherji and Karplus,[5] using experimental values for the dipole moment and the electric field gradient of the hydrofluoric acid, proposed a perturbation approach to constrain a single Slater determinant wave function. Similar techniques were subsequently exploited to perform calculations on lithium hydride,[6,7] while Chong and Rasiel tried to extend the constraint procedure through nonperturbative strategies.[8]

Nevertheless, a milestone in this area of interest is represented by the series of pioneering papers published by Clinton and coworkers since 1969.[9–15] In those studies, they have devised original methods to mainly extract idempotent one-particle density matrices from theoretically generated density data. These techniques have been improved over the years, for example, extending the original strategy to the case of open-shell single determinant wave functions[16] or introducing a steepest descent algorithm to solve the problem of simultaneously fulfilling the idempotency conditions and the desired agreement with the experimental measurements.[17] In this context, Frishberg and Massa[16] have obtained interesting results using high-quality theoretically determined X-ray structure factors to fit the wave functions for simple atomic and molecular systems. Their main conclusion was that the determination of one-electron properties improved over variational calculations, although that was not observed for the two-electron ones. However, concerning this family of techniques, the most significant result was obtained by Massa et al.[18] extracting the idempotent density matrix of beryllium by directly exploiting experimental X-ray diffraction data for the first time.

A more recent variant of the Clinton's strategy was introduced by Howard and coworkers.[19] They resorted to a simulated annealing technique to fit single determinant wave functions of much larger systems, such as methylamine and formamide. In this context, it is also important to cite the study conducted by Snyder and Stevens[20] who obtained the experimental idempotent density matrix for the azide ion in crystalline potassium azide. In particular, they have shown that the electron deformation density maps constructed from the fitted density matrices present some features that can be found in the corresponding theoretical densities but that are not generally observable in multipole model charge distributions. Finally, in this class of strategy, another noteworthy technique is the X-ray atomic orbital method proposed by Tanaka[21,22] that mainly aims at modeling the distortion of the heavy-atom atomic orbitals due to the crystal-field effects.

Nevertheless, an intrinsic and important limitation of all the previous Clinton-like techniques is their only focus on the use of idempotent density matrices (namely, single Slater determinant wave functions). To overcome this serious drawback, Hibbs, Waller, and coworkers[23,24] have recently devised a new interesting method in which the occupation numbers of predetermined (occupied and virtual) molecular orbitals are optimized through a fitting to experimental structure factors. The MOON (Molecular Orbitals Occupation Numbers) approach is potentially linear scaling and it is also characterized by an advantageous ease of properties calculation, even if the precomputation and selection of molecular orbitals could be a serious limit for quite large systems.

In more advanced approaches, the simple idempotency requirement has been replaced by the *N*-representability conditions. As well known, these conditions guarantee that the density operators come from a valid antisymmetric wave function and, for a one-particle density matrix, they simply correspond to the fact that the density matrix eigenvalues must lie between 0 and 1. In this context, among the *N*-representable strategies a prominent role is occupied by the methods developed by Weyrich and coworkers[25–27] and by Gillet and collaborators.[28–30] In fact, these researchers have interestingly demonstrated over the years that the wealth of information about the chemical bonding effects contained in the inelastic Compton scattering data[31,32] can be efficiently used to reconstruct one-particle density matrices. Finally, it is worth mentioning the studies performed by Cassam-Chenaï[33,34] that pointed out the importance of using ensemble

N-representable density matrices. In particular, he has proposed to express these one-particle density matrices in terms of a small number of precomputed *ab initio* wave functions and then to determine a few coupling coefficients through a fitting to collected experimental data. The strategy has been applied in connection with polarized neutron diffraction data and it allowed analysis of the magnetization density for the $CoCl_4^{2-}$ ion in the Cs_3CoCl_5 crystal.

Unfortunately, even introducing *N*-representability conditions may not be enough. In fact, in 1975 Gilbert[35] showed that, in principle, an infinite number of *N*-representable one-particle density matrices can reproduce a desired electron density distribution. Therefore, fitting to density data alone, as done in all the previously described strategies, is a necessary but not sufficient criterion to guarantee the determination of physically meaningful experimentally constrained wave functions or density matrices. To overcome this drawback, Henderson and Zimmerman[36] have suggested that of all the possible single Slater determinants that are compatible with a given electron density, the optimal one is that which minimizes the Hartree–Fock energy. This idea led to another modified Clinton approach that was applied to lithium hydride using theoretical structure factors. An alternative was subsequently proposed by Levy and Goldstein[37] and by Gritsenko and Zhidomirov,[38] according to whom, when a given wave function is not univocally determined from the data, the single Slater determinant to be chosen is that simply minimizing the kinetic energy. Following this philosophy, Parr and coworkers[39–41] have devised original strategies to extract Kohn–Sham orbitals using electron density data obtained from theoretical wave functions or quantum Monte-Carlo computations. These strategies are also currently used to generate new DFT functionals.

However, supported by the Levy's constrained search formalism,[42] according to which the exact wave function is that minimizing the total energy while reproducing a given electron distribution, Jayatilaka[43–50] has quite recently proposed a more practical implementation of Henderson and Zimmerman's idea. This novel technique, which is known as X-ray constrained wave function fitting, consists in determining the single Slater determinant that minimizes the corresponding energy but that simultaneously reproduces a set of experimental structure factor amplitudes to a desired accuracy. In other words, using the Lagrange multipliers technique, the strategy variationally minimizes a new functional given by the energy associated with the single Slater determinant and an additional term represented by the statistical agreement with the experimental diffraction data.

In particular, the Lagrange multiplier is iteratively increased until the desired agreement between experimental and calculated values is reached.

The Jayatilaka approach is probably the most promising strategy among all the experimentally constrained methods developed so far. In fact, in addition to giving physically meaningful "experimental" electron distributions, the technique has also been successful in the study of chemical and physical properties such as in-crystal molecular dipole moments, in-crystal polarizabilities, and refractive indices.[51,52] Unfortunately, due to the complete delocalization of the molecular orbitals obtained, the simple chemical interpretation of the resulting charge distributions is not obvious, and for this reason, Jayatilaka and coworkers have already tried to recover it, extracting quantities such as the Electron Localization Functions (ELFs),[53] the Electron Localizability Indicators (ELIs),[54,55] and the Roby indices[56] directly from X-ray constrained wave functions.

In this context, we have recently proposed an alternative technique[57–59] to directly include the traditional Lewis picture of molecules in the X-ray constrained wave function fitting, without using *a posteriori* localization techniques or analyses. In particular, the new method enables extraction of a single Slater determinant from X-ray diffraction data, constructed with molecular orbitals strictly localized on small molecular fragments (e.g., atoms, bonds, or functional groups) and it can be considered as the combination of Jayatilaka's approach with the technique developed by Stoll et al.[60] for the *a priori* determination of Extremely Localized Molecular Orbitals (ELMOs). Our previous studies have shown that, provided good-quality crystallographic data are available, the extraction of X-ray constrained ELMOs is straightforward and that the XC-ELMO strategy is a new and potentially useful tool for the reconstruction and analysis of electron densities from experimental diffraction data.[58,59]

Concerning the ELMOs, it is worth pointing out that, unlike the traditional *a posteriori* localization techniques (e.g., Pipek-Mezey,[61] Foster-Boys,[62,63] or Edimiston-Ruedenberg[64,65]) and analyses (e.g., ELF[66] or ELI[67]), the usual chemical picture of molecules is obtained only introducing a localization scheme before performing the calculations. If, on the one hand, this user-defined prelocalization can be considered as a strong empirical bias that makes the ELMO technique less general than all the *a posteriori* methods, on the other hand the definition of a localization pattern allows obtaining molecular orbitals that are really strictly localized on molecular subunits, which make them easily transferable from one molecule to another and, consequently, suitable to be collected in proper libraries to instantaneously reconstruct wave functions or electron distributions of large systems. On the contrary, all the traditional localization procedures provide molecular orbitals that keep orthogonalization tails beyond the main localization region, which prevents a direct and harmless transferability. In fact, although small, these tails must be deleted before the transfer and this unfortunately corresponds to a nonnegligible perturbation that results in a large increase of the energy of the system.[68,69]

Although, as just mentioned, the introduction of a predefined localization is a strong empirical constraint in electronic structure calculations, the effects of using ELMOs in the framework of the X-ray constrained wave function fitting have not been studied yet. Nevertheless, shedding light on this aspect is crucial for the recently proposed XC-ELMO technique since it will allow evaluating if the experimental diffraction data are able to mitigate the bias represented by the user-defined prelocalization. Therefore, the objective of this work is dual. On the one hand, we will present in detail the basic theory and the capabilities of Jayatilaka's approach and of its recent extension into the ELMOs framework in order to further spread these original methods among the quantum chemistry community. On the other hand, we will study the effects of the *a priori* localization, on the experimentally reconstructed wave functions and electron densities, analyzing both the convergence toward the desired statistical agreement with the collected X-ray diffraction data and the main topological properties of the obtained electron distributions.

2. THEORY

In this section, the theory of the X-ray constrained wave function strategies will be described in detail. At first, after focusing on the main assumptions of Jayatilaka's approach, we will present the fundamental equations of the original X-ray constrained Hartree–Fock (XC-HF) method. Afterward, we will briefly introduce the theory of the recently developed X-ray constrained ELMO (XC-ELMO) technique, and finally, we will discuss important details strictly related to the X-ray constrained calculations.

2.1 X-Ray Constrained Wave Function Fitting: Basic Assumptions

Assuming we have molecular crystals, let us consider a fictitious crystalline system in which (a) each crystal unit does not interact with the other ones and (b) the global electron density is identical to that of the real interacting system.

The first assumption allows us to write the global wave function of the crystal as:

$$|\Psi_{\text{cryst}}\rangle = \prod_k |\Psi_k\rangle \tag{1}$$

where all the crystal-unit wave functions $|\Psi_k\rangle$ are formally identical and related to each other through the crystal symmetry operations. The choice of the functional form for $|\Psi_k\rangle$ is completely arbitrary, and it can be considered as a further assumption of the technique since it will eventually determine the type of X-ray constrained wave function (e.g., Hartree–Fock, ELMO, CI, MCSCF) that is desired.

Considering all the noninteracting units as symmetry-unique portions of the crystal unit cell, the unit-cell electron distribution can be simply computed as sum of N_m crystal-unit charge densities $\rho_k(\mathbf{r})$ that are related to the reference electron density $\rho_0(\mathbf{r})$ through the unit-cell symmetry operations $\{\mathbf{R}_k, \mathbf{r}_k\}$:

$$\rho_{\text{cell}}(\mathbf{r}) = \sum_{k=1}^{N_m} \rho_k(\mathbf{r}) = \sum_{k=1}^{N_m} \rho_0\left(\mathbf{R}_k^{-1}(\mathbf{r} - \mathbf{r}_k)\right) \tag{2}$$

The charge distribution $\rho_0(\mathbf{r})$ is associated with the wave function $|\Psi_0\rangle$ for the reference unit. Following Jayatilaka, this wave function is obtained not only minimizing the energy of the reference crystal unit but also reproducing a set of experimental structure factor amplitudes. In other words, $|\Psi_0\rangle$ is the wave function that minimizes the following functional:

$$\begin{aligned} J[\Psi_0] &= \langle \Psi_0 | \hat{H}_0 | \Psi_0 \rangle + \lambda(\chi^2 - \Delta) \\ &= E_0[\Psi_0] + \lambda(\chi^2[\Psi_0] - \Delta) \end{aligned} \tag{3}$$

with $\hat{H}_0$ as the Hamiltonian operator for the reference crystal unit, λ as the Lagrange multiplier associated with the experimental constraint, Δ as the desired agreement between theoretical and experimental values, and χ^2 as the measure of the statistical agreement between the calculated and the observed structure factor amplitudes $|F_{\mathbf{h}}^{\text{calc}}|$ and $|F_{\mathbf{h}}^{\text{exp}}|$, namely:

$$\chi^2 = \frac{1}{N_r - N_p} \sum_{\mathbf{h}} \frac{\left(\eta |F_{\mathbf{h}}^{\text{calc}}| - |F_{\mathbf{h}}^{\text{exp}}|\right)^2}{\sigma_{\mathbf{h}}^2} \tag{4}$$

where N_r is the number of considered experimental scattering data, N_p is the number of adjustable parameters (namely, only the Lagrange multiplier λ), $\mathbf{h}$ is the term of Miller indices labeling the reflection, $\sigma_{\mathbf{h}}$ is the error associated with each measure, and η is an $\mathbf{h}$-independent scale factor that is determined to minimize χ^2.

It is crucial to point out that since the experimental constraint is present in the functional (3), we are able to fulfill the initial assumption that the global electron density of the fictitious noninteracting system is identical to that of the real interacting crystal. Of course, the accuracy of the assumption is strictly related to the quality and completeness of the experimental X-ray diffraction data used in the calculations.

2.2 The X-Ray Constrained Hartree–Fock Method

Since the XC-HF strategy has recently been extended also to the unrestricted case,[49] for the sake of generality, in this section we will derive the X-ray constrained Hartree–Fock equations in their spin-orbital form. To accomplish this task, consider an N-electron system and assume that the reference wave function $|\Psi_0\rangle$ is a single Slater determinant built up with a set of N spin orbitals:

$$|\Psi_0\rangle = \frac{1}{\sqrt{N!}} \hat{A}\left[\phi_1^0 \ \phi_2^0 \ \phi_3^0 \ldots \phi_N^0\right] \tag{5}$$

The XC-HF equations can be simply derived through the minimization of the functional (3) with respect to the spin orbitals, subject to the constraint of spin-orbital orthonormality. Therefore, omitting any subscripts and superscripts corresponding to the reference crystal unit, the new functional to be minimized becomes

$$L[\boldsymbol{\phi}] = E[\boldsymbol{\phi}] + \lambda(\chi^2[\boldsymbol{\phi}] - \Delta) - \sum_{i,k=1}^{N} \epsilon_{ki}(\langle \phi_i | \phi_k \rangle - \delta_{ik}) \tag{6}$$

where $\{\epsilon_{ki}\}$ constitutes a set of Lagrange multipliers and where the symbol $\boldsymbol{\phi}$ stresses the functional dependence on the spin orbitals.

Now, assuming that real spin orbitals are used and considering the arbitrary variation of $L[\boldsymbol{\phi}]$ with respect to the occupied spin orbital $|\phi_j\rangle$, we obtain

$$\delta_{(\phi_j)} L = \delta_{(\phi_j)} E + \lambda \delta_{(\phi_j)} \chi^2 - 2\sum_{k=1}^{N} \epsilon_{kj} \left\langle \delta\phi_j | \phi_k \right\rangle \tag{7}$$

where, as well known, $\delta_{(\phi_j)}E$ is

$$\delta_{(\phi_j)}E = 2\left\langle \delta\phi_j|\hat{F}|\phi_j\right\rangle \tag{8}$$

with $\hat{F}$ as the usual Fock operator for the spin-orbital case, while, exploiting the relation $\left|F_{\mathbf{h}}^{\text{calc}}\right| = \left[F_{\mathbf{h}}^{\text{calc}}\left(F_{\mathbf{h}}^{\text{calc}}\right)^*\right]^{\frac{1}{2}}$, $\delta_{(\phi_j)}\chi^2$ is

$$\delta_{(\phi_j)}\chi^2 = \frac{\eta}{N_r - N_p} \sum\nolimits_{\mathbf{h}} \frac{\eta\left|F_{\mathbf{h}}^{\text{calc}}\right| - \left|F_{\mathbf{h}}^{\text{exp}}\right|}{\sigma_{\mathbf{h}}^2\left|F_{\mathbf{h}}^{\text{calc}}\right|}\left\{\left(F_{\mathbf{h}}^{\text{calc}}\right)^*\delta_{(\phi_j)}F_{\mathbf{h}}^{\text{calc}} + F_{\mathbf{h}}^{\text{calc}}\delta_{(\phi_j)}\left(F_{\mathbf{h}}^{\text{calc}}\right)^*\right\} \tag{9}$$

Since the structure factors are Fourier transforms of the unit-cell electron density and given that the structure factor operator is defined as

$$\hat{I}_{\mathbf{h}} = \sum_{k=1}^{N_m} e^{\,i2\pi(\mathbf{R}_k\mathbf{r}+\mathbf{r}_k)\cdot(\mathbf{Bh})} = \hat{I}_{\mathbf{h},R} + i\ \hat{I}_{\mathbf{h},C} \tag{10}$$

where $\mathbf{B}$ is the reciprocal lattice matrix and both $\hat{I}_{\mathbf{h},R}$ and $\hat{I}_{\mathbf{h},C}$ (real and imaginary part of $\hat{I}_{\mathbf{h}}$, respectively) are Hermitian, we get

$$F_{\mathbf{h}}^{\text{calc}} = \sum\nolimits_{i=1}^{N}\langle\phi_i|\hat{I}_{\mathbf{h}}|\phi_i\rangle \tag{11}$$

and

$$\left(F_{\mathbf{h}}^{\text{calc}}\right)^* = \sum\nolimits_{i=1}^{N}\left\langle\phi_i\middle|\hat{I}_{\mathbf{h}}^{\dagger}\middle|\phi_i\right\rangle \tag{12}$$

Furthermore, bearing in mind that we are using real spin orbitals, from Eqs. (11) and (12), it is easy to obtain

$$\delta_{(\phi_j)}F_{\mathbf{h}}^{\text{calc}} = 2\left\langle\delta\phi_j\middle|\hat{I}_{\mathbf{h}}\middle|\phi_j\right\rangle \tag{13}$$

and

$$\delta_{(\phi_j)}\left(F_{\mathbf{h}}^{\text{calc}}\right)^* = \left(\delta_{(\phi_j)}F_{\mathbf{h}}^{\text{calc}}\right)^* \tag{14}$$

Using Eq. (14) in (9), $\delta_{(\phi_j)}\chi^2$ becomes

$$\delta_{(\phi_j)}\chi^2 = \frac{2\eta}{N_r - N_p} \sum\nolimits_{\mathbf{h}} \frac{\eta\left|F_{\mathbf{h}}^{\text{calc}}\right| - \left|F_{\mathbf{h}}^{\text{exp}}\right|}{\sigma_{\mathbf{h}}^2\left|F_{\mathbf{h}}^{\text{calc}}\right|}\ \text{Re}\left\{\left(F_{\mathbf{h}}^{\text{calc}}\right)^*\delta_{(\phi_j)}F_{\mathbf{h}}^{\text{calc}}\right\} \tag{15}$$

and, exploiting the definition (10), relation (13) can be written as

$$\delta_{(\phi_j)} F_{\mathbf{h}}^{\text{calc}} = 2 \left[\left\langle \delta\phi_j \middle| \hat{I}_{\mathbf{h},R} \middle| \phi_j \right\rangle + \mathrm{i} \left\langle \delta\phi_j \middle| \hat{I}_{\mathbf{h},C} \middle| \phi_j \right\rangle \right] \tag{16}$$

Now, substituting Eq. (16) into (15), we have

$$\begin{aligned} \delta_{(\phi_j)} \chi^2 = 2 \Big\{ & \sum\nolimits_{\mathbf{h}} K_{\mathbf{h}} \mathrm{Re}\{F_{\mathbf{h}}^{\text{calc}}\} \left\langle \delta\phi_j \middle| \hat{I}_{\mathbf{h},R} \middle| \phi_j \right\rangle \\ & + \sum\nolimits_{\mathbf{h}} K_{\mathbf{h}} \mathrm{Im}\{F_{\mathbf{h}}^{\text{calc}}\} \left\langle \delta\phi_j \middle| \hat{I}_{\mathbf{h},C} \middle| \phi_j \right\rangle \Big\} \end{aligned} \tag{17}$$

with

$$K_{\mathbf{h}} = \frac{2\eta}{N_r - N_p} \quad \frac{\eta \left| F_{\mathbf{h}}^{\text{calc}} \right| - \left| F_{\mathbf{h}}^{\text{exp}} \right|}{\sigma_{\mathbf{h}}^2 \left| F_{\mathbf{h}}^{\text{calc}} \right|} \tag{18}$$

and the arbitrary variation of the functional L given by Eq. (7) becomes

$$\begin{aligned} \delta_{(\phi_j)} L = 2 \Bigg\{ & \left\langle \delta\phi_j \,|\, \hat{F} \,|\, \phi_j \right\rangle \\ & + \lambda \sum\nolimits_{\mathbf{h}} K_{\mathbf{h}} \ \mathrm{Re}\left\{ F_{\mathbf{h}}^{\text{calc}} \right\} \left\langle \delta\phi_j \middle| \hat{I}_{\mathbf{h},R} \middle| \phi_j \right\rangle \\ & + \lambda \sum\nolimits_{\mathbf{h}} K_{\mathbf{h}} \ \mathrm{Im}\left\{ F_{\mathbf{h}}^{\text{calc}} \right\} \left\langle \delta\phi_j \middle| \hat{I}_{\mathbf{h},C} \middle| \phi_j \right\rangle \\ & - \sum_{k=1}^{N} \epsilon_{kj} \left\langle \delta\phi_j \middle| \phi_k \right\rangle \Bigg\} \end{aligned} \tag{19}$$

Since $\left\langle \delta\phi_j \right|$ is arbitrary, the minimum of L, which corresponds to $\delta_{(\phi_j)} L = 0$, is reached when the following relation is fulfilled for each j:

$$\hat{F}^{\text{exp}} \ \left| \phi_j \right\rangle = \sum\nolimits_{k=1}^{N} \epsilon_{kj} \left| \phi_k \right\rangle \tag{20}$$

where the new "experimental" Fock operator, which takes into account the constraint of the experimental structure factor amplitudes, is given by

$$\hat{F}^{\text{exp}} = \hat{F} + \lambda \sum\nolimits_{\mathbf{h}} K_{\mathbf{h}} \mathrm{Re}\{F_{\mathbf{h}}^{\text{calc}}\} \hat{I}_{\mathbf{h},R} + \lambda \sum\nolimits_{\mathbf{h}} K_{\mathbf{h}} \ \mathrm{Im}\{F_{\mathbf{h}}^{\text{calc}}\} \hat{I}_{\mathbf{h},C} \tag{21}$$

Finally, performing a unitary transformation of the spin orbitals among themselves and considering the invariance of $\hat{F}^{\text{exp}}$ to that transformation, Eq. (20) transforms into the canonical X-ray constrained Hartree–Fock equations:

$$\hat{F}^{\exp}\left|\phi'_j\right\rangle = \epsilon'_j \left|\phi'_j\right\rangle \tag{22}$$

2.3 X-Ray Constrained ELMOs

Unlike the XC-HF method, the more recent X-ray constrained ELMO strategy has been developed only in the restricted case. For this reason, from now on, we will assume that each crystal unit is a $2N$-electron closed-shell system described by an ELMO wave function, namely, by a single Slater determinant built up with ELMOs.

For the sake of simplicity, let us consider the reference crystal unit and let us subdivide it into f fragments (e.g., atoms, bonds, or functional groups) that may overlap. Due to this partition, each subunit is associated with a local basis-set that is constituted by the only basis functions centered on the atoms belonging to the fragment. Therefore, if $\left\{\left|\chi^0_{i\mu}\right\rangle\right\}_{\mu=1}^{M_i^0}$ is the set of atomic orbitals for the ith subunit, the αth ELMO of the fragment can be written like this

$$\left|\varphi^0_{i\alpha}\right\rangle = \sum_{\mu=1}^{M_i^0} C^0_{i\mu,\,i\alpha} \left|\chi^0_{i\mu}\right\rangle \tag{23}$$

and the ELMO wave function of the reference unit as

$$\left|\Psi^0_{\mathrm{ELMO}}\right\rangle = \frac{1}{\det[\mathbf{S}]\sqrt{(2N)!}} \hat{A}\left[\prod_{i=1}^{f}\prod_{\alpha=1}^{n_i}\varphi^0_{i\alpha}\ \overline{\varphi}^0_{i\alpha}\right] \tag{24}$$

where $\mathbf{S}$ is the overlap matrix between the occupied ELMOs, $\hat{A}$ is the antisymmetrizer, n_i is the number of occupied ELMOs for the ith subunit, and $\overline{\varphi}^0_{i\alpha}$ is a spin orbital with spatial part $\varphi^0_{i\alpha}$ and spin part β. The overlap matrix arises from the fact that ELMOs localized on different subunits are generally nonorthogonal due to the partial sharing of basis functions between predefined overlapping fragments.

Now, in order to obtain the equations for the XC-ELMO technique, let us minimize the functional (3) with respect to the occupied ELMOs. In particular, let us consider the variation of J associated with the arbitrary variation of the occupied ELMO $\left|\varphi^0_{j\beta}\right\rangle$. Omitting all the subscripts and superscripts corresponding to the reference unit, we obtain

$$\delta_{(\varphi_{j\beta})}J = \delta_{(\varphi_{j\beta})}E + \lambda \ \delta_{(\varphi_{j\beta})}\chi^2 \tag{25}$$

where, similar to the XC-HF equations, we get

$$\delta_{(\varphi_{j\beta})}\chi^2 = \frac{\eta}{N_r - N_p} \ \sum\nolimits_{\mathbf{h}} \frac{\eta \left|F_{\mathbf{h}}^{\text{calc}}\right| - \left|F_{\mathbf{h}}^{\text{exp}}\right|}{\sigma_{\mathbf{h}}^2 \left|F_{\mathbf{h}}^{\text{calc}}\right|} \left\{ \left(F_{\mathbf{h}}^{\text{calc}}\right)^* \delta_{(\varphi_{j\beta})} F_{\mathbf{h}}^{\text{calc}} + F_{\mathbf{h}}^{\text{calc}} \delta_{(\varphi_{j\beta})} \left(F_{\mathbf{h}}^{\text{calc}}\right)^* \right\} \tag{26}$$

while, following Stoll et al.,[60] we have

$$\delta_{(\varphi_{j\beta})}E = 4\left\langle \delta\varphi_{j\beta} \middle| (1 - \hat{\rho})\hat{F} \middle| \widetilde{\varphi}_{j\beta} \right\rangle \tag{27}$$

with the dual orbital $|\widetilde{\varphi}_{j\beta}\rangle$ defined as

$$|\widetilde{\varphi}_{j\beta}\rangle = \sum_{k=1}^{f} \sum_{\gamma=1}^{n_k} [\mathbf{S}^{-1}]_{k\gamma,\, j\beta} |\varphi_{k\gamma}\rangle \tag{28}$$

and the global density operator $\hat{\rho}$ as

$$\hat{\rho} = \sum_{j=1}^{f} \sum_{\beta=1}^{n_j} \ |\widetilde{\varphi}_{j\beta}\rangle\langle\varphi_{j\beta}| = \sum_{j=1}^{f} \sum_{\beta=1}^{n_j} |\varphi_{j\beta}\rangle\langle\widetilde{\varphi}_{j\beta}|. \tag{29}$$

As already shown in our previous studies,[57,58] through a simple mathematical derivation that exploits the definition (10), we obtain

$$\begin{aligned}\delta_{(\varphi_{j\beta})}J = 4\Big\{ &\left\langle \delta\varphi_{j\beta} \middle| (1 - \hat{\rho})\hat{F} \middle| \widetilde{\varphi}_{j\beta} \right\rangle \\ &+ \lambda \sum\nolimits_{\mathbf{h}} K_{\mathbf{h}} \ \text{Re}\left\{F_{\mathbf{h}}^{\text{calc}}\right\} \left\langle \delta\varphi_{j\beta} \middle| (1 - \hat{\rho})\hat{I}_{\mathbf{h},R} \middle| \widetilde{\varphi}_{j\beta} \right\rangle \\ &+ \lambda \sum\nolimits_{\mathbf{h}} K_{\mathbf{h}} \ \text{Im}\left\{F_{\mathbf{h}}^{\text{calc}}\right\} \left\langle \delta\varphi_{j\beta} \middle| (1 - \hat{\rho})\hat{I}_{\mathbf{h},C} \middle| \widetilde{\varphi}_{j\beta} \right\rangle \Big\}\end{aligned} \tag{30}$$

with $K_{\mathbf{h}}$ given by Eq. (18). Furthermore, since in this case $\left\langle \delta\varphi_{j\beta} \right|$ is also arbitrary, the functional minimum (corresponding to $\delta_{(\varphi_{j\beta})}J = 0$) is reached when, for all j and β, we have

$$\left\{(1-\hat{\rho})\hat{F}+\lambda\ \sum_{\mathbf{h}}K_{\mathbf{h}}\ \mathrm{Re}\left\{F_{\mathbf{h}}^{\mathrm{calc}}\right\}(1-\hat{\rho})\hat{I}_{\mathbf{h},R}\right.$$
$$\left.+\lambda\sum_{\mathbf{h}}K_{\mathbf{h}}\ \mathrm{Im}\left\{F_{\mathbf{h}}^{\mathrm{calc}}\right\}(1-\hat{\rho})\hat{I}_{\mathbf{h},C}\right\}\left|\widetilde{\varphi}_{j\beta}\right\rangle=0 \tag{31}$$

Now, considering the definition of fragment density operator

$$\hat{\rho}_j=\sum_{\beta=1}^{n_j}\ |\widetilde{\varphi}_{j\beta}\rangle\langle\varphi_{j\beta}| \tag{32}$$

and proceeding as in our original paper,[57] we add

$$|Q\rangle=\left\{\hat{\rho}_j^{\dagger}\hat{F}+\lambda\sum_{\mathbf{h}}K_{\mathbf{h}}\mathrm{Re}\left\{F_{\mathbf{h}}^{\mathrm{calc}}\right\}\hat{\rho}_j^{\dagger}\hat{I}_{\mathbf{h},R}+\lambda\sum_{\mathbf{h}}K_{\mathbf{h}}\mathrm{Im}\left\{F_{\mathbf{h}}^{\mathrm{calc}}\right\}\hat{\rho}_j^{\dagger}\hat{I}_{\mathbf{h},C}\right\}\left|\widetilde{\varphi}_{j\beta}\right\rangle \tag{33}$$

to both sides of Eq. (31) and we obtain

$$\hat{F}^{j,\exp}\left|\varphi_{j\beta}\right\rangle=\sum_{\gamma=1}^{n_j}\left\langle\varphi_{j\gamma}\left|\hat{F}^{j,\exp}\right|\varphi_{j\beta}\right\rangle\left|\varphi_{j\gamma}\right\rangle \tag{34}$$

where $\hat{F}^{j,\exp}$ is the modified "experimental" Fock operator for the fragment j:

$$\hat{F}^{j,\exp}=\hat{F}^{j}+\lambda\sum_{\mathbf{h}}\ \hat{I}^{j}_{\mathbf{h},R}+\hat{I}^{j}_{\mathbf{h},C} \tag{35}$$

with

$$\hat{F}^{j}=(1-\hat{\rho}+\hat{\rho}_j^{\dagger})\hat{F}(1-\hat{\rho}+\hat{\rho}_j) \tag{36}$$

$$\hat{I}^{j}_{\mathbf{h},R}=K_{\mathbf{h}}\ \mathrm{Re}\left\{F_{\mathbf{h}}^{\mathrm{calc}}\right\}(1-\hat{\rho}+\hat{\rho}_j^{\dagger})\hat{I}_{\mathbf{h},R}(1-\hat{\rho}+\hat{\rho}_j) \tag{37}$$

$$\hat{I}^{j}_{\mathbf{h},C}=K_{\mathbf{h}}\ \mathrm{Im}\left\{F_{\mathbf{h}}^{\mathrm{calc}}\right\}(1-\hat{\rho}+\hat{\rho}_j^{\dagger})\hat{I}_{\mathbf{h},C}(1-\hat{\rho}+\hat{\rho}_j) \tag{38}$$

Finally, given the invariance of the $\hat{F}^{j,\exp}$ operator to a transformation that mixes only the occupied ELMOs of each fragment among themselves, Eq. (34) can be rewritten in the following canonical form:

$$\hat{F}^{j,\exp}\left|\varphi'_{j\beta}\right\rangle=\epsilon_{j\beta}\left|\varphi'_{j\beta}\right\rangle \tag{39}$$

The XC-ELMO equations (39) are solved self-consistently for all the fragments. However, although solved separately, these equations are coupled due to the presence of the global density operator $\hat{\rho}$ in the definition of $\hat{F}^{j,\exp}$. Finally, note that in case of convergence problems associated with the intrinsic nonorthogonality of the ELMOs,[60,70,71] we have also devised an alternative strategy for the direct minimization of the functional J. This strategy relies on a quasi-Newton procedure that directly exploits first- and second-order derivatives of J with respect to the XC-ELMO coefficients.[57,58,71]

2.4 Details About the X-Ray Constrained Calculations

We point out that the set of theoretical structure factor amplitudes $\left\{\left|F_{\mathbf{h}}^{\mathrm{calc}}\right|\right\}$ is obtained taking into account the effects of the nuclear thermal motion. In particular, as already observed in Jayatilaka's and our previous work,[43,44,57,58] since the structure factors are Fourier transforms of the unit-cell electron distribution, we can write

$$F_{\mathbf{h}}^{\mathrm{calc}} = \mathrm{Tr}[\boldsymbol{\rho}_0 \mathbf{I}_{\mathbf{h}}] \tag{40}$$

with $\boldsymbol{\rho}_0$ as the one-particle LCAO density matrix corresponding to the wave function of the reference crystal unit and with $\mathbf{I}_{\mathbf{h}}$ as a matrix whose elements, in the atomic orbital basis, are given by this expression:

$$[\mathbf{I}_{\mathbf{h}}]_{\mu\nu} = \sum_{k=1}^{N_m} \mathrm{e}^{\mathrm{i}2\pi \mathbf{r}_k \cdot (\mathbf{Bh})} T_{\mu\nu}\left[\mathbf{B}^{-1}\mathbf{R}_k^T \mathbf{Bh}\right] \int \mathrm{d}\mathbf{r}\ \chi_\mu(\mathbf{r})\ \chi_\nu(\mathbf{r})\ \mathrm{e}^{\mathrm{i}2\pi(\mathbf{R}_k \mathbf{r})\cdot(\mathbf{Bh})} \tag{41}$$

where $T_{\mu\nu}[\mathbf{B}^{-1}\mathbf{R}_k^T \mathbf{Bh}]$ is an empirical term that can have different functional forms[21,72,73] and that depends on the thermal vibration parameters (namely, the Anisotropic Displacement Parameters (ADPs)) obtained experimentally for the atoms on which the basis functions $\chi_\mu(\mathbf{r})$ and $\chi_\nu(\mathbf{r})$ are centered.

Another important point is that, due to the experimental errors associated with the X-ray diffraction measurements, the χ^2 statistical agreement (see Eq. 4) is not usually forced to 0, but the desired accuracy Δ (see Eq. 3) is usually chosen equal to 1.0, so that the final computed structure factor amplitudes are on average within one standard deviation of the experimental data. Nevertheless, a common issue associated with the XC calculations is that they do not always rapidly converge toward $\chi^2 = 1.0$. Different criteria have been proposed and adopted to halt the X-ray constrained procedure[51,52,58] but, unfortunately, this remains an unsolved problem and an open field of research. In this chapter, we will use a strategy that we have introduced along with the XC-ELMO technique[58] and that consists in stopping the XC calculations when one of the following three conditions is fulfilled:

$$\begin{cases} \chi^2 < 1 \\ \dfrac{\chi_i^2 - \chi_{i-1}^2}{\lambda_i - \lambda_{i-1}} > -0.5 \\ \left| \dfrac{E_{\lambda_i}^{\mathrm{el}} - E_{\lambda=0}^{\mathrm{el}}}{E_{\lambda=0}^{\mathrm{el}}} \right| > 5 \times 10^{-4} \end{cases} \tag{42}$$

While it is obvious that the first condition is the traditional one, the second and the third ones take into account the incremental ratio of χ^2 with respect to the Lagrange multiplier λ and the percentage change of the electronic energy with respect to the unconstrained value, respectively. They have been proposed to prevent situations in which minimal improvements in the statistical agreement correspond to large values of the Lagrange multiplier and to large unphysical changes in the energy.

3. COMPUTATIONAL RESULTS

3.1 Preliminary Information

To show how the techniques discussed in the previous section perform and to study the effects of the molecular orbitals localization in the framework of the X-ray constrained wave function fitting, we have considered the crystallographic structure of the zwitterionic L-alanine determined by Destro et al. at 23 K.[74] In particular, using the molecular geometry obtained from the diffraction experiment (see Fig. 1), we have performed single-point calculations at Restricted Hartree–Fock (RHF), ELMO, XC-RHF, and XC-ELMO levels with six different basis-sets of increasing size and flexibility, namely the 3-21G, 6-31G(d), cc-pVDZ, 6-311G(d,p), aug-cc-pVDZ, and 6-311++G(2d,2p) basis-sets. Furthermore, for all the ELMO and XC-ELMO computations, we have adopted a localization scheme almost corresponding to the Lewis structure of the system in question. In fact, core and lone-pair electrons have been described considering atomic fragments, while almost all the bond electron pairs have been treated resorting to bond subunits; the only exceptions are represented by the σ and π electrons of the carboxylic group for which the three-atom fragment C1–O5–O6 (see Fig. 1) has been taken into account. For the "experimental" RHF and

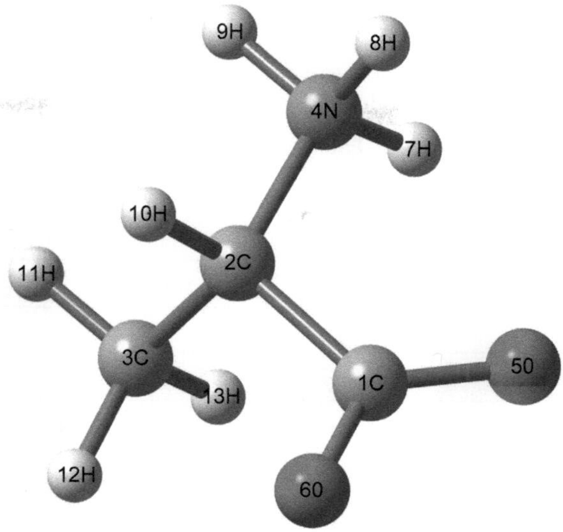

Figure 1 L-Alanine molecular geometry obtained from the X-ray diffraction experiment.

ELMO calculations, we have also exploited the unit-cell parameters, the ADPs, and the experimental structure factor amplitudes deposited by Destro and coworkers with their paper. Concerning the experimental diffraction data, which are unaffected by serious secondary extinction effects and which had been previously corrected for scan-truncation losses, note that only those characterized by $|F_{\mathbf{h}}^{\text{exp}}| > 3\sigma_{\mathbf{h}}$ have been selected for our computations, resulting in a set of 2435 structure factor amplitudes.

All the RHF and XC-RHF calculations have been performed using the TONTO package[75] developed by Jayatilaka and coworkers, while each ELMO and XC-ELMO computation has been carried out exploiting version 8 of the GAMESS-UK quantum chemistry suite of programs[76] that has been suitably modified to implement both the original ELMO technique proposed by Stoll and the new X-ray constrained ELMO approach described in the previous section. Finally, the topological analyses of all electron distributions obtained have been performed by means of version 13.11.04 of the AIMAll package.[77]

3.2 Performance and Convergence of the Methods

In Table 1, we have reported the χ^2 and energy values obtained from calculations on the L-alanine system. As expected, it is easy to observe that the X-ray constrained techniques always provide better statistical agreements

Table 1 χ^2 Agreement Statistics and Energy Values Corresponding to All the Unconstrained and Constrained Calculations Performed on the L-Alanine System

	Unconstrained Calculations[a]		X-Ray Constrained Calculations		
Method and Basis-Set	χ^2	**Energy (a.u.)**	λ_{max}	χ^2	**Energy (a.u.)**
3-21G					
RHF	4.43	−320.001	0.442	1.71	−319.841
ELMO	5.66	−319.850	0.580	2.29	−319.609
6-31G(d)					
RHF	2.53	−321.798	0.248	1.26	−321.747
ELMO	2.94	−321.696	0.320	1.30	−321.621
cc-pVDZ					
RHF	2.52	−321.831	0.240	1.23	−321.782
ELMO	3.02	−321.711	0.300	1.29	−321.640
6-311G(d,p)					
RHF	2.48	−321.891	0.230	1.21	−321.846
ELMO	2.94	−321.778	0.280	1.27	−321.712
aug-cc-pVDZ					
RHF	2.52	−321.863	0.230	1.20	−321.815
ELMO	2.74	−321.769	0.260	1.23	−321.712
6-311++G(2d,2p)					
RHF	2.47	−321.918	0.226	1.17	−321.873
ELMO	2.78	−321.830	0.280	1.19	−321.769

The λ_{max} value is shown for the constrained wave functions computations.
[a]The scale factors η have been optimized using the density matrices obtained from the unconstrained calculations.

with respect to the experimental data and that this agreement improves when larger and more flexible basis-sets are taken into account.

Nevertheless, despite the improved descriptions, even with the large 6-311++G(2d,2p) basis-set, we are unable to reach $\chi^2 = 1.0$. As already noted in our previous papers,[58,59] the difficulties in converging toward the desired agreement may be associated with errors in the collected experimental data. To better investigate this problem, we have analyzed the normalized

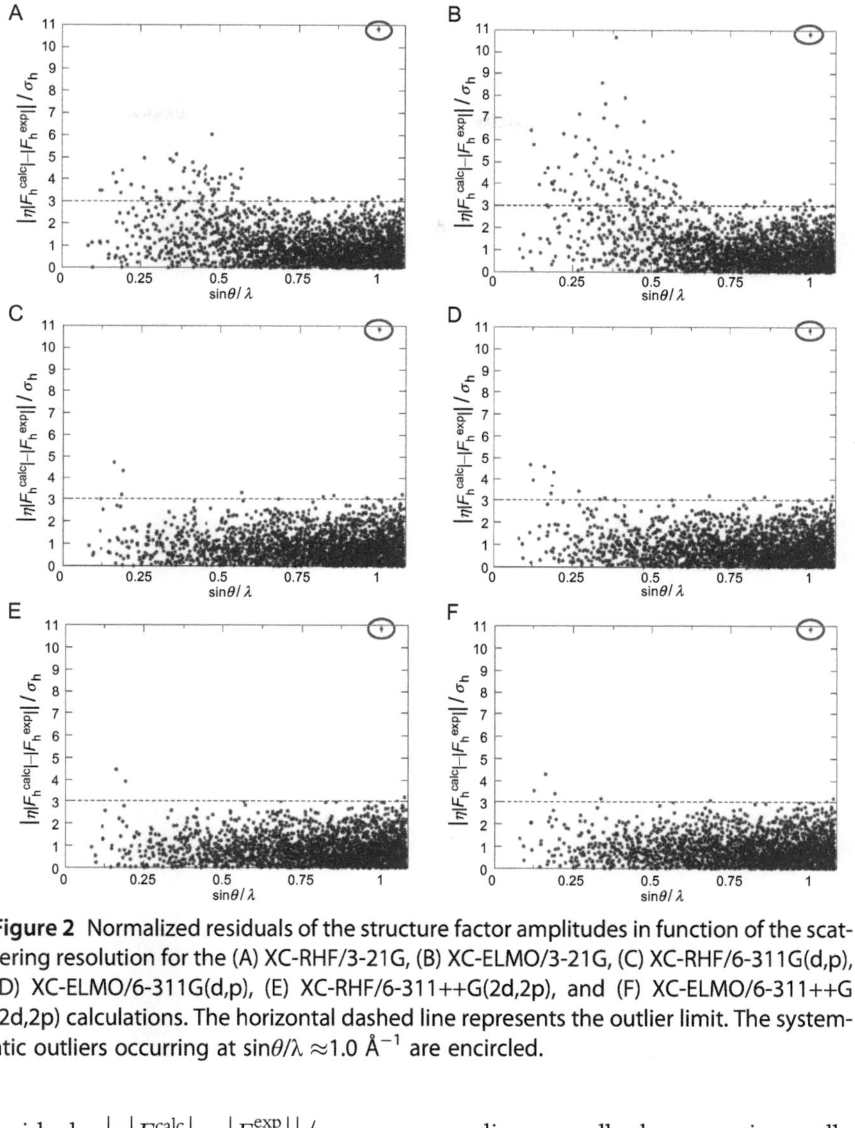

Figure 2 Normalized residuals of the structure factor amplitudes in function of the scattering resolution for the (A) XC-RHF/3-21G, (B) XC-ELMO/3-21G, (C) XC-RHF/6-311G(d,p), (D) XC-ELMO/6-311G(d,p), (E) XC-RHF/6-311++G(2d,2p), and (F) XC-ELMO/6-311++G (2d,2p) calculations. The horizontal dashed line represents the outlier limit. The systematic outliers occurring at $\sin\theta/\lambda \approx 1.0$ Å^{-1} are encircled.

residuals $\left|\eta\left|F_{\mathbf{h}}^{\text{calc}}\right| - \left|F_{\mathbf{h}}^{\text{exp}}\right|\right|/\sigma_{\mathbf{h}}$ corresponding to all the experimentally constrained calculations. As we can observe in Fig. 2, where we have reported the normalized residuals as a function of scattering resolution only for the 3-21G, 6-311G(d,p), and 6-311++G(2d,2p) basis-sets, a significant and systematic outlier occurs for all the XC-RHF and XC-ELMO computations at $\sin\theta/\lambda \approx 1.0$ Å^{-1}, i.e., normalized residual approximately equal to 10.8 in all the cases. Excluding the corresponding experimental value in the calculation of the statistical agreements for the basis-set 6-311++G(2d,2p), χ^2 further decreases to 1.12 and 1.14 for the XC-HF and XC-ELMO wave functions, respectively. This result further confirms that the impossibility of approaching the desired agreement is strictly related to the quality of diffraction data used in the X-ray constrained calculations.

In Table 1, we can also note that all the RHF computations provide χ^2 values lower than those corresponding to the ELMO calculations. This is particularly evident in the unconstrained cases, whereas the discrepancy significantly reduces when X-ray constrained computations are performed. The observed trend can be ascribed to the effects of the experimental data that probably compensate for the quite strong approximation represented by the introduction of a predefined localization scheme. Furthermore, we can easily see that the difference between the XC-RHF and XC-ELMO statistical agreements constantly decreases as more and more flexible basis-sets are used in the calculations, until a discrepancy of 0.02 is observed in the case of the 6-311++G(2d,2p) basis. This tendency can be also observed in Fig. 3, where we have reported χ^2 trends corresponding to X-ray constrained RHF and X-ray constrained ELMO calculations carried out with three basis-sets of different size and quality: the quite small 3-21G basis-set (68 atomic orbitals (AOs) for the system in exam), the medium-size 6-311G(d,p) set of basis functions (156 AOs), and the large and flexible 6-311++G(2d,2p) basis (244 AOs). In fact, while in the 3-21G case the XC-RHF and XC-ELMO calculations converge to two different values (see Fig. 3A), in the other two situations the XC-Hartree–Fock and XC-ELMO curves progressively approach the same limit (see Fig. 3B and, especially, Fig. 3C), which is not far from 1.0.

Further insights into the trends just discussed above can be obtained from a detailed analysis of the normalized residuals reported in Fig. 2. Indeed, for the three basis sets taken into account, the number of outliers (i.e., normalized residuals greater than 3) is always lower for the XC-RHF wave functions: 65 against 116, 10 against 18, and 5 against 8 for the 3-21G, 6-311G (d,p), and 6-311++G(2d,2p) basis-sets, respectively. Moreover, it is easy to observe that the discrepancy in the number of outliers between corresponding XC-RHF and XC-ELMO computations gradually reduces as the basis-set quality increases.

Therefore, all the results shown and discussed in this section seem to suggest that introducing a strict *a priori* localization in experimentally constrained calculations can have a strong influence on the final results if small basis-sets are used. However, it also seems that, when higher-quality basis-sets are considered, the bias coming from the predefined localization

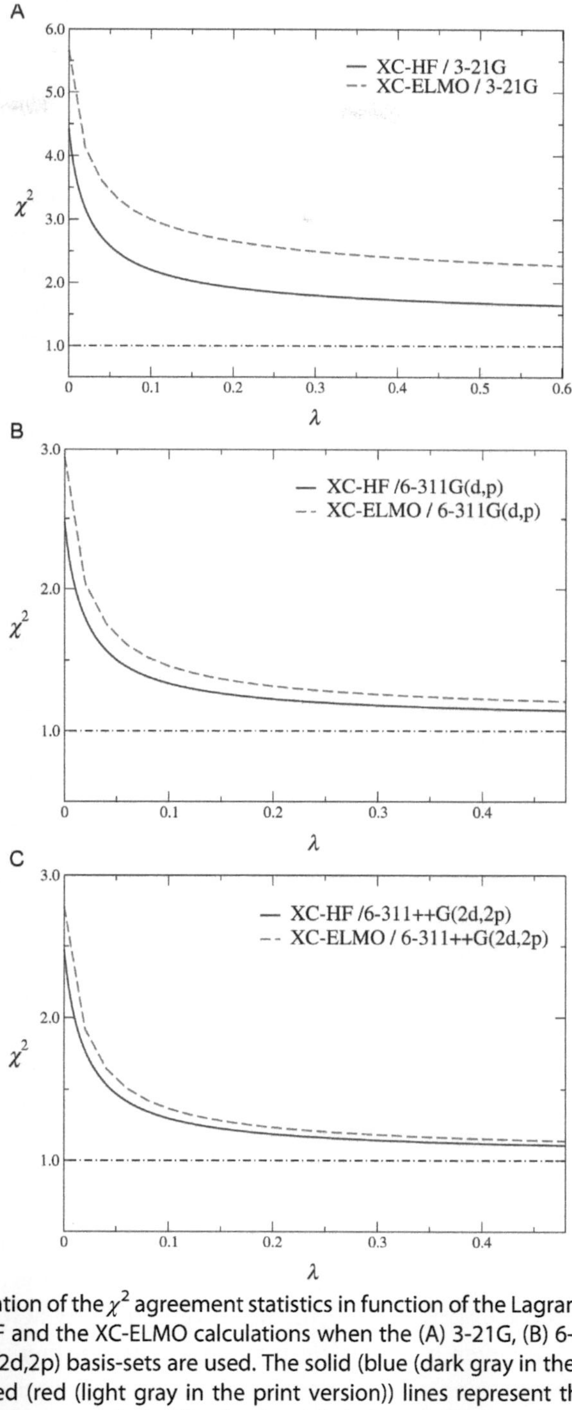

Figure 3 Variation of the χ^2 agreement statistics in function of the Lagrange multiplier λ for the XC-RHF and the XC-ELMO calculations when the (A) 3-21G, (B) 6-311G(d,p), and (C) 6-311++G(2d,2p) basis-sets are used. The solid (blue (dark gray in the print version)) and the dashed (red (light gray in the print version)) lines represent the XC-RHF and XC-ELMO trends, respectively. The horizontal dash-and-dot line shows the desired statistical agreement.

becomes less important. In fact, in those cases, the more flexible basis-sets probably allow to capture most of information coming from the experimental X-ray diffraction data, thus reducing the influence of the strict localization constraints (namely, the localization scheme) initially imposed on the electronic structure.

Before concluding this section, for the sake of completeness, we want to point out that all the energies associated with the X-ray constrained wave functions are higher than the energies obtained from to the corresponding unconstrained computations. As already explained,[44,49,57] this is the result of adding a constraint to the variational procedure without providing a new variational parameter.

3.3 Effects of the Localization on the Charge Distributions

To further investigate the effects of introducing an *a priori* molecular orbitals localization into the framework of the X-ray constrained wave function fitting, we have also analyzed the electron densities obtained from the previously discussed calculations. To accomplish this task, we have exploited the Quantum Theory of Atoms in Molecules (QTAIM),[78] and for the sake of simplicity, we have considered only the charge distributions computed using the 3-21G, 6-311G(d,p), and 6-311++G(2d,2p) basis-sets. Furthermore, we have mainly focused on two of the main topological properties currently used in QTAIM charge density analyses: the electron density and its Laplacian at the bond critical points (BCPs). In particular, since our main goal was to study the influence of the imposed prelocalization on the final electron distributions, we have determined the relative variations of the topological properties considered when the ELMO and XC-ELMO techniques have been used, taking the RHF and XC-RHF values as references, respectively.

The relative variations of the electron density values at the BCPs ($\rho(\mathbf{r}_b)$) are collected in Table 2. It is easy to observe that, both for unconstrained and X-ray constrained computations, the use of ELMOs does not entail significant changes, being the largest relative variation slightly larger than 4%. However, analyzing the data obtained from the experimentally constrained calculations, it is worth noting that, in most cases, the relative variation decreases as the quality of the basis-set increases. Moreover, the lowest values are generally those corresponding to the flexible 6-311++G(2d,2p) basis. These remarks are even more interesting considering the fact that the same trends are no longer seen when we examine the relative variations associated

Table 2 Unconstrained and X-Ray Constrained Relative Variations (U-RV and XC-RV, Respectively) of the Electron Density Values at the L-Alanine Bond Critical Points (BCPs)

	U-RV (%)[a]			XC-RV(%)[a]		
BCPs	3-21G	6-311G(d,p)	6-311++ G(2d,2p)	3-21G	6-311G(d,p)	6-311++ G(2d,2p)
N4–H7	−0.25	0.26	0.40	2.20	−1.55	−0.86
N4–H8	−0.19	0.16	0.33	−0.05	−0.30	−0.05
N4–H9	−0.65	0.29	0.29	3.82	1.99	0.69
N4–C2	0.48	−0.01	0.99	0.66	−0.19	0.23
C2–H10	0.36	1.85	1.50	−0.57	−0.08	−0.55
C2–C3	−0.96	1.56	1.33	−0.38	0.46	0.00
C3–H11	−0.16	0.26	0.39	−0.23	0.14	−0.11
C3–H12	−0.05	0.16	0.49	1.53	0.69	−0.03
C3–H13	0.04	0.23	0.41	1.15	−0.03	−0.36
C2–C1	−0.42	3.54	2.81	−0.56	1.33	0.82
C1–O5	1.95	2.02	1.84	4.22	2.38	2.04
C1–O6	1.73	1.97	1.93	2.74	2.05	1.89

[a]U-RV(%) = 100.0(ELMO − RHF)/RHF; XC-RV(%) = 100.0(XC-ELMO − XC-RHF)/XC-RHF.

with unconstrained computations. In fact, in those situations, the more flexible basis-sets 6-311G(d,p) and 6-311++G(2d,2p) are those for which we mainly observe the largest changes due to the introduction of the ELMOs.

In Table 3, we have reported the relative variations of the Laplacian values at the BCPs. Unlike the previous case, it is possible to note a larger variability that was actually expected due to the extreme sensitivity of the Laplacian to the position of the BCPs. Nevertheless, also for this topological property, we observe the same trends of Table 2. In fact, considering the data corresponding to the X-ray constrained calculations, the lowest relative variations are mainly obtained when the 6-311++G(2d,2p) basis-set is taken into account while, focusing on the unconstrained computations, the more flexible basis-sets (i.e., 6-311G(d,p) and 6-311++G(2d,2p)) generally provide the largest relative changes.

The trends observed for the relative variations of the topological properties corroborate the results discussed in the previous section. In fact, from Tables 2 and 3, we can infer that the use of ELMOs in X-ray constrained

Table 3 Unconstrained and X-Ray Constrained Relative Variations (U-RV and XC-RV, Respectively) of the Laplacian Values of the Electron Density at the L-Alanine Bond Critical Points (BCPs)

	U-RV (%)[a]			XC-RV(%)[a]		
BCPs	**3-21G**	**6-311G(d,p)**	**6-311++ G(2d,2p)**	**3-21G**	**6-311G(d,p)**	**6-311++ G(2d,2p)**
N4–H7	−0.38	0.54	0.50	2.14	0.56	0.97
N4–H8	0.03	0.37	0.38	−1.03	1.52	0.75
N4–H9	−0.94	1.56	0.74	−1.87	2.84	0.44
N4–C2	54.37	−2.26	31.29	9.28	0.21	3.49
C2–H10	0.81	3.88	2.82	0.66	2.00	0.30
C2–C3	0.03	7.88	6.66	0.13	5.50	3.62
C3–H11	−0.68	0.95	1.19	−0.18	1.31	1.04
C3–H12	−0.71	0.09	1.27	1.46	1.52	0.81
C3–H13	−0.45	0.60	1.21	1.62	0.60	−0.60
C2–C1	8.26	14.89	11.01	9.41	8.56	6.58
C1–O5	27.64	93.78	50.44	44.53	29.90	28.79
C1–O6	16.90	47.87	37.20	20.93	18.95	20.44

[a]U-RV(%) = 100.0(ELMO −RHF)/RHF; XC-RV(%) = 100.0(XC-ELMO −XC-RHF)/XC-RHF.

calculations determines larger changes in the electron density properties only when small basis-sets are used. On the contrary, when we employ larger and more flexible sets of basis functions, the influence of the predefined localization on the electron distribution becomes less significant.

4. CONCLUSIONS AND PERSPECTIVES

Nowadays, the experimentally constrained wave function approaches represent an alternative to the very well known *ab initio* and DFT methods. Although several researchers have been working on this subject for almost five decades, the field is largely unexplored and new interesting developments are still possible. New advances in this area of research will be beneficial not only to the quantum chemistry community but also to experimental researchers. In fact, if, on the one hand, new strategies could help to shed light on crucial theoretical problems (e.g., the Hohenberg and Kohn mapping between electron densities and wave functions), on the other hand new methods could also become valuable tools for the direct interpretation of experimental data.

In this chapter, we focused our attention on the X-ray constrained wave function fitting proposed by Jayatilaka. This technique is probably the most promising of all the "experimental" wave function strategies, and it is currently considered as a quantum mechanically rigorous method to reconstruct charge distributions in crystals from experimental diffraction data. Nevertheless, due to the complete delocalization of the molecular orbitals obtained, this method is not characterized by an immediate chemical interpretation in terms of the traditional Lewis molecular picture. To recover it, the strategy has been recently extended in order to obtain molecular orbitals strictly localized on small molecular fragments (e.g., atoms, bonds, or functional groups) from experimental X-ray diffraction data by defining *a priori* localization schemes that, however, represent strong initial assumptions about the electronic structure of the systems in exam.

In order to evaluate to what extent these assumptions impact the reconstructed electron densities, here we have compared results obtained through "traditional" X-ray constrained Hartree–Fock calculations to results of X-ray constrained ELMO computations. Analyzing statistical agreements with the experimental data, convergence trends, and also topological properties of charge distributions, we have observed that, if small basis-sets are used in the calculations, the *a priori* localization of the molecular orbitals indeed has a strong influence on the final electron densities. Nevertheless, employing larger sets of basis functions, the XC-HF and XC-ELMO calculations tend to more similar results. This can be explained considering that the flexibility of larger and larger basis-sets allows to increasingly exploit the information provided by the experimental data and, consequently, to gradually mitigate the initial bias represented by the predefined localization. Therefore, when adequate basis-sets are used, the introduction of a localization scheme represents a less severe approximation and it allows us to directly include traditional chemical concepts in the X-ray constrained wave function fitting.

However, in order to further improve the X-ray constrained wave function approaches, some open questions must be still addressed. For instance, as already mentioned in Section 2, a definitive and sound criterion to halt the X-ray constrained procedure should be defined. Furthermore, it will be crucial to perform detailed studies to understand whether X-ray constrained wave functions are really able to capture both the electron correlation and the crystal field effects, which are intrinsically included in the experimental diffraction data. Besides these important problems, in our research group we are also considering the possibility of extending Jayatilaka's approach to a multideterminant wave function ansatz since, until now, only single Slater determinants have been considered.

Finally, concerning the more recent XC-ELMO technique, we are currently exploiting the reliable transferability of the ELMOs[79–81] to construct original libraries of XC-ELMOs. They could become an alternative to the existing pseudo-atom databases[82–87] that have already been successfully used in the framework of the multipole model techniques to refine crystallographic structures and to determine electrostatic properties of macromolecular systems.

ACKNOWLEDGMENTS

The CNRS and the University of Lorraine (PEPS Mirabelle project "KiteComb") and the Institute Jean Barriol of the University of Lorraine (Axe-R12 project) are gratefully acknowledged for financial support.

REFERENCES

1. Hohenberg, P.; Kohn, W. Inhomogeneous Electron Gas. *Phys. Rev. B* **1964**, *36*, 864–871.
2. Hirshfeld, F. L. Difference Densities by Least-Squares Refinement: Fumaramic Acid. *Acta Cryst. B* **1971**, *27*, 769–781.
3. Stewart, R. F. Electron Population Analysis with Rigid Pseudoatoms. *Acta Cryst. A* **1976**, *32*, 565–574.
4. Hansen, N. K.; Coppens, P. Testing Aspherical Atom Refinements on Small-Molecule Data Sets. *Acta Cryst. A* **1978**, *34*, 909–921.
5. Mukherji, A.; Karplus, M. Constrained Molecular Wavefunctions: HF Molecule. *J. Chem. Phys.* **1963**, *38*, 44–48.
6. Rasiel, Y.; Whitman, D. R. Constrained-Variation Method in Molecular Quantum Mechanics. *J. Chem. Phys.* **1965**, *42*, 2124–2131.
7. Chong, D. P.; Byers-Brown, W. Perturbation Theory of Constraints: Application to a Lithium Hydride Calculation. *J. Chem. Phys.* **1966**, *45*, 392–395.
8. Chong, D. P.; Rasiel, Y. Constrained-Variation Method in Molecular Quantum Mechanics. Comparison of Different Approaches. *J. Chem. Phys.* **1966**, *44*, 1819–1823.
9. Clinton, W. L.; Nakhleh, J.; Wunderlich, F. Direct Determination of Pure-State Density Matrices. I. Some Simple Introductory Calculations. *Phys. Rev.* **1969**, *177*, 1–6.
10. Clinton, W. L.; Galli, A. J.; Massa, L. J. Direct Determination of Pure-State Density Matrices. II. Construction of Constrained Idempotent One-Body Densities. *Phys. Rev.* **1969**, *177*, 7–13.
11. Clinton, W. L.; Henderson, G. A.; Prestia, J. V. Direct Determination of Pure-State Density Matrices. III. Purely Theoretical Densities via an Electrostatic-Virial Theorem. *Phys. Rev.* **1969**, *177*, 13–18.
12. Clinton, W. L.; Lamers, G. B. Direct Determination of Pure-State Density Matrices. IV. Investigation of Another Constraint and Another Application of the P Equations. *Phys. Rev.* **1969**, *177*, 19–27.

13. Clinton, W. L.; Galli, A. J.; Henderson, G. A.; Lamers, G. B.; Massa, L. J.; Zarur, J. Direct Determination of Pure-State Density Matrices. V. Constrained Eigenvalue Problems. *Phys. Rev.* **1969**, *177*, 27–33.
14. Clinton, W. L.; Massa, L. J. Determination of the Electron Density Matrix from X-Ray Diffraction Data. *Phys. Rev. Lett.* **1972**, *29*, 1363–1366.
15. Clinton, W. L.; Frishberg, C. A.; Massa, L. J.; Oldfield, P. A. Methods for Obtaining an Electron-Density Matrix from X-Ray Diffraction Data. *Int. J. Quantum Chem. Symp.* **1973**, 7, 505–514.
16. Frishberg, C.; Massa, L. J. Idempotent Density Matrices for Correlated Systems from X-Ray Diffraction Structure Factors. *Phys. Rev. B* **1981**, *24*, 7018–7024.
17. Pecora, L. M. Determination of the Quantum Density Matrix from Experiment: An Application to Positron Annihilation. *Phys. Rev. B* **1986**, *33*, 5987–5993.
18. Massa, L.; Goldberg, M.; Frishberg, C.; Boehme, R. F.; Placa, S. J. L. Wave Functions Derived by Quantum Modeling of the Electron Density from Coherent X-Ray Diffraction: Beryllium Metal. *Phys. Rev. Lett.* **1985**, *55*, 622–625.
19. Howard, S. T.; Huke, J. P.; Mallinson, P. R.; Frampton, C. S. Density Matrix Refinement for Molecular Crystals. *Phys. Rev. B* **1994**, *49*, 7124–7136.
20. Snyder, J. A.; Stevens, E. D. A Wavefunction and Energy of the Azide Ion in Potassium Azide Obtained by a Quantum-Mechanically Constrained Fit to X-Ray Diffraction Data. *Chem. Phys. Lett.* **1999**, *313*, 293–298.
21. Tanaka, K. X-Ray Analysis of Wavefunctions by the Least-Squares Method Incorporating Orthonormality. I. General Formalism. *Acta Cryst. A* **1988**, *44*, 1002–1008.
22. Tanaka, K.; Makita, R.; Funahashi, S.; Komori, T.; Win, Z. X-Ray Atomic Orbital Analysis. I. Quantum-Mechanical and Crystallographic Framework of the Method. *Acta Cryst. A* **2008**, *64*, 437–449.
23. Hibbs, D. E.; Howard, S. T.; Huke, J. P.; Waller, M. P. A New Orbital-Based Model for the Analysis of Experimental Molecular Charge Densities: An Application to (Z)-N-Methyl-C-Phenylnitrone. *Phys. Chem. Chem. Phys.* **2005**, 7, 1772–1778.
24. Waller, M. P.; Howard, S. T.; Platts, J. A.; Piltz, R. O.; Willock, D. J.; Hibbs, D. E. Novel Properties from Experimental Charge Densities: An Application to the Zwitterionic Neurotransmitter Taurine. *Chem. Eur. J.* **2006**, *12*, 7603–7614.
25. Schmider, H.; Smith, V. H.; Weyrich, W. Determination of Electron Densities and One-Matrices from Experimental Information. *Trans. Am. Cryst. Assoc.* **1990**, *26*, 125–140.
26. Schmider, H.; Smith, V. H.; Weyrich, W. Reconstruction of the One-Particle Density Matrix from Expectation Values in Position and Momentum Space. *J. Chem. Phys.* **1992**, *96*, 8986–8994.
27. Weyrich, W. An Electronic Position and Momentum Density Study of Chemical Bonding in TiO2 (Rutile). *Lect. Ser. Comput. Sci.* **2006**, *5*, 1–3.
28. Gillet, J. M.; Becker, P. J. Position and Momentum Densities. Complementarity at Work: Refining a Quantum Model from Different Data Sets. *J. Phys. Chem. Sol.* **2004**, *65*, 2017–2023.
29. Gillet, J. M. Determination of a One-Electron Reduced Density Matrix Using a Coupled Pseudo-Atom Model and a Set of Complementary Scattering Data. *Acta Cryst. A* **2007**, *63*, 234–238.
30. Gillet, J. M.; Becker, P. J.; Cortona, P. Joint Refinement of a Local Wave-Function Model from Compton and Bragg Scattering Data. *Phys. Rev. B* **2001**, *63*, 235115.
31. Cooper, M. Compton Scattering and Electron Momentum Determination. *Rep. Prog. Phys.* **1985**, *48*, 415–481.
32. Loupias, G.; Chomilier, J. Electron Momentum Density and Compton Profiles: An Accurate Check of Overlap Models. *Z. Phys. D Atoms Mol. Clusters* **1986**, *2*, 297–308.
33. Cassam-Chenaï, P. Ensemble Representable Densities for Atoms and Molecules. I. General Theory. *Int. J. Quantum Chem.* **1995**, *54*, 201–210.

34. Cassam-Chenaï, P.; Wolff, S.; Chandler, G.; Figgis, B. Ensemble-Representable Densities for Atoms and Molecules. II. Application to $CoCl_4^{2-}$. *Int. J. Quantum Chem.* **1996**, *60*, 667–680.
35. Gilbert, T. L. Hohenberg-Kohn Theorem for Nonlocal External Potentials. *Phys. Rev. B* **1975**, *12*, 2111–2120.
36. Henderson, G. A.; Zimmerman, R. K. One-Electron Properties as Variational Parameters. *J. Chem. Phys.* **1976**, *65*(2), 619–622.
37. Levy, M.; Goldstein, J. A. Electron Density-Functional Theory and X-Ray Structure Factors. *Phys. Rev. B* **1987**, *35*, 7887–7890.
38. Gritsenko, O. V.; Zhidomirov, G. M. On the Determination of the 1st-Order Density-Matrix from the Diffraction Experiment Data. *Dokl. Akad. Nauk. SSSR* **1987**, *293*, 1162–1166.
39. Zhao, Q.; Parr, R. G. Quantities $T_s[n]$ and $T_c[n]$ in Density Functional Theory. *Phys. Rev. A* **1992**, *46*, 2337–2343.
40. Zhao, Q.; Parr, R. G. Constrained-Search Method to Determine Electronic Wave Functions from Electronic Densities. *J. Chem. Phys.* **1993**, *98*, 543–548.
41. Zhao, Q.; Morrison, R. C.; Parr, R. G. From Electron Densities to Kohn-Sham Kinetic Energies, Orbital Energies, Exchange-Correlation Potentials, and Exchange Correlation Energies. *Phys. Rev. A* **1994**, *50*, 2138–2142.
42. Levy, M. Universal Variational Functionals of Electron Densities, First-Order Density Matrices, and Natural Spin-Orbitals and Solution of the v-Representability Problem. *Proc. Natl. Acad. Sci. U.S.A.* **1979**, *76*, 6062–6065.
43. Jayatilaka, D. Wave Function for Beryllium from X-Ray Diffraction Data. *Phys. Rev. Lett.* **1998**, *80*, 798–801.
44. Jayatilaka, D.; Grimwood, D. J. Wavefunctions Derived from Experiment. I. Motivation and Theory. *Acta Cryst. A* **2001**, *57*, 76–86.
45. Grimwood, D. J.; Jayatilaka, D. Wavefunctions Derived from Experiment. II. A Wavefunction for Oxalic Acid Dihydrate. *Acta Cryst. A* **2001**, *57*, 87–100.
46. Bytheway, I.; Grimwood, D.; Jayatilaka, D. Wavefunctions Derived from Experiment. III. Topological Analysis of Crystal Fragments. *Acta Cryst. A* **2002**, *58*, 232–243.
47. Bytheway, I.; Grimwood, D. J.; Figgis, B. N.; Chandler, G. S.; Jayatiaka, D. Wavefunctions Derived from Experiment. IV. Investigation of the Crystal Environment of Ammonia. *Acta Cryst. A* **2002**, *58*, 244–251.
48. Grimwood, D. J.; Bytheway, I.; Jayatilaka, D. Wavefunctions Derived from Experiment. V. Investigation of Electron Densities, Electrostatic Potentials, and Electron Localization Functions for Noncentrosymmetric Crystals. *J. Comput. Chem.* **2003**, *24*, 470–483.
49. Hudák, M.; Jayatilaka, D.; Peraínova, L.; Biskupic, S.; Kozísek, J.; Bucinský, L. X-Ray Constrained Unrestricted Hartree-Fock and Douglas-Kroll-Hess Wavefunctions. *Acta Cryst. A* **2010**, *66*, 78–92.
50. Jayatilaka, D. Using Wavefunctions to Get More Information Out of Diffraction Experiments. In *Modern Charge-Density Analysis*; Gatti, C., Macchi, P., Eds.; Springer: Berlin, 2012; pp 213–257.
51. Whitten, A. E.; Jayatilaka, D.; Spackman, M. Effective Molecular Polarizabilities and Crystal Refractive Indices Estimated from X-Ray Diffraction Data. *J. Chem. Phys.* **2006**, *125*, 174505.
52. Hickstein, D. D.; Cole, J. M.; Turner, M. J.; Jayatilaka, D. Modeling Electron Density Distributions from X-Ray Diffraction to Derive Optical Properties: Constrained Wavefunction Versus Multipole Refinement. *J. Chem. Phys.* **2013**, *139*, 064108.
53. Jayatilaka, D.; Grimwood, D. Electron Localization Functions Obtained from X-Ray Constrained Hartree-Fock Wavefunctions for Molecular Crystals of Ammonia, Urea and Alloxan. *Acta Cryst. A* **2004**, *60*, 111–119.

54. Grabowsky, S.; Jayatilaka, D.; Mebs, S.; Luger, P. The Electron Localizability Indicator from X-Ray Diffraction Data—A First Application to a Series of Epoxide Derivatives. *Chem. Eur. J.* **2010**, *16*, 12818–12821.
55. Grabowsky, S.; Weber, M.; Jayatilaka, D.; Chen, Y. S.; Grabowski, M. T.; Brehme, R.; Hesse, M.; Schirmeister, T.; Luger, P. Reactivity Differences Between α,β-Unsaturated Carbonyls and Hydrazones Investigated by Experimental and Theoretical Electron Density and Electron Localizability Analyses. *J. Phys. Chem. A* **2011**, *115*, 12715–12732.
56. Grabowsky, S.; Luger, P.; Buschmann, J.; Schneider, T.; Schirmeister, T.; Sobolev, A. N.; Jayatilaka, D. The Significance of Ionic Bonding in Sulfur Dioxide: Bond Orders from X-Ray Diffraction Data. *Angew. Chem. Int. Ed. Engl.* **2012**, *51*, 6776–6779.
57. Genoni, A. Molecular Orbitals Strictly Localized on Small Molecular Fragments from X-Ray Diffraction Data. *J. Phys. Chem. Lett.* **2013**, *4*, 1093–1099.
58. Genoni, A. X-Ray Constrained Extremely Localized Molecular Orbitals: Theory and Critical Assessment of the New Technique. *J. Chem. Theory Comput.* **2013**, *9*, 3004–3019.
59. dos Santos, L. H. R.; Genoni, A.; Macchi, P. Unconstrained and X-Ray Constrained Extremely Localized Molecular Orbitals: Analysis of the Reconstructed Electron Density. *Acta Cryst. A* **2014**, *70*, 532–551.
60. Stoll, H.; Wagenblast, G.; Preuss, H. On the Use of Local Basis Sets for Localized Molecular Orbitals. *Theor. Chim. Acta* **1980**, *57*, 169–178.
61. Pipek, J.; Mezey, P. G. A Fast Intrinsic Localization Procedure Applicable for Ab Initio and Semiempirical Linear Combination of Atomic Orbital Wave Functions. *J. Chem. Phys.* **1989**, *90*, 4916–4926.
62. Boys, S. F. Construction of Some Molecular Orbitals to be Approximately Invariant for Changes from One Molecule to Another. *Rev. Mod. Phys.* **1960**, *32*, 296–299.
63. Foster, J. M.; Boys, S. F. Canonical Configurational Interaction Procedure. *Rev. Mod. Phys.* **1960**, *32*, 300–302.
64. Edmiston, C.; Ruedenberg, K. Localized Atomic and Molecular Orbitals. *Rev. Mod. Phys.* **1963**, *35*, 457–465.
65. Edmiston, C.; Ruedenberg, K. Localized Atomic and Molecular Orbitals. II. *J. Chem. Phys.* **1965**, *43*, S97–S116.
66. Becke, A. D.; Edgecombe, K. E. A Simple Measure of Electron Localization in Atomic and Molecular Systems. *J. Chem. Phys.* **1990**, *92*, 5397–5403.
67. Kohout, M. A Measure of Electron Localizability. *Int. J. Quantum Chem.* **2004**, *97*, 651–658.
68. Genoni, A.; Sironi, M. A Novel Approach to Relax Extremely Localized Molecular Orbitals: The Extremely Localized Molecular Orbital – Valence Bond Method. *Theor. Chem. Acc.* **2004**, *112*, 254–262.
69. Genoni, A.; Fornili, A.; Sironi, M. Optimal Virtual Orbitals to Relax Wave Functions Built Up with Transferred Extremely Localized Molecular Orbitals. *J. Comput. Chem.* **2005**, *26*, 827–835.
70. Smits, G. F.; Altona, C. Calculation and Properties of Non-Orthogonal, Strictly Local Molecular Orbitals. *Theor. Chim. Acta* **1985**, *67*, 461–475.
71. Fornili, A.; Sironi, M.; Raimondi, M. Determination of Extremely Localized Molecular Orbitals and Their Application to Quantum Mechanics/Molecular Mechanics Methods and to the Study of Intramolecular Hydrogen Bonding. *J. Mol. Struct. (Theochem)* **2003**, *632*, 157–172.
72. Stewart, R. F. Generalized X-Ray Scattering Factors. *J. Chem. Phys.* **1969**, *51*, 4569–4577.
73. Coppens, P.; Willoughby, T. V.; Csonka, L. N. Electron Population Analysis of Accurate Diffraction Data. I. Formalisms and Restrictions. *Acta Cryst. A* **1971**, *27*, 248–256.

74. Destro, R.; Marsh, R. E.; Bianchi, R. A Low-Temperature (23-K) Study of L-Alanine. *J. Phys. Chem.* **1988**, *92*, 966–973.
75. Jayatilaka, D.; Grimwood, D. J. Tonto: A Fortran Based Object-Oriented System for Quantum Chemistry and Crystallography. Computational Science—ICCS 2003. *Lect. Notes Comput. Sci.* **2003**, *2660*, 142–151.
76. Guest, M. F.; Bush, I. J.; van Dam, H. J. J.; Sherwood, P.; Thomas, J. M. H.; van Lenthe, J. H.; Havenith, R. W. A.; Kendrick, J. The GAMESS-UK Electronic Structure Package: Algorithms, Developments and Applications. *Mol. Phys.* **2005**, *103*, 719–747.
77. Keith, T. A. *AIMAll (Version 13.11.04)*; TK Gristmill Software: Overland Park, KS, 2012. URL: aim,tkgristmill.com.
78. Bader, R. F. W. *Atoms in Molecules: A Quantum Theory*; Oxford University Press: Oxford, 1990.
79. Genoni, A.; Ghitti, M.; Pieraccini, S.; Sironi, M. A Novel Extremely Localized Molecular Orbitals Based Technique for the One-Electron Density Matrix Computation. *Chem. Phys. Lett.* **2005**, *415*, 256–260.
80. Sironi, M.; Genoni, A.; Civera, M.; Pieraccini, S.; Ghitti, M. Extremely Localized Molecular Orbitals: Theory and Applications. *Theor. Chem. Acc.* **2007**, *117*, 685–698.
81. Sironi, M.; Ghitti, M.; Genoni, A.; Saladino, G.; Pieraccini, S. DENPOL: A New Program to Determine Electron Densities of Polypeptides Using Extremely Localized Molecular Orbitals. *J. Mol. Struct. (Theochem)* **2009**, *898*, 8–16.
82. Pichon-Pesme, V.; Lecomte, C.; Lachekar, H. On Building a Data Bank of Transferable Experimental Electron Density Parameters: Application to Polypeptides. *J. Phys. Chem.* **1995**, *99*, 6242–6250.
83. Zarychta, B.; Pichon-Pesme, V.; Guillot, B.; Lecomte, C.; Jelsch, C. On the Application of an Experimental Multipolar Pseudo-Atom Library for Accurate Refinement of Small-Molecule and Protein Crystal Structures. *Acta Cryst. A* **2007**, *63*, 108–125.
84. Koritsanszky, T.; Volkov, A.; Coppens, P. Aspherical-Atom Scattering Factors from Molecular Wave Functions. 1. Transferability and Conformation Dependence of Atomic Electron Densities of Peptides Within the Multipole Formalism. *Acta Cryst. A* **2002**, *58*, 464–472.
85. Volkov, A.; Li, X.; Koritsanszky, T.; Coppens, P. Ab Initio Quality Electrostatic Atomic and Molecular Properties Including Intermolecular Energies from a Transferable Theoretical Pseudoatom Databank. *J. Phys. Chem. A* **2004**, *108*, 4283–4300.
86. Dittrich, B.; Koritsanszky, T.; Luger, P. A Simple Approach to Nonspherical Electron Densities by Using Invarioms. *Angew. Chem. Int. Ed. Engl.* **2004**, *43*, 2718–2721.
87. Dittrich, B.; Hübschle, C. B.; Luger, P.; Spackman, M. A. Introduction and Validation of an Invariom Database for Amino-Acid, Peptide and Protein Molecules. *Acta Cryst. D* **2006**, *62*, 1325–1335.

CHAPTER EIGHTEEN

Electron Impact Atomic and Ionic Ionization: Analytical, Semiempirical, and Semiclassical Methods

A.K. Fazlul Haque*, M. Atiqur R. Patoary*, M. Alfaz Uddin*, Arun K. Basak*, Bidhan C. Saha†,1

*Department of Physics, University of Rajshahi, Rajshahi, Bangladesh
†Department of Physics, Florida A & M University, Tallahassee, Florida, USA
1Corresponding author: e-mail address: bidhan.saha@famu.edu

Contents

Abstract

Calculations of electron impact ionization cross sections (EIICS) for H to Sc (with atomic number $Z=1-21$) and their corresponding isoelectronic series over incident energies up to 10^6 keV have been reviewed. This work reports the results of calculations using our various easy-to-use models reproducing closely the experimental EIICS data. We also report how these models can easily be implemented to generate accurate data as demanded by various model calculations. The selection of the range of atomic number Z for different isoelectronic series was made possible by the wealth of the EIICS data either from experiments or from rigorous quantum calculations. We have furnished the detailed description of our various models with comparison of the outcomes of these models with experimental and other theoretical results. Our models describe the experimental results with reasonable success for various atomic and ionic targets belonging to the H to Sc isoelectronic series over a wide range of atomic number in each series.

Advances in Quantum Chemistry, Volume 73
ISSN 0065-3276
http://dx.doi.org/10.1016/bs.aiq.2015.07.004

1. INTRODUCTION

Electron impact ionization (EII) is of fundamental importance in understanding the physics of the collision process associated with many electron problems involving atoms, neutral, and ions. A knowledge of electron impact ionization cross sections, noted here as EIICS, is needed in various applied physics areas such as chemical-, biological-, plasma-, laser-, astrophysics, etc., for many different targets over a wide range of incident energy. In addition to many practical applications, the electron impact inelastic collision processes involving both atomic and ionic targets create a basic testing ground for a deeper understanding of the quantum collision theory.

The purpose of this chapter is to review the recent progress on ionization problems. It is shown explicitly how the limitations of previous models in the literature have been resolved, making them valid for a wider range of applications involving both the energies and targets. By now, various quantum mechanical (QM) methods have been developed for generating accurate cross sections. Of these, notable calculations are from the plane-wave Born approximation (PWBA) of Omidvar et al.[1]; distorted-wave Born approximation (DWBA) with variants including exchange (DWBAX) due to Younger[2–8]; relativistic plane-wave Born approximation (RPWBA) due to Scofield,[9] and Ndefru and Malik[10]; relativistic DWBA (RDWBA) due to Segui et al.,[11] Fontes et al.,[12,13] and Moores and Reed[14]; DWBA with R-matrix (DWBARM) due to Laghdas et al.[15]; Born approximation of Peach[16]; Coulomb–Born approximation (CBA) due to Moores, and Nussbaumer[17] and Moores[18]; CBA with exchange (CBAX) due to Stingl,[19] and Jacubowicz and Moores[20]; two-potential DWBA (TPDWBA) due to Kuo and Huang,[21] and Kao et al.[22]; convergent closed coupling (CCC) methods due to Bray[23]; and exterior complex scaling techniques[24] and others with variations from these QM methods.

However, many QM calculations are, in practice, arduous and time consuming and based on various approximations. They are also limited in terms of atomic species and incident energies. Experimental techniques produce data at discrete energy values and for some suitable selected targets. Thus the great demand in applications for EIICS can be fulfilled neither by generating a database of experimental measurements nor by *ab initio* QM calculations.

In literature, there are a few good reviews on the electron impact single ionization cross sections (EISICS) calculations,[25–27] employing simple-to-use models with limited reliability at higher energies, where relativistic effects become important, and hence their use remains constrained. We follow the acronyms used in our previous review[27] on EII of neutral atomic targets. Thus, in this review, the experimental data are labeled by EIICS to distinguish from the theoretical ionization values, which are calculated by adopting a single-step collision mechanism.

Several successful and simple models for calculating EIICS have been proposed since the late twentieth century. These calculations include the several versions of Deutsch and Märk (DM) models[26,28–30] starting with a semiclassical approach[28] which combines the classical binary encounter approximation (BEA) of Gryzinski[31–33] and the weighted sum of squared radius for maximum charge density of the outer electron shells from the Born–Bethe approximation[34]; a simplified version of improved binary-encounter dipole (SIBED) model of Huo[35]; fitting formulas of Bell et al.[36] with polynomial series and Godunov and Ivanov[37] with nonlinear exponential form; Kolbenstvedt model (KLV) of Kolbenstvedt[38,39]; empirical model (BRY) of Bernshtam et al.[40]; and analytic expression of Cengiz (CN),[41] CN based on distant and close interactions using the generalized oscillator strength.

The quantum aspect in the DM model is considered in the modified version of Ref. 29 through radius dependence of the electronic shells on the principal and orbital quantum numbers *nl*. Neither the nonrelativistic nor relativistic DM models before the version of Ref. 30 satisfies the asymptotic Bethe condition.[34] The revised DM model (RDM) of Ref. 30 was made to satisfy the Bethe asymptotic condition to extend the range of validity of the DM to higher impact energies. The DM models, however, have been applied mostly to neutral atoms and molecules. These have restricted applications on ionic targets.[42,43] However, the DM and BRY models have important bearing on the development of models, considered in this review. Uddin et al.[44–48] and Haque et al.[49–54] have extended these models in terms of applicability over wide range of incident energies and atomic targets. In achieving this, they accounted for the following effects:

i. *Correlation*: Effective energy seen by a bound electron is not merely the incident projectile energy T, but the Burgess denominator[55] $T+\zeta_i+I_i$, where ζ_i and I_i are, respectively, the kinetic and binding energies of the atomic electron in the ith orbit. This results in a considerable reduction in the ionization cross sections at the low energy region.

ii. *Ionic*: The incident electron approaching the inner-shell electron feels the atom as having an effective charge $q=Z-N_{q_i}$, N_{q_i} being the total number of electrons in the atomic/ionic target (with atomic number Z) from the innermost 1s-orbital to the relevant ionizing i^{th} orbital. This, however, leads to a greater overlap in the charge clouds of the projectile and target, resulting in an enhancement of the cross sections.

iii. *Relativistic*: For application of a model in the relativistic domain, an appropriate form of relativistic factor conformable to the model structure has to be incorporated. Gryzinski relativistic factor[32,33] is a widely used form. If it does not work well in the model structure, other simple forms, as in Uddin et al.[48,56] and Haque et al.[53] are envisaged.

iv. *Generalization of parameters*: Here, the dependence of the model parameters only on the principal and orbital angular momentum quantum numbers *nl* of the ionized orbital is sought as in Haque et al.[50,51,53,57] following the *i*-dependent generalization of the parameters of Deutsch et al.[29,30,58] and Bernshtam et al.[40]

v. *Asymptotic behavior in incident energy*: Energy-dependent factor should obey the asymptotic condition.

Consideration of the above-mentioned effects creates a window for the development of simple models with profound success in fast generation of reliable EIICS data over wide ranges of energy and atomic targets, both neutral and ionic. These include, to name a few, RQIBED of Uddin et al.[44–46] with ionic and relativistic corrections in siBED of Huo[35]; MRIBED of Uddin et al.,[47] the modified RQIBED[44–46] with the inclusion of *Z*-dependent factor; MBELL of Haque et al.,[49–52] the modified BELL[36] with incorporation of ionic and relativistic factors; MKLV of Uddin et al.,[48] the modified KLV[38,39] with ionic and relativistic corrections; GKLV in Haque et al.,[53] the modified version of MKLV[48] with generalized parameters; NMBELL of Patoary et al.,[59] the nonlinear version of modified MBELL[49–51]; and MCN of Haque et al.,[60] the modified CN model[41] including both the ionic and relativistic corrections.

This chapter reviews recent progress in predicting EISICS results from the MCN, RQIBED/ MRIBED, MBELL, NMBELL, and GKLV models on both neutral and ionic targets in 21 isoelectronic series from H- to Sc-like atoms. In various model applications, use of these reliable EISICS results will ease the massive data generation needs.

2. OUTLINE OF THE MODELS

2.1 MCN Model

In the Cengiz CN model,[41] the correlation effect is included via the focusing term, which is used to mitigate the overestimation of cross sections at the low energy region. But for ionic targets this causes too much reduction.

These issues are best addressed using the following steps inside the MCN model:

i. *Ionic correction*: It remedies the underestimation of cross sections in the low energy region.

ii. *Relativistic correction*: The relativistic enhancement of cross sections is taken care of by considering a relativistic factor of Gryzinski type.[32,33]

iii. *Modulation of the focusing term*: The focusing term in CN, modulated in the MCN model by considering the correlation effect and ionic corrections becomes a function of $T+\zeta_i+I_i$ and q_i, which is appropriate to describe the EIICS data at the low energy region.

iv. *Removal of the oscillatory part*: Incorporation of the ionic and relativistic corrections in the original structure of the CN model results in an overestimation of cross sections in the threshold region in addition to a superfluous fluctuation in the cross section in the relativistic domain of the incident energies. It is found that both these problems emanate from the oscillator part of the CN model (see Ref. 60 for details). As a remedy, this term is discarded altogether before embodying the ionic and relativistic corrections in the MCN model.

Denoting the velocity of light in a vacuum, electron charge, rest mass, incident projectile velocity and number of electrons in the ith atomic orbital, respectively, by c, e, m, v and N_i and defining the reduced incident energy, reduced velocity and reduced rest energy, respectively, by $U_i = T/I_i$, $\beta = v/c$ and $J_i = mc^2/I_i$, the expression for EISICS for ionization of the ith orbital in the MCN model is given[60] by

$$\sigma_{\text{MCN}} = 2\tau N_i P_i^{\text{C}} \; R_{\text{DM}} \; R_{\text{G}}\left[(T-I_i)\left(\frac{1}{TI_i}+\frac{w}{2T^2}\right)+\frac{(1-w)}{T+I_i}\ln\left(\frac{I_i}{T}\right)\right], \tag{1}$$

where $w = \left(\frac{T}{T+mc^2}\right)^2$ and $\tau = \frac{2\pi e^4}{mc^2\beta^2}$. The focusing term, corrected for the correlation and ionic effects, is

$$P_i^{\text{C}} = \frac{T}{T+(I_i+\zeta_i)/Q_{\text{f}}} \tag{2}$$

with the charge factor Q_{f}[60] defined by

$$Q_{\text{f}} = \left(1+\xi_1\left(\frac{q}{U_i}\right)^{\xi_2}\right). \tag{3}$$

R_G and R_{DM} are, respectively, the relativistic factors due to Gryzinski[33] in a modified from and a factor following Deutsch et al.[61] suited to the EIICS data at ultra-relativistic region given by

$$R_G(U_i) = \left(\frac{1+2J_i}{1.1U_i+2J_i}\right)\left(\frac{U_i+J_i}{1+J_i}\right)^{2.07}\left[\frac{(1+U_i)(U_i+2J_i)(1+J_i)^2}{J_i^2(1+2J_i)+U_i(U_i+2J_i)(1+J_i)^2}\right]^{1.5}, \tag{4}$$

and

$$R_{DM}(U_i) = \left(1+\lambda_1 U_i^{\lambda_2}\right). \tag{5}$$

The parameters ξ_1 and ξ_2 in the charge factor Q_f in Eq. (3) and those λ_1 and λ_2 associated with the relativistic factor R_{DM} in Eq. (5) have been determined through fitting the experimental EIICS data on the neutral and ionic species of atomic targets. Optimization of the two parameters has been accomplished through the minimization of χ^2 defined by

$$\chi^2 = \frac{1}{N_p}\sum_{j=i}^{N_p}\left(\frac{\sigma_{Th}^j - \sigma_{Ex}^j}{\sigma_{Ex}^j}\right)^2. \tag{6}$$

Here, N_p is the total number of data points on different species and incident energies. σ_{Th}^j refers to the predicted EISCS, summed over all the contributing shells, and σ_{Ex}^j is the experimental value of EIICS at the jth point. Nonlinear least-square fitting code MINUIT[62] has been used in both local and global modes to achieve the optimization of the model parameters. The optimization procedure using the χ^2 minimization is applicable to the individual analysis with all the models, considered herein.

The minimization approach of χ^2 on the MCN model parameters leads to their estimated values as $\xi_1=0.10$, $\xi_2=0.05$, $\lambda_1=0.30$, and $\lambda_2=0.17$. These parameters are finally used to calculate the EISICS values using the MCN model.

2.2 RQIBED/MRIBED Model

For an ionized atomic shell, the ionization cross section in the RQIBED model of Uddin et al.[45] is

$$\sigma_{RQIBED} = \sum_i \left(S_i^R H_i + F_i^R G_i\right). \tag{7}$$

The term S_i^{R} in Eq. (7) are given as

$$S_i^{\mathrm{R}} = \frac{4\pi N_i \alpha^2}{\beta_t^2 + (\beta_{bi}^2 + \beta_{ai}^2)/(q_i + 1)} \quad . \tag{8}$$

Here, the summation over i is taken over all the contributing orbitals each with quantum number nl in the considered shell. N_i is the number of electrons in the contributing i^{th}-orbital, q_i, denotes the effective charge as seen by the incident electron, and α is the fine structure constant.

The terms H_i, F_i^{R}, and G_i is given in Ref. 45 as

$$H_i = \left[\frac{k_0^2 - \alpha_{0i}^2}{k_0^2 \alpha_{0i}^2} - \frac{\ln(k_0^2/\alpha_{0i}^2)}{k_0^2 + \alpha_{0i}^2}\right], \tag{9}$$

$$F_i^{\mathrm{R}} = \frac{64\beta_{ai}^3 N_i}{\alpha \beta_t^2}, \tag{10}$$

and

$$G_i = \int_0^{(k_0^2 - \alpha_{i0}^2)/2} k_{pi}(k_{pi}^2 + \alpha_{0i}^2)\mathrm{d}E_{pi} \times \int_{K_{\min}}^{K_{\max}} \frac{1 + d_1 t_i + d_2 t_i^2}{K\left[(K + k_{pi})^2 + \alpha_{0i}^2\right]^3 \left[(K - k_{pi})^2 + \alpha_{0i}^2\right]^3} \mathrm{d}K, \tag{11}$$

where

$$t_i = \frac{K^4}{\left[(K + k_{pi})^2 + \alpha_{0i}^2\right]\left[(K - k_{pi})^2 + \alpha_{0i}^2\right]}. \tag{12}$$

Using atomic units in the above Eqs. (8–12), $k_0^2/2$ is the energy of the incident electron; $k_{bi}^2/2$, the kinetic energy of the bound electron; $I_i = \alpha_{0i}^2/2$, the ionization potential of the relevant i^{th} orbital electron and $k_{pi}^2/2$, the energy of the ejected electron with k_{pi} having the dimension of momentum.[35] $\vec{K} = \vec{k_0} - \vec{k_f}$ denotes the momentum transfer vector with k_f as the momentum of the incident electron after a collision. The values of $K_{\max}$ and $K_{\min}$ are given in Ref. 63. The parameters d_1 and d_2 in Eq. (11) determine, respectively, the EISICS peak and EISICS values at higher energies.[45] The EISICS data on most atomic targets favor $d_1 = 0.0$[44–46] and, with the inclusion of relativistic factors, d_2 is also found to be $d_2 = 0.0$.[45] The quantities in Eqs. (8) and (10) are defined in terms of $t' = k_0^2/(2c^2)$, $b_i' = k_{bi}^2/(2c^2)$, and $a_i' = \alpha_{0i}^2/(2c^2)$, respectively, as

$$\beta_t^2 = 1 - \frac{1}{(1+t')^2}, \tag{13}$$

$$\beta_{bi}^2 = 1 - \frac{1}{(1+b_i')^2}, \tag{14}$$

and

$$\beta_{ai}^2 = 1 - \frac{1}{(1+a_i')^2}. \tag{15}$$

In the MRIBED model of Uddin et al.,[47] the K-shell ionization cross section after including a Z-dependent function

$$F(Z) = 1 + \xi Z^{\lambda}, \tag{16}$$

reduces to

$$\sigma_{\text{MRIBED}} = F(Z)\sigma_{\text{RQIBED}}. \tag{17}$$

For filled 1s-orbitals $\xi = 0.0$, $\lambda = 0.0$, and $F(Z) = 1$; and for vacant orbitals, the values of the parameters are $\xi = 0.365$ and $\lambda = 0.05$.

2.3 MBELL Model

The total cross section of electron impact single ionization, according to Haque et al.,[49,51,52] contributed from different ionized orbitals is given by

$$\sigma_{\text{MBELL}}(T) = \sum_i N_i F_{\text{ion}}^i G_{\text{R}}^i \sigma_{\text{BELL}}^i(T), \tag{18}$$

where the summation over the index i refers to different ionized nl-orbitals and $\sigma_{\text{BELL}}(T)$ having the form[36]:

$$\sigma_{\text{BELL}}^i(T) = \frac{1}{I_i T}\left[A_i \ \ln(U_i) + \sum_{k=1}^{7} B_{ik}\left(1 - \frac{1}{U_i}\right)^k\right]. \tag{19}$$

Here, A and B_k are the so-called BELL parameters. G_{R}^i is the relativistic factor of the Gryzinski type[31–33] given by

$$G_{\text{R}}^i = \left(\frac{1+2J_i}{U_i+2J_i}\right)\left(\frac{U_i+J_i}{1+J_i}\right)^2\left[\frac{(1+U_i)(U_i+2J_i)(1+J_i)^2}{J_i^2(1+2J_i)+U_i(U_i+2J_i)(1+J_i)^2}\right]^{1.5}. \tag{20}$$

F_{ion}^i is the ionic correction factor involving the ionic parameters ξ_i and λ_i having the form

$$F_{\text{ion}}^{i} = 1 + \xi_i \left(\frac{q_i}{ZU_i} \right)^{\lambda_i}. \tag{21}$$

The form of the above ionic factor is to account for its effects from two different origins:

i. The cross section enhancement via the greater overlap of the charge clouds of the incident and target electrons.

ii. The decrease of the enhancement with the increase of incident energy due to sorter interaction time with the atomic field.

For an ionized orbital *nl*, the Bell parameters A and B_k with $k=1$–7, shown in Eq. (19) above, are obtained by fitting the EIICS data of atomic targets having their outermost orbital with the configuration *nl*. Following the prescription of Ref. 52, the ionic parameters ξ_i and λ_i are then extracted from the experimental data of the ionic targets. Having obtained the values of ξ_i and λ_i for a particular *nl*, these parameters are then employed for all orbitals with the same l and different n. This means that the ionic parameters for the 1s, 2s, and 3s orbitals remain the same and likewise, those for the 2p and 3p are identical. The detailed procedure is outlined in Ref. 52. The optimized values of the parameters of the MBELL model are listed in Table 1.

2.4 NMBELL Model

Godunov and Ivanov[37] have shown that a better fit to the experimental EIICS data can be achieved if the polynomial series in the Bell model[36] or its modified version MBELL[49,51,52] is replaced by a nonlinear exponential term that results in a fewer number of parameters. Patoary et al.[59] incorporated this nonlinear feature and the quantum aspect of atomic radius[29] in the MBELL model to give rise to what is referred to as the NMBELL model. The total cross section for electron impact single ionization in the NMBELL model is expressed[59] as

$$\sigma_{\text{NMBELL}}(T) = \sum_i \pi r_i^2 N_i g_i F_{\text{ion}}^{i} \frac{R_{\text{G}}^{i}}{Z^{\alpha} U_i} \left[A_i \ \ln(U_i) + \exp\left\{ \sum_{k=1}^{2} B_{ik} \left(1 - \frac{1}{U_i} \right)^{k} \right\} \right], \tag{22}$$

where the summation over the index i refers to different ionized *nl*-orbitals. The weighting factor g_i, in line with Ref. 59, is parameterized in terms of D_i and b parameters as

Table 1 The MBELL Parameters A_i and B_{ik}'s and the Ionic Parameters ξ_i and λ_i for the 1s, 2s, 2p, 3s, 3p, 3d, and 4s Orbitals

	Parameter Values									
Orbits	A_i	B_{i1}	B_{i2}	B_{i3}	B_{i4}	B_{i5}	B_{i6}	B_{i7}	ξ_i	λ_i
1s	0.525	−0.5100	0.2000	0.0500	−0.025	−0.100	0.00	0.00	3.0	1.270
2s	0.530	−0.4100	0.1500	0.1500	−0.200	−0.150	0.00	0.00	3.0	1.270
2p	0.600	−0.4000	−0.7100	0.6550	0.425	−0.750	0.00	0.00	3.0	0.542
3s	0.130	0.2500	−1.5000	2.4000	3.220	−3.667	0.00	0.00	3.0	1.270
3p	0.388	−0.2000	−0.2356	0.5355	3.150	−8.500	5.05	0.370	3.0	0.542
3d	0.350	1.6000	−3.000	4.000	2.000	−5.000	−1.50	3.50	3.0	0.95
4s	0.130	0.2300	−1.6000	2.4000	3.220	−3.667	0.20	0.20	3.0	0.542

The parameters are in the unit of 10^{-13} eV^2 cm^2, excepting ξ_i and λ_i.

Table 2 The Parameters of the NMBELL Model for the Ionization from the 1s, 2s, 2p, 3s, 3p, 3d, and 4s Orbitals

Parameters	**Ionization (Target Status)**	**Orbits**						
		1s	**2s**	**2p**	**3s**	**3p**	**3d**	**4s**
B_{i1}	Neutral	0.50	0.55	1.10	0.45	0.4	0.150	0.35
	Ion	0.50	0.50	0.15	0.15	0.16	0.150	0.35
B_{i2}	Neutral	0.50	0.55	1.10	0.45	0.40	0.150	0.35
	Ion	0.50	0.50	0.15	0.15	0.15	0.150	0.35
D_i	Neutral	5.20	1.50	1.50	2.25	4.5	6.000	2.0
	Ion	4.00	1.20	6.40	2.00	2.0	12.00	2.0

$$g_i = \frac{D_i}{I_i}(1 + bq_i/Z), \tag{23}$$

to accommodate cases for various ionic targets. The ionic correction factor F^i_{ion} for the ith ionize orbital has already been expressed in Eq. (21). The modified relativistic factor R^i_{G} is expressed as given in terms of the Gryzinsky type correction factor G^i_{R}[33] in Eq. (20) as

$$R^i_{\text{G}}(U_i) = G^i_{\text{R}}(U_i)[1 + 2\ \ U_i^{0.2}/J_i^2]. \tag{24}$$

The NMBELL model fixes $A_i = 2.0$, $\alpha = 0.16$ in Eq. (22); $b = 0.4$ in Eq. (23); and $\xi_i = 1.6$ and $\lambda_i = 1.10$ in ionic correction factor (21), which are identical for all the contributing orbitals. All required values for the parameters B_{ik}'s and D_i are also listed in Table 2.

2.5 GKLV Model

In GKLV model,[53,57] the total EISIC is expressed as

$$\sigma_{\text{GKLV}} = \sum_i N_i R^i_{\text{F}}(\sigma^i_{\text{M}} + F^i_{\text{S}}\sigma^i_{\text{ph}}). \tag{25}$$

Here, the summation runs over all the ionized *nl*-orbitals. σ^i_{M} and σ^i_{ph} denote,[38,39] respectively, the EII cross sections due to close encounters that are dominated by the Møller interaction and due to distant collisions, which are primarily determined by the exchange effects of virtual photons.

The Møller part σ^i_{M} (barn unit) after introducing a first orbital-dependent generalized parameter η_i is given[38,39] by

$$\sigma_{\mathrm{M}}^{i} = \frac{\eta_i (T+1)^2}{I_i T(T+2)} \left[1 - \frac{I_i}{T} \left(1 - \frac{T^2}{2(T+1)^2} + \frac{2T+1}{(T+1)^2} \ln \frac{T}{I_i} \right) \right]. \tag{26}$$

The part σ_{ph}^{i} via the virtual exchange effect of photons in Refs. 38, 39 is modified following the procedure of Kim.[64] Here, the total effect of correlation and ionic effects is considered leading to replacement of the Burgess denominator $T(T+1)$[55] by the corresponding term in the effective kinetic energy[53,57] defined as

$$T_i' = T + \frac{\epsilon_i (\zeta_i + I_i)}{(q+1)^{1/2}}. \tag{27}$$

Here, ζ_i is again the kinetic energy of the bound electron and ϵ_i represents the second orbital-dependent generalized parameter for ionized orbital. In terms of T_i', the σ_{ph}^{i} part again in barn unit can be expressed[53,57] as

$$\sigma_{\mathrm{ph}}^{i} = \frac{0.141(T+1)^2}{I_i T_i' (T_i' + 2)} \left[\ln \frac{1.243(T+2)}{I_i} - \frac{T(T+2)}{(T+1)^2} \right]. \tag{28}$$

In Eqs. (26–28), T, I_i and ζ_i are expressed in natural unit i.e., in terms of the electron rest-mass energy. F_{S}^{i} in Eq. (25) is the scale down factor to correct for the overestimation of the cross sections from threshold to peak region caused by the σ_{ph}^{i} part. R_{F}^{i} also in Eq. (25) refers to the relativistic correction factor for both σ_{M}^{i} and σ_{ph}^{i} at ultrahigh energies. F_{S}^{i} is expressed[48] as

$$\begin{aligned} F_{\mathrm{S}}^{i} &= 2.5(1 - 1/U_i), \quad \text{for} \quad U_i \leq 1.70, \\ &= 1, \quad \text{for} \quad U_i > 1.70. \end{aligned} \tag{29}$$

The relativistic factor is given[53,57] by

$$R_{\mathrm{F}}^{i} = 1 + \xi U_i^{\lambda}, \tag{30}$$

with $\xi = 0.054$ and $\lambda = 0.067$.[48,53,57]

Values of the generalized parameters ϵ_i and η_i of the GKLV model for different ionized *nl*-orbitals are optimized by fitting the experimental EIICS data from various sources on the several chosen neutral and ionic targets using the criterion of reliability, discussed in Section 3. The resulting optimized parameters are listed in Table 3.

Table 3 The Values of Parameters ϵ_i and η_i, where *i* denotes *nl*, of the GKLV Model for the Inner and Outer Shell (Only for Neutral Target(NT)) Ionization in the 1s, 2s, 2p, 3s, 3p, 3d, and 4s Orbitals

		Orbits						
Parameters	**Ionization**	**1s**	**2s**	**2p**	**3s**	**3p**	**3d**	**4s**
ϵ_i	Inner-shell	2.00	1.00	1.00	2.00	0.30	0.10	1.50
	Outer-shell (NT)	2.50	2.00	2.00	0.50	1.50	–	1.50
η_i	Inner-shell	0.499	0.40	0.57	0.65	0.70	1.15	0.01
	Outer-shell (NT)	0.470	0.470	0.450	0.270	0.77	–	0.45

The parameters for an inner-shell also apply to the outer-shell of ionic targets.[53]

3. EXPERIMENTAL DATA SOURCES

Various techniques of measuring the EII cross sections have been provided in the excellent reviews by Powell.[65–67] Three techniques are usually employed for measuring K-, L-, and M-inner-shell ionization cross sections. These are (i) X-ray fluorescence, (ii) Auger electron spectroscopy and (iii) transmission electron energy loss spectroscopy. In the X-ray spectroscopy technique, cross sections are obtained from the measurement of fluorescence yield from X-rays emission due to occupying vacancies created by the EII in the inner shells. Auger electron spectroscopy however, involves measurements of yields of characteristic Auger electrons for determination of EIICS. Inner-shell ionization cross sections are also derived from the transmission electron energy loss experiments.

The experimental data are chosen with the following criteria: (i) the systematic error associated with the data collection is less than 20%, and (ii) in the case of multiple data sets for a given target, the data sets do not differ by more than 12% within the error bars. The data sources, used in this review, are from Refs. 68,69,88–191.

4. DISCUSSION

The calculated EISICS of the H to Sc isoelectronic series obtained from our MCN, GKLV, RQIBED/ MRIBED, MBELL and NMBELL models are compared in Figs. 1–26 with the experimental EIICS data, selected using the criteria of Section 2.1 and the predicted results from other useful theories available concerning the neutral and ionic targets up to U.

In Figs. 1 and 2, we compare the EIICS data and the calculated EISICS of the H-isoelectronic series with the DM results of Margreiter et al.[29] for H;

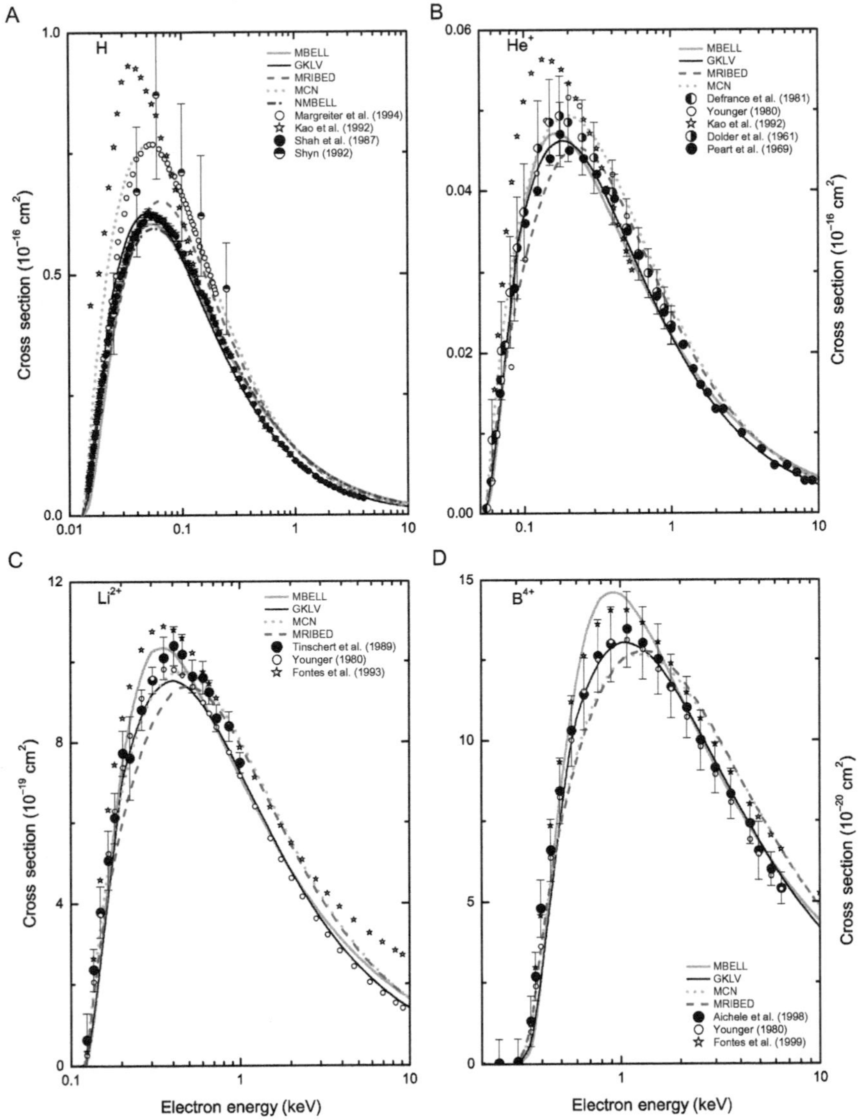

Figure 1 Electron impact ionization cross sections for neutral atoms and ions are plotted as a function of incident electron energy. EII cross sections for atomic targets: (A) H, (B) He^+, (C) Li^{2+}, and (D) B^{4+} in H-isoelectronic series. The sources of experimental EIICS data in solid symbols and half-solid symbols are from Refs. 68,88-93. Thick solid (green (gray in the print version)), thin solid (black), dash (red (dark gray in the print version)), short dash (cyan (light gray in the print version)), and dash dot (blue (black in the print version)) lines represent the calculated results using the MBELL, GKLV, MRIBED/RQIBED, MCN, and NMBELL models, respectively. Predicted EISICS of Younger,[2] Fontes et al.,[13] Margreiter et al.,[29] and Kao et al.[22] are also demonstrated.

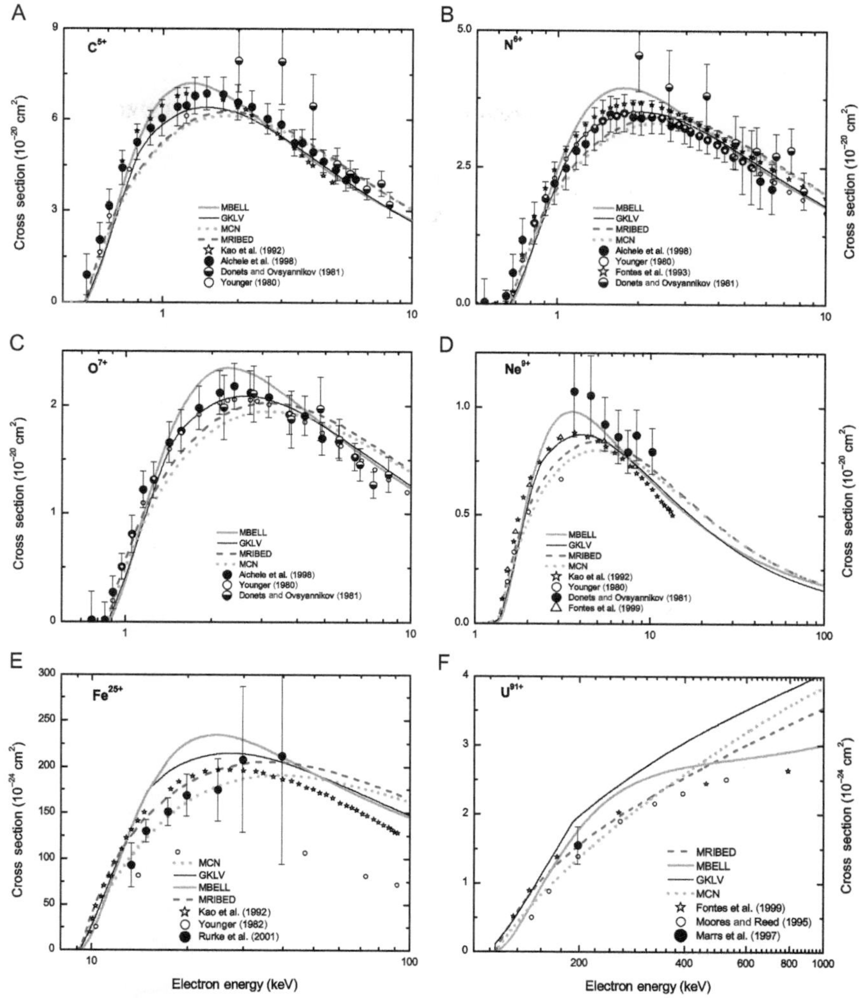

Figure 2 Same as Fig. 1 for (A) C^{5+}, (B) N^{6+}, (C) O^{7+}, (D) Ne^{9+}, (E) Fe^{25+}, and (F) U^{91+}. *The experimental data are from Refs. 93–96 and QM results from Refs. 2,7,12-14,22.*

the DWBA results of Younger[2,7] for He^{+}, Li^{2+}, B^{4+}, C^{5+}, N^{6+}, O^{7+}, Ne^{9+} and Fe^{25+}; the TPDWBA calculations of Kao et al.[22] for H, He^{+}, C^{5+}, Ne^{9+} and Fe^{25+}; the RDWBA calculations of Fontes et al.[12,13] for Li^{2+}, B^{4+}, N^{6+}, Ne^{9+} and U^{91+}, and of Moores and Reed[14] for U^{91+}.

In Figs. 3 and 4, we present the EIICS data and the EISICS findings on the He-isoelectronic series from calculations of DM[29] for He; of DWBA[3,4,70] for Li^{+}, B^{3+}, N^{5+}, O^{6+}, Ne^{8+}, and Fe^{24+}; of TPDWBS of Ref. 21 for He, Li^{+}, B^{3+}, N^{5+}, and Fe^{24+}; the RDWBA of Ref. 13 for Fe^{24+} and U^{90+}.

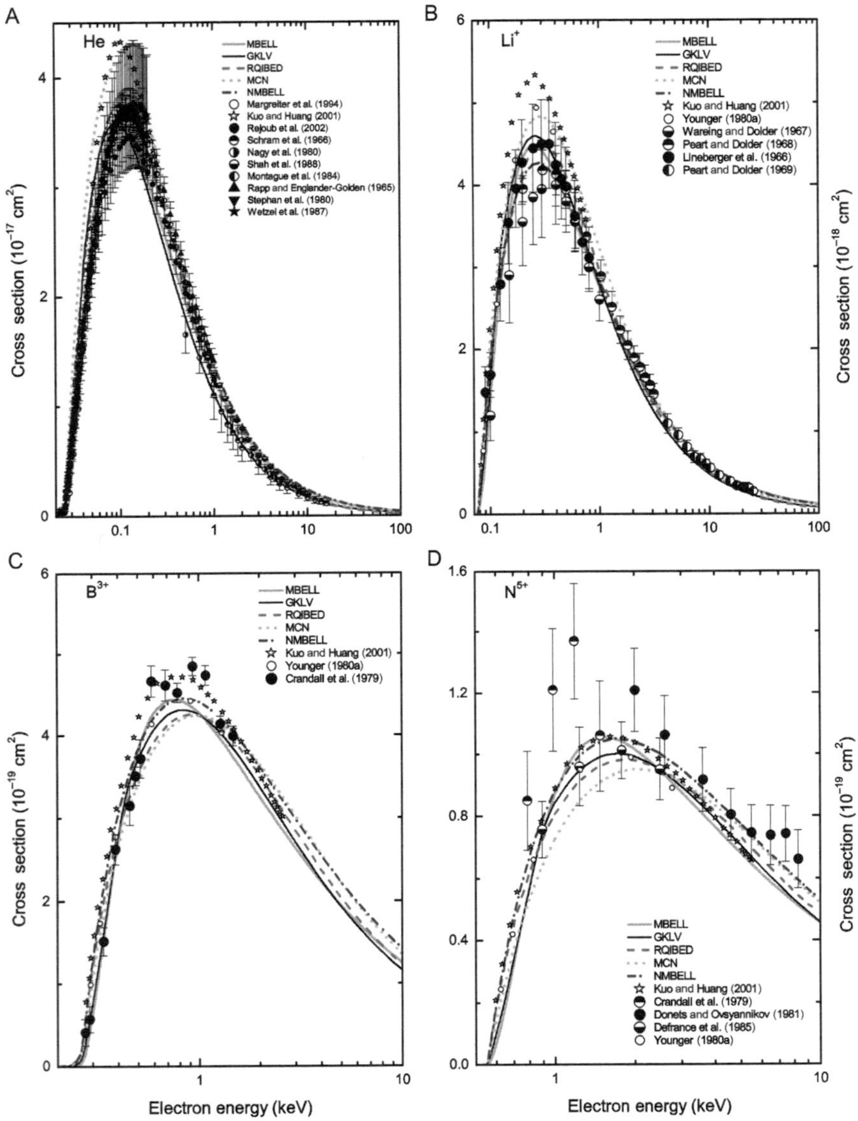

Figure 3 Same as Fig. 1 for EIICS in He-isoelectronic series of (A) He, (B) Li^+, (C) B^{3+}, and (D) N^{5+}. *The experimental data are from Refs. 94,97–110 and QM results from Refs. 21,29,70.*

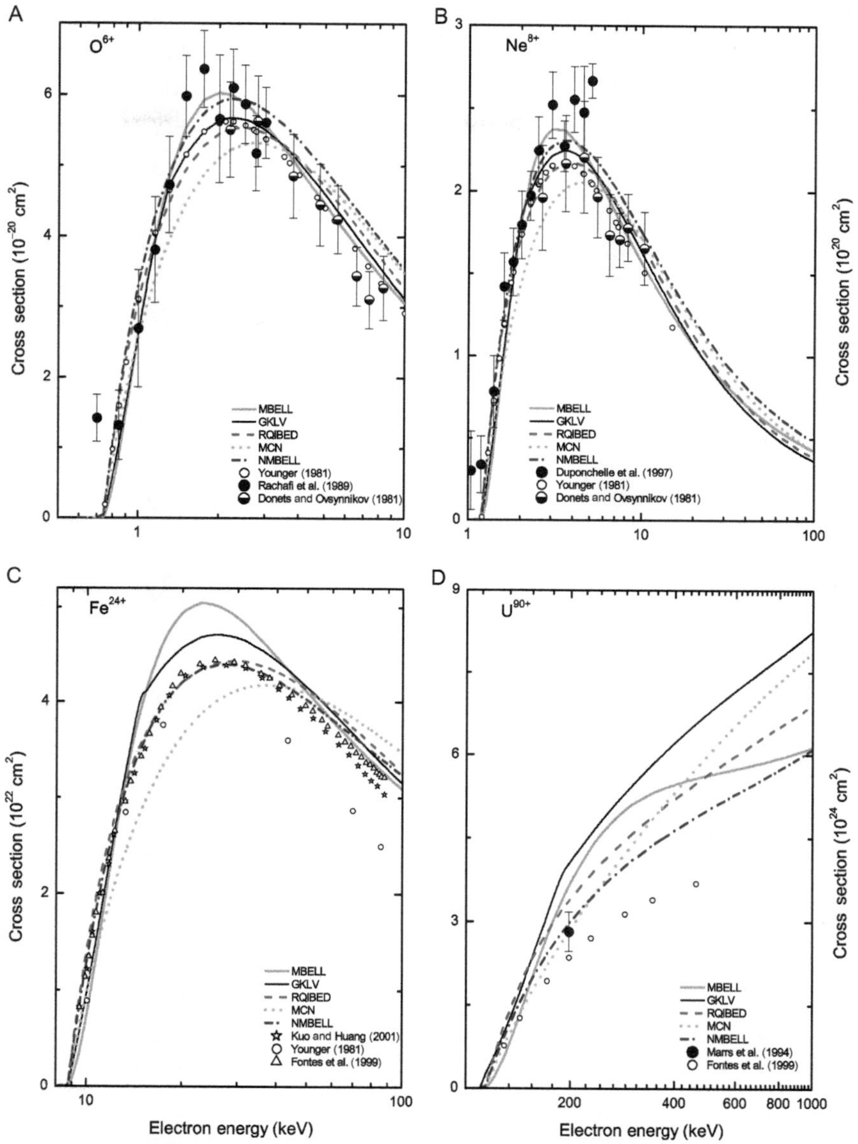

Figure 4 Same as Fig. 1 for EIICS in He-isoelectronic series of (A) O^{6+}, (B) Ne^{8+}, (C) Fe^{24+}, and (D) U^{90+}. *The experimental data are from Refs. 94,111–113 and QM results from Refs. 3,13.*

Figs. 5 and 6 show the comparison of the EIICS and EISICS for Li-isoelectronic series with the results of DM[29] for Li; of CCC[23] for Li, Be^{+}, N^{4+}, O^{5+}, and Ar^{15+}; of DWBA[2] for Be^{+} and O^{5+}; of CBAX[20] for N^{4+}, O^{5+}, and Ne^{7+}; of RDWBA[71] for U^{89+}.

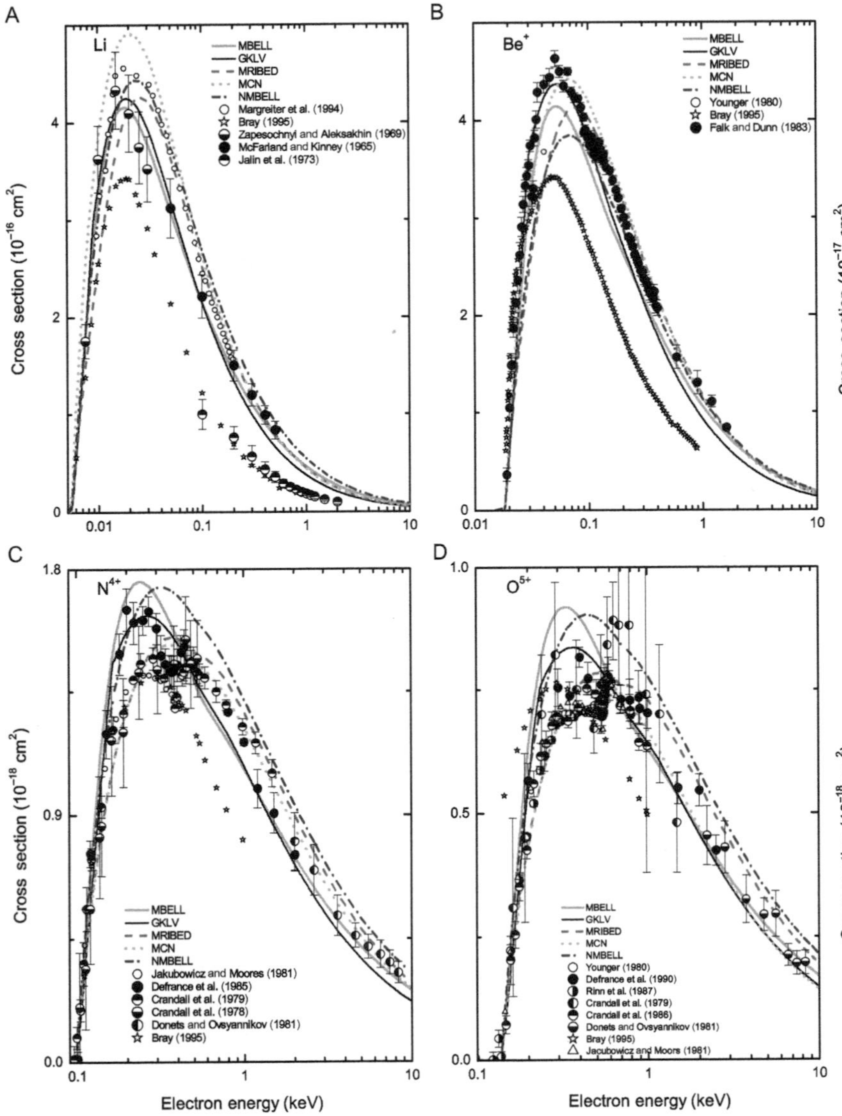

Figure 5 Same as Fig. 1 for EIICS in Li-isoelectronic series of (A) Li, (B) Be^{+}, (C) N^{4+}, and (D) O^{5+}. *The experimental data are from Refs. 94,107,108,114–121 and QM results from Refs. 2,20,23,29.*

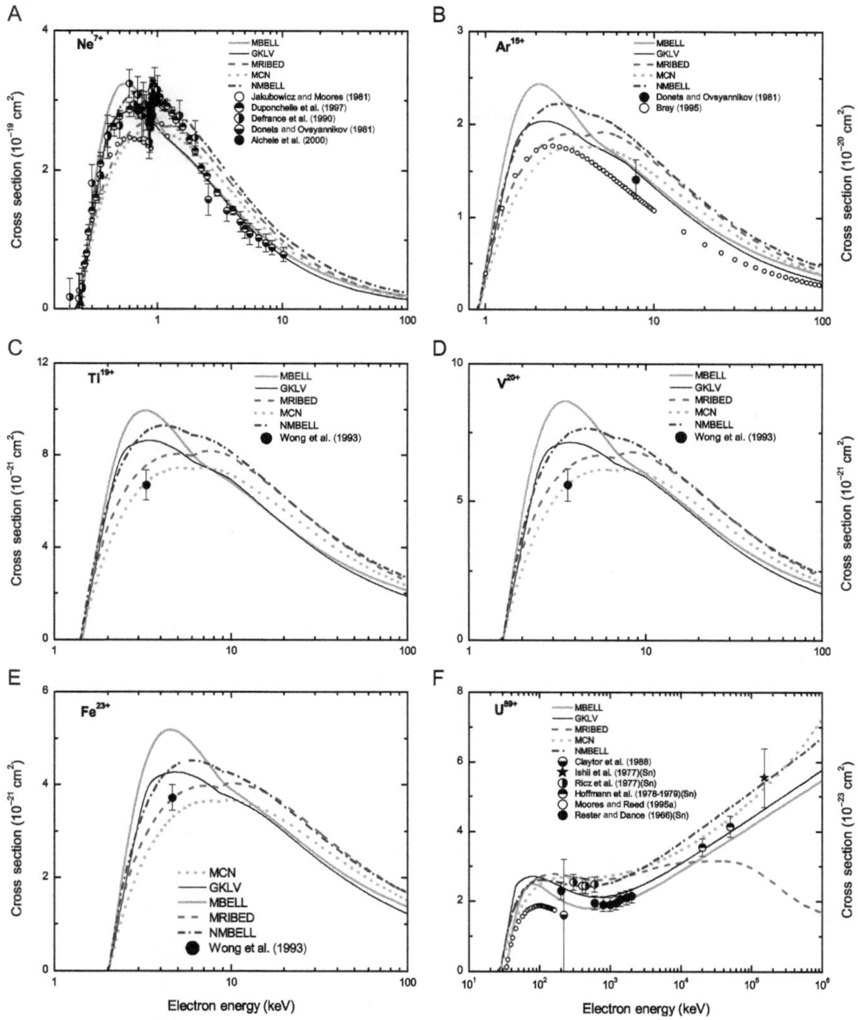

Figure 6 Same as Fig. 1 for EIICS in Li-isoelectronic series of (A) Ne^{7+}, (B) Ar^{15+}, (C) Ti^{19+}, (D) V^{20+}, (E) Fe^{23+}, and (F) U^{89+}. The experimental data in (F) for U^{89+} are evaluated from the normalized K-shell ionization cross sections of Sn (for details see the Ref. 51). *The experimental data are from Refs. 94,112,119,122–128 and QM results from Refs. 20,23,71.*

In Figs. 7 and 8, we compare the EIICS data and EISICS results of Be-isoelectronic series with the TPDWBA results of Chang et al.[72] for B^{+}, C^{2+}, N^{3+}, O^{4+}, Ne^{6+}, and Fe^{22+}; the DWBA results of Ref. 3 for B^{+}, C^{2+}, N^{3+}, and O^{+}; the CBAX results of Ref. 20 for C^{+}, N^{3+}, O^{4+}, Ne^{6+}, and Fe^{22+}; and the RDWBA results of Ref. 71 for U^{88+}.

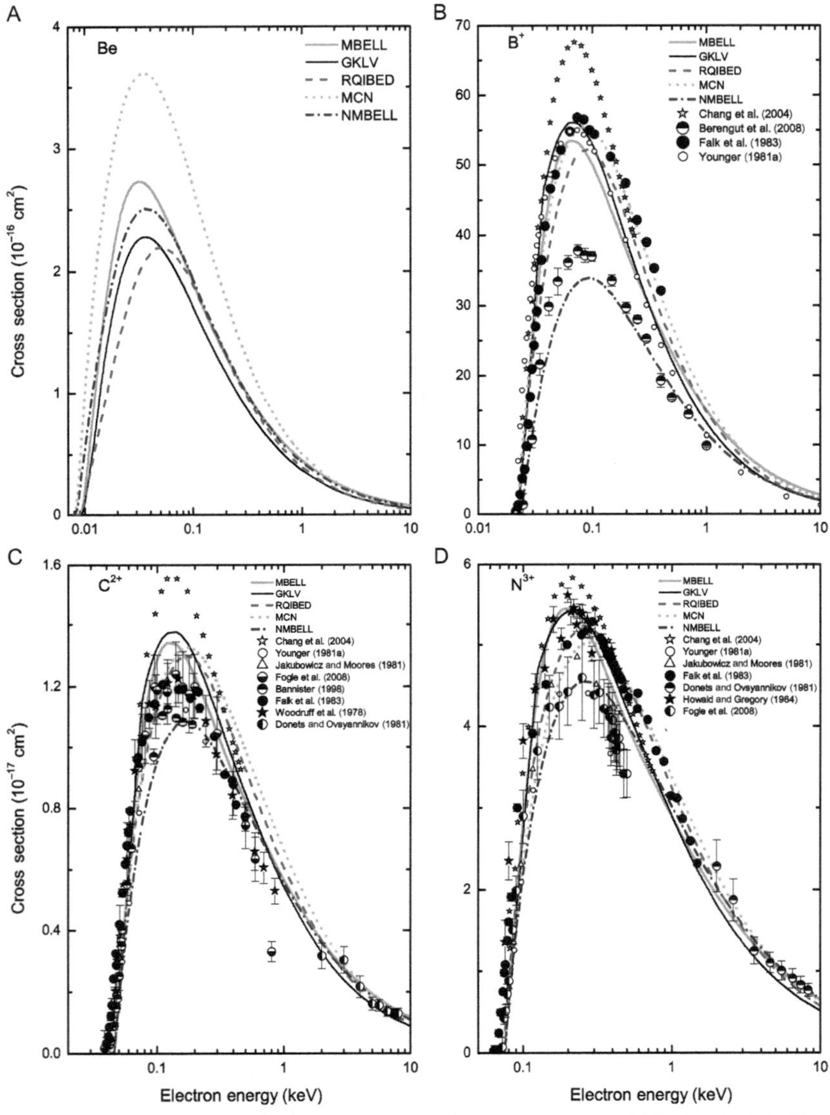

Figure 7 Same as Fig. 1 for EIICS in Be-isoelectronic series of (A) Be, (B) B^+, (C) C^{2+}, and (D) N^{3+}. *The experimental data are from Refs. 94,117,129–133 and QM results from Refs. 4,20,72.*

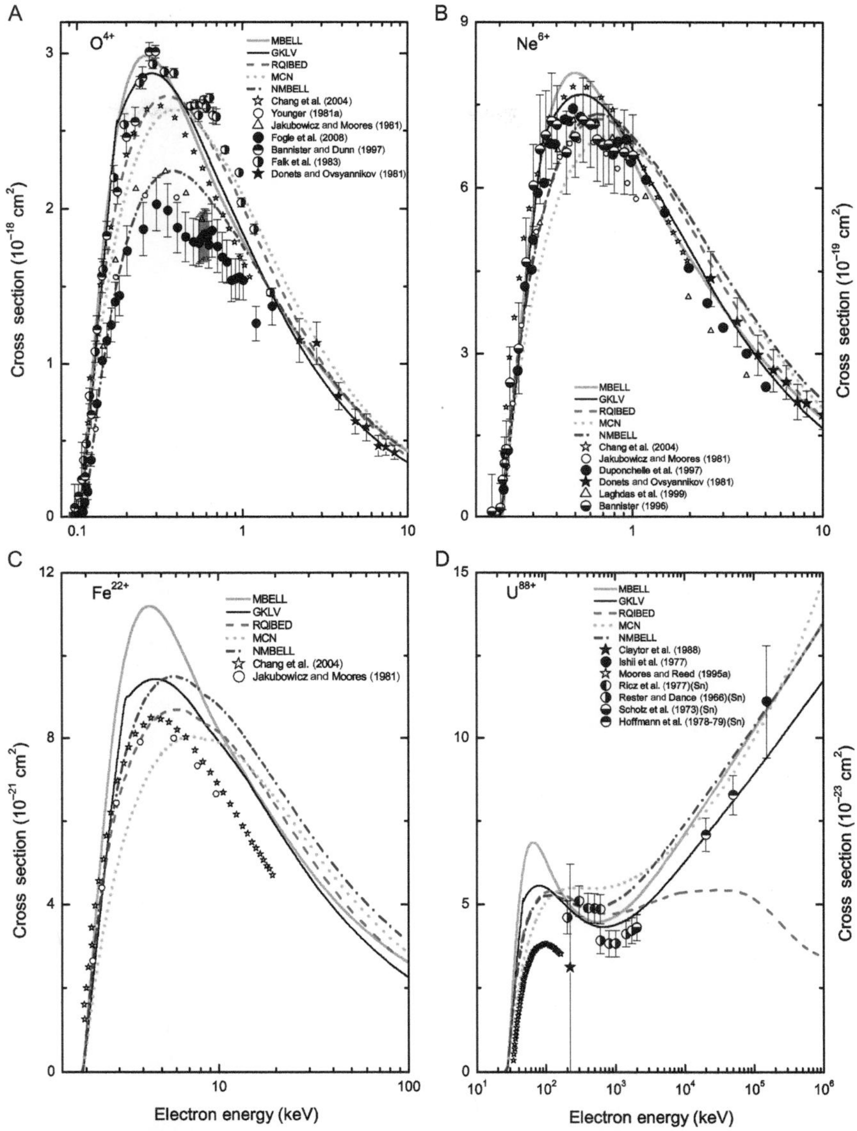

Figure 8 Same as Fig. 1 for EIICS in Be-isoelectronic series of (A) O^{4+}, (B) Ne^{6+}, (C) Fe^{22+}, and (D) U^{88+}. The experimental data in (D) for U^{88+} are evaluated from the normalized K-shell ionization cross sections of Sn (for details see Ref. 51). *The experimental data are from Refs. 94,112,117,124–128,130,134–136 and QM results from Refs. 4,15,20,71,72.*

In Fig. 9, the EIICS and EISICS data of B-isoelectronic series are displayed along with the calculations from CBAX[19] for B; from the scaled PWBA

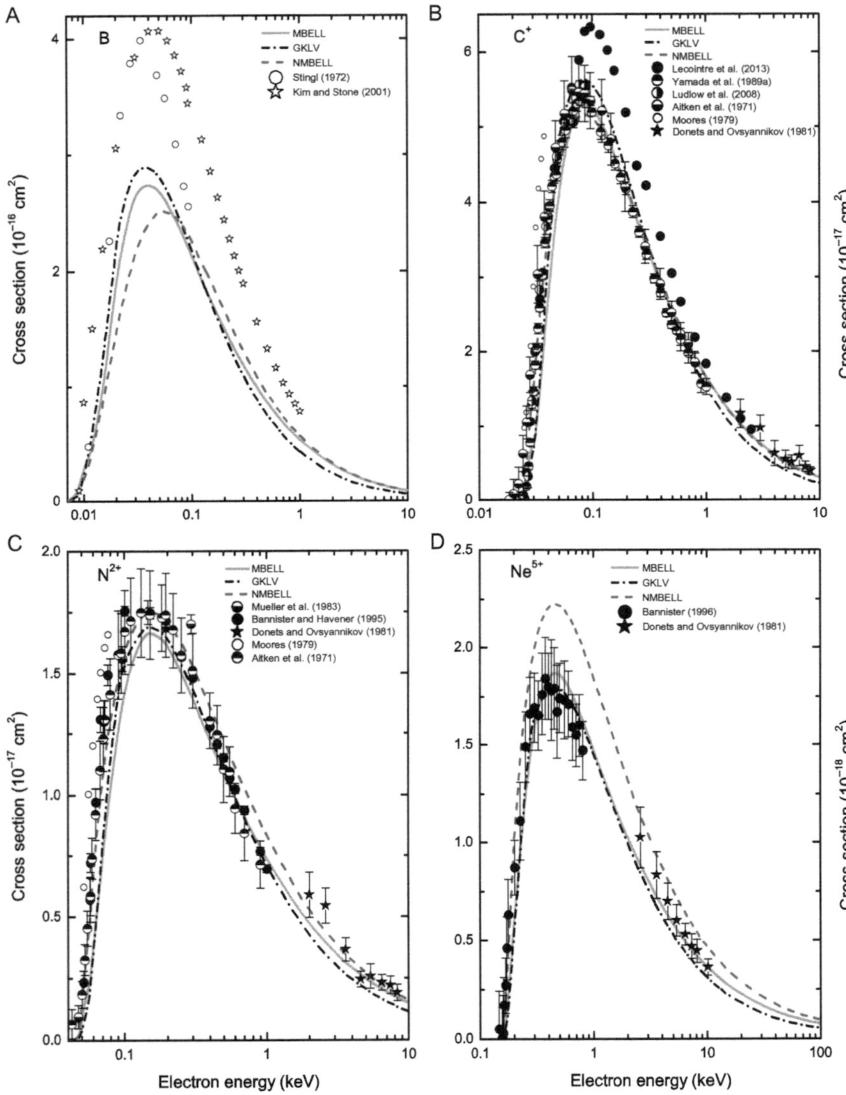

Figure 9 Electron impact ionization cross sections for neutral atoms and ions are plotted as a function of incident electron energy. EII cross sections for atomic targets: (A) B, (B) C^+, (C) N^{2+}, and (D) Ne^{5+} in B-isoelectronic series. The sources of experimental EIICS data in solid symbols and half-solid symbols are from Refs. 94,135,137–142 and QM results from Refs. 19,73,74. Solid (green (gray in the print version)), dash dot (black), and dash (red (dark gray in the print version)) lines represent the calculated results using the MBELL, GKLV, and NMBELL models, respectively.

including autoionization contribution (SPWBA) of Kim and Stone[73] for B; and from autoionization contribution of Moores[74] for C^+ and N^{2+}.

In Fig. 10, the EIICS and EISICS data of C-isoelectronic series are compared with the results from DM[29] for C; from CBA[18] for N^+; and from

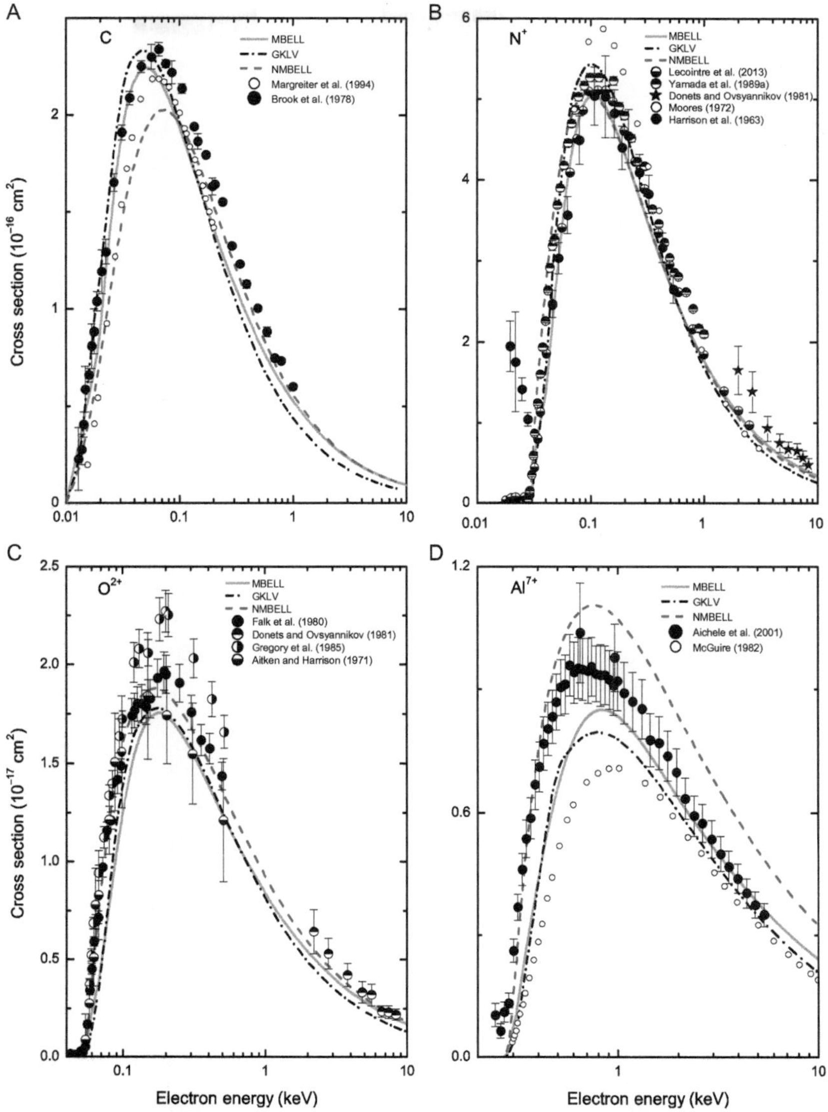

Figure 10 Same as Fig. 9 for EIICS in C-isoelectronic series of (A) C, (B) N^+, (C) O^{2+}, and (D) Al^{7+}. *The experimental data are from Refs. 94,137,138,140,143–147 and QM results from Refs. 18,29,75.*

generalized oscillator strength formulation of the Born approximation (GOSBA) of McGuire[75] for Al^{7+}.

In Fig. 11, the EIICS data and EISICS results of N-isoelectronic series are compared with those from DM[28,29] for N; from CBA[18] for O^+; and GOSBA[75] for Al^{6+}.

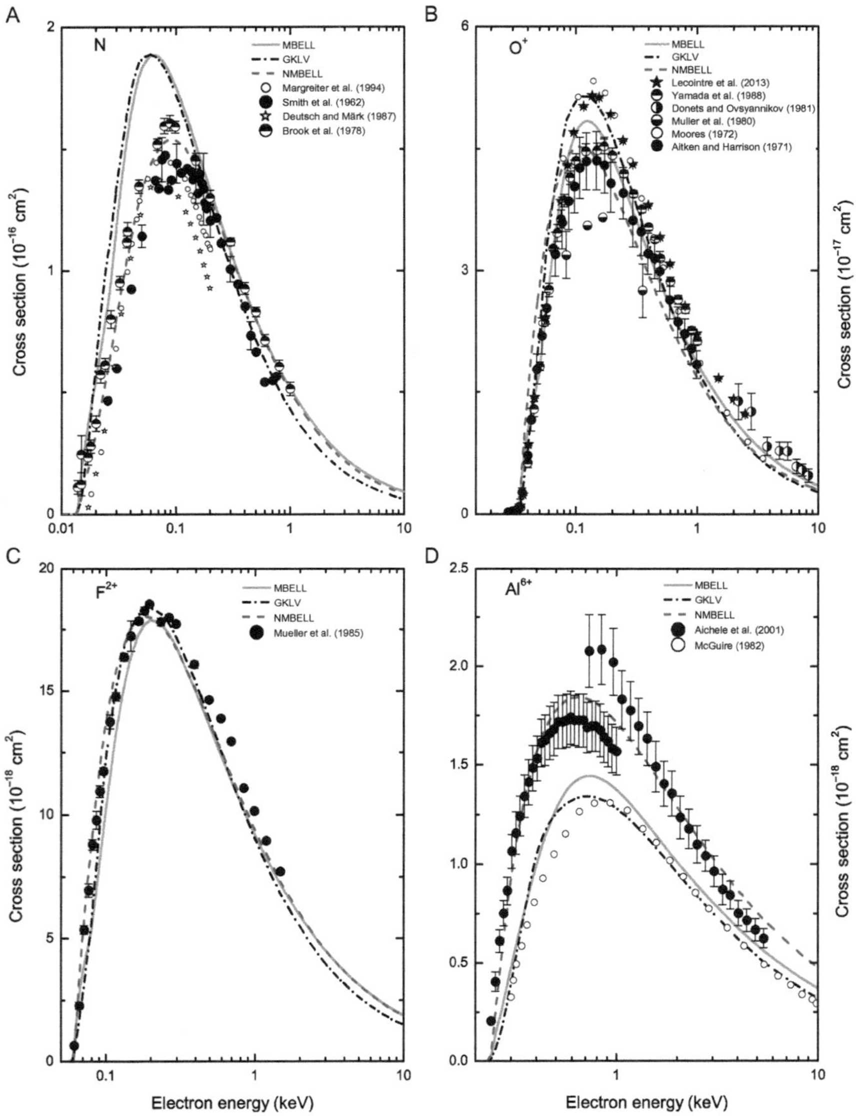

Figure 11 Same as Fig. 9 for EIICS in N-isoelectronic series of (A) N, (B) O^+, (C) F^{2+}, and (D) Al^{6+}. *The experimental data are from Refs. 94,137,140,143,147–150 and QM results from Refs. 18,28,29,75.*

The EIICS data and EISICS results are displayed and compared with the predicted values from calculations with DM[29] for O in Fig. 12; and with DM[28,29] for F, CBA[18] for Ne^+, and GOSBA[75] for Al^{4+} inside the F-isoelectronic series in Fig. 13.

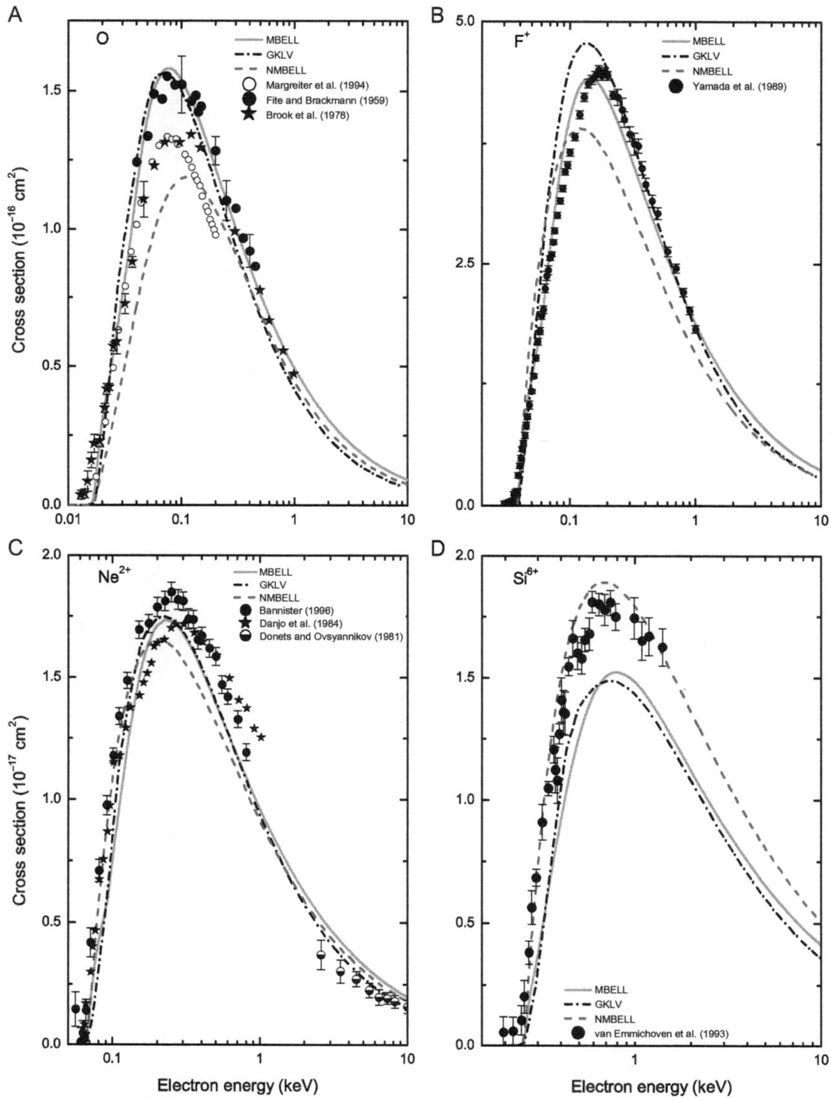

Figure 12 Same as Fig. 9 for EIICS in O-isoelectronic series of (A) O, (B) F^+, (C) Ne^{2+}, and (D) Si^{6+}. *The experimental data are from Refs. 94,135,143,151–154 and QM results from Ref. 29.*

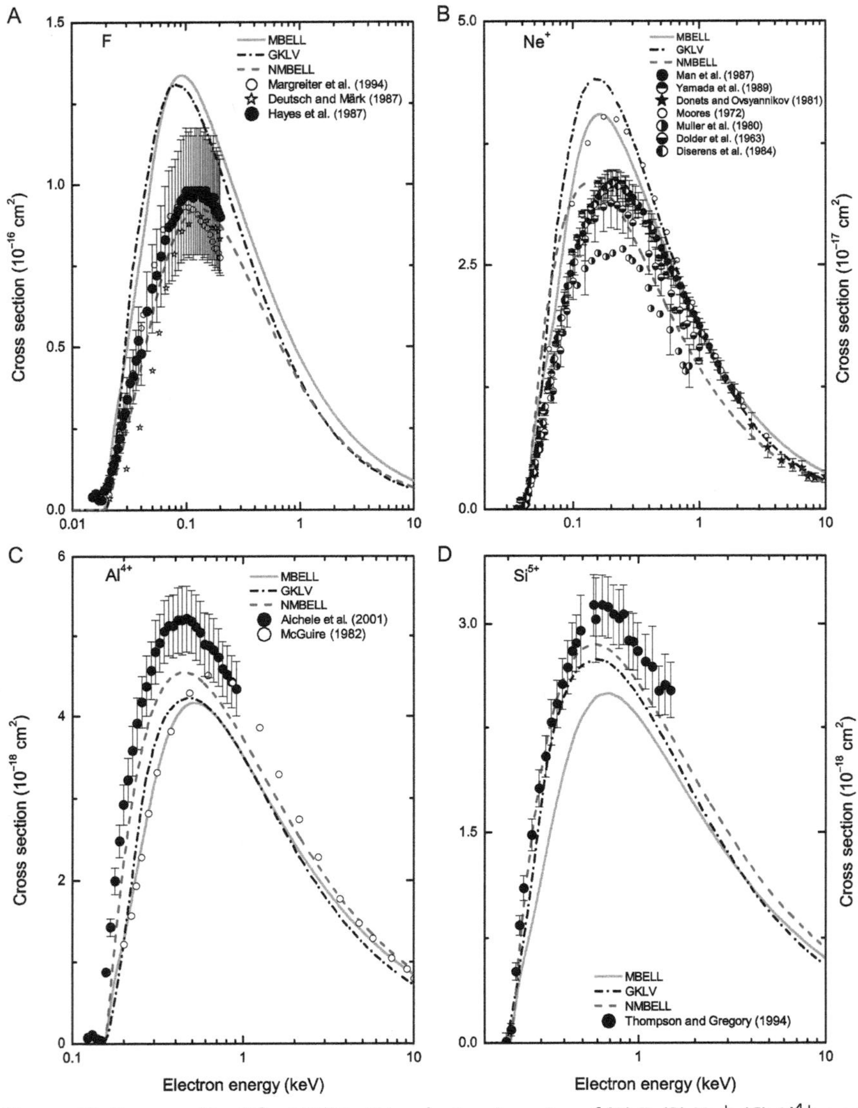

Figure 13 Same as Fig. 9 for EIICS in F-isoelectronic series of (A) F, (B) Ne^+, (C) Al^{4+}, and (D) Si^{5+}. *The experimental data are from Refs. 94,147,151,155–160 and QM results from Refs. 18,28,29,75.*

In Figs. 14 and 15, we compare the EIICS and EISICS data of Ne-isoelectronic series with the results of DM[18] and PWBA[1] for Ne; DWBA[5,7] for Na^+, Mg^{2+}, Ar^{8+}, and Fe^{16+}; GOSBA[75] for Al^{3+}; CBA[18] for Na^+ and Mg^{2+}; and DWBA of Hahn[76,77] for Fe^{16+}.

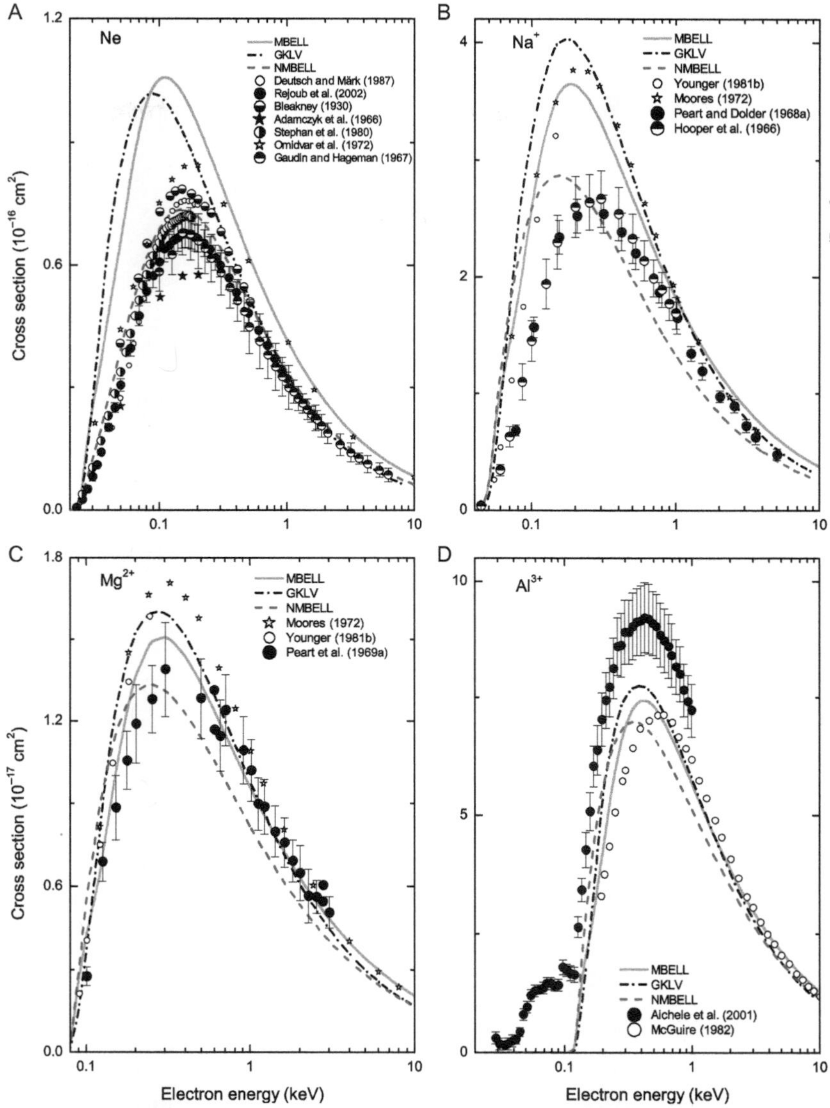

Figure 14 Same as Fig. 9 for EIICS in Ne-isoelectronic series of (A) Ne, (B) Na^+, (C) Mg^{2+}, and (D) Al^{3+}. *The experimental data are from Refs. 97,103,147,161–166 and QM results from Refs. 1,5,18,28,75.*

The EIICS data and EISICS results are presented for comparison with the findings from the calculations of DM,[29] PWBA,[1] and CCC[78] for Na, DWBA,[6] and CBAX of Moores and Nussbaumer[17] for Mg^+ and Al^{2+},

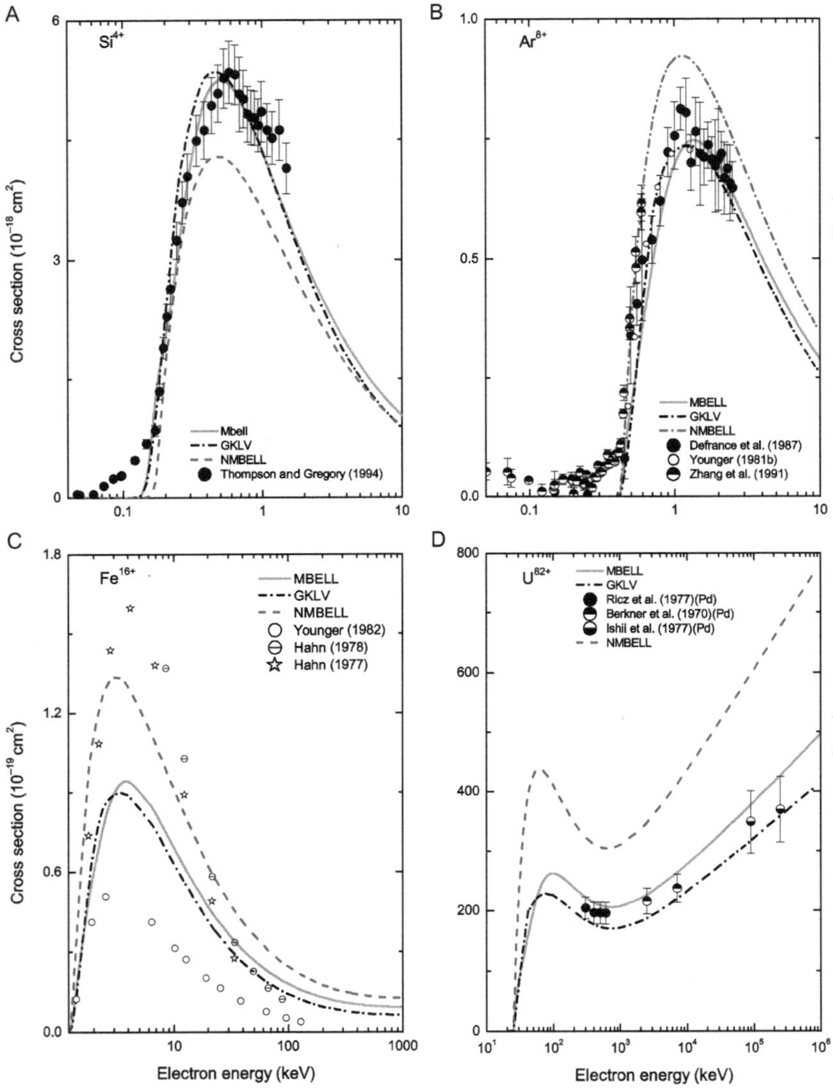

Figure 15 Same as Fig. 9 for EIICS in Ne-isoelectronic series of (A) Si^{4+}, (B) Ar^{8+}, (C) Fe^{16+}, and (D)U^{82+}. The experimental data in (D) for U^{82+} are evaluated from the normalized K-shell ionization cross sections of Pd (details in Ref. 51). *The experimental data are from Refs. 125,126,160,167–169 and QM results from Refs. 5,7,76,77.*

and Coulomb–Born–Oppenheimer approximation (CBOA) of Golden and Sampson[79] for Al^{2+} and Si^{3+} ions belonging to Na-isoelectronic series in Fig. 16; DM[29] for Mg, and DWBA of Pindzola et al.[80] for S^{4+}, Cl^{5+} and Ar^{6+} ions belonging to Na-isoelectronic series in Fig. 17; and SPWBA[73]

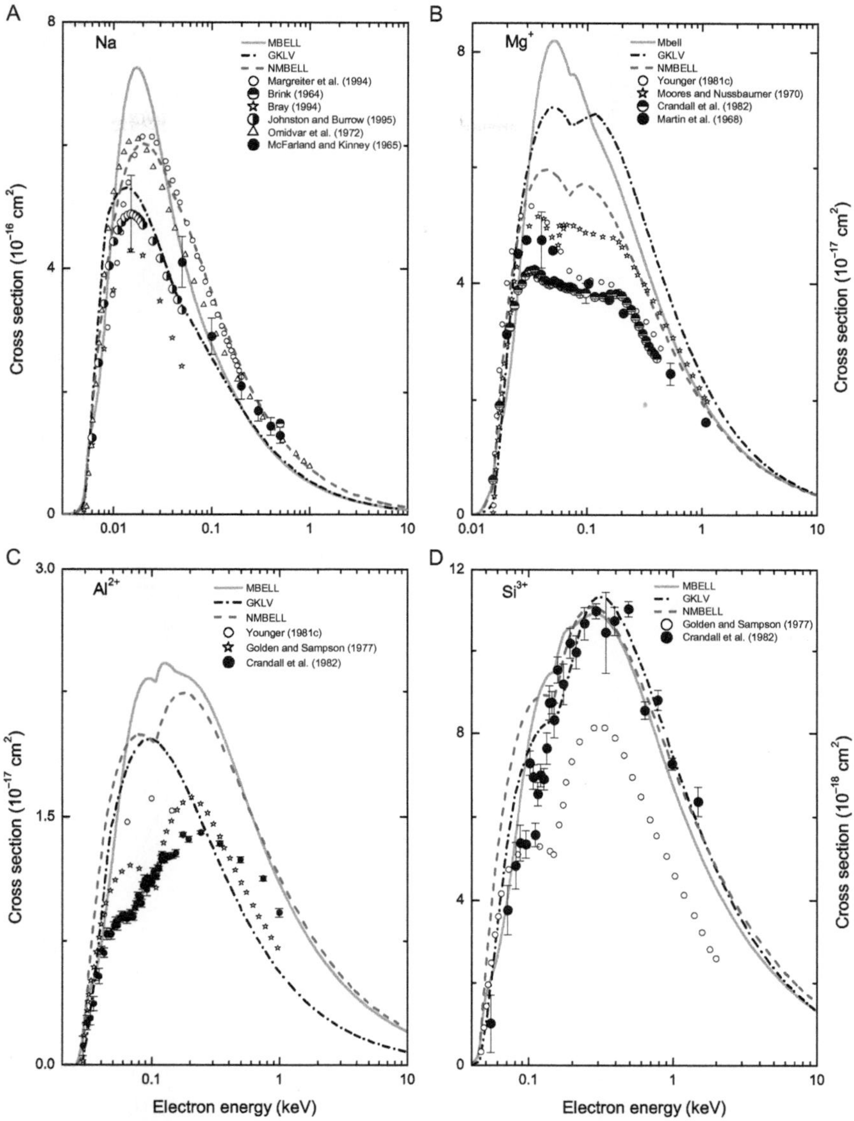

Figure 16 Same as Fig. 9 for EIICS in Na-isoelectronic series of (A) Na, (B) Mg^+, (C) Al^{2+}, and (D) Si^{3+}. *The experimental data are from Refs. 115,170–173 and QM results from Refs. 1,6,17,29,78,79.*

for Al; GOSBA[75] for Al; and DWBA[80] for Fe^{13+} ions belonging to Al-isoelectronic series in Fig. 18.

We compare the EIICS and EISICS results with the available calculations of DM[29] for Si belonging to the Si-isoelectronic series in Fig. 19;

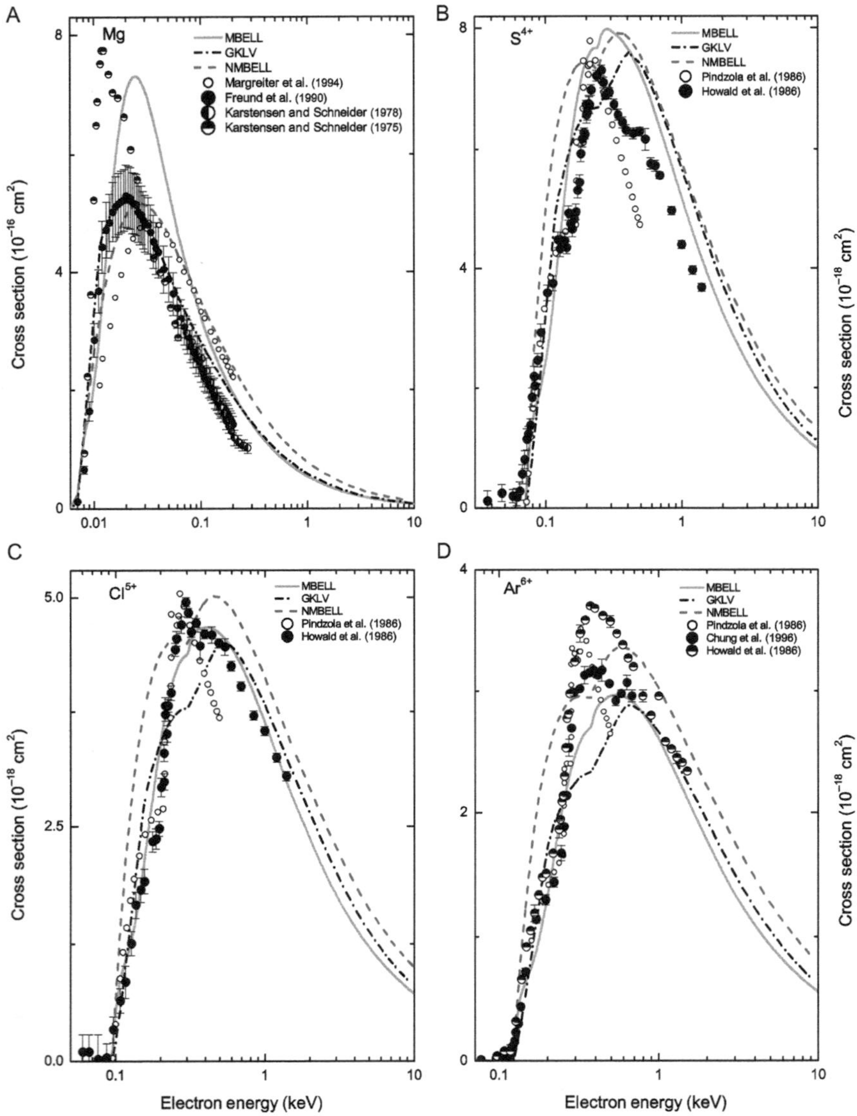

Figure 17 Same as Fig. 9 for EIICS in Mg-isoelectronic series of (A) Mg, (B) S^{4+}, (C) Cl^{5+}, and (D) Ar^{6+}. *The experimental data are from Refs. 174–178 and QM results from Refs. 29,87.*

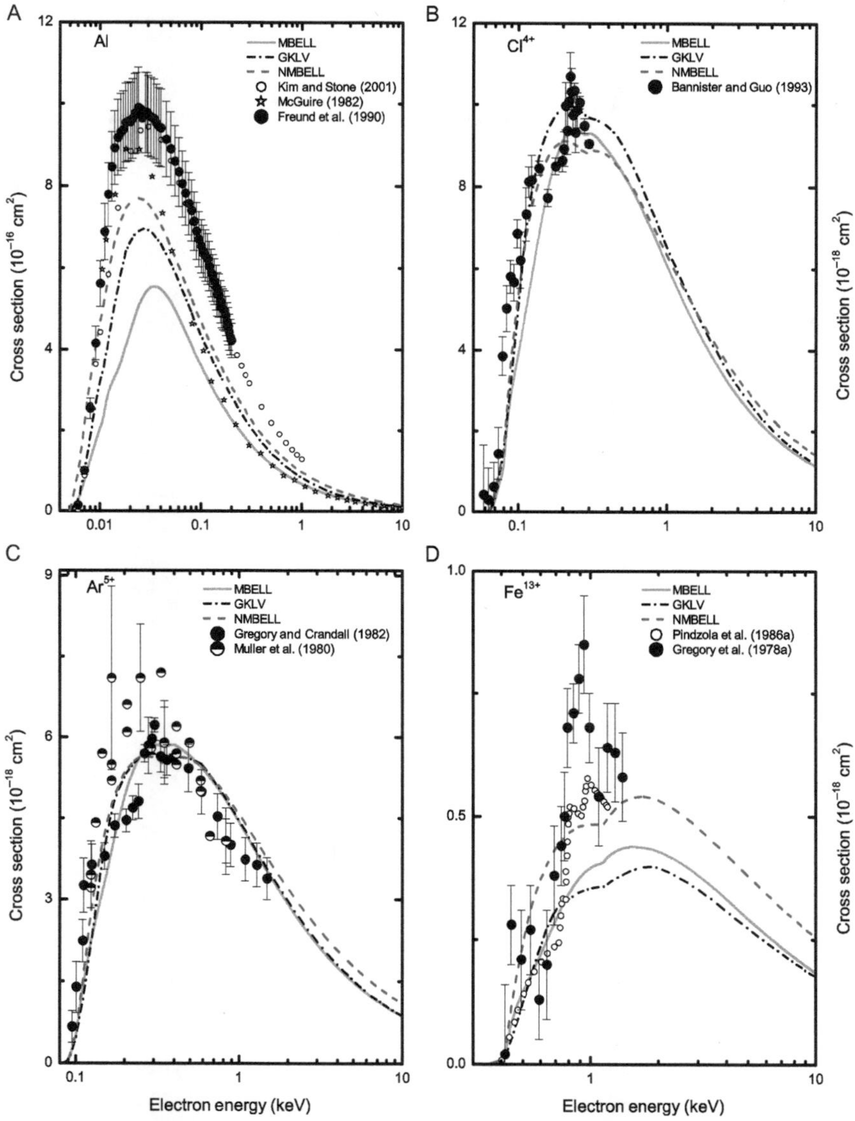

Figure 18 Same as Fig. 9 for EIICS in Al-isoelectronic series of (A) Al, (B) Cl^{4+}, (C) Ar^{5+}, and (D) Fe^{13+}. *The experimental data are from Refs. 155,174,179–181 and QM reesults from Refs. 73,75,80.*

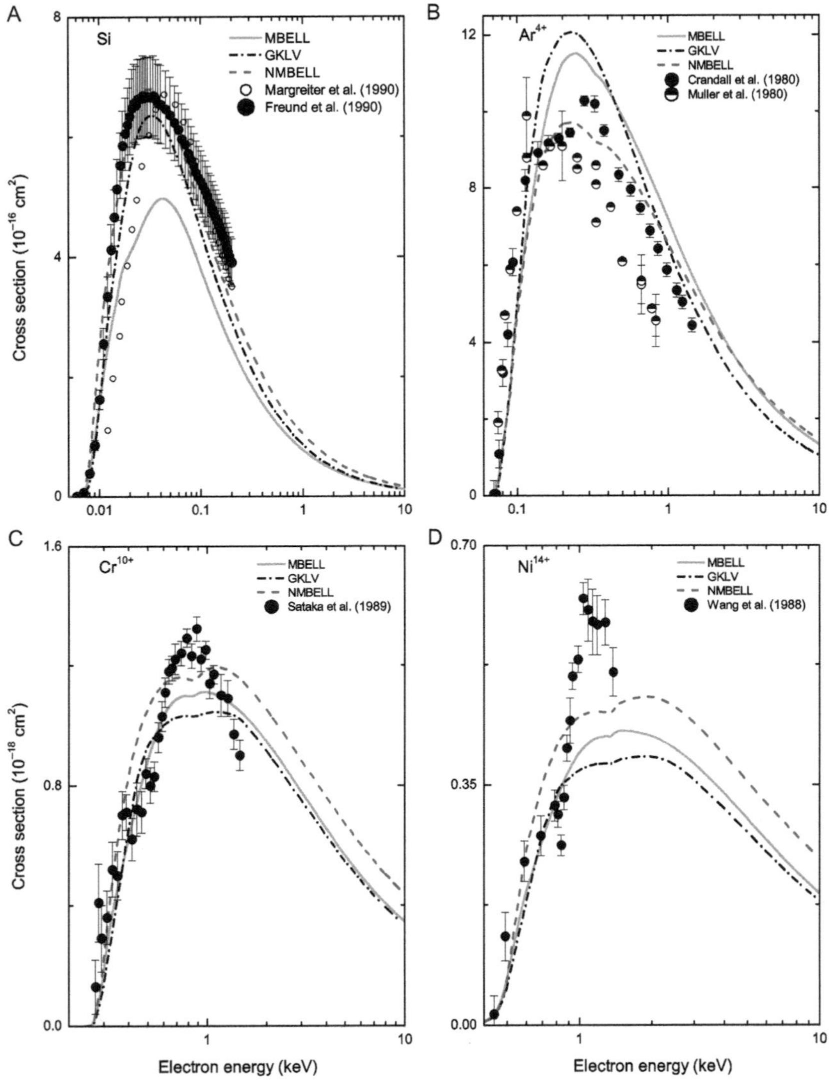

Figure 19 Same as Fig. 9 for EIICS in Si-isoelectronic series of (A) Si, (B) Ar^{4+}, (C) Cr^{10+}, and (D) Ni^{14+}. *The experimental data are from Refs. 107,155,174,182,183 and QM results from Ref. 29.*

DM[29] for P, and DWBA for Fe^{11+} belonging to the P-isoelectronic series in Fig. 20; DM[29] for S belonging to the S-isoelectronic series in Fig. 21; and DM[29] for Cl, DWBA for Fe^{9+} belonging to the Cl-isoelectronic series in Fig. 22.

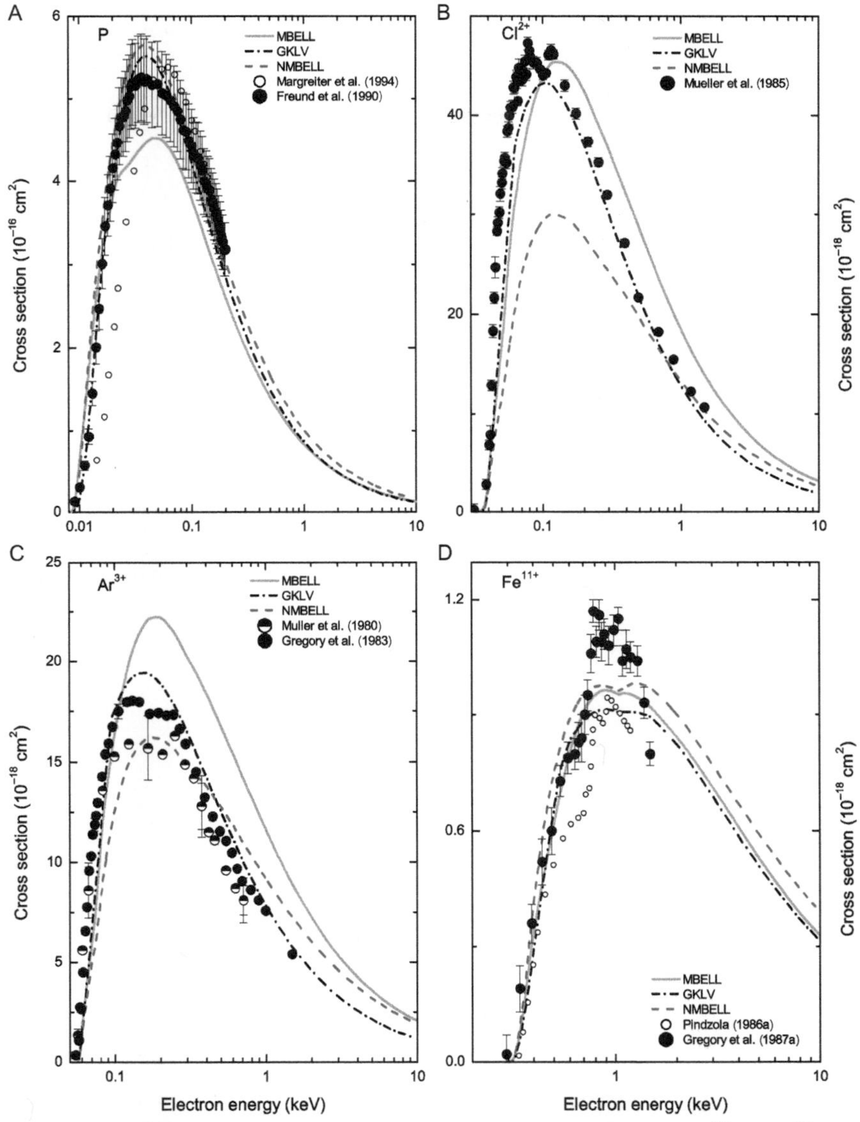

Figure 20 Same as Fig. 9 for EIICS in P-isoelectronic series of (A) P, (B) Cl^{2+}, (C) Ar^{3+}, and (D) Fe^{11+}. *The experimental data are from Refs. 150,155,174,181,184 and QM results from Refs. 29,80.*

The EIICS and EISICS results are compared with the available calculations of DM,[29] the BEA of Jha et al.[81] for Ar, DWBA of Younger,[8] BA of Peach[16] and BEA of Kumar and Roy[82] for K^+ belonging to the Ar-isoelectronic series in Fig. 23; DM,[29] PWBA,[1] and GOSBA[83] for K,

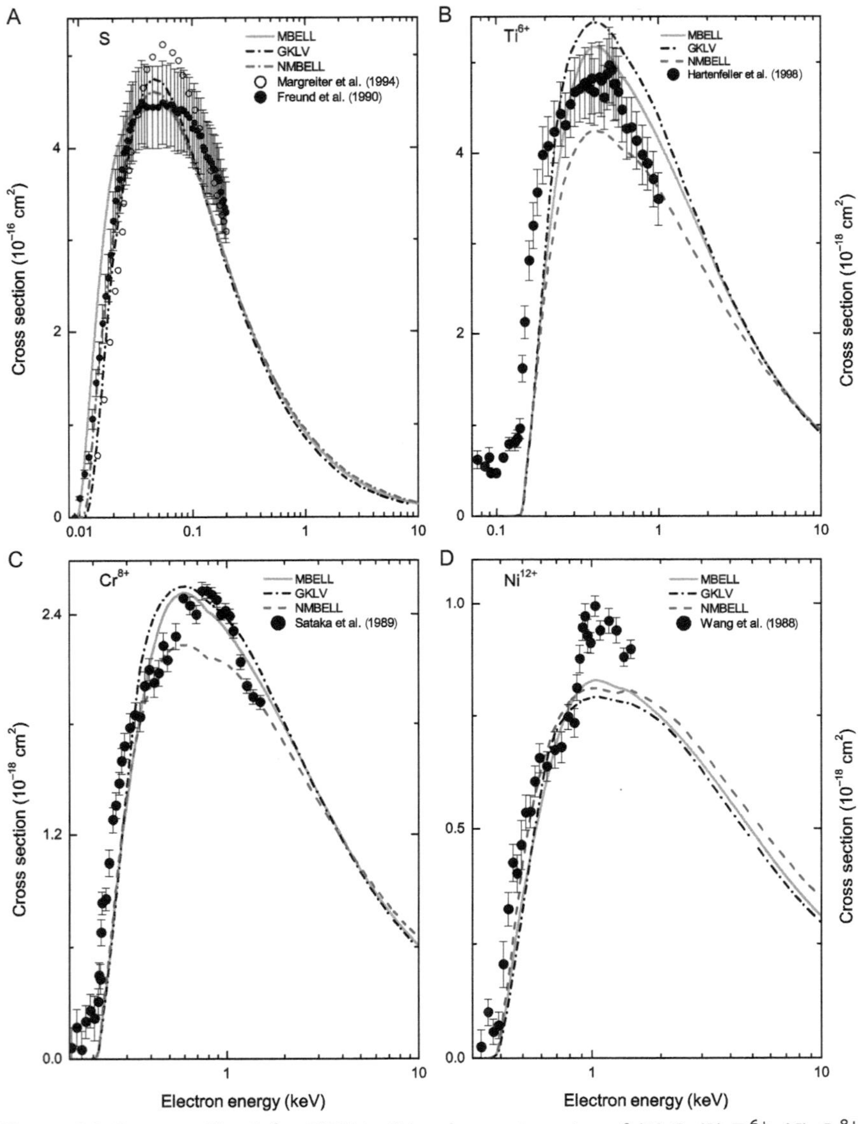

Figure 21 Same as Fig. 9 for EIICS in S-isoelectronic series of (A) S, (B) Ti^{6+}, (C) Cr^{8+}, and (D) Ni^{12+}. *The experimental data are from Refs. 174,182,183,185 and QM results from Ref. 29.*

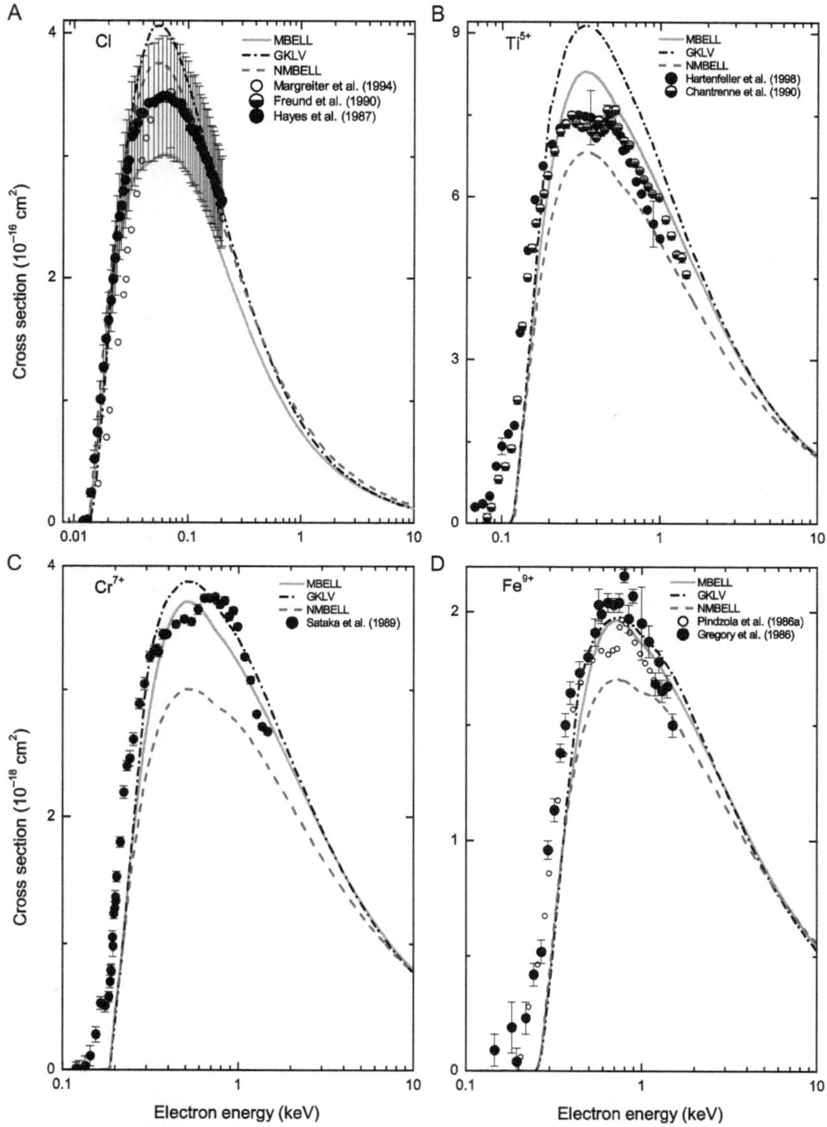

Figure 22 Same as Fig. 9 for EIICS in Cl-isoelectronic series of (A) Cl, (B) Cr^{7+}, (C) Mn^{8+}, and (D) Fe^{9+}. *The experimental data are from Refs. 156,174,182,185–187 and QM results from Refs. 29,80.*

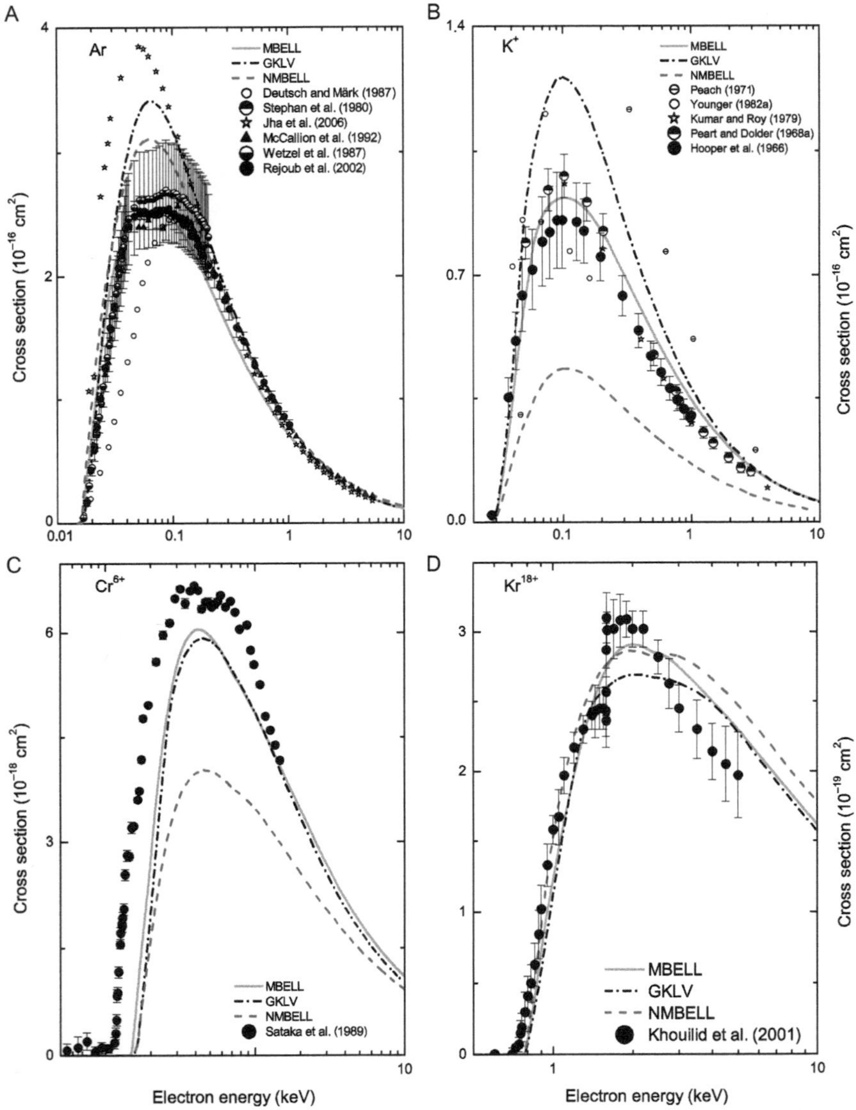

Figure 23 Same as Fig. 9 for EIICS in Ar-isoelectronic series of (A) Ar, (B) K^+, (C) Cr^{6+}, and (D) Kr^{18+}. *The experimental data are from Refs. 97,103,104,161,166,182,188,189 and QM results from Refs. 8,16,28,81,82.*

DWBA of Colgan et al.[84] for Mn^{6+}, and DWBA of Pindzola et al.[85] for Fe^{7+} belonging to the K-isoelectronic series in Fig. 24; DM[29] and McGuire et al. [83] for Ca belonging to the Ca-isoelectronic series in Fig. 25; and none belonging to the Sc-isoelectronic series in Fig. 26.

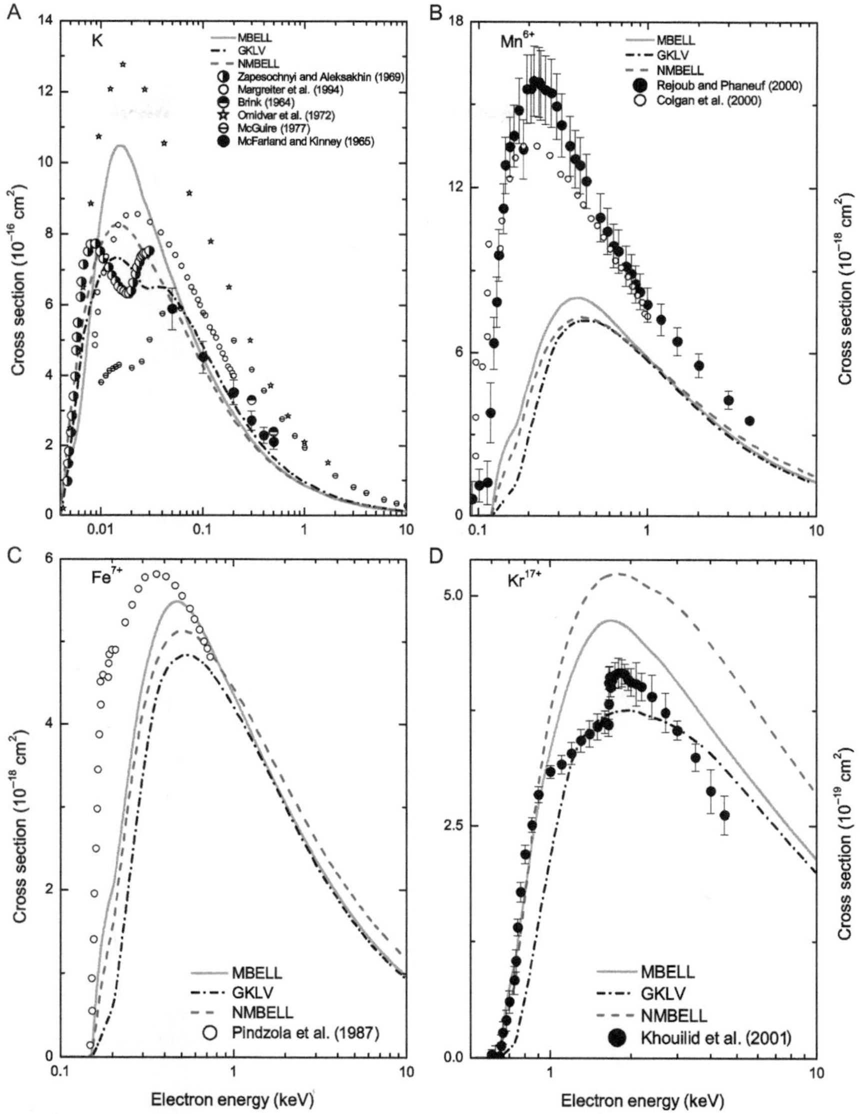

Figure 24 Same as Fig. 9 for EIICS in K-isoelectronic series of (A) K, (B) Mn^{6+}, (C) Fe^{7+}, and (D) Kr^{17+}. *The experimental data are from Refs. 114,115,170,188,190 and QM results from Refs. 1,29,83–85.*

Before discussing the relative performance of our MCN, RQIBED/MRIBED, MBELL, NMBELL, and GKLV models, let us delve into the effect of different corrections incorporated in the respective parent model to wide their applications. The consideration of the correlation

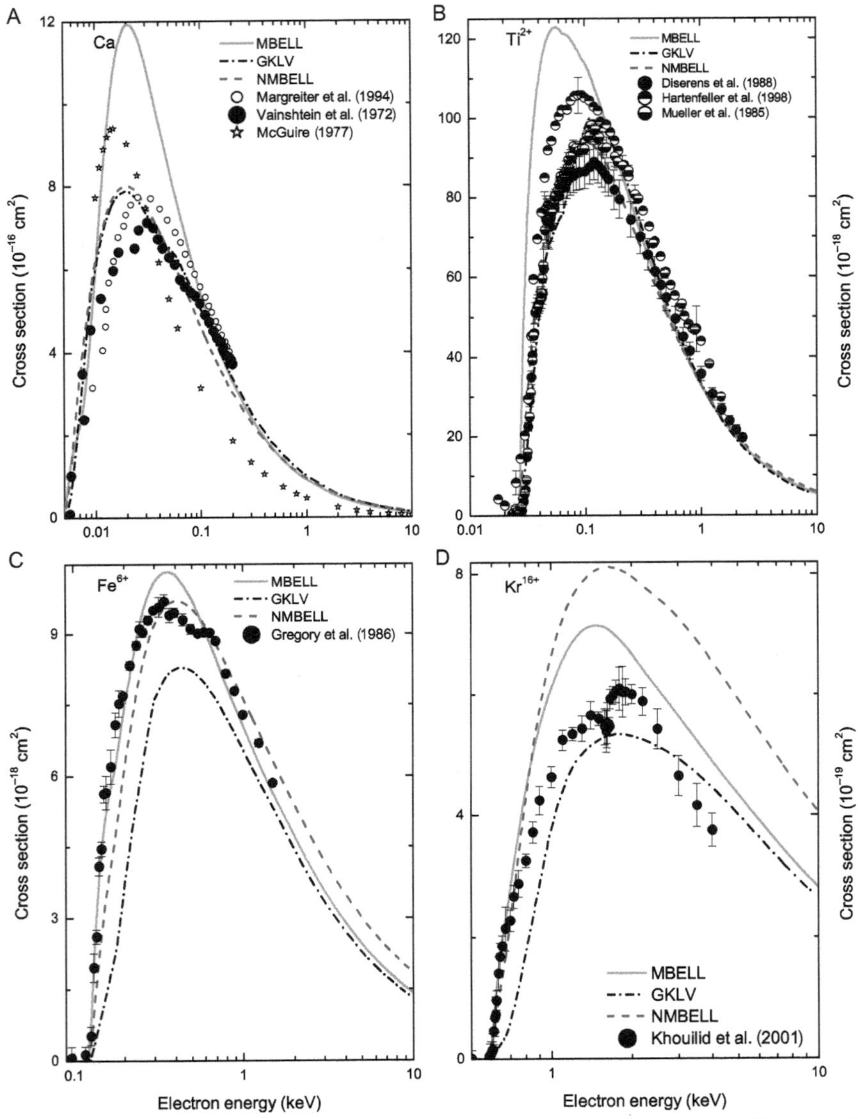

Figure 25 Same as Fig. 9 for EIICS in Ca-isoelectronic series of (A) Ca, (B) Ti^{2+}, (C) Fe^{6+}, and (D) Kr^{16+}. *The experimental data are from Refs. 69,150,185,187,188,191 and QM results from Refs. 29,75.*

effect and ionic and relativistic corrections lead to the generalization of the relevant model parameters in terms of the quantum numbers *nl* of the ionized orbital. Apart from embodying the corrections, the individual performance also depends on the inherent structure as well as the incident

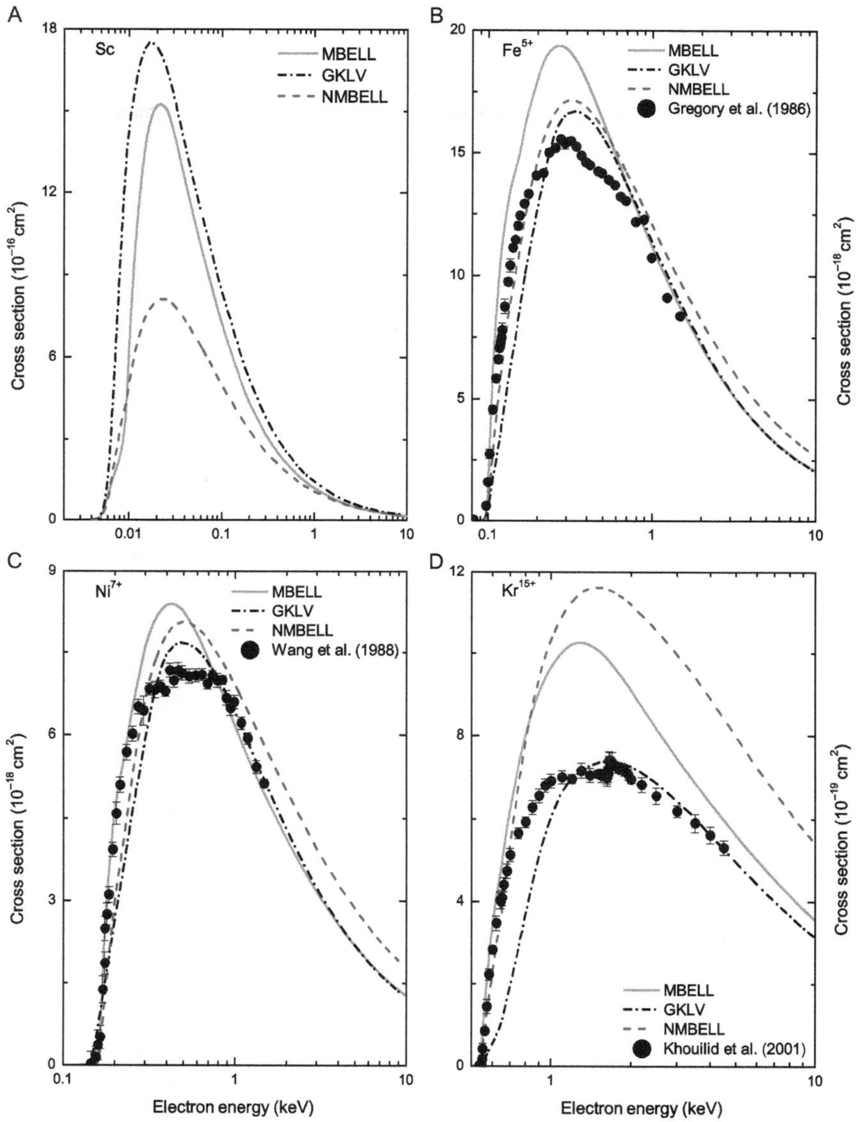

Figure 26 Same as Fig. 9 for EIICS in Sc-isoelectronic series of for (A) Sc, (B) Fe^{5+}, (C) Ni^{7+}, and (D) Kr^{15+}. *The experimental data are from Refs. 183,187,188.*

energies and atomic species considered. Their basic structures are as follows:

1. The simplest MCN[60] considers the ionization effect only from the close part of interactions of its parent CN[41] assuming equal contributions from the distant and close collisions for *s-orbital* ionization. This assumption,

however, limits the model application to the atomic targets ranging from the H-like to Be-like species with no species- and shell-dependent parameter even after the implementation of correlation, ionic and relativistic corrections. These corrections produce further simplicity in making the contribution from the oscillator part of the CN model (see eq. (41) in Ref. 41 and eq. (1) in Ref. 60) negligible compared to the other terms.

2. The RQIBED/MRIBED model, through the parent siBED,[35] replaces the Bethe part of the BED model of Kim and Rudd[86] with a parametric Born term involving d_1 and d_2 parameters in Eq. (11). d_1 and d_2 in siBED are the generic parameters, chosen in Ref. 35 to represent the shielding effect of a dipole potential. The parameter d_1 determines the cross sections at the peak region while d_2 optimizes the fit to the data at the higher incident energies. Our implementation of the above-mentioned corrections for the *s-shell* ionization in the atomic targets from H- to Be-isoelectronic series up to incident energies of $\sim$ 1 MeV leads to a further simplification with $d_1 = d_2 = 0.0$ with a constant set of other parameters.
3. The MBELL combines the Bethe asymptotic condition and the low energy behavior of collisions through the first and series terms in Eq. (19) with up to *nine* shell-dependent parameters, shown in Table 1, generalized in *nl* quantum numbers. This approach is profoundly useful for applications to all species up to Sc in the full energy domain up to ultrahigh values. One of the model parameters λ is involved with the ionic correction factor in Eq. (21) to take care of the residual effect from the relevant ionized orbital.
4. The NMBELL model replaces the series term in MBELL by a nonlinear exponential form[37] and introduces the quantum aspect of orbital radius.[29] This approach reduces the shell-dependent parameters to *three*, as shown in Table 2, generalized in *nl* orbitals, and listed in Table 2. This generalization also improves the description of the experimental EIICS data of the atomic species excluding the ionic H-like atoms. The parameter D in Eq. (23) takes care of the specific ionic effect from the individual orbitals.
5. The GKLV model, based on the Kolbenstvedt model,[38,39] has its roots in quantum electrodynamics with its profound predictive power with just *two* shell-dependent parameters, η_{nl} and ϵ_{nl} in Table 3. The former parameter in Eq. (26) determines the ionization arising from the close collisions through the Møller interaction while the latter in Eq. (27) controls the contribution in Eq. (28) through the exchange of virtual

photons. Set of η_{nl} and ϵ_{nl} values for the individual ionizing inner orbits are identical for both neutral and ionic species. However, the two sets of values are different for ionization in case of the outer shell as the effective charges q's for neutral and ionic targets differ by unity.

In this review, we report the predictions of EISICS on 112 atomic targets including the experimental EIICS data of 105 species in 21 isoelectronic series up to Sc for examining the ensuring individual performance of the five models considered herein. The performance index of the models is summarized in Table 4.

As can be noted from Table 4, the RQIBED/MRIBED and MCN have restrictive application on targets up to Be-isoelectronic series. Considering the simplicity MCN competes better of the two models. The MBELL model is applicable to all species its success index 75.2% obtained with up to nine shell-independent parameters. Its improved version NMMELL has its limitation in application to the H-like ionic targets where the predicted EISICS are too far from the EIICS data to be displayed in Figs. 1 and 2. A good description of the experimental data can be achieved with just *three* shell-dependent parameters, shown in Table 2 for the neutral H and all atomic targets in the range of He- to Sc-isoelectronic sequences with the performance rate of 70.3%. The GKLV model with its *two* shell-parameters enjoys the performance level of 81.2% with the highest number of 25 for the excellent fits. Best performance of GKLV is the outcome of the underlying basic feature of its parent KLV.[38,39]

Table 4 Presentation of Performance Index for Fitting Quality

		Number of Species with Fitting Quality					
Model	**Species**	**Excellent**	**Good**	**Reasonable**	**Total**	**Poor**	**Success Index**
GKLV	All	25	30	27	82	19	82/101=81.2%
NMBELL	All but H-like ionic target	12	23	36	71	21	71/101=70.3%
MBELL	All	11	42	23	76	25	76/101=75.2%
RQIBED/ MRIBED	H- to Be-like	0	18	10	28	4	28/32=87.5%
MCN	H- to Be-like	1	13	15	29	3	29/32=90.6%

Simultaneous poor performances of MBELL, NMBELL and GKLV in describing the experimental EIICS data, in general, provides a strong signal of significant contributions from indirect processes. Glaring examples are cases of Mg^{+} in Fig. 16B and Mn^{6+} in Fig. 24B. In the former, Moores and Nussbaumer[17] noted a significant autoionization contribution from the inner-shell. Their CBA calculation with inclusion of the autoionization contribution produces feature of the ionization cross sections amazingly similar to the experimental EIICS data. On the other hand, Colgan et al.[84] observed that contributions of excitation-autoionization (EA) in multiply charged atomic ions like Mn^{6+} is key to the complete understanding of dynamics of EII. Their DWBA calculations with consideration of EA produced remarkable QM results in close agreement with the experimental EIICS data for Mn^{6+} in Fig. 24B. Successes of other QM calculations can be observed in the DWBA predictions of Ref. 87 including EA for Ar^{6+} in Fig. 17D and those of Ref. 80 for Fe^{13+} in Fig. 18D where the feature of the EIICS data is reproduced. Indirect EA processes are dominant for Fe-, Ni-, Mn-, and Kr-iso-nuclear sequences near the peak region. In most cases, however, at least one of these considered models: MCN, RQIBED/MRIBED, MBELL, NMBELL and GKLV, represents the EIICS data reasonably well.

5. CONCLUSIONS

This review presents the predicted results using five models: MCN, RQIBED/MRIBED, MBELL, NMBELL and GKLV. The finding from these results suggests that our models are complementary in providing a satis factory description of the experimental data. The calculated EISICS values from the five considered models yield the upper and lower bounds for the EII cross sections. To be specific, GKLV and NMBELL for C^{2+} (Fig. 7), Al^{7+} (Fig. 10), Si^{6+} (Fig. 12), Mg^{2+} (Fig. 14), Ti^{8+} (Fig. 21), K^{+} (Fig. 23); MCN and NMBELL for U^{89+} GKLV and MBELL for U^{82+} (Fig. 15) predict the limiting values of EIICS in various modeling applications.

ACKNOWLEDGMENTS

This chapter is dedicated to the fond memories of our coworker and a great icon, late Professor F. Bary Malik, Southern Illinois University at Carbondale, IL. Professor B. C. Saha would like to thank the National Nuclear Security Administration (NNSA), Award Number(s) DE-NA0002630, for partial support.

REFERENCES

1. Omidvar, K.; Kyle, H. L.; Sullivan, E. C. Ionization of Multi-Electron Atoms by Fast Charged Particles. *Phys. Rev. A* **1972**, *5*, 1174.
2. Younger, S. M. Electron-Impact Ionization Cross Sections for Highly Ionized Hydrogen- and Lithium-Like Atoms. *Phys. Rev. A* **1980**, *22*, 111.
3. Younger, S. M. Electron Impact Ionization Cross Sections and Rates for Highly Ionized Atoms. *J. Quant. Spectrosc. Radiat. Transf.* **1981**, *26*, 329.
4. Younger, S. M. Electron-Impact-Ionization Cross Sections and Rates for Highly Ionized Beryllium Like Ions. *Phys. Rev. A* **1981**, *24*, 1278.
5. Younger, S. M. Distorted-Wave Electron-Impact-Ionization Cross Sections for Highly Ionized Neon like Atoms. *Phys. Rev. A* **1981**, *23*, 1138.
6. Younger, S. M. Cross Sections and Rates for Direct Electron-Impact Ionization of Sodium Like Ions. *Phys. Rev. A* **1981**, *24*, 1272.
7. Younger, S. M. Electron Impact Ionization Rate Coefficients and Cross Sections for Highly Ionized Iron. *J. Quant. Spectrosc. Radiat. Transf.* **1982**, *27*, 54.
8. Younger, S. M. Distorted-Wave Electron-Impact Ionization Cross Sections for the Argon Isoelectronic Sequence. *Phys. Rev. A* **1982**, *26*, 3177.
9. Scofield, J. H. K- and L-Shell Ionization of Atoms by Relativistic Electrons. *Phys. Rev. A* **1978**, *18*, 963.
10. Ndefru, J. T.; Malik, F. B. Relativistic Effects in the K-Shell Ionization of Au and Hg by Electron Impact. *Phys. Rev. A* **1982**, *25*, 2407.
11. Segui, S.; Dingfelder, M.; Salvat, F. Distorted-Wave Calculation of Cross Sections for Inner-Shell Ionization by Electron and Positron Impact. *Phys. Rev. A* **2003**, *67*, 062710.
12. Fontes, C. J.; Sampson, D. H.; Zhang, H. L. Method for Calculating the Electron-Impact Ionization of Ions of Any Complexity. *Phys. Rev. A* **1993**, *48*, 1975.
13. Fontes, C. J.; Sampson, D. H.; Zhang, H. L. Fully Relativistic Calculations of and Fits to 1S Ionization Cross Sections. *Phys. Rev. A* **1999**, *59*, 1329.
14. Moores, D. L.; Reed, K. J. Effect of the Møller Interaction on Electron-Impact Ionization of High-Z Hydrogen like Ions. *Phys. Rev. A* **1995**, *51*, R9.
15. Laghdas, K.; Reid, R. H. G.; Joachain, C. J.; Burke, P. G. Electron-Impact Ionization of Ne^{6+}. *J. Phys. B: At. Mol. Opt. Phys.* **1999**, *32*, 1439.
16. Peach, G. Ionization of Atoms and Positive Ions by Electron and Proton Impact. *J. Phys. B: At. Mol. Phys.* **1971**, *4*, 1670.
17. Moores, D. L.; Nussbaumer, H. The Ionization of Li^+ and Mg^+ by Electron Impact. *J. Phys. B: At. Mol. Phys.* **1970**, *3*, 161.
18. Moores, D. L. Electron Impact Ionization of Positive Ions with Configurations $1s^2 2s^2 2p^q$. *J. Phys. B: At. Mol. Phys.* **1972**, *5*, 286.
19. Stingl, E. The Ionization of Boron and the Isoelectronic Ions Carbon II, Nitrogen III, and Oxygen IV by Electron Impact. *J. Phys. B: At. Mol. Phys.* **1972**, *5*, 1160.
20. Jakubowicz, H.; Moores, D. L. Electron Impact Ionization of Li-Like and Be-Like Ions. *J. Phys. B: At. Mol. Phys.* **1981**, *14*, 3733.
21. Kuo, T. Y.; Huang, K. N. Relativistic Cross Sections of the Electron-Impact Ionization of Heliumlike Ions. *Phys. Rev. A* **2001**, *64*, 032710.
22. Kao, H. C.; Kuo, T. Y.; Yen, H. P.; Wei, C. M.; Huang, K. N. Relativistic Cross Sections of Electron-Impact Ionization of Hydrogenic Ions. *Phys. Rev. A* **1992**, *45*, 4646.
23. Bray, I. Calculation of Electron-Impact Ionization of Lithium-Like Targets. *J. Phys. B: At. Mol. Opt. Phys.* **1995**, *28*, L247.
24. Rescigno, T. N.; Baertschy, M.; Isaacs, W. A.; McCurdy, C. W. Collision Breakup in a Quantum System of Three Charged Particles. *Science* **1999**, *286*, 2474; Baertschy, M.; Rescigna, T. N.; McCurdy, C. W. Accurate Amplitudes for Electron-Impact Ionization. *Phys. Rev. A* **2001**, *64*, 022709; Bartlett, P. L.; Stelbovics, A. T.; Bray, I.

Iteratively-Coupled Propagating Exterior Complex Scaling Method for Electron-Hydrogen Collisions. *J. Phys. B* **2004**, *37*, L69.
25. Moores, D. L.; Reed, K. J. Indirect Processes in Electron Impact Ionization of Positive Ions. *Adv. At. Mol. Opt. Phys.* **1994**, *34*, 301.
26. Deutsch, H.; Becker, K.; Matt, S.; Märk, T. D. Theoretical Determination of Absolute Electron-Impact Ionization Cross Sections of Molecules. *Int. J. Mass Spectrom.* **2000**, *197*, 37.
27. Haque, A. K. F.; Uddin, M. A.; Shahajahan, M.; Talukder, M. R.; Basak, A. K.; Saha, B. C. Electron Impact Inner-Shell Ionization of Atoms. *Adv. Quantum Chem.* **2011**, *61*, 317.
28. Deutsch, H.; Märk, T. D. Calculation of Absolute Electron Impact Ionization Cross Section Functions for Single Ionization of He, Ne, Ar, Kr, Xe, N and F. *Int. J. Mass Spectrom. Ion Process.* **1987**, *79*, R1.
29. Margreiter, D.; Deutsch, H.; Märk, T. D. A Semi-Empirical Approach to the Calculation of Electron Impact Ionization Cross Sections of Atoms: from Hydrogen to Uranium. *Int. J. Mass Spectrom. Ion Process.* **1994**, *139*, 127.
30. Deutsch, H.; Scheier, P.; Becker, K.; Märk, T. D. Revised High Energy Behavior of the Deutsch-Märk (DM) Formula for Calculation of Electron Impact Cross Sections of Atoms. *Int. J. Mass Spectrom.* **2004**, *233*, 13.
31. Gryzinski, M. Two-Particle Collisions I. General Realations for Collision in the Laboratory Systems. *Phys. Rev.* **1965**, *138*, A305.
32. Gryzinski, M. Two-Particle Collisions. II. Coulomb Collisions in the Laboratory System of Coordinates. *Phys. Rev.* **1965**, *138*, A322.
33. Gryzinski, M. Classical Theory of Atomic Collisions: I. Theory of Inelastic Collisions. *Phys. Rev.* **1965**, *138*, A336.
34. Bethe, H. Zur Theorie des Durchgangs Schneller Korpuskularstrahlen Durch Materie. *Ann. Phys. (Leipzig)* **1930**, *5*, 325.
35. Huo, W. M. Convergent Series Representation for the Generalized Oscillator Strength of Electron-Impact Ionization and an Improved Binary-Encounter Dipole Model. *Phys. Rev. A* **2001**, *64*, 042719.
36. Bell, K. L.; Gilbody, H. B.; Hughes, J. G.; Kingston, A. E.; Smith, F. J. Recommended Data on the Electron Impact Ionization of Light Atoms and Ions. *J. Phys. Chem. Ref. Data* **1983**, *12*, 891.
37. Godunov, A. L.; Ivanov, P. B. On Analytical Fits for Electron Impact Ionization Cross Sections. *Phys. Scr.* **1999**, *59*, 277.
38. Kolbenstvedt, H. Simple Theory for K-Ionization by Relativistic Electrons. *J. Appl. Phys.* **1967**, *38*, 4785.
39. Kolbenstvedt, H. Asymptotic Expression for K-Shell Ionization Cross Section with Electrons. *J. Appl. Phys.* **1975**, *46*, 2771.
40. Bernshtam, V. A.; Ralchenko, Yu. V.; Maron, Y. Empirical Formula for Cross Section of Direct Electron-Impact Ionization of Ions. *J. Phys. B: At. Mol. Opt. Phys.* **2000**, *33*, 5025.
41. Cengiz, A. Approximate Inelastic Scattering Cross Sections of Electrons. *Rad. Phys. Chem.* **2002**, *65*, 33.
42. Deutsch, H.; Scheier, P.; Becker, K.; Märk, T. D. Calculated Cross-Sections for the Electron-Impact Detachment from Negative Ions Using the Deutsch-Märk (DM) Formalism. *Chem. Phys. Lett.* **2003**, *382*, 26.
43. Deutsch, H.; Becker, K.; Defrance, P.; Onthong, U.; Probst, M.; Matt, S.; Scheier, P.; Märk, T. D. Calculated Absolute Cross Section for the Electron-Impact Ionisation of Simple Molecular Ions. *Int. J. Mass Spectrom.* **2003**, *223-224*, 639.
44. Uddin, M. A.; Haque, A. K. F.; Basak, A. K.; Saha, B. C. Calculations of Electron-Impact Single-Ionization Cross Sections of Helium Isoelectronic Systems. *Phys. Rev. A* **2004**, *70*, 032706.

45. Uddin, M. A.; Haque, A. K. F.; Billah, M. M.; Basak, A. K.; Karim, K. R.; Saha, B. C. Computation of Electron-Impact K-Shell Ionization Cross Sections of Atoms. *Phys. Rev. A* **2005**, *71*, 032715.
46. Uddin, M. A.; Haque, A. K. F.; Mahbub, M. S.; Karim, K. R.; Basak, A. K.; Saha, B. C. Electron Impact Ionization of Beryllium Isoelectronic Ions. *Int. J. Mass Spectrom.* **2005**, *244*, 76.
47. Uddin, M. A.; Haque, A. K. F.; Basak, A. K.; Karim, K. R.; Saha, B. C. Electron-Impact Ionization of Hydrogen and Lithiumlike Systems. *Phys. Rev. A* **2005**, *72*, 032715.
48. Uddin, M. A.; Haque, A. K. F.; Karim, K. R.; Basak, A. K.; Malik, F. B. Modified Kolbensvedt Model for the Electron Impact K-Shell Ionization Cross-Sections of Atoms and Ions. *Eur. Phys. J. D* **2006**, *37*, 361.
49. Haque, A. K. F.; Uddin, M. A.; Basak, A. K.; Karim, K. R.; Saha, B. C. Empirical Model for the Electron-Impact K-Shell-Ionization Cross Section. *Phys. Rev. A* **2006**, *73*, 012708.
50. Haque, A. K. F.; Uddin, M. A.; Basak, A. K.; Karim, K. R.; Saha, B. C.; Malik, F. B. Relativistic Effects in Electron Impact Ionization from the P-Orbital. *Phys. Lett. A* **2006**, *354*, 449.
51. Haque, A. K. F.; Uddin, M. A.; Basak, A. K.; Karim, K. R.; Saha, B. C.; Malik, F. B. Electron-Impact Ionization of L-Shell Atomic Species. *Phys. Rev. A* **2006**, *73*, 052703.
52. Haque, A. K. F.; Uddin, M. A.; Basak, A. K.; Karim, K. R.; Saha, B. C.; Malik, F. B. Electron Impact Ionization of M-Shell Atoms. *Phys. Scr.* **2006**, *74*, 377.
53. Haque, A. K. F.; Uddin, M. A.; Patoary, M. A. R.; Basak, A. K.; Talukder, M. R.; Saha, B. C.; Karim, K. R.; Malik, F. B. Generalized Kolbenstvedt Model for Electron Impact Ionization of K-, L- and M-Shell Atoms. *Eur. Phys. J. D* **2007**, *42*, 203.
54. Haque, A. K. F.; Sarker, M. S. I.; Patoary, M. A. R.; Shahjahan, M.; Hossain, M. I.; Uddin, M. A.; Basak, A. K.; Saha, B. Modified Version of Revised Deutsch-Märk Model for Electron Impact K-Shell Ionization Cross Sections of Atoms at Relativistic Energies. *Int. J. Quantum Chem.* **2009**, *109*, 1442.
55. Burgess, A.; Percival, I. C. Classical Theory of Atomic Scattering. *Adv. At. Mol. Phys.* **1968**, *4*, 109.
56. Uddin, M. A.; Haque, A. K. F.; Karim, K. R.; Basak, A. K. Empirical Model for the Ionization Cross Sections of H- and He-Like Ions. *Phys. Scr.* **2005**, *72*, 389.
57. Haque, A. K. F.; Shahjahan, M.; Uddin, M. A.; Patoary, M. A. R.; Basak, A. K.; Saha, B. C.; Malik, F. B. Generalized Kolbenstvedt Model for Electron Impact Ionization of K-, L- and M-Shell Ions. *Phys. Scr.* **2010**, *81*, 045301.
58. Deutsch, H.; Margreiter, D.; Märk, T. D. A Semi-Empirical Approach to the Calculation of Absolute Inner-Shell Electron Impact Ionization Cross Sections. *Z. Phys. D* **1994**, *29*, 31.
59. Patoary, M. A. R.; Uddin, M. A.; Haque, A. K. F.; Shahjahan, M.; Basak, A. K.; Talukder, M. R.; Saha, B. C. Electron Impact Ionization in K-, L- and M-Shells of Atomic Targets. *Int. J. Quantum. Chem.* **2011**, *111*, 923.
60. Haque, A. K. F.; Hossain, M. I.; Talukder, M. R.; Hasan, M.; Uddin, M. A.; Basak, A. K.; Saha, B. C.; Malik, F. B. Electron Impact Ionization Cross-Section of K-Shell and H- to Be-Isoelectronic Series: An Empirical Model. *Rad. Phys. Chem.* **2013**, *91*, 50.
61. Deutsch, H.; Becker, K.; Gstir, B.; Märk, T. Calculated Electron Impact Cross Sections for the K-Shell Ionization of Fe, Co, Mn, Ti, Nb and Mo Atoms Using the DM Formalism. *Int. J. Mass Spectrom.* **2002**, *213*, 5.
62. James, F.; Roos, M. MINUIT-A System for Function Minimization and Analysis of the Parameter Errors and Correlations. *Comput. Phys. Commun.* **1975**, *10*, 343.
63. Inokuti, M. Inelastic Collisions of Fast Charged Particles with Atoms and Molecules–The Bethe Theory Revisited. *Rev. Mod. Phys.* **1971**, *43*, 297.

64. Kim, Y. K. Electron Correlation in Electron-Impact Ionization of Hydrogen-Like Atoms. *Phys. Essays* **2000**, *13*, 473.
65. Powell, C. J. Cross-Sections for Ionization of Inner-Shell Electrons by Electrons. *Rev. Mod. Phys.* **1976**, *48*, 33.
66. Powell, C. J. Inner-Shell Ionization Cross Sections. In: Märk, T. D., Dunn, D. H., Eds.; Electron Impact Ionization; Springer: Berlin, 1985; p 198.
67. Powell, C. J. Inner-Shell Ionization Cross Sections. In: Michael, J. R., Ingram, P., Eds.; Microbeam Analysis; SunFrancisco Press: San Francisco, CA, 1990; p 13.
68. Shah, M. B.; Elliott, D. S.; Gilbody, H. B. Pulsed Crossed-Beam Study of the Ionisation of Atomic Hydrogen by Electron Impact. *J. Phys. B: At. Mol. Phys.* **1987**, *20*, 3501.
69. Diserens, M. J.; Smith, A. C. H.; Harrison, M. F. A. Ionisation of Ti^{+}, Ti^{2+} and Ar^{2+} by Electron Impact. *J. Phys. B: At. Mol. Opt. Phys.* **1988**, *21*, 2129.
70. Younger, S. Electron-Impact-Ionization Cross Sections for Highly Ionized Heliumlike Atoms. *Phys. Rev. A* **1980**, *22*, 1425.
71. Moores, D. L.; Reed, K. J. Electron Impact Ionization of Highly-Charged Uranium Ions. *J. Phys. B: At. Mol. Opt. Phys.* **1995**, *28*, 4861.
72. Chang, J.; Sun, H.; Cheng, W.; Huang, K. Electron-Impact Ionization of Be-Like Ions. *Phys. Rev. A* **2004**, *69*, 052713.
73. Kim, Y. K.; Stone, P. M. Ionization of Boron, Aluminum, Gallium, and Indium by Electron Impact. *Phys. Rev. A* **2001**, *64*, 052707.
74. Moores, D. L. The Autoionisation Contribution to the Electron Impact Ionisation Cross Sections of C^{+} and N^{2+} at Low Energies. *J. Phys. B: At. Mol. Phys.* **1979**, *12*, 4171.
75. McGuire, E. J. Born-Approximation Electron Ionization Cross Sections for Al^{N+} ($0 \leq N \leq 11$) and Some Ions of the Na Isoelectronic Sequence. *Phys. Rev. A* **1982**, *26*, 125.
76. Hahn, Y. Distorted-Wave Theory of Electron-Ion Collisions. I. Direct Excitation and Ionisation. *Phys. Rev. A* **1977**, *16*, 1964.
77. Hahn, Y. Distorted-Wave Theory of Electron-Ion Collisions. II. Auger Ionisation and Excitation Fluorescence. *Phys. Rev. A* **1978**, *18*, 1028.
78. Bray, I. Calculation of Electron Impact Total, Ionization, and Nonbreakup Cross Sections from the 3S and 3P States of Sodium. *Phys. Rev. Lett.* **1994**, *73*, 1088.
79. Golden, L. B.; Sampson, D. H. Ionization of Highly Charged Ions by Electron Impact. *J. Phys. B: At. Mol. Phys.* **1977**, *10*, 2229.
80. Pindzola, M. S.; Griffin, D. C.; Bottcher, C. Electron-Impact Ionization in the Iron Isonuclear Sequence. *Phys. Rev. A* **1986**, *34*, 3668.
81. Jha, L. K.; Kumar, S.; Roy, B. N. Electron Impact Single and Double Ionization of Argon. *Eur. Phys. J. D* **2006**, *40*, 101.
82. Kumar, A.; Roy, B. N. Electron Impact Ionisation of Alkali Metal Ions. *J. Phys. B: At. Mol. Phys.* **1979**, *12*, 3979.
83. McGuire, E. J. Electron Ionization Cross Sections in the Born Approximation. *Phys. Rev. A* **1977**, *16*, 62.
84. Colgan, J.; Mitnik, D. M.; Pindzola, M. S. Electron-Impact Ionization of Multiply Charged Manganese Ions. *Phys. Rev. A* **2000**, *63*, 012712.
85. Pindzola, M. S.; Griffin, D. C.; Bottcher, C.; Younger, S. M.; Hunter, H. T. Electron Impact Ionization Data for the Iron Isonuclear Sequence. *Nucl. Fusion* **1987**, *27*, 21.
86. Kim, Y. K.; Rudd, M. E. Binary-Encounter-Dipole Model for Electron-Impact Ionization. *Phys. Rev. A* **1994**, *50*, 3954.
87. Pindzola, M. S.; Griffin, D. C.; Bottcher, C. Electron-Impact Ionization in the Magnesium-Isoelectronic Sequence. *Phys. Rev. A* **1986**, *33*, 3787.
88. Shyn, T. W. Doubly Differential Cross Sections of Secondary Electrons Ejected from Atomic Hydrogen by Electron Impact. *Phys. Rev. A* **1992**, *45*, 2951.

89. Defrance, P.; Brouillard, F.; Claeys, W.; Wassenhove, G. V. Crossed Beam Measurement of Absolute Cross Sections: An Alternative Method and Its Application to the Electron Impact Ionisation of He^{+}. *J. Phys. B: At. Mol. Phys.* **1981**, *14*, 103.
90. Dolder, K. T.; Harrison, M. F. A.; Thonemann, P. C. A Measurement of the Ionization Cross-Section of Helium Ions by Electron Impact. *Proc. R. Soc. Lond.* **1961**, *264*, 367.
91. Peart, B.; Walton, D. S.; Dolder, K. T. The Ranges of Validity of the Born and Bethe Approximations for the Single Ionization of He^{+} and Li^{+} Ions by Electron Impact. *J. Phys. B: At. Mol. Phys.* **1969**, *2*, 1347.
92. Tinschert, K.; Muller, A.; Hofmann, G.; Huber, K.; Becker, R.; Gregory, D. C.; Salzborn, E. Experimental Cross Sections for Electron Impact Ionisation of Hydrogen-Like Li^{2+} Ions. *J. Phys. B: At. Mol. Opt. Phys.* **1989**, *22*, 531.
93. Aichele, K.; Hartenfeller, U.; Hathiramani, D.; Hofmann, G.; Schäfer, V.; Steidl, M.; Stenke, M.; Salzborn, E.; Pattard, T.; Rost, J. M. Electron Impact Ionization of the Hydrogen-Like Ions B^{4+}, C^{5+}, N^{6+} and O^{7+}. *J. Phys. B: At. Mol. Opt. Phys.* **1998**, *31*, 2369.
94. Donets, E. D.; Ovsyannikov, V. P. Investigation of Ionization of Positive Ions by Electron Impact. *Zh. Eksp. Teor. Fiz.* **1981**, *80*, 916.
95. O'Rourke, B.; Currell, F.; Kuramoto, H.; Li, Y.; Ohtani, S.; Tong, X. M.; Watanabe, H. Electron-Impact Ionization of Hydrogen-Like Iron Ions. *J. Phys. B: At. Mol. Opt. Phys.* **2001**, *34*, 4003.
96. Marrs, R. E.; Elliott, S. R.; Scofield, J. H. Measurement of Electron-Impact Ionization Cross Sections for Hydrogenlike High-Z Ions. *Phys. Rev. A* **1997**, *56*, 1338.
97. Rejoub, R.; Lindsay, B. G.; Stebbings, R. F. Determination of the Absolute Partial and Total Cross Sections for Electron-Impact Ionization of the Rare Gases. *Phys. Rev. A* **2002**, *65*, 042713.
98. Schram, B. L.; Moustafa, H. R.; Schutten, J.; de Heer, F. J. Ionization Cross Sections for Electrons (100-600 Ev) in Noble and Diatomic Gases. *Physica* **1966**, *32*, 734.
99. Nagy, P.; Skutlartz, A.; Schmidt, V. Absolute Ionisation Cross Sections for Electron Impact in Rare Gases. *J. Phys. B: At. Mol. Phys.* **1980**, *13*, 1249.
100. Shah, M. B.; Elliott, D. S.; McCallion, P.; Gilbody, H. B. Single and Double Ionisation of Helium by Electron Impact. *J. Phys. B: At. Mol. Opt. Phys.* **1988**, *21*, 2751.
101. Montague, R. G.; Harrison, M. F.; Smith, A. C. H. A Measurement of the Cross Section for Ionisation of Helium by Electron Impact Using a Fast Crossed Beam Technique. *J. Phys. B: At. Mol. Phys.* **1984**, *17*, 3295.
102. Rapp, D.; Englander-Golden, P. Total Cross Sections for Ionization and Attachment in Gases by Electron Impact. I. Positive Ionization. *J. Chem. Phys.* **1965**, *43*, 1464.
103. Stephan, K.; Helm, H.; Märk, T. D. Mass Spectrometric Determination of Partial Electron Impact Ionization Cross Sections of He, Ne, Ar and Kr from Threshold up to 180 Ev. *J. Chem. Phys.* **1980**, *73*, 3763.
104. Wetzel, R. C.; Baiocchi, F. A.; Hayes, T. R.; Freund, R. S. Absolute Cross Sections for Electron-Impact Ionization of the Rare-Gas Atoms by the Fast-Neutral-Beam Method. *Phys. Rev. A* **1987**, *35*, 559.
105. Wareing, J. B.; Dolder, K. T. A Measurement of the Cross Section for Ionization of Li^{+} to Li^{2+} by Electron Impact. *Proc. Phys. Soc.* **1967**, *91*, 887.
106. Lineberger, W. C.; Hooper, J. W.; McDaniel, E. W. Absolute Cross Sections for Single Ionization of Alkali Ions by Electron Impact. I. Description of Apparatus and Li^{+} Results. *Phys. Rev.* **1966**, *141*, 151.
107. Crandall, D. H.; Phaneuf, R. A.; Gregory, D. C. *Electron Impact Ionization of Multicharged Ions*; 1979. ORNL/TM-7020, Oak Ridge National Laboratory.
108. Defrance, P.; Chantrenne, S.; Brouillard, F.; Rachafi, S.; Belic, D.; Gregory, J. J. D. Electron Impact Ionization of N^{4+} and N^{5+}. *Nucl. Instrum. Methods Phys. Res. B* **1985**, *9*, 400.

109. Peart, B.; Dolder, K. T. Measurements of Cross Sections for the Ionization of Li^+ and Ba^+ Ions by Electron Impact. *J. Phys. B: At. Mol. Phys.* **1968**, *1*, 872.
110. Peart, B.; Dolder, K. T. The Ionization of Li^+ to Li^{3+} by Electron Impact. *J. Phys. B: At. Mol. Phys.* **1969**, *2*, 1169.
111. Rachafi, S.; Zambar, M.; Hui, Z.; Duponchelle, M.; Jureta, J.; Defrance, P. Electron Impact Ionization of C^{4+} and O^{6+}. *Phys. Scr.* **1989**, *T28*, 12.
112. Duponchelle, M.; Khouilid, M.; Oualim, E. M.; Zhang, H.; Defrance, P. Electron-Impact Ionization of Neon Ions (q = 4-8). *J. Phys. B: At. Mol. Opt. Phys.* **1997**, *30*, 729.
113. Marrs, R. E.; Elliott, S. R.; Knapp, D. A. Production and Trapping of Hydrogenlike and Bare Uranium Ions in an Electron Beam Ion Trap. *Phys. Rev. Lett.* **1994**, *72*, 4082.
114. Zapesochnyi, I. P.; Aleksakhin, I. S. Ionization of Alkali Metal Atoms by Slow Electrons. *Sov. Phys. JETP* **1969**, *28*, 41 *Zh. Eksp. Teor. Fiz.* **July, 1968**, *55*, 76–85.
115. McFarland, R. H.; Kinney, J. D. Absolute Cross Sections of Lithium and Other Alkali Metal Atoms for Ionization by Electrons. *Phys. Rev.* **1965**, *137*, A1058.
116. Jalin, R.; Hagemann, R.; Botter, R. Absolute Electron Impact Ionization Cross Sections of Li in the Energy Range from 100 to 2000 Ev. *J. Chem. Phys.* **1973**, *59*, 952.
117. Falk, R. A.; Dunn, G. H. Electron-Impact Ionization of Be^+. *Phys. Rev. A* **1983**, *27*, 754.
118. Crandall, D. H.; Phaneuf, R. A.; Taylor, P. O. Electron-Impact Ionization of C^{3+} and N^{4+}. *Phys. Rev. A* **1978**, *18*, 1911.
119. Defrance, P.; Chantrenne, S.; Rachafi, S.; Belic, D. S.; Jureta, J.; Gregory, D.; Brouillard, F. Absolute Cross-Section Measurements for Electron Impact Ionisation of Li-Like N^{4+}, o^{5+} and Ne^{7+} Ions. *J. Phys. B: At. Mol. Opt. Phys.* **1990**, *23*, 2333.
120. Rinn, K.; Gregory, D. C.; Wang, L. J.; Phaneuf, R. A.; Müller, A. Electron-Impact Ionization of O^{5+}: Improved Measurements. *Phys. Rev. A* **1987**, *36*, 595.
121. Crandall, D. H.; Phaneuf, R. A.; Gregory, D. C.; Howald, A. M.; Mueller, D. W.; Morgan, T. J.; Dunn, G. H.; Griffin, D. C.; Henry, R. J. W. Electron-Impact Ionization of B^{2+} and O^{5+}: Excitation-Autoionization in Li-Like Ions. *Phys. Rev. A* **1986**, *34*, 1757.
122. Aichele, K.; Shi, W.; Scheuermann, F.; Teng, H.; Salzborn, E.; Müller, A. Resonant Indirect Processes in Electron-Impact Single Ionization of Ne^{7+} Ions. *Phys. Rev. A* **2000**, *63*, 014701.
123. Wong, K. L.; Beiersdorfer, P.; Chen, M. H.; Marrs, R. E.; Reed, K. J.; Scofield, J. H.; Vogel, D. A.; Zasadzinski, R. Electron-Impact Ionization of Lithiumlike Ions: Ti^{19+}, V^{20+}, Cr^{21+}, Mn^{22+}, and Fe^{23+}. *Phys. Rev. A* **1993**, *48*, 2850.
124. Claytor, N.; Feinberg, B.; Gould, H.; Bemis, C.; Campo, J. D.; Ludemann, C.; Vane, C. Electron Impact Ionization of U^{88+}-U^{91+}. *Phys. Rev. Lett.* **1988**, *61*, 2081.
125. Ishii, K.; Kamiya, M.; Sera, K.; Morita, S.; Tawara, H.; Oyamada, M.; Chu, T. Inner-Shell Ionization by Ultrarelativistic Electrons. *Phys. Rev. A* **1977**, *15*, 906.
126. Ricz, S.; Schlenk, B.; Berenyi, D.; Hock, G.; Valek, A. K-Shell Ionization Cross Sections of Pd, Ag, in and Sn for Relativistic Electrons. *Acta Physiol. Acad. Sci. Hung.* **1977**, *42*, 269.
127. Hoffmann, D. H. H.; Genz, H.; Löw, W.; Richter, A. Z and E Dependence and Scaling Behaviour of the K-Shell Ionization Cross Section for Relativistic Electron Impact. *Phys. Lett. A* **1978**, *65*, 304; Hoffmann, D. H. H.; Brendel, C.; Genz, H.; Löw, W.; Müller, S.; Richter, A. Inner-Shell Ionization by Relativistic Electron Impact. *Z. Phys. A* **1979**, *293*, 187.
128. Rester, D. H.; Dance, W. E. K-Shell Ionization of Ag, Sn, and Au from Electron Bombardment. *Phys. Rev.* **1966**, *152*, 1.
129. Berengut, J. C.; Loch, S. D.; Pindzola, M. S.; Ballance, C. P.; Griffin, D. C.; Fogle, M.; Bannister, M. E. Electron-Impact Ionization of B^+. *Phys. Rev. A* **2008**, *78*, 012704.
130. Fogle, M.; Bahati, E. M.; Bannister, M. E.; Vane, C. R.; Loch, S. D.; Pindzola, M. S.; Balance, C. P.; Thomas, R. D.; Zhaunerchyk, V.; Bryans, P.; Mitthumsiri, W.;

Savin, D. W. Electron-Impact Ionization of Be-Like C III, N IV, and O V. *Astrophys. J. Suppl. Ser.* **2008**, *175*, 543.

131. Woodruff, P. R.; Hublet, M. C.; Harrison, M. F. A.; Brook, E. A Measurement of the Cross Section for Electron Impact Ionisation of C^{2+}. *J. Phys. B: At. Mol. Phys.* **1978**, *11*, L679.
132. Howald, A. M.; Gregory, D. C. *Absolute Cross Sections for Electron-Impact Single Ionization of* N^{3+}. 1984. Unpublished ORNL data, http://www-cfadc.phy.ornl.gov/xbeam/n3-4tbl_howald.html.
133. Bannister, M. *Absolute Cross Sections for Electron-Impact Single Ionization of* C^{3+}. 1998. Unpublished ORNL data, http://www-cfadc.phy.ornl.gov/xbeam/c3-4tbl_bannister.html.
134. Bannister, M.; Dunn, G. *Absolute Cross Sections for Electron-Impact Single Ionization of* O^{4+}. 1997. Unpublished ORNL data, http://www-cfadc.phy.ornl.gov/xbeam/o4-5tbl_bannister.html.
135. Bannister, M. E. Absolute Cross Sections for Electron-Impact Single Ionization of Ne^{q+} (q=2,4-6) Ions. *Phys. Rev. A* **1996**, *54*, 1435.
136. Scholz, A. L.; Colle, R.; Preiss, I. L.; Scholz, W. Cross Sections for K-Shell Ionization by 2-Mev-Electron Impact,. *Phys. Rev. A* **1973**, 7, 1957.
137. Lecointre, J.; Kouzakov, K. A.; Belic, D. S.; Defrance, P.; Yu, V.; Shevelko, V. P. Popov Multiple Ionization of C^+, N^+ and O^+ Ions by Fast Electron Impact. *J. Phys. B: At. Mol. Opt. Phys.* **2013**, *46*, 205201.
138. Yamada, I.; Danjo, A.; Hirayama, T.; Matsumoto, A.; Othani, S.; Suzuki, H.; Takayanagi, T.; Tawara, H.; Wakiya, K.; Yoshino, M. Electron Impact Ionization of C^+, N^+ and P^+ Ions. *J. Phys. Soc. Jpn.* **1989**, *58*, 1585.
139. Ludlow, J. A.; Loch, S. D.; Pindzola, M. S.; Ballance, C. P.; Griffin, D. C.; Bannister, M. E.; Fogle, M. Electron-Impact Ionization of C^+ in Both Ground and Metastable States. *Phys. Rev. A* **2008**, *78*, 052708.
140. Aitken, K. L.; Harrison, M. F.; Rundel, R. D. Measurement of the Cross Section for Electron Impact Ionization of Multi-Electron Ions. II. N^{2+} to N^{3+} and C^+ to C^{2+}. *J. Phys. B: At. Mol. Phys.* **1971**, *4*, 1189.
141. Mueller, D. *Absolute Cross Sections for Electron-Impact Single Ionization of* N^{2+}; 1983. Unpublished ORNL data, http://www-cfadc.phy.ornl.gov/xbeam/n2-3tbl_mueller.html.
142. Bannister, M.; Havener, C. C. *Absolute Cross Sections for Electron-Impact Single Ionization of* N^{2+}; 1995. Unpublished ORNL data, http://www-cfadc.phy.ornl.gov/xbeam/n2-3tbl_bannister.html.
143. Brook, E.; Harrison, M. F.; Smith, A. C. H. Measurements of the Electron Impact Ionisation Cross Sections of He, C, O and N Atoms. *J. Phys. B: At. Mol. Phys.* **1978**, *11*, 3115.
144. Harrison, M. F.; Dolder, K. T.; Thonemann, P. C. A Measurement of the Cross Section for the Ionization of N^+ to N^{2+} by Electron Impact. *Proc. Phys. Soc.* **1963**, *82*, 368.
145. Falk, R. *Absolute Cross Sections for Electron-Impact Single Ionization of* O^{2+}; 1980. Unpublished ORNL data, http://www-cfadc.phy.ornl.gov/xbeam/o2-3tbl_falk.html.
146. Gregory, D. C.; Crandall, D. H.; Phaneuf, R. A.; Howald, M. A.; Dunn, G. H.; Falk, R. A.; Muller, D. W.; Morgan, T. J. *Electron Impact Ionization of Multi-Charged Ions at ORNL: 1980-1984*; 1985. ORNL/TM-9501, Oak Ridge National Laboratory.
147. Aichele, K.; Steidl, M.; Hartenfeller, U.; Hathiramani, D.; Scheuermann, F.; Westermann, M.; Salzborn, E.; Pindzola, M. S. Electron-Impact Single Ionization of Multiply Charged Aluminium Ions. *J. Phys. B: At. Mol. Opt. Phys.* **2001**, *34*, 4113.
148. Smith, A. C. H.; Caplinger, E.; Neynaber, R. H.; Rothe, E. W.; Trujillo, S. M. Electron Impact Ionization of Atomic Nitrogen. *Phys. Rev.* **1962**, *127*, 1647.

149. Yamada, I.; Danjo, A.; Hirayama, T.; Matsumoto, A.; Othani, S.; Suzuki, H.; Tawara, H.; Takayanagi, T.; Wakiya, K.; Yoshino, M. Electron Impact Ionization of O^+, S^+ and S^{2+} Ions. *J. Phys. Soc. Jpn.* **1988**, *57*, 2699.
150. Mueller, D. W.; Morgan, T. J.; Dunn, G. H.; Gregory, D. C.; Crandall, D. H. Absolute Cross-Section Measurements for Electron-Impact Ionization of Doubly Charged Ions Ti^{2+}, Fe^{2+}, Ar^{2+}, Cl^{2+}, and F^{2+}. *Phys. Rev. A* **1985**, *31*, 2905.
151. Yamada, I.; Danjo, A.; Hirayama, T.; Matsumoto, A.; Othani, S.; Suzuki, H.; Takayanagi, T.; Tawara, H.; Wakiya, K.; Yoshino, M. Electron Impact Ionization of F^+, Ne^+, Cl^+ and Ar^+ Ions. *J. Phys. Soc. Jpn.* **1989**, *58*, 3151.
152. Fite, W. L.; Brackmann, R. T. Ionization of Atomic Oxygen on Electron Impact. *Phys. Rev.* **1959**, *113*, 815.
153. Danjo, A.; Matsumoto, A.; Ohtani, S.; Suzuki, H.; Tawara, H.; Wakiya, K.; Yoshino, M. Electron Impact Single Ionization of Ne^{2+}, Ar^{2+}, Kr^{2+} and Xe^{2+} Ions. *J. Phys. Soc. Jpn.* **1984**, *53*, 4091.
154. Van Emmichoven, P. A. Z.; Bannister, M. E.; Gregory, D. C.; Havener, C. C.; Phaneuf, R. A.; Bell, E. W.; Guo, X. Q.; Thompson, J. S.; Sataka, M. Electron-Impact Ionization of Si^{6+} and Si^{7+} Ions. *Phys. Rev. A* **1993**, *47*, 2888.
155. Muller, A.; Salzborn, E.; Frodl, R.; Becker, R.; Klein, H.; Winter, H. Absolute Ionization Cross Sections for Electrons Incident on O^+, Ne^+, Xe^+ and Ar^{i+} (i = 1,.,5) Ions. *J. Phys. B: At. Mol. Phys.* **1980**, *13*, 1877.
156. Hayes, T. R.; Wetzel, R. C.; Freund, R. Absolute Electron-Impact-Ionization Cross-Section Measurements of the Halogen Atoms. *Phys. Rev. A* **1987**, *35*, 578.
157. Man, K. F.; Smith, A. C.; Harrison, M. F. A Measurement of the Cross Section for Electron Impact Ionisation of Ne^+, Ar^+, Kr^+ and Xe^+. *J. Phys. B: At. Mol. Phys.* **1987**, *20*, 5865.
158. Dolder, K. T.; Harrison, M.; Thonemann, P. A Measurement of the Ionization Cross-Section of Ne^+ to Ne^{2+} by Electron Impact. *Proc. R. Soc. Lond. A Math. Phys. Sci.* **1963**, *274*, 546.
159. Diserens, M. J.; Harrison, M. F. A.; Smith, A. C. A Measurement of the Cross Section of Electron Impact Ionisation of Ne^+. *J. Phys. B: At. Mol. Phys.* **1984**, *17*, L621.
160. Thompson, J. S.; Gregory, D. C. Absolute Cross-Section Measurements for Electron-Impact Single Ionization of Si^{4+} and Si^{5+}. *Phys. Rev. A* **1994**, *50*, 1377.
161. Peart, B.; Dolder, K. T. Measurements of Cross Sections for the Ionization of Na^+ to Na^{2+} and K^+ to K^{2+} by Electron Impact. *J. Phys. B: At. Mol. Phys.* **1968**, *1*, 240.
162. Peart, B.; Martin, S. O.; Dolder, K. T. Measurements of Electron Impact Ionization Cross Sections of Mg^{2+} Ions and a Note on the Classical Scaling Law. *J. Phys. B: At. Mol. Phys.* **1969**, *2*, 1176.
163. Bleakney, W. Ionization Potentials and Probabilities for the Formation of Multiply Charged Ions in Helium, Neon and Argon. *Phys. Rev.* **1930**, *36*, 1303.
164. Adamczyk, B.; Boerboom, A. J. H.; Schram, B. L.; Kistemaker, J. Partial Ionization Cross Sections of He, Ne, H_2, and Ch_4 for Electrons from 20 to 500 Ev. *J. Chem. Phys.* **1966**, *44*, 4640.
165. Gaudin, A.; Hageman, R. Determinations Absolues Des Sections Efficacies Totales Et Partielles Dionisation De Lhelium, Du Neon, De Largon Et De Lacetylene, Pour Des Electrons De 100 A. *J. Chim. Phys.* **1967**, *64*, 1209.
166. Hooper, J. W.; Lineberger, W. C.; Bacon, F. M. Absolute Cross Sections for Single Ionization of Alkali Ions by Electron Impact. II. Na^+ and K^+ Results and Comparisons with Theory. *Phys. Rev.* **1966**, *141*, 165.
167. Defrance, P.; Rachafi, S.; Jureta, J.; Meyer, F.; Chantrenne, S. Electron Impact Ionization of Ar^{8+}. *Nucl. Instrum. Methods Phys. Res. B* **1987**, *23*, 265.

168. Zhang, Y.; Reddy, C. B.; Smith, R. S.; Golden, D. E.; Mueller, D. W.; Gregory, D. C. Measurement of Electron-Impact Single-Ionization Cross Sections of Ar^{8+}. *Phys. Rev. A* **1991**, *44*, 4368.
169. Berkner, K. H.; Kaplan, S. N.; Pyle, R. V. Cross Sections for K-Shell Ionization of Pd and Au by 2.5 and 7.1 Mev Electrons. *Bull. Am. Phys. Soc.* **1970**, *15*, 786.
170. Brink, G. O. Absolute Ionization Cross Sections of the Alkali Metals. *Phys. Rev.* **1964**, *134*, A345.
171. Johnston, A. R.; Burrow, P. D. Electron-Impact Ionization of Na. *Phys. Rev. A* **1995**, *51*, R1735(R).
172. Crandall, D. H.; Phaneuf, R. A.; Falk, R. A.; Belic, D. S.; Dunn, G. H. Absolute Cross-Section Measurements for Electron-Impact Ionization of Na-Like Ions Mg^{+}, Al^{2+}, and Si^{3+}. *Phys. Rev. A* **1982**, *25*, 143.
173. Martin, S. O.; Peart, B.; Dolder, K. T. Measurements of Cross Sections for the Ionization of Mg^{+} to Mg^{2+} by Electron Impact. *J. Phys. B: At. Mol. Phys.* **1968**, *1*, 537.
174. Freund, R. S.; Wetzel, R. C.; Shul, R. J.; Hayes, T. R. Cross-Section Measurements for Electron-Impact Ionization of Atoms. *Phys. Rev. A* **1990**, *41*, 3575.
175. Karstensen, F.; Schneider, M. Absolute Cross Sections for Single and Double Ionisation of Mg Atoms by Electron Impact. *J. Phys. B: At. Mol. Phys.* **1978**, *11*, 167.
176. Karstensen, F.; Schneider, M. Ionization Function of Magnesium for Electron Impact. *Z. Phys. A* **1975**, *273*, 321.
177. Howald, A. M.; Gregory, D. C.; Meyer, F. W.; Phaneuf, R. A.; Müller, A.; Djuric, N.; Dunn, G. H. Electron-Impact Ionization of Mg-Like Ions: S^{4+}, Cl^{5+}, and Ar^{6+}. *Phys. Rev. A* **1986**, *33*, 3779.
178. Chung, Y. S. *Electron-Impact Ionization Cross Sections for Ar^{6+}*; 1996. Unpublished ORNL data, http://www-cfadc.phy.ornl.gov/xbeam/ar6-7tbl_chung.html.
179. Bannister, M.; Guo, X. *Electron-Impact Single Ionization of Cl^{4+}*; 1993. Unpublished ORNL data, http://www-cfadc.phy.ornl.gov/xbeam/cl4-5tbl_bannister.html.
180. Gregory, D. C.; Crandall, D. H. *Experimental Electron-Impact Cross Sections for Single Ionization of Ar^{5+}*; 1982. Unpublished ORNL data, http://www-cfadc.phy.ornl.gov/xbeam/ar5-6tbl_gregory.html.
181. Gregory, D. C.; Wang, L. J.; Meyer, F. W.; Rinn, K. Electron-Impact Ionization of Iron Ions: Fe^{11+}, Fe^{13+}, and Fe^{15+}. *Phys. Rev. A* **1987**, *35*, 3256.
182. Sataka, M.; Ohtani, S.; Swenson, D.; Gregory, D. C. Experimental Cross Sections for Electron-Impact Ionization of Chromium Ions: Cr^{6+}, Cr^{7+}, Cr^{8+}, and Cr^{10+}. *Phys. Rev. A* **1989**, *39*, 2397.
183. Wang, L. J.; Rinn, K.; Gregory, D. C. Electron Impact Ionisation of Nickel Ions. *J. Phys. B: At. Mol. Opt. Phys.* **1988**, *21*, 2117.
184. Gregory, D. C.; Dittner, P. F.; Crandall, D. H. Absolute-Cross-Section Measurements for Electron-Impact Ionization of Triply Charged Inert-Gas Ions: Ne^{3+}, Ar^{3+}, Kr^{3+}, and Xe^{3+}. *Phys. Rev. A* **1983**, *27*, 724.
185. Hartenfeller, U.; Aichele, K.; Hathiramani, D.; Hofmann, G.; Schäfer, V.; Steidl, M.; Stenke, M.; Salzborn, E.; Pindzola, M. S. Electron-Impact Single Ionization of Low-Charged Titanium Ions. *J. Phys. B: At. Mol. Opt. Phys.* **1998**, *31*, 2999.
186. Chantrenne, S. J.; Gregory, D. C.; Buie, M. J.; Pindzola, M. S. Experimental and Theoretical Cross Sections for Electron-Impact Ionization of Ti^{5+}. *Phys. Rev. A* **1990**, *41*, 140.
187. Gregory, D. C.; Meyer, F. W.; Müller, A.; Defrance, P. Experimental Cross Sections for Electron-Impact Ionization of Iron Ions: Fe^{5+}, Fe^{6+}, and Fe^{9+}. *Phys. Rev. A* **1986**, *34*, 3657.
188. Khouilid, M.; Cherkani-Hassani, S.; Rachafi, S.; Teng, H.; Defrance, P. Electron-Impact Single Ionization of Krypton Ions (q = 12–18). *J. Phys. B: At. Mol. Opt. Phys.* **2001**, *34*, 1727.

189. McCallion, P.; Shah, M. B.; Gilbody, H. B. A Crossed Beam Study of the Multiple Ionization of Argon by Electron Impact. *J. Phys. B: At. Mol. Opt. Phys.* **1992**, *25*, 1061.
190. Rejoub, R.; Phaneuf, R. A. Electron-Impact Single Ionization of Multiply Charged Manganese Ions. *Phys. Rev. A* **2000**, *61*, 032706.
191. Vainshtein, L. A.; Ochkur, V. I.; Rakhovskii, V. I.; Stepanov, A. M. Absolute Values of Electron Impact Ionization Cross Section for Magnesium, Calcium, Strontium and Barium. *Zh. Eksp. Teor. Fiz.* **1972**, *61*, 511 *Sov. Phys. JETP Fiz.* **1972**, *34*, 271.

INDEX

Note: Page numbers followed by "*f*" indicate figures, "*t*" indicate tables, and "*np*" indicate footnotes.

F

N

O

FSC
www.fsc.org
MIX
Papier | Fördert gute Waldnutzung
FSC® C083411

Kolibri 360 GmbH
Ferdinand-Jühlke-Straße 7
99095 Erfurt, Deutschland
produktsicherheit@kolibri360.de